Electrokinetic Phenomena

Electrokinetic Phenomena

Principles and Applications in Analytical Chemistry and Microchip Technology

edited by

Anurag S. Rathore
Amgen, Inc.
Thousand Oaks, California, U.S.A.

András Guttman
Diversa Corporation
San Diego, California, U.S.A.

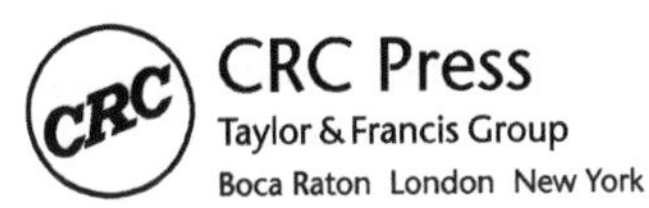

CRC Press is an imprint of the
Taylor & Francis Group, an **informa** business

First published 2004 by Marcel Dekker, Inc.

Published 2019 by CRC Press
Taylor & Francis Group
6000 Broken Sound Parkway NW, Suite 300
Boca Raton, FL 33487-2742

First issued in paperback 2019

ISBN 13: 978-0-367-44666-6 (pbk)
ISBN 13: 978-0-8247-4306-2 (hbk)

Visit the Taylor & Francis Web site at
http://www.taylorandfrancis.com

and the CRC Press Web site at
http://www.crcpress.com

Library of Congress Cataloging-in-Publication Data
A catalog record for this book is available from the Library of Congress.

Foreword

The field of electrokinetic phenomena has had a long history. For over a century, it has been recognized that ions in solution move in the presence of electric fields. It is also well known that bulk flow can occur when electric fields are applied and when the surfaces containing the fluid are charged (electroosmosis). These two phenomena, ion and bulk flow movement in the presence of electric fields, and their applications in both analytical chemistry and microfluidics are covered in depth in this book.

The book is an outgrowth of the field of electrophoresis, and as such it owes a debt of gratitude to all the pioneers who brought the science to where it is today. While I do not wish to provide a historical survey, it is nevertheless useful to explore briefly the past so that the reader will understand and appreciate the advances outlined in this book.

Electrophoresis, a method of differential migration based on the size-to-charge ratio of ions in the presence of an applied electric field, had from its beginning a strong biological focus. It took this direction because many biological substances are charged, an effective means for solubilization in the predominantly aqueous environment of the biological milieu. The Uppsala School, under the leadership of Nobel Laureate Arne Tiselius, enumerated many of the important first principles of the method before and after World War II. The process of moving boundary electrophoresis in free solution, for example, was developed by Tiselius and was widely used for the study of proteins during this era (1). Later, in order to contain the bulk fluids, electrophoresis was practiced using support (anticonvective) media, e.g., paper or starch. From these early choices, gels made of cross-linked polyacrylamide and agarose became widely used as support media. When the pore sizes of such gels are optimally controlled, the electrophoretic migration of proteins, DNA, and other biopolymers are retarded as a function of their size, due to interaction with the gel. It was later discovered that denatured proteins in the presence of 0.1% SDS or higher yielded migration based on molecular weight, as a result of a constant mass-to-charge ratio of the SDS protein complexes. Similar phenomena occur in the gel separation of single- and double-stranded DNA, in which each base or base pair yields a constant increment of mass and charge. Today, separations based on slab gel electrophoresis are widely practiced, particularly in biological laboratories—for example, 2-D gel electrophoresis separation of proteins.

Slab gel electrophoresis has limitations, perhaps the most important being that it is difficult to automate. Thus, column or free zone electrophoresis, in analogy to column liquid chromatography, was a field that early on attracted attention. Particularly noteworthy was the 1967 publication of the Ph.D. thesis of Stellan Hjerten, where many of the major principles of the method were described (2). Unfortunately, in those days relatively wide tubes (3 mm) were required in order to inject sufficient material to be observed with the detectors available. It was the development of the appropriate detection and other instrumentation that led to capillary electrophoresis (CE) in 1981 by Jim Jorgenson (3). As demonstrated in that first paper, very high-efficiency separation using bulk flow electroosmosis was achieved. As discussed in this book, the sharp band was in large part the result of the plug-like flow of electroosmosis in contrast to the parabolic flow profile generated by pressure-driven laminar flow.

The spectacular efficiencies generated by capillary electrophoresis drew others, such as myself, to enter the field. Fortunately, the Human Genome Project began in 1990, and capillary electrophoresis was one of the technologies to be developed for DNA sequencing. With linear polymer solutions, enormous column efficiencies were found, up to 3×10^7 plates/meter (4). Such performance was required to separate consecutive Sanger fragments for which the mobility differences were extremely small. The rest is history, as capillary electrophoresis using replaceable polymer matrices in a multiplexed capillary array format was used to sequence the human genome and is presently being applied to sequence the genomes of many other organisms.

Capillary electrophoreses also began the drive toward miniaturization and lab-on-a-chip. At its inception was the μTAS concept of Michael Widmer (5). This lab-on-a-chip method, in which microfluidic devices were manufactured using a methold similar to that for semiconductor devices, first focused on the transfer of CE methods from the capillary to the chip. Today, electrophoresis utilizing electric fields for bulk flow, injection, gradient mixing, and other applications, is well advanced, as described in this book. In addition, other innovative approaches to the use of electric fields have been introduced. For example, capillary electrochromatography (CEC), in which electric fields are employed to drive fluids across chromatographic packed beds, has been advanced and is also discussed in this book.

Electrokinetic phenomena is no longer merely a laboratory curiosity. It is a key component of present-day microscale analysis. Looking toward the future, it may well be that some of the current techniques will be supplemented by nanotechnology approaches, but fluid control and ion movement will likely still involve electric fields. This book will afford the reader an appreciation of the power of electrokinetic phenomena. I congratulate the editors on the breadth of coverage and the timeliness of the topic.

Barry L. Karger
Barnett Institute
Northeastern University
Boston, Massachusetts, U.S.A.

REFERENCES

1. Tiselius A. A new apparatus for electrophoretic analysis of colloidal mixtures, Trans Faraday Soc 1937; 33: 524.
2. Hjerten S. Free zone electrophoresis. Chromatogr Rev 1967; 9:122–219.
3. Jorgenson JW, Lucas KD. High-performance separations based on electrophoresis and electroosmosis. J Chromotogr 1981; 218:209–216.
4. Guttman A, Cohen AS, Heiger DN and Karger BL. Analytical and micropreparative ultrahigh resolution of oligonucleotides by polyacrylamide gel high-performance capillary electrophoresis. Anal Chem 1990; 62:137–141.
5. Manz A, Graber N, and Widmer HM. Sens. Actuators B 1990; 1:244–248.

Preface

Electrokinetic phenomena represent the basis of different electric-field-mediated separation techniques in use today. These analytical tools, including capillary zone electrophoresis (CZE), micellar electrokinetic chromatography (MEKC), capillary electrochromatography (CEC), capillary gel electrophoresis (CGE), and other forms of capillary electrophoresis, are gaining more and more popularity in the analytical labs of pharmaceutical and biotechnology companies as orthogonal techniques to traditional chromatographicly based methods of analysis. The recent great interest in microfabricated separation devices has led to newer challenges in the areas of system design, as well as application of the various electrokinetic phenomena for optimized and efficient separations. The goal of this book is to address both the various underlying fundamentals in this area and the numerous applications of these electro-driven separation tools. In this spirit, the book includes contributions from active researchers in both academia and industry to present a complete picture of where this field is at present and where and how it is evolving.

The first chapter discusses the various electrokinetic interactions that govern migration and separation of different sample components in two of the most commonly used electrodriven analytical techniques—CZE and CEC—in order to provide a platform of fundamental understanding for more detailed and specific chapters that follow.

The next four chapters discuss the basic principles underlying operation and method development of the most common electrodriven analytical techniques: CE, capillary isoelectric focusing (cIEF), capillary gel electrophoresis (CGE), and affinity capillary electrophoresis (ACE). Weinberger presents a comprehensive approach for method development in CE with an emphasis on small-molecule applications. This is followed by Kilár's chapter describing the principles of and method development in cIEF, as well as recent innovations and applications. Guttman's chapter covers the theoretical and practical aspects of capillary gel electrophoresis and reviews the key application areas of nucleic acid, protein and complex carbohydrate analysis, affinity-based methodologies, and related microseparation methods such as ultra-thin-layer gel electrophoresis and electric-field-mediated separations on microchips. Okun and Kenndler review two approaches to performing ACE—the equilibrium case for the study of weak to moderate affinities and the nonequilibrium case for moderate to high affinities. The practical advantages and limitations of both modes of ACE are discussed.

The next three chapters discuss the various fundamental and practical aspects of separations via CEC. Wen et al. review the theoretical basis of the generation and control of electroosmotic flow in CEC, followed by a discussion of various factors that can influence the separation speed, including novel column designs. Bartle and Myers review the progress made toward understanding some of the factors that influence the performance of silica-based columns in CEC. Various practical aspects of CEC are addressed, including column packing, frit formation, and column repeatability and reproducibility. The impact of properties of stationary and mobile phases on CEC separation is also discussed. Stol and Kok present a novel method for measuring electroosmotic pore flow in packed CEC columns. Theoretical and experimental results are presented and used as the basis for guidelines for optimizing the stationary-phase pore size and the ionic strength of the mobile phase.

The next four chapters discuss the extension of electric-field-mediated separation tools to the nano-world. Tsuda's chapter describes design and performance of an ultra-short column CEC system. Column preparation and design of the pumping system, injector, and detector are discussed. Laurell et al. discuss the impact of microstructure developments on the new era of proteomics. Issues associated with design of a lab-on-a-chip system are discussed, and it is shown that microfluidic modeling and simulation by in silico technology are highly useful in the first attempts at hypothesis testing. Kikutani and Kitamori report recent developments in the fields of microunit operations (MUOs), continuous-flow chemical processing (CFCP), and thermal lens microscopy (TLM) detection for microchips. The application of these methodologies to integrate a variety of chemical and analytical processes on a monolithic microchip is examined, and their use for highly sensitive wet analysis of heavy metal ions, diagnostic immunoassays, and rapid and high-yield chemical synthesis is discussed. Khandurina's chapter presents a survey of the most recent advances in CE-based preparative methods with special attention to microscale fraction collection and on-line sample preparation. These chapters suggest that emerging novel microfabricated devices will further increase the level of integration and multifunctionality in chemical and biochemical processing technology.

The last three chapters discuss some of the most important applications of the techniques mentioned above. Sweedler and coworkers describe a novel combination of CE and CEC with nuclear magnetic resonance (NMR) spectroscopy (CE-NMR and CEC-NMR). The chapter presents the theory and instrumentation of this hyphenated technique, also examining issues regarding design of the interface and finally, applications. Remcho reviews various CEC applications, including analysis of inorganic anions and cations, pollutants, amino acids, peptides, proteins, carbohydrates, nucleotides and their derivatives, pharmaceuticals, and different chiral compounds. Landers' group reviews clinical uses of microfluidic devices. These include analysis of clinically relevant analytes such as small molecules (drugs, ions, neurotransmitters, carbohydrates, and amino acids), proteins and peptides, and nucleic acids. Further, challenges of integrating the individual processes into a single device are explored.

The book is intended to provide a useful summary of the various electrokinetic phenomena that play a fundamental role in the various methods and applications mentioned above. In addition to serving as a reference book for analytical chemists in pharmaceutical and biotechnology companies, the material should be helpful to graduate students in many disciplines including chemistry, pharmacy, biochemistry, chemical engineering, and microfluidics.

Anurag S. Rathore
András Guttman

Contents

Contributors

Keith D. Bartle University of Leeds, Leeds, United Kingdom

Joan M. Bienvenue University of Virginia, Charlottesville, Virginia, U.S.A.

Gabriela S. Chirica Sandia National Laboratories, Livermore, California, U.S.A.

Stacey L. Clark Oregon State University, Corvallis, Oregon, U.S.A.

Angela Doneau Oregon State University, Corvallis, Oregon, U.S.A.

Jerome P. Ferrance University of Virginia, Charlottesville, Virginia, U.S.A.

András Guttman* Torrey Mesa Research Institute, San Diego, California, U.S.A.

Csaba Horváth Yale University, New Haven, Connecticut, U.S.A.

Dimuthu A. Jayawickrama University of Illinois, Urbana, Illinois, U.S.A.

James Karlinsey University of Virginia, Charlottesville, Virginia, U.S.A.

Ernst Kenndler University of Vienna, Vienna, Austria

**Current affiliation*: Diversa Corporation, San Diego, California, U.S.A.

Julia Khandurina Torrey Mesa Research Institute, San Diego, California, U.S.A.

Yoshikuni Kikutani Kanagawa Academy of Science and Technology, Kanagawa, Japan

Ferenc Kilár University of Pécs, Pécs, Hungary

Takehiko Kitamori The University of Tokyo, Tokyo, and Kanagawa Academy of Science and Technology, Kanagawa, Japan

Wim Th. Kok University of Amsterdam, Amsterdam, The Netherlands

James P. Landers University of Virginia, Charlottesville, Virginia, U.S.A.

Thomas Laurell Lund University, Lund, Sweden

György Marko-Varga Lund University, Lund, Sweden

Peter Myers University of Leeds, Leeds, United Kingdom

Johan Nilsson Lund University, Lund, Sweden

Vadim M. Okun University of Vienna, Vienna, Austria

Anurag S. Rathore* Pharmacia Corporation, North Chesterfield, Missouri, U.S.A.

V. T. Remcho Oregon State University, Corvallis, Oregon, U.S.A.

Remco Stol Organon, Oss, The Netherlands

Jonathan V. Sweedler University of Illinois, Urbana, Illinois, U.S.A.

Takao Tsuda Nagoya Institute of Technology, Nagoya, Japan

Robert Weinberger CE Technologies, Inc., Chappaqua, New York, U.S.A.

Emily Wen Merck Research Laboratories, West Point, Pennsylvania, U.S.A.

Andrew M. Wolters University of Illinois, Urbana, Illinois, U.S.A.

**Current affiliation*: Amgen, Inc., Thousand Oaks, California, U.S.A.

1

Migration of Sample Components in Capillary Analytical Techniques

Chromatography, Electrophoresis, and Electrochromatography

Anurag S. Rathore*
Pharmacia Corporation, Chesterfield, Missouri, U.S.A.

1. INTRODUCTION

This chapter discusses the various interactions that govern migration and separation of various sample components in some of the most commonly used analytical techniques: high-performance liquid chromatography (HPLC), capillary zone electrophoresis (CZE), and capillary electrochromatography (CEC). The first few sections of the chapter discuss the relatively simpler cases of HPLC and CZE, while the last few sections delve into the complexities of "hybrid" techniques, such as CEC. Even though the focus is on capillary techniques, understanding of the concepts discussed here plays a central role in the emerging area of microchip separations. The goal of this chapter is to provide a platform of fundamental understanding for more detailed and specific chapters that come later.

2. SEPARATIVE AND NONSEPARATIVE MIGRATION

In HPLC and CZE, separation occurs due to differences in the retention volumes at constant flow, which is conveniently represented by retention times and the electrophoretic migration velocities of the separands, respectively. However, the

**Current affiliation*: Amgen, Inc., Thousand Oaks, California, U.S.A.

time the components spend in the mobile phase or the distance they move with electroosmotic flow (EOF) does not contribute directly to the separation, because these are the same for all sample components. This distinction between separative and nonseparative migration leads to additivity relationships involving the mobile- and stationary-phase holdup times in HPLC and the additivity of the electrophoretic and electroendoosmotic velocities in CZE.

2.1. High-Performance Liquid Chromatography

In HPLC the mobile-phase flow and retention by the stationary phase take place in series so that the total migration time, t_R, is the sum of the times the eluite spends in the stationary and mobile phases [2,3]:

$$t_R = t_O + t'_R = t_O + k' t_O \tag{1}$$

where t_O is the retention time of an unretained tracer, i.e., the holdup time in the mobile phase, t'_R is the adjusted retention time, i.e., the holdup time in the stationary phase, and k' is the chromatographic retention factor, which, among others, expresses the mass distribution of the eluite between the two phases.

In view of Eq. (1), the time domain offers the natural framework for such an additivity relationship in chromatography [1].

2.2. Capillary Zone Electrophoresis

The overall migration velocity in electrophoresis, u_m, is the sum of the electrophoretic velocity of the charged separand, u_{ep}, and the electrosmotic velocity measured by a neutral tracer, u_{eo}, as both migration processes take place simultaneously. Thus

$$u_m = u_{ep} + u_{eo} \tag{2a}$$

where each velocity is a signed quantity and can be either positive or negative. The velocity of a migrant in electrophoresis is the product of its mobility and the applied electric field, so that Eq. (2a) can be expressed as

$$\mu_m = \mu_{ep} + \mu_{eo} \tag{2b}$$

where μ stands for the mobility of a migrant and the subscripts have the same meanings as in Eq. (2a). It follows from Eq. (2a) that in CZE the velocity is considered the natural frame for the additivity relationship [1].

3. VIRTUAL MIGRATION DISTANCES

An additivity relationship for the separative and nonseparative parts of the overall migration process that is equally applicable to HPLC and CZE arises naturally upon introduction of "virtual migration distances," which represent the product of the migration time and velocity [1]. These distances are obtained by dividing the real migration distance, i.e., the appropriate length of the column, into two virtual distances, ℓ_s and ℓ_o. The first, ℓ_s, represents the separative component of the overall migration process, and thus it reflects the magnitude of the adjusted retention time, t_R', in HPLC, and that of the electrophoretic velocity in CZE. The other virtual migration distance, ℓ_o, represents that part of the overall migration process that does not lead directly to separation and thus reflects the retention time of an inert tracer, i.e., the mobile-phase holdup time, t_o, in HPLC and the velocity of an uncharged tracer, i.e., that of EOF, in CZE [1]. By definition, the sum of the two virtual lengths, ℓ_s and ℓ_o, always equals the actual migration distance.

3.1. High-Performance Liquid Chromatography

In HPLC, the virtual distance, ℓ_s, representing separative migration, is defined as [1]

$$\ell_s = u_b t_R' = L_m\left(\frac{k'}{1+k'}\right) \tag{3}$$

where u_b the migration velocity of the eluite band (u_o is the velocity of an inert tracer) and L_m is the actual migration distance, i.e., the column length. On the other hand, ℓ_o, the virtual migration distance accounting for the holdup time of an inert tracer, is given by [1]

$$\ell_o = u_b t_o = L_m\left(\frac{1}{1+k'}\right) \tag{4}$$

Here u_b is the migration velocity of the eluite band and u_o is the velocity of an inert tracer. In HPLC both virtual lengths, ℓ_s and ℓ_o, are positive quantities, and therefore neither of them can be greater than the total migration distance.

3.2. Capillary Zone Electrophoresis

In CZE, the virtual distance associated with separative migration, ℓ_s, is [1]

$$\ell_s = u_{ep} t_m \tag{5}$$

whereas the virtual migration distance, ℓ_o, arising from EOF is [1]

$$\ell_o = u_{eo} t_m \tag{6}$$

It is recalled that u_{eo} is the velocity of a neutral tracer that migrates with the EOF velocity.

Both ℓ_s and ℓ_o are signed quantities, and their sign depends on the direction of the migration. When electrorheophoresis is co-directional, i.e., the electrophoretic velocity of the separands and the EOF have the same direction, both ℓ_s and ℓ_o have the same sign. In counterdirectional modes of CZE the signs of the two are opposite.

Evidently, the virtual distances, ℓ_s, have to be different for the sample components separated by HPLC or CZE, and as a consequence the corresponding ℓ_o values also have to be different. Such an interdependence of ℓ_o on ℓ_s may be counterintuitive since it arises from the fictitiousness of the virtual length domain used here to unify the natural additivity relationships of times and velocities in HPLC and CZE, respectively.

4. FUNDAMENTAL MIGRATION PARAMETERS IN HPLC AND CZE

4.1. Chromatographic Retention Factor

In view of Eq. (1), the retention factor in HPLC can be expressed in terms of the two virtual length components as [1]

$$k' = \frac{t_R - t_o}{t_o} = \frac{\ell_s}{\ell_o} \tag{7}$$

As mentioned before, the retention factors are always positive in chromatography, so both ℓ_s and ℓ_o have to be positive.

4.2. Electrophoretic Velocity Factor

In CZE, the ratio of the two virtual lengths defines a migration parameter that is analogous to the chromatographic retention factor and is conveniently called the electrophoretic velocity factor, k_e'. It is given by [1]

$$k_e' = \frac{\ell_s}{\ell_o} = \frac{u_{ep}}{u_{eo}} = \frac{t_{eo} - t_m}{t_m} \tag{8}$$

Since the direction of EOF can be the opposite of the electrophoretic migration of the separand, the velocity factor k_e' can be negative.

5. THE FOUR MAIN OPERATIONAL MODES OF CZE

Unlike in HPLC, where the sample components and the mobile phase move in the same direction, in CZE we encounter four fundamentally different operational modes as far as the relative directions and magnitudes of the electrophoretic and electroendoosmotic migration velocities are concerned. The four modes and their characteristics, in terms of the sign and the magnitude of the corresponding key migration parameters, including the virtual migration distances, are given in Table 1. Figures 1 and 2 are complementary to Table 1 and provide graphical illustration of the various features associated with the four main modes A–D, of capillary zone electrorheophoresis [1].

The interrelationships among the various operational modes is revealed by Figure 1, which shows the effect of changing EOF at a fixed value and sign of the electrophoretic velocity of the sample components. Our starting point is K, where the EOF velocity is co-directional and much higher than that of the electrophoretic migration ($u_{eo} \approx \infty$). As we move clockwise from point K along the chart that encompasses the four operational modes, the EOF diminishes and becomes zero at point L. Thus, the right-hand side of the chart is the domain of mode B, which encompasses co-directional CZE with the velocity factor increasing from zero to infinity. This point represents mode A when there is no EOF, i.e., arheic conditions prevail. It is also the point of transition from co-directional to counterdirectional CZE, where the EOF velocity and hence the velocity factor change sign.

Moving further in the clockwise direction, on the left side of the graph we encounter sectors of modes C and D, the two counterdirectional modes of CZE. The magnitude of EOF, the direction of which is now opposite to that of the electrophoretic velocity, increases in mode C as we move from point L to point M, where the magnitude of the EOF becomes the same as that of the electrophoretic velocity and the velocity factor equals −1. Thus at M, which is the point of transition between modes C and D, there is no migration due to the counterveiling effect of the two velocities. The EOF velocity increases as we move along the sector of mode D in the clockwise direction, and since it is higher than the electrophoretic velocity, the separands migrate in the direction

Table 1 The Four Main Modes of Capillary Zone Electrophoresis, Characterized by the Signs and Limiting Values of the Key Migration Parameters[a]

	Velocity factor, K_e		Virtual migration distances				
			ℓ_s(separative)		ℓ_o(nonseparative)		
Mode	Sign	Magnitude	Sign	Magnitude	Sign	Magnitude	Sample electropherogram
A	+	∞	+	L_m	+	0	Figure 2a
B	+	$[0, \infty]$	+	$[0, L_m]$	+	$[L_m, 0]$	Figure 2b
C	−	$[1, \infty]$	+	$[\infty, L_m]$	−	$[\infty, 0]$	Figure 2c
D	−	$[0, 1]$	−	$[0, \infty]$	+	$[L_m, \infty]$	Figure 2d

[a]Mode A, arheic CZE; mode B, co-directional CZE; mode C, counterdirectional CZE with the mobility of the neutral tracer being the smallest; mode D, counterdirectional CZE with the mobility of the neutral tracer being the greatest. The notation $[a, b]$ means: from a to b, including both a and b.

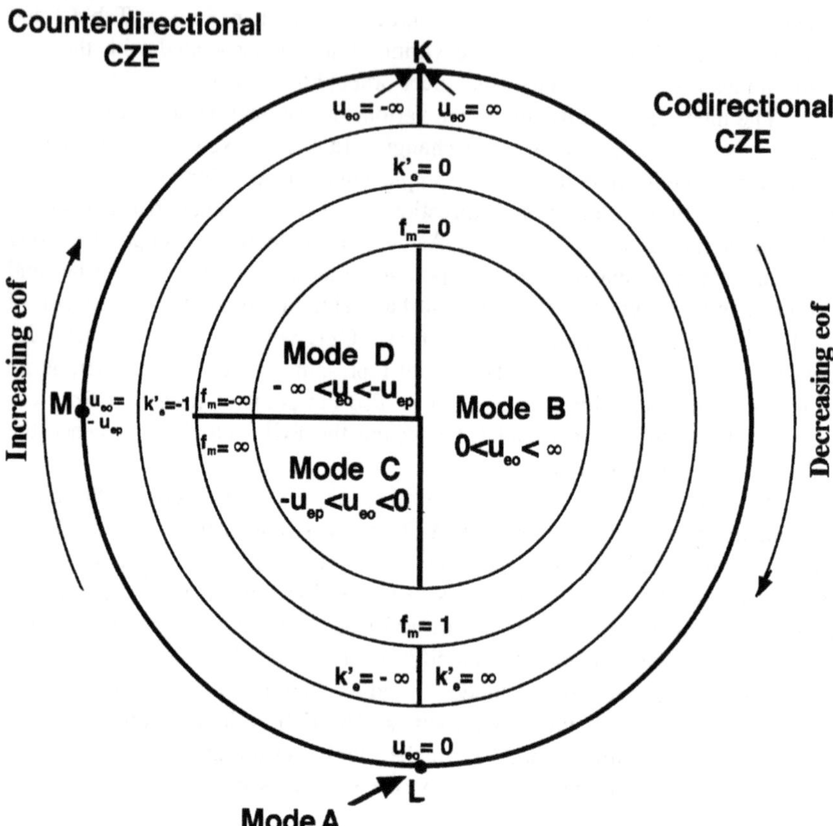

FIGURE 1 Schematic illustration of the four main operational modes of CZE and the pertinent values of some key migration parameters. The right-hand side of the chart represents the counterdirectional mode B and the left-hand side shows the two counterdirectional modes C and D. Mode A lies at point *L* at the bottom. At a fixed electrophoretic velocity of the sparand, the EOF decreases as we go clockwise from *K* to *L*. At point *L* a transition occurs from the co-directional to the counterdirectional CZE, and as we move farther clockwise the magnitude of the EOF increases in the opposite direction in mode C. Point *M*, where both EOF and electrophoretic migration have the same velocity, is the transition point to mode D, the other counterdirectional mode of CZE. Finally, at point *N*, the relative magnitude of the EOF becomes very large in the direction opposite to that of electrophoretic migration.

of EOF, i.e., in the direction opposite to that of the electrophoretic velocity. Finally, we reach point *N*, where EOF becomes so high that the velocity factor diminishes to zero. This corresponds to a situation where no separation occurs, either because the separands are neutral or because their electrophoretic velocity is negligibly small in comparison to the EOF velocity.

Another way to illustrate the four operational modes of CZE is by means of representative electropherograms as shown in Figure 2 [1]. In mode A, arheic conditions prevail and the sample components migrate solely by their electrophoretic velocity and appear in the order of their decreasing electrophoretic

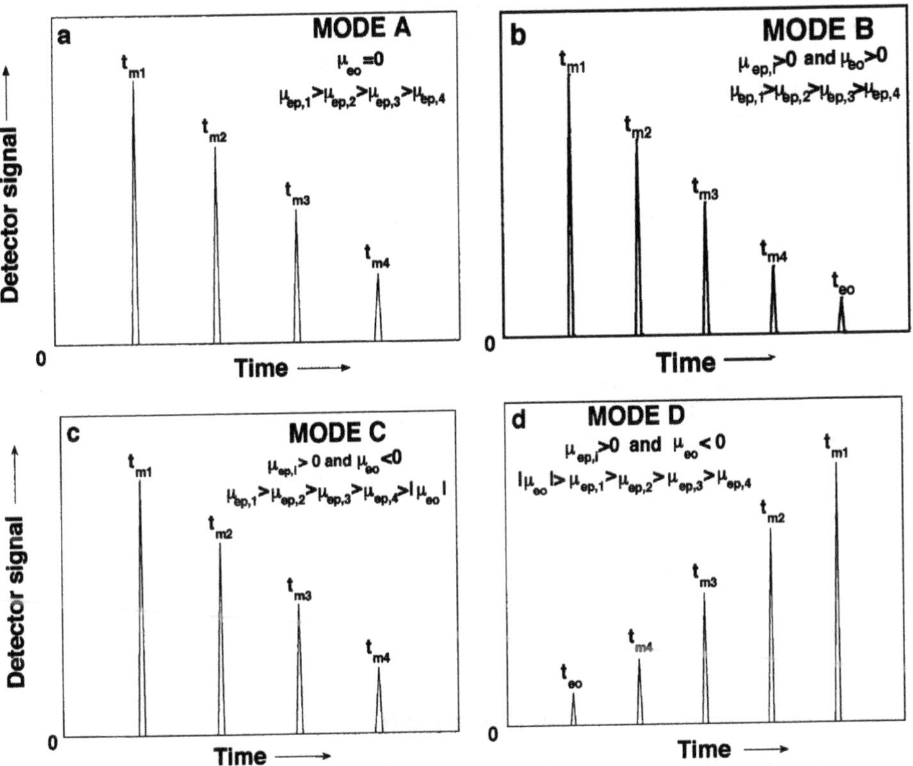

FIGURE 2 Schematic illustration of electropherograms typical for the four main modes of CZE listed in Table 1. The conditions are stated on charts a to d for each mode: A, arheic CZE; B, co-directional CZE; C, counterdirectional CZE, when the neutral tracer migrates slower than any of the charged sample components; D, counterdirectional CZE with the neutral tracer being the fastest-migrating component.

mobilities as depicted in Figure 2a. In the absence of EOF, neutral sample components do not migrate, so the velocity factor, k'_e, is infinitely large and the nonseparative component of the virtual length, ℓ_o, is zero as shown in Table 1.

Co-directional CZE is encompassed by mode B, where the migration velocities of charged sample components are always greater than that of the neutral tracer for EOF [1]. As seen in Figure 2b, the components appear on the electropherogram in the order of decreasing mobility as in mode A. Both ℓ_s and ℓ_o are of the same sign in mode B, each of them is smaller than the total capillary length, and as a result the k'_e is always positive and ranges between zero and infinity. In mode B, k'_e is always positive, and this behavior is analogous to that of the factors k' and $k'/(1 + k')$ in HPLC. Thus mode B of CZE, which is most widely used in free-solution capillary zone electrophoresis, exhibits among the four modes the closest formal resemblance to chromatography.

Modes C and D both represent counterdirectional CZE, with the restriction that all charged sample components exit at the same end of the capillary [1]. As expected, the velocity factors are negative in both modes as shown in Table 1. In mode C, the EOF velocity is lower than the electrophoretic velocity of any charged separand, so the order of their appearance on the representative electropherogram (see Figure 2) is the same as in modes A and B. However, no neutral tracer can traverse the column, and therefore its peak is absent on the electropherogram in Figure 2c. In mode C, as seen in Table 1, k'_e is greater than unity and negative.

In mode D, as illustrated in Figure 2d, the EOF velocity is higher than the electrophoretic velocity and the sample components appear in an order reversed with respect to that found in all other operational modes [1]. As seen, they appear at the other end of the capillary in the order of increasing electrophoretic mobilities, with the peak of the neutral tracer being the first. Table 1 shows that in mode D the k'_e is negative and less than unity.

The discussion above covers the four main modes of CZE [1]. Special situations, e.g., modes with migrants carrying opposite charge and moving in opposite directions, can be also treated as combinations of two or more of these modes.

6. MIGRATION OF SAMPLE COMPONENTS IN CEC

For the sake of simplicity, it has been assumed in the ensuing discussion of HPLC, CZE, and CEC that u_m, u_o, u_p, t_m, and t_o denote the overall migration flow velocity of the migrant, the velocity of the mobile phase, the electrophoretic velocity of the migrant, the migration time of the analyte, and the migration time of an inert and neutral tracer, respectively.

The k' defined by Eq. (7) is the dimensionless measure of retention in chromatography and is widely used in three different ways as follows [1,4].

First, it is a dimensionless measure of the position of an eluite peak on the chromatogram with respect to t_o and the starting point ($t = 0$). As such, it plays an important role in expressions for separation parameters such as selectivity and resolution. Second, since log k' is proportional to the free-energy change for the chromatographic retention process at a given temperature, k' is employed in thermodynamic and mechanistic studies of the chromatographic process [1–3]. Third, k' measures the mass distribution of the eluite between the stationary and the mobile phases and $1/(1 + k')$ and $k'/(1 + k')$ express the respective probabilities that the analyte is in the mobile or stationary phase during the time it traverses the column.

The velocity factor, k'_e, as defined in Eq. (8), serves as a peak locator and can be used for the evaluation of various separation parameters such as selectivity and resolution [1,4]. The velocity factor can either be positive or negative when the electrophoretic migration of the analyte is co- or counterdirectional to the EOF, respectively. For this reason, and since k'_e is void of thermodynamic or any other kind of fundamental significance, the velocity factor finds very limited applications.

We have shown recently that the different separation mechanisms of HPLC and CZE preclude the definition of any retention or velocity factor for CEC that would have comparable significance to that of k' in HPLC [4]. However, as shown in the following section, the CEC system was defined by a combination of a "retention factor" that measures the retention in CEC and a velocity factor that characterizes the electrophoretic migration. Further, expressions were proposed for possible peak locators that, just like k' and k'_e in HPLC and CZE, respectively, can serve as useful indicators of peak position on the electrochromatogram and facilitate estimation of various separation parameters such as selectivity and resolution.

The migration velocity, u_m, of a charged sample component in CEC is expressed in the literature by the sum of the velocity of the mobile phase, u_o, and the electrophoretic velocity of the migrant of interest, u_p, multiplied by the retardation factor, $1/(1 + k'')$, as [4–8]

$$u_m = \frac{u_p + u_o}{1 + k''} = \frac{u_o(1 + k''_e)}{1 + k''} \tag{9}$$

where k'' is the measure of chromatographic retention under conditions of the CEC experiments, i.e., the retention factor in CEC, and k''_e is the velocity factor. They are given by [1,4]

$$k''_e = \frac{\mu_p}{\mu_o} \tag{10}$$

and

$$k'' = \frac{t_m(1 + k_e'') - t_o}{t_o} \tag{11}$$

The double prime distinguishes the symbols k'' and k_e'' from their respective counterparts, k' and k'_e, which have been defined earlier in Eqs. (7) and (8) for HPLC and CZE, respectively.

It should be emphasized that both k'' and k_e'' need to be evaluated under conditions used in the CEC experiments. First, the electrophoretic mobility of the sample component is obtained from separate CZE measurements using the mobile phase used in CEC. Then electroosmotic mobility, which is the interstitial EOF mobility in the packing, is evaluated from the results of measuring the currents and the EOF with CEC columns [9,10]. This allows for calculation of k_e'' according to Eq. (16) followed by calculation of k'' from Eq. (11) using the migration times of the different sample components in the CEC column.

The k'' measures the magnitude of retention in CEC due to reversible binding of the analyte to the stationary phase. Inspection of Eq. (11) shows that for all components (neutral or charged), k'' is always positive, as a chromatographic "retention factor" should be. Further, while the retention factor in HPLC and the velocity factor in CZE are able to characterize the respective differential migration processes alone, both of them are required to characterize CEC.

7. PEAK LOCATOR

It follows from the discussion above that knowledge of both k'' and k_e'' is required to assess the chromatographic and electrophoretic contributions to migration of charged sample components in CEC. However, both of them cannot be determined from CEC data alone and require electrophoretic measurements, as described above.

Often, however, a dimensionless peak locator suffices. For such a case, a much simpler approach can be followed and a peak locator, k_{cc}'', can be defined in terms of k'' and k_e'' by following a chromatographic or an electrophoretic formalism, respectively. It should be emphasized that the peak locator is devoid of any mechanistic insight offered by k'' and k_e'' and so has limited utility. However, the peak locators can be evaluated directly from the electrochromatograms without any additional measurements. Both the chromatographic and the electrophoretic formalisms can be used for defining the peak locator [4].

7.1. Chromatographic Formalism

A peak locator, k_{cc}'', following this formalism can be expressed in a manner similar to Eq. (7) for chromatography as follows [4]:

$$k''_{cc} = \frac{t_m - t_o}{t_o} \tag{12}$$

where t_m and t_o denote the migration time of the analyte and migration time of an inert and neutral tracer in a CEC column, respectively.

The use of Eq. (12) has been proposed in the literature for CEC as well [5–8,11,12]. As in HPLC, k''_{cc} in CEC can be conveniently evaluated from the migration times on the electrochromatogram. However, since $t_m < t_o$ may be possible for positively charged sample components migrating faster than the EOF, k''_{cc} would be negative for those cases.

7.2. Electrophoretic Formalism

Another method is to define a peak locator, k''_{ce}, by using the electrophoretic formalism. In a manner similar to Eq. (8), it can be expressed as follows [4]:

$$k''_{ce} = \frac{t_o - t_m}{t_m} \tag{13}$$

Table 2 Display of the Retention Factor, the Velocity Factor, and the Two Peak Locators and Their Uses in CEC

Name	Symbol	Formula	Measure of chromatographic retention	Measure of electrophoretic migration	Measure of peak position
Retention factor	k''	$\frac{t_m(1 + k''_e) - t_o}{t_o}$	Yes		
Velocity factor	k''_e	$\frac{\mu_p}{\mu_o}$		Yes	
Peak locator[a] (LC formalism)	k''_{cc}	$\frac{t_m - t_o}{t_o}$			Yes
Peak locator[a] (CE formalism)	k''_{ce}	$\frac{t_o - t_m}{t_m}$			Yes

[a]The peak locators have been derived following the chromatography [Eq. (12)] or electrophoretic [Eq. (13)] formalisms. A more detailed comparison of the two peak locators is given in Table 3.

TABLE 3 Merits and Shortcomings of Using the Chromatographic and Electrophoretic Formalisms for the Peak Locators

	Peak locator, k''_{cc}, by chromatographic formalism	Peak locator, k''_{ce}, by electrophoretic formalism
Formula	$k''_{cc} = \frac{t_m - t_o}{t_o}$	$k''_{ce} = \frac{t_o - t_m}{t_m}$
Merits	• Useful for calculating resolution • Enjoys the familiarity of the retention factor as defined in chromatography • Increases with the migration time, like k' in chromatography	• Useful for calculating resolution • Equivalence of the retention process to an electrophoretic mobility counterdirectional to the EOF is easy to understand
Shortcomings	• Is negative for components migrating faster than the EOF	• Lacks the familiarity that k' has in chromatography • Is negative for components migrating more slowly than the EOF

where t_m and t_o denote the migration time of the analyte and the migration time of an inert and neutral tracer in a CEC column, respectively. Equation (13) is now very similar to the corresponding equation for electrophoresis. k''_{ce} can also be calculated from the migration times on the electrochromatogram. However, it follows from Eq. (13) that, like the velocity factor in CZE, k''_{ce} is negative for components moving more slowly than the EOF.

Table 2 shows the different expressions that have been proposed in Sections 6 and 7 for characterizing the migration of charged sample components in CEC. Further, Table 3 lists the merits and shortcomings of the different definitions and it is evident that, due to the dual separation mechanisms at work during the CEC separation process, none of the two definitions can treat all possible cases that may arise in CEC with the same simplicity as the k' offers to do in HPLC.

8. CONCLUSIONS

In this chapter, definitions of the various migration parameters are offered for the capillary analytical techniques of HPLC, CZE, and CEC. It is shown that in both HPLC and CZE with EOF, the differential migration process can be divided into a separative component, which involves selective interactions with the stationary phase or differences in the electrophoretic migration velocities, and a nonseparative component representing migration by convection that does not contribute directly to separation. The separation system in HPLC and CZE can be characterized by retention factor and velocity factor, respectively.

For CEC, no single parameter equivalent to k' in HPLC has been found to be applicable to the wide range of mechanistic changes in CEC. However, it has been shown that the CEC system can be defined by a combination of a "retention factor" that measures the retention in CEC and a velocity factor that characterizes electrophoretic migration. These two parameters together can be very useful in a complete characterization of the CEC system.

REFERENCES

1. Rathore AS, Horváth Cs. Separation parameters via virtual migration distances in HPLC and CZE and electrokinetic chromatography. J Chromatogr A 1996; 743: 231–246.
2. Horváth Cs, Melander WR. In: Heftmann E, ed. Theory of Chromatography. Amsterdam: Elsevier, 1983:A27.
3. Giddings JC. Dynamics of Chromatography, Part I: Principles and Theory. New York: Marcel Dekker, 1965.
4. Rathore AS, Horváth Cs. Electrophoresis 2002; 23:1211–1216.
5. Wu J-T, Huang P, Li MX, Lubman DM. Anal Chem. 1987; 69:2908–2913.
6. Dittmann MM, Masuch K, Rozing GP. J Chromatogr A 2000; 887:209–221.
7. Cikalo MG, Bartle KD, Myers P. J Chromatogr A 1999; 836:35–51.
8. Eimer T, Unger KK, Greef Jvd. Trends Anal Chem 1996; 15:463–468.
9. Rathore AS, Wen E, Horváth Cs. Anal Chem 1999; 71:2633–2641.
10. Rathore AS, Horváth Cs. In: Deyl Z, Svec F, eds. Capillary Electrochromatography. Amsterdam: Elsevier, 2001: Chapter 1.
11. Crego AL, González A, Marina ML. Crit Rev Anal Chem 1996; 26:261–304.
12. Bartle KD, Myers P. J Chromatogr A 2001; 916:3–23.

2

Methods Development for Capillary Electrophoresis with Emphasis on Small Molecules

Robert Weinberger
CE Technologies, Inc., Chappaqua, New York, U.S.A.

1. INTRODUCTION

When deciding whether to write this chapter, I was faced with an unusual dilemma. Being a consultant and developing methods on a contract basis using capillary electrophoresis, I asked myself the question, "Do I want to give away my trade secrets in this chapter?" Well, dear reader, it turns out that methods development for capillary electrophoresis (CE), particularly for small molecules, is not very difficult. Of course, this must be qualified with the caution: for most applications. There are no real secrets here, only simple and logical approaches that rapidly produce reliable methods. So please forgive me if this chapter seems relatively simple.

There are several good reasons why CE has not captured the lion's share of small-molecule separations. High-performance liquid chromatography (HPLC) has a 23-year head start over capillary electrophoresis, and most of the problems have been worked out. HPLC is rugged, sensitive, scales up to preparative and commercial modes, and scales down to the capillary format. Poorly developed CE methods by ill-trained chromatographers are another contributing factor to the slow acceptance of CE in the world of small molecules.

The role of CE in the human genome project is the stuff that legends are made of. The predominant feature of CE that made that possible was the automation of the slab-gel processes using batteries of highly multiplexed 96-capillary-array CE instruments. It seems that both HPLC and CE have separate strengths that clearly define noncompeting applications areas.

CE's strengths lie in several areas. Separations that are difficult or impossible by HPLC are often easily performed by CE. The speed of separation usually favors CE. Multiplexed HPLC is a nightmare, whereas CE multiplexes more easily. Miniaturization of CE using microfluidics is a natural evolution of the field, and this can be multiplexed as well for high-throughput applications.

This brings me back to methods development. Running a 96-capillary instrument with a poorly constructed method presents the opportunity to perform 96 horrible separations simultaneously.

There are many books devoted to capillary electrophoresis, including my own [1]. Being redundant to already published material is sometimes necessary but seldom fulfilling, so I will try to be different, within the constraints of technological necessities. This chapter is not an exhaustive review of methods development. What follows is a personal road map describing my intellectual processes that go into methods development, with emphasis on small-molecule separations. I have been fortunate to have been involved in CE since 1987. During the past 15 years, I enjoyed watching the technique mature and in playing my own small role in its development.

2. PRELIMINARY CONSIDERATIONS

2.1. Define the Analytical Problem

This is the place to start, since most often, analytical chemists are trying to help solve someone else's problem. We need to define the solute and its matrix as well as the nature of the analytical problem. For example, in the world of pharmaceuticals, there are raw material identification and purity determinations, in-process testing, dosage-form determinations, content uniformity, dissolution testing, stability studies, bioavailability, pharmacokinetics, and drug metabolism, to name a few. Each of these analytical problems has its own specific requirements. The matrix can be a raw material, granulation, tablet, capsule, solution, lotion, cream, syrup, dissolution medium, blood serum, urine, or various body tissues and fluids. Similar definitions can be described for virtually any industrial area and problem set. These definitions will help select sample preparation, separation, and detection techniques.

Whatever the sample matrix, ensure that blank matrices are available for recovery and selectivity studies. For impurity determinations, it is best to have impurity standards and degradation products available for selectivity studies and quantitative validation. For quantitative analysis of individual major components or impurities, internal standards are usually necessary to ensure precise quantitation.

2.2. Define the Analytical Figures of Merit

The analytical figures of merit are the limit of detection (LOD), limit of quantitation (LOQ), linear dynamic range, percent recovery, peak area, and migration time precision on a run-to-run, day-to-day, and capillary-to-capillary basis. Some of these parameters are generated on a best-efforts basis. Other parameters, such as LOD, LOQ, and percent recovery, become fundamental requirements for the method.

2.3. Select the Detection Technique

The inability to detect precludes the ability to develop a separation. The selected technique is defined by the required limit of detection. If low-pg/mL levels are needed, it is fruitless to use a UV/visible absorbance detector. Laser-induced fluorescence (LIF) is usually appropriate, provided derivatization reagents are available if the solute does not have significant native fluorescence [2]. Limits of detection of 10^{-10} M are easily achieved using LIF, provided the solute absorbs at a laser emission wavelength and has a reasonable fluorescence quantum yield.

For separations in the ng/mL-to-µg/mL range and solutes that absorb in the UV/visible portion of the spectrum, optimize the detector wavelength. The optimal wavelength is usually in the low-UV portion of the spectrum. Using LC, the UV cutoff of the mobile phase often prevents such low wavelengths from being employed. Limits of detection usually approach 10^{-6} M without heroic measures. The downside of low-UV detection is a loss of selectivity, since more solutes will absorb there. This is countered in part by the high peak capacity of CE. In some cases, appropriate sample preparation may be required for selectivity.

If the solute lacks UV absorption but is charged, the appropriate indirect detection technique must be selected [3]. If the solute is neutral and does not absorb in the UV, derivatization is required to add charge and a chromophore/fluorophore to the molecule. For this case, alternative separation techniques are usually advantageous.

2.4. Given the Choice, Separate the Solutes as Anions

This is particularly important for large polymeric molecules. Anions are repelled from the negatively charged capillary wall, simplifying methods development since buffer additives or coated capillaries are not required to minimize wall effects [4].

There are exceptions to this, so do not rule out the separation of zwitterionic biomolecules as cations. Some proteins and peptides separate well at low pH on bare silica or coated capillaries.

2.5. Is the Application for Large or Small Molecules?

Polycations are strongly retained by the bare silica capillary wall, since they can ion-pair at multiple sites on the anionic wall. Since all electrostatic bonds must be broken to free the solute, this problem becomes increasingly severe as the number of positive charges increases. Polycations may include small molecules such as tetrazoles and even small peptides. Either permanently coated or charged-reversed capillaries are usually required to separate polyvalent cations. Monovalent cations show some wall effects, but these do not preclude the bare silica capillary from being used. Polyanions and monovalent anions can usually be separated on bare silica.

Peptides and proteins with high pI's usually show wall effects at most pHs, since they are positively charged. If the peptide is sufficiently small (<10 amino acids), low-pH buffers (<pH 2.5) may be suitable, since the capillary wall approaches neutrality. For larger peptides, a wall effect may still be observed at low pH.

2.6. Is the Sample Soluble in Aqueous Solutions?

The entire sample must be soluble to be separated. If it is not, sample preparation can be performed to remove insoluble endogenous sample components. If the solute is not soluble, the appropriate additives must be used to solubilize it. So long as wavelengths less than 220 nm are not required, 6 M urea is an excellent solvent. Both ionic and nonionic surfactants are also useful in this regard. Acetate buffers provide better solubility compared to phosphate buffers. These approaches maintain a totally aqueous system, which is most robust. Organic solvents are often used in CE, but migration time precision is usually worse compared to aqueous systems.

If the samples are poorly soluble in aqueous media, then nonaqueous CE [5], capillary electrochromatography (CEC), or HPLC should be considered.

2.7. How Many Solutes Are Present?

For simple mixtures containing only a few components, prepare each solute at a different concentration. This in combination with diode-array UV detection simplifies peak tracking. For complex mixtures computerized experimental design speeds methods development [6].

2.8. Use the Short End of the Capillary for Scouting Runs

Methods development is speeded using this approach. First of all, the direction of migration is quickly determined. If you see broad peaks, wall effects are certainly occurring. The frustration of waiting for what seems like forever for peaks to appear is avoided. Sometimes, the short end is sufficient in length to run simple separations. It is useful to run short-end separations using both positive and negative polarity. The direction of migration is quickly found, and you can determine if any solutes are migrating in opposite directions.

2.9. Two or Three Electrolyte Reservoirs?

The selection of the background electrolyte (BGE) fill process can affect the separation, depending on the characteristics of the electrolyte. The two-reservoir method uses inlet and outlet vials. The capillary is filled using the inlet vial. Multiple runs from a single set of vials are possible, but sometimes migration time drift is observed due to buffer depletion [7]. This limits the number of runs from a single set of vials.

Another method uses three buffer vials, an inlet, an outlet, and a fill vial. Instead of filling the capillary using the inlet vial, the capillary is filled using BGE contained in the fill vial. In this manner, the capillary is starting out with BGE that has not been electrophoresed. Migration time precision is often improved compared to the two-vial method. A problem occurs when the BGE has some UV absorbance and there is strong electroosmotic flow (EOF). Since the fill BGE and the inlet BGE now differ somewhat, a baseline shift may be noted somewhere in the electropherogram, depending on the mobility of the UV-absorbing ions. This stair-step shift can be significant and can only be avoided by using the two-vial method or by refreshing the buffers for each run.

During methods development or when you wish or need to stretch the buffer, set up two methods where the inlet and outlet vials are reversed. If you alternate between the two methods, the pH-changing depletion effects are effectively canceled and you can get many runs out of a set of buffers. Eventually the vials become sufficiently contaminated that baseline upsets occur.

3. DEVELOPING THE SEPARATION

3.1. Basic Concepts

The fundamental separation mechanism of capillary zone electrophoresis (CZE) is based on differences in the mobilities of solutes. Mobility is defined as the charge/mass ratio for each solute. Since the charge is often a function of pH, the pH is the most important adjustable parameter for control of resolution. The order of elution on bare silica at high pH is cations, unseparated neutrals, and anions. At low pH, where the EOF is very low, the anions may migrate toward the positive electrode and may not be seen using normal polarity.

The efficiency of a separation is defined as

$$N = \frac{\mu_{ep} E L}{2 D_m}$$

where N = the number of theoretical plates; μ_{ep} = solute mobility; E = field strength; L = capillary length, and D_m = solute diffusion coefficient. Since time-related diffusion is a fundamental limiting factor for efficient separations, the most efficient separations occur rapidly at high field strengths. The limiting factor for the field strength is the production of heat. Currents should generally be kept below 100 μA (lower is better) or band broadening due to a thermal (viscosity, mobility) gradient will occur. Loss of linearity of the Ohm's law plot (voltage/current) indicates that the instrument cooling system is becoming inadequate for the removal of heat [8]. Do not work too far away from the linear part of the plot. High current also accelerates buffer depletion, limiting the number of runs from a set of buffers.

The general resolution equation is

$$R_s = 0.177\ \Delta\mu_{ep}\sqrt{\frac{EL}{(\mu_{ep} + \mu_{eo})D_m}}$$

where R_s = the resolution; $\Delta\mu$ = the differences in mobilities between two solutes; E = field strength; L = capillary length; μ_{ep} = average electrophoretic mobility; μ_{eo} = electroosmotic mobility, and D_m = average solute diffusion coefficient. This equation explains why it is best to optimize resolution by adjusting $\Delta\mu$, since it is the only adjustable parameter that falls outside the square root sign. Adjusting parameters within the square-root function is less fruitful. For example, doubling the capillary length results in a maximum 41% increase in resolution. The separation time doubles only when the voltage is doubled to maintain a constant field strength. If 30 kV (maximum voltage) is already being applied, doubling the capillary length quadruples the separation time.

When the EOF and solute mobility are in the same direction, this is known as co-migration. If the EOF and solute mobility are in opposing directions, this is known as countermigration. Co-migration produces separations that are more rapid. Countermigration is particularly useful because the solutes spend more time on the capillary and thus resolution is improved. The gain in resolution occurs without a reduction in field strength, since the capillary length is kept constant. You still only realize a maximum of 41% improvement in resolution with doubling of the migration time.

3.2. Capillary Zone Electrophoresis

Since the first reports in 1981 [9], there have been at least 500 English-language papers on CZE of small molecules using simple buffers. The following approach will rapidly determine if this mode of CE is applicable to the separation problem at hand.

Set the detector to the known UV absorption maximum of the solute. If it is unknown, set the wavelength to 200 nm. If you have a diode array detector, collect all spectra. Using the diode array detector, you can also collect many channels simultaneously. For example, you can monitor at 200, 220, 240, 260, and 280 nm.

If you have reference standard material, use a solute concentration of 1 mg/mL dissolved in water, if possible. For acidic solutes, use a background electrolyte consisting of 50 mM borate buffer, pH 9.3. For basic solutes, use 50 mM phosphate buffer, pH 2.5. This and yet higher buffer concentrations optimizes sample stacking and minimizes electromigration dispersion [1]. Migration time data for hundreds of drug substances using simple buffers have been reported [10,11].

To obtain other pHs, a listing of common buffers used in my laboratory is given in Table 1. Phosphate is an excellent buffer over a wide range of pHs. It binds to the capillary wall and generally produces reproducible electroosmotic flow. Should phosphate bind to the solute, a wall effect is produced that lowers efficiency [12]. This tends to occur with some biopolymers. If the effect is noted, select an alternative buffer.

Install a 50- or 72-cm × 50-μm-i.d. capillary into the instrument. Condition the capillary for 15 min with 1 N sodium hydroxide followed by water for 2 min and BGE for 3 min, using a flush pressure of about 1 atm. Note that some separations function better without sodium hydroxide conditioning. It may be necessary to use longer buffer equilibration times to optimize precision. These parameters are optimized as part of methods development.

TABLE 1 Typical Simple Buffer Compositions

Buffer	pH	Concentration (mM)
Citrate	2.5	50
Phosphate	2.5	50
Acetate	4.0	50
MES	6.0	100
Phosphate	7.0	50
MOPS	7.0	100
Tris	8.0	100
Borate	9.3	50
CAPS	11.0	100
CABS	11.7	100

CAPS = 3-(cyclohexylamino)-1-propanesulfonic acid; CABS = 4-(cyclohexylamino)-1-butanesulfonic acid; MES = 2-(N-morpholino)ethanesulfonic acid; MOPS = 3-(N-morpholino)propanesulfonic acid.

Perform the first run on the short end of the capillary. Use an injection size of 50 millibar seconds (mbs) or 0.5 pounds per square inch seconds (psi-s), depending on which type instrument you have. Set the voltage polarity to negative. Use a voltage ramp from 0 to 30 kV over 0.3 min. If the peak does not elute within 5 min, reverse the polarity and repeat the separation. Even if a peak elutes using negative polarity, reverse the polarity and perform another run in any case. In this fashion, you can determine if any sample components are going in a direction opposite to one another. This is more likely to occur at low pH when the EOF is low or when coated capillaries are employed.

Chances are, a peak or peaks will elute within a few minutes. Since the length of the short end of the capillary is known, you can calculate the time of separation on the long end. Add 10 min to the calculated time, reverse the voltage polarity, double the injection time, and perform a run.

If a partial separation is found, adjust the pH ± 0.5 pH units and perform the additional runs. If the separation improves, continue adjusting the pH in the direction in which improvement was noted. If the pK_a's of the solutes are known, the pH equal to the average pK_a may provide the best separation. In other cases, the fully ionized solutes yield the best separation. A mobility (or migration time) plot covering a wide range of pH can be performed if desired. The short end can be used to minimize run times. The most promising pH values can then be studied on the long end of the capillary.

An illustration of a pH versus migration time plot is shown in Figure 1 [13]. While the separation continues to improve as the pH is increased, pH 9.4 with 20 mM borate buffer provides an adequate separation of purines as shown in Figure 2. Note that very small differences in migration times produce good separations, since the peak widths are narrow.

Even if the separation is still incomplete, increasing the buffer concentration from 20 mM to 50 or 75 mM and even higher may provide for a complete separation. If the currents get too high (>100 μA), consider the use of a 25-μm-i.d. capillary. Once a good separation is obtained, try shortening the capillary length.

If broad peaks are noted during these scouting runs, the solute may be adhering to the capillary wall. This is commonplace when polycations are being separated. Before any further optimization can be performed, this problem must be dealt with. For low-pH separations, a coated capillary can be tried. Alternatively, the charge on the capillary wall can be reversed with a cationic surfactant [14] or a polycation [15]. Under these conditions, the electroosmotic flow is reversed and separations are performed under conditions of reversed polarity. Usually, one of these techniques will eliminate or reduce the wall effect.

Not all separations can be accomplished with simple buffers. For example, chiral recognition and separation of neutral molecules always require additives. Even charged-molecule separations can often be improved via secondary equi-

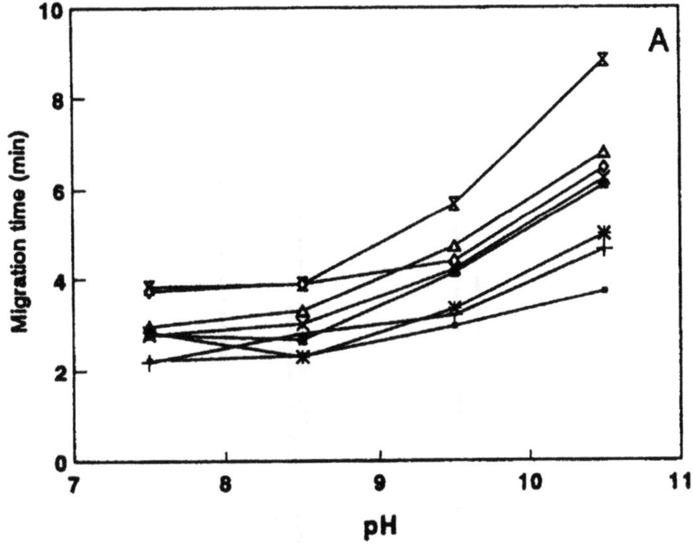

FIGURE 1 Influence of pH on migration times of purine compounds. Conditions: capillary, 44 cm (37 cm to detector) × 75 μm i.d.; buffer, 20 mM borate; voltage, 20 kV; temperature, 37°C; detection, UV, 254 nm. Key: ■ = adenine, + = adenosine, * = guanine, × = guanosine, □ = hypoxanthine, △ = xanthine, X = uric acid. (Reprinted from Ref. 13 with permission of Elsevier Science Publishers.)

librium. There are many powerful tools at our disposal to engineer selectivity into the separation. A list of typical reagents used in my laboratory is given in Table 2. This list is far from all inclusive but is sufficient to solve most small-molecule separation problems. These techniques will be described in the forthcoming sections.

3.3. Micellar Electrokinetic Capillary Chromatography

The first reports of micellar electrokinetic capillary chromatography (MECC or MEKC) appeared in the literature in 1984 [16]. Through April 2002, there have been over 1100 English-language papers published in the field. The subject is covered in all general textbooks on capillary electrophoresis. While a vast number of surfactants and related reagents can be employed, most separations can be accomplished with a few simple recipes.

A high percentage of published papers use sodium dodecyl sulfate (SDS) as the micelle-forming surfactant. Most separations are performed using 20 mM phosphate buffer, pH 7, or 20 mM borate buffer, pH 9.3, containing 20–150

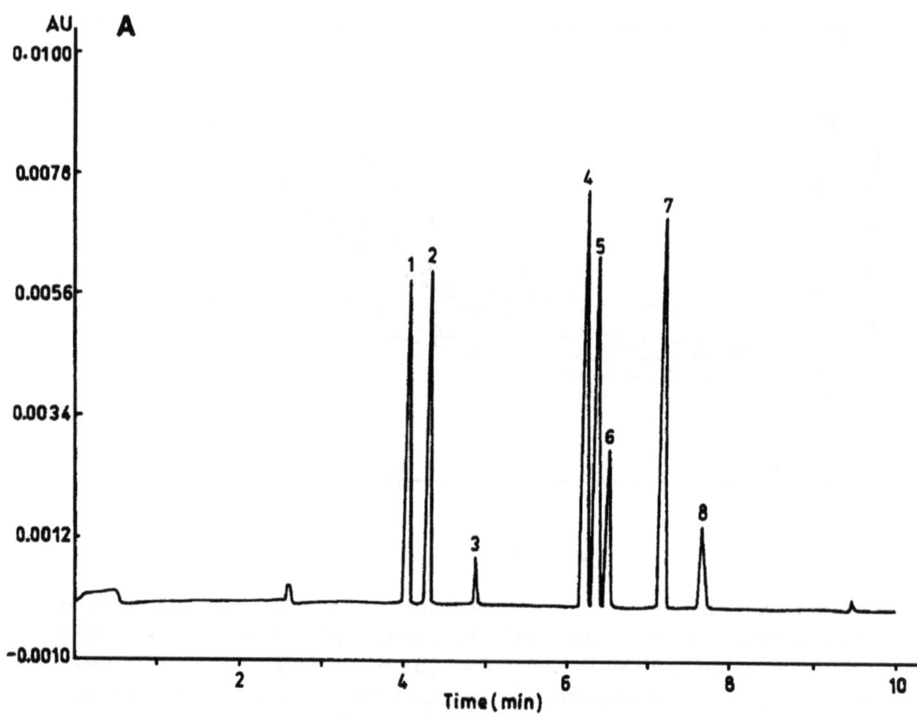

Figure 2 Optimized electropherogram for purine compounds at pH 9.4. Other conditions as per Figure 1. (Reprinted from Ref. 13 with permission of Elsevier Science Publishers.)

Table 2 Reagents for Secondary Equilibrium

Reagent	Concentration (mM)	Application
Boric acid	100–200	Carbohydrates
Brij 35	10–100	MECC
CETAB	20–100	MECC
Cyclodextrins	5–50	Chiral
SDS	20–150	MECC
Sodium cholate	20–100	MECC
Silver ion	1–5	Alkenes

SDS = sodium dodecyl sulfate; CETAB = cetyltrimethylammonium bromide.

mM SDS. Many papers report a blend of phosphate/borate buffers, but I find that either buffer is sufficient by itself.

Select the detection wavelength as described in Section 3.2, install a 50- or 72-cm × 50-µm-i.d. capillary, prepare aqueous solutions of solutes at a concentration of 1 mg/mL* and perform a scouting run on the short end of the capillary using a 50-mbs injection with 50 mM SDS/20 mM borate, pH 9.3, as the BGE. In some instances a complete separation may be found. If not, repeat the separation of the long end. The next experiments are based on your observations. These are summarized in Figure 3 and discussed in more detail below.

1. If migration times are short (< 10 min) and separation is incomplete, increase the SDS concentration in 25 mM increments.

2. If migration times are long (> than 30 min), lower the SDS concentration to 25 mM. If cations are being separated, raise the pH to above the pK_a to neutralize or partially neutralize the solutes. For runs above pH 10.3, use CAPS (or CABS buffer) in place of borate.

3. If migration times are short and anions are being separated, lower the pH, but never lower it below pH 6. If the pH is lower than 6, the EOF may not be sufficiently strong enough to sweep the surfactant toward the cathode (negative electrode). You can run at pH 2.5, but a coated capillary is advised to completely suppress the EOF [17]. This separation is run using reverse polarity, as the micelle now migrates toward the anode. In this example, solutes that spend more time attached to the micelle now elute first.

4. If migration times are long and neutral hydrophobic solutes are being separated, they are spending too much time attached to the micelle. Conventional practice calls for adding an organic solvent such as acetonitrile or methanol [18], but this can lead to migration-time imprecision due to solvent evaporation. It is my personal preference to use totally aqueous systems. If the solutes absorb above 220 nm, prepare the electrolyte using 6 M urea and repeat the run. Mixed-mode separations employing cyclodextrins can be employed here as well [19] (see step 6, below). If the separation is not forthcoming, small amounts of organic solvents may be required. Acetonitrile is the preferred solvent, since it does not affect the EOF. Use 0–20% of the solvent mixed with BGE. It is best to try the solutions offered below prior to using organic solvents.

The next series of experiments is empirical in nature, but they seem to work when separations are troublesome:

5. Add 20–50 mM Brij 35 to the electrolyte. Other nonionic surfactants can also be considered, as long as they do not absorb in the UV. This combination forms a mixed micelle with a lower mobility. The migration time is short-

*Do not dissolve the solute in an organic solvent. This will lead to bandbroadening and possibly peak splitting. Try dissolving the solute in BGE for scouting runs if it is insoluble in water.

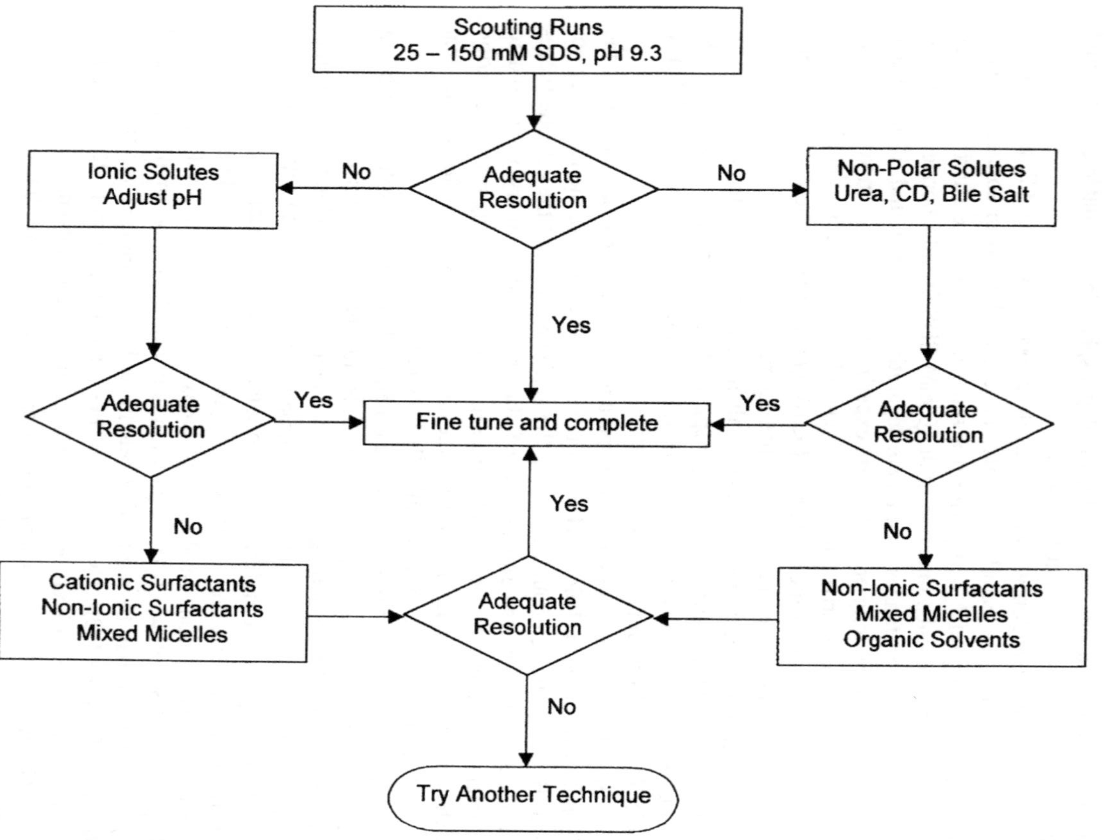

FIGURE 3 Flow chart for methods development by MECC.

ened and substantial changes selectivity may be observed [20]. For charged solutes, try 50 mM Brij 35 without SDS in the appropriate buffer [21].

6. Add a cyclodextrin to the electrolyte [19]. The neutral cyclodextrin serves as a hydrophobic vehicle to carry hydrophobic solutes through the aqueous bulk phase. Charged cyclodextrins can also be employed in the mixed mode in conjunction with SDS [22].

7. If the solutes are planar, such as a steroid [23] or macrocyclic antibiotic [24], try a bile salt micelle. Bile salts can be used in conjunction with SDS as well [25].

8. If the migration times are too long for cations or too short for anions, use a cationic surfactant such as 50 mM cetyltrimethylammonium bromide in the appropriate buffer [26,27]. Since the EOF is reversed, be sure to run scouting separations to ensure the voltage polarity is set properly. Be aware that most cationic surfactants absorb in the low UV, below 220 nm. This can affect the LOD of the method. You can also follow the experiments described in the above sections if necessary. These include pH variation and the use of additives such as Brij 35, β-CD, and urea. Once a capillary has been used with a cationic surfactant, it often cannot be used for any other application.

While the numbers of possible experiments seem enormous, I seldom get past step 6 and usually steps 1–4 do the job. It is easy to prepare these reagents, and the capillary equilibrates quickly. Even for difficult separations, only a few days of testing are usually required to establish feasibility of separation.

Please note that I have not covered the whole of MECC. One notable omission involves the use of surfactants that generate in-situ charge [28]. Another related omission is microemulsion electrokinetic capillary chromatography (MEEKC) [29]. I have not had the opportunity or the need to try these systems.

3.4. Cyclodextrins

While chiral recognition and cyclodextrins (CDs) are synonymous, CDs are also alternative reagents for generating secondary equilibrium. They can be used in place of and in combination with SDS and bile salts. A particular advantage favoring CDs is that they do not form micelles and are largely unaffected by the overall composition of the BGE. Applications include chiral recognition, structural isomer separation, impurity separations, and general applications of secondary equilibria to enhance selectivity.

Despite the seemingly small market size, a wide and diverse selection of CDs is available.* These include the naturally occurring and neutral CDs such

*The widest variety of CDs is available from Cyclolab (Budapest, Hungary). Beckman Instruments and J&W Scientific have highly sulfated CDs. The sulfobutylether variety are available from Cydex (Overland Park, KS).

as α-, β-, and γ-CD (6, 7, or 8 glucopyranose units). These materials are produced by bacterial digestion of starches. The most useful CD for separating compounds containing one or two aromatic rings is β-CD, but this material is not very soluble in aqueous solution. Hydroxypropyl-β-CD is a functionalized material that is much more soluble. Other materials such as heptakis(2,6-di-O-methyl)-β-CD and carboxymethyl-β-CD may be used as well.

In my lab I always start with sulfobutylether-β-CD, particularly for chiral and structural isomer separation. This negatively charged CD produces the countermigration mode of CE. If the solute does not separate using this material, chances are it will not using the simpler CDs. The same can be said for highly sulfated CDs. If separations do occur with this material, it is worthwhile to try the simpler materials to determine if a lower-cost reagent can do the job.

The scheme for methods development is quite simple and is similar to that given in Section 3.1. When charged CDs are used, it is important to perform scouting runs on the short end of the capillary using both positive and negative polarity. At low pH, the negatively charged CD may migrate toward the positive electrode and carry a neutral or cationic solute in that direction.

Select the appropriate buffer based on the charge of the solute, phosphate, pH 2.5, or borate, pH 9.3. Since the sulfated CDs are highly charged, start with a 20 mM buffer to minimize current. Prepare both a 5 mM and a 25 mM CD solution in the buffer. While most separations require the more concentrated reagent, some solutes bind tightly to the CD, and for those, the lower concentration is necessary. If it is necessary to use higher sulfated CD concentrations or if a short capillary is employed, a 25-μm-i.d. capillary can be used to lower the current. Alternatively, the field strength can be lowered, but this increases the time of separation. To minimize both current and capillary length, separations using highly sulfated CDs are often run at 15°C.

Dual cyclodextrin systems can be developed to provide impressive selectivity of separation [30—32]. A typical recipe uses 0.5–1 mM of a charged CD and 5–10 mM of a neutral CD. The development of a dual-CD system chiral separation of secondary amine drug substances is illustrated in Figure 4 using various blends of CDs. A complete separation is only found in run (d), where the buffer contained 5 mM of a neutral and 1 mM of a charged CD.

3.5. Transition Metals

Only a few references on the use of transition metals appear in the literature [33,34]. The separations can be so striking, so one is included here. Silver ion is very valuable for separating structural isomers of alkenes. The BGE consists of 2.5 mM silver(I), 5 mM SDS, and 32.5 mM borate buffer, pH 9.3. SDS is

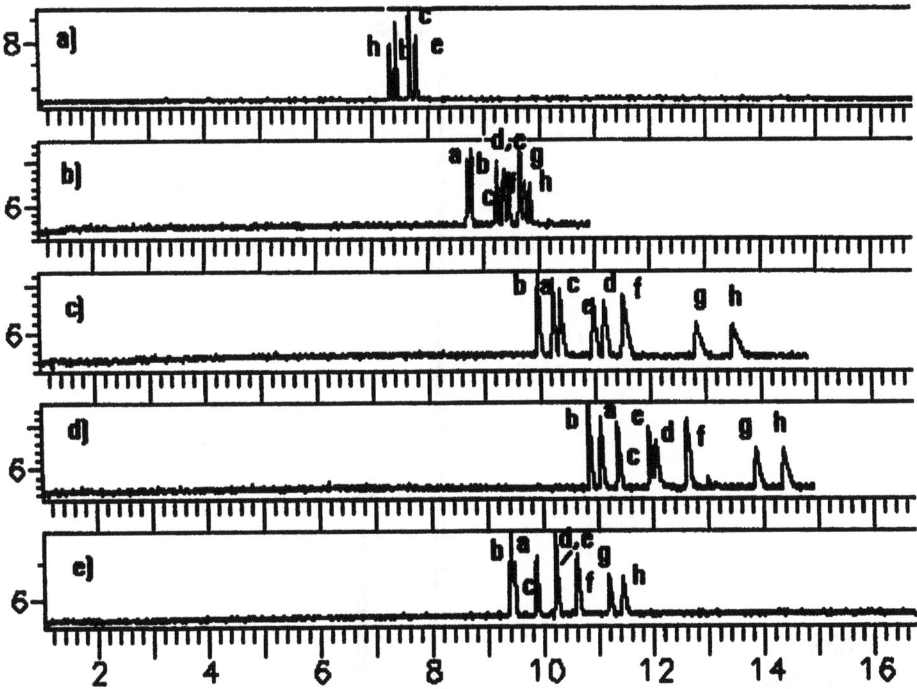

FIGURE 4 Development of the separation of secondary amine drug standards using dual cyclodextrins. Conditions: capillary, 82 cm (60 cm effective length) × 50 μm i.d.; injection, 0.5 psi vacuum, 0.5 s; temperature, 30°C; voltage, 30 kV; detection, UV, 210 nm. Buffers: (a) 25 mM tris-phosphate, pH 2.1; (b) 98.5% 5 mM dimethyl-β-CD (DM-β-CD) in buffer (a), 1.5% methanol; (c) as in (b) except 1 mM sulfobutylether(IV)-β-CD (SBE-β-CD); (d) as in (b) except 5 mM DM-β-CD and 1 mM SBE-β-CD; (e) as in (b) except 5 mM DM-β-CD and 0.5 mM SBE-β-CD. Key: (a) (+)-methcathinone, (b) (−) methcathinone, (c) (−)-pseudoephedrine, (d) (−)-ephedrine, (e) (+)-ephedrine, (f) (+)-pseudoephedrine, (g) (−)-methamphetamine, (h) (+)-methamphetamine. (Reprinted from Ref. 30 with permission of the American Chemical Society.)

used at a concentration below the critical micelle concentration to prevent silver ion from binding to the capillary wall. In the absence of SDS, a moderate wall effect was found. The separation was complete in less than 4 min (Figure 5). Without silver ion, no separation occurs. In addition to silver, consider the use of other transition metals to adjust selectivity for metal-binding proteins, peptides, or other polyvalent ions.

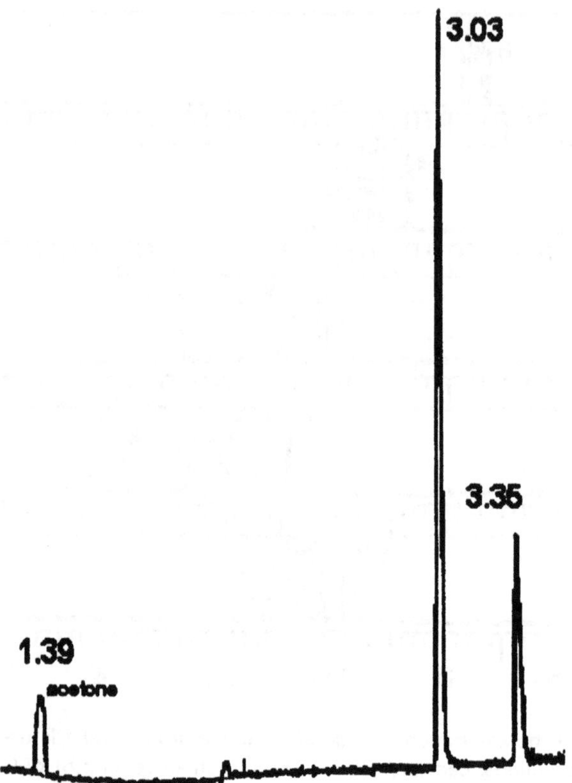

FIGURE 5 Separation of cis/trans isomers of an alkene using silver ion. Capillary, 26 cm (20 cm to detector) × 50 μm i.d.; BGE, 32.5 mM borate, pH 9, 2.5 mM silver nitrate, 5.0 mM SDS; voltage, 30 kV; injection, 0.5 psi for 5 s; temperature, 40°C, detection, UV, 213 nm.

4. COMPLETING THE METHOD

4.1. Is a Capillary Wash Procedure Needed?

The best wash procedure is no wash procedure, but this implies that the migration times are stable. If migration times increase with each ensuing run, add a 1–2-min 0.1 N sodium hydroxide wash. If the trend shows a decrease, add a 1–2-min acid wash. While 0.1 N phosphoric acid is often used, be careful if phosphate buffer is not used for the separation. Wall effects can occur if phosphate binds to both the wall and the solute. If protein or other cations are sticking to the capillary wall, add a 2-min 0.1 M SDS wash step [35]. Before you launch wash procedure studies, ensure that the buffer is effective at the selected pH. Change the buffer after each run and see if depletion effects are causing the drift. If migration-time drift is persistent, try a dynamic coating such as CElixir (Microsolve Technologies, Long Branch, NJ) [36,37]. Even for well-controlled methods, dynamic coatings usually improve the migration time and peak area precision on a run-to-run, day-to-day, and capillary-to-capillary basis [36].

4.2. Is Sample Preparation Necessary?

While it is wonderful to be able to inject neat samples directly, sample preparation can often improve selectivity and sensitivity. If the resolution is poor, the salt content of the sample too high, or the capillary fouls, consider a sample cleanup. This can include liquid–liquid extraction, solid-phase extraction, supercritical fluid extraction, protein precipitation, or dialysis, depending on the solutes and application [38]. The final sample diluent should be a solution that is CE-friendly. That usually means low ionic strength compared to the BGE.

Aqueous extraction procedures for water-soluble molecules are especially advantageous in CE. For example, toluene extraction of nicotine from tobacco prior to gas chromatography gives low recoveries compared to aqueous extraction and CE [39].

When solvent extraction is required, evaporation to dryness followed by dissolution of the residue in 100 μM phosphoric acid (for bases) provides for sample concentration as well as the ability to perform field-amplified stacking [40]. This is discussed in Section 4.4.

The determination of drug substances in blood serum and plasma is often carried out by protein precipitation with acetonitrile/trifluoroacetic acid mixtures. This combination produces impressive sample stacking as well [41].

4.3. Must the Sample Be Desalted?

CE is quite tolerant to salt, provided only small injections are made. If large injections and sample stacking is required, it is necessary to desalt the sample prior to injection. High concentrations of salts increase conductivity and cause antistacking (band broadening) to occur.

Figure 6 illustrates some separations of a 20-mer antisense oligonucleotide with a neutral backbone. With the sample diluted in water, a symmetrical peak is obtained for a small 100-mbs injection (data not shown). The buffer is 100 mM CAPS, pH 11.7, with 6 M urea. The solute is separated as an anion, with thymine and guanine being ionized at the high pH. Figure 6 shows the separations at expanded scale. Some subtle but surprising observations are made. In 1 M salt diluent, two impurity peaks on the tail of the main component show improved resolution, while the impurity peak that elutes just before the main

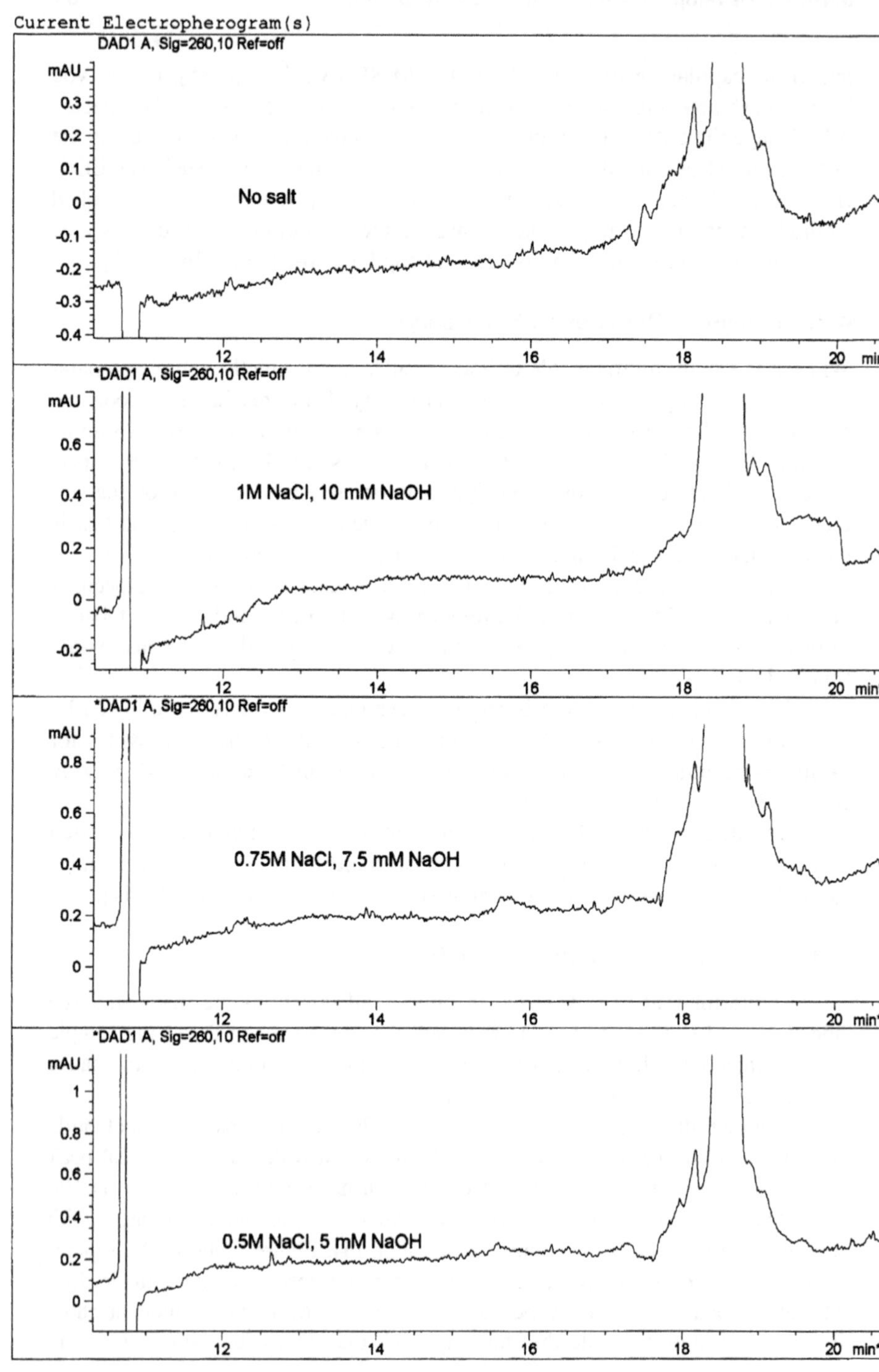

Current Electropherogram(s)
DAD1 A, Sig=260,10 Ref=off
mAU
0.3
0.2
0.1
0
-0.1
-0.2
-0.3
-0.4
No salt
12
14
16
18
20 min
*DAD1 A, Sig=260,10 Ref=off
mAU
0.6
0.4
0.2
0
-0.2
1M NaCl, 10 mM NaOH
12
14
16
18
20 min*
*DAD1 A, Sig=260,10 Ref=off
mAU
0.8
0.6
0.4
0.2
0
0.75M NaCl, 7.5 mM NaOH
12
14
16
18
20 min*
*DAD1 A, Sig=260,10 Ref=off
mAU
1
0.8
0.6
0.4
0.2
0
0.5M NaCl, 5 mM NaOH
12
14
16
18
20 min*

component is no longer resolved. In 0.75 M salt, the early-eluting impurity peak begins to resolve as the resolution of the later-eluting impurities begins to deteriorate. The separation in 0.5 M salt is virtually identical to that of the sample dissolved in water.

One explanation for the sharpening of the later-eluting impurities is a phenomenon called sample self-stacking [42,43]. This is a variant of transient isotachophoresis (ITP) [44]. In this case, the leading electrolyte is chloride ion, while the main component acts as the terminating electrolyte. The requirement for ITP is fulfilled since the impurity's mobility is bracketed by leader and terminator. While you may not understand all of this, accidental ITP is not rare [45]. When it does occur, the sharpened peaks are often incorrectly considered to be artifacts.

4.4. Is Sample Stacking Necessary?

Ionic strength-mediated stacking can improve detectability by a factor of 10. Use as a final sample diluent a 10-fold dilution of the background electrolyte [46]. This will enable large injections, upwards of 5–10% of the capillary volume, to be made. Do not inject large quantities of water into the capillary. If the conductivity of the sample plug is very low, the field strength becomes enormous over the point of injection. The generated heat can decompose labile solutes or even fracture the capillary. Band broadening may also be found for large water injections, since the high field over the sample zone may cause electroosmotic pumping [46].

For yet larger injections, other, more specialized techniques such as pH mediated stacking [47] (for zwitterions) or transient isotachophoresis (difficult to implement) can be considered [44]. Neutral molecules can also be effectively stacked using charged surfactants or CDs [48,49].

Field-amplified or electrokinetic injection stacking can provide impressive LODs, so long as injections are made from solutions with very low conductivity [50]. For small anions and cations, LODs easily reach the low-ppb range when the sample is ultrapure water. This technique is commonly employed to monitor water quality in the semiconductor industry.

For small molecules and peptides, it is necessary to add a small amount of acid or base to ionize the solutes. The impact of acid addition to some peptide

---

FACING PAGE

Figure 6 Separation of an oligonucleotide dissolved in the specified matrices. Conditions: capillary, 72 cm × 50 μm i.d. bubble factor 3; BGE, 100 mM CAPS, pH 11.7, 6 M urea; voltage, 30 kV; temperature, 20°C; injection, 100 mbs; detection, UV, 260 nm.

samples is shown in Figure 7. The solutes are small peptides at a concentration of 1 μg/mL. When no acid is added, only a small signal is obtained for a 15-s, 10-kV injection. As the acid in the sample is increased, the signal increases accordingly. At this concentration, the optimal acid concentration is 100 μM.

When the sample size is small (10–20 μL), it becomes possible to effectively inject most of the ions in the sample in less than 30 s at 5 kV. A small water plug injected hydrodynamically has been shown to further improve sensitivity [51], though this was not the case for the above-mentioned application. Since the sample conductivity severely affects the amount of solute injected, internal standards are important for this mode of stacking. For quantitative analysis, make only a single injection out of each vial, as the sample becomes ion-depleted.

4.5. What Is the Required Linear Dynamic Range of the Method?

Most CE methods have a linear dynamic range of about 10^4. At the high-concentration portion of the calibration curve, electrodispersion may cause substantial band broadening, as shown in Figure 8 [52]. If the separation can be performed using charged micelles or cyclodextrins at a pH at which the solute is neutral, no such band broadening will occur. In this case, the solute concentration can approach 10 mg/mL. This makes low-level impurities simpler to detect [53]. This is illustrated with separations of impurities found in the manufacture of cresylic acid, a mixture of *o*-, *m*-, and *p*-cresol. Figure 9 (top) shows an MECC separation at pH 7 of a 1% solution of cresylic acid. The cresols give an amazing 2-AU signal and exhibit slight band broadening due to micellar saturation. When the scale is expanded (Figure 9, bottom), perfectly symmetrical and sharp peaks appear for the catechol and phenol impurities. This would never be observed if charged solutes were employed. The technique should allow the determination of trace impurities in pharmaceuticals with LODs approaching 0.01%.

4.6. Are Internal Standards Necessary?

Internal standards usually improve quantitative precision because injection bias is compensated for. If the internal standard elutes at the correct migration time

---

FACING PAGE

Figure 7 Impact of added acid on peptides using field-amplified electrokinetic injection. Conditions: capillary, 40 cm (effective length) × 50 μm BF3; buffer, 50 mM phosphate, pH 2.5; buffer equilibration, 3-min rinse between runs, dip inlet in water prior to injection; voltage, 30 kV; injection, 10 kV for 15 s; temperature, 30°C; detection, UV, 200 nm; sample, peptide GGR, 0.54 μg/mL; GGYR, 0.92 μg/mL.

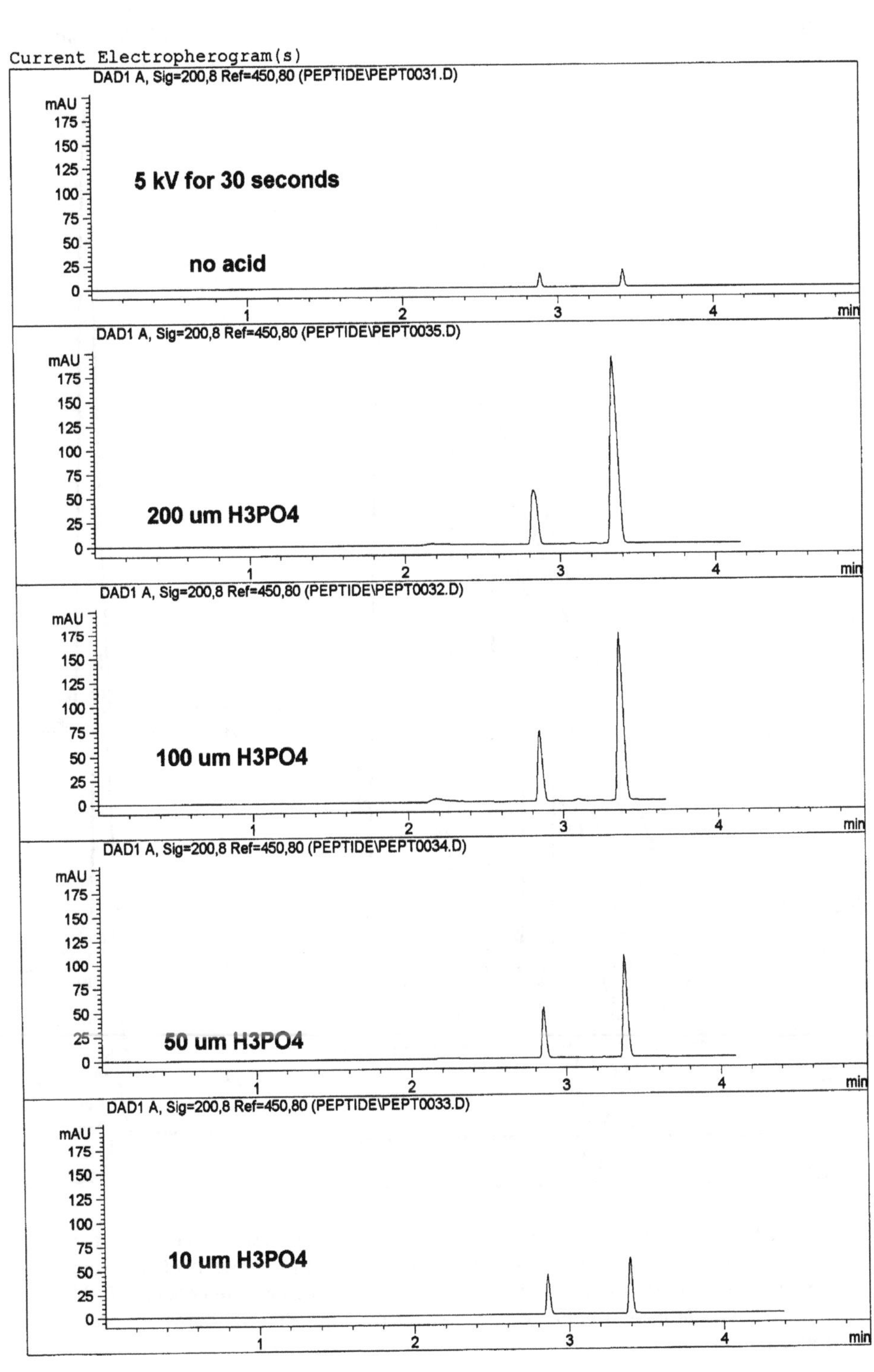

Current Electropheregram(s)
DAD1 A, Sig=200,8 Ref=450,80 (PEPTIDE\PEPT0031.D)
mAU
5 kV for 30 seconds
no acid
min
DAD1 A, Sig=200,8 Ref=450,80 (PEPTIDE\PEPT0035.D)
mAU
200 um H3PO4
min
DAD1 A, Sig=200,8 Ref=450,80 (PEPTIDE\PEPT0032.D)
mAU
100 um H3PO4
min
DAD1 A, Sig=200,8 Ref=450,80 (PEPTIDE\PEPT0034.D)
mAU
50 um H3PO4
min
DAD1 A, Sig=200,8 Ref=450,80 (PEPTIDE\PEPT0033.D)
mAU
10 um H3PO4
min

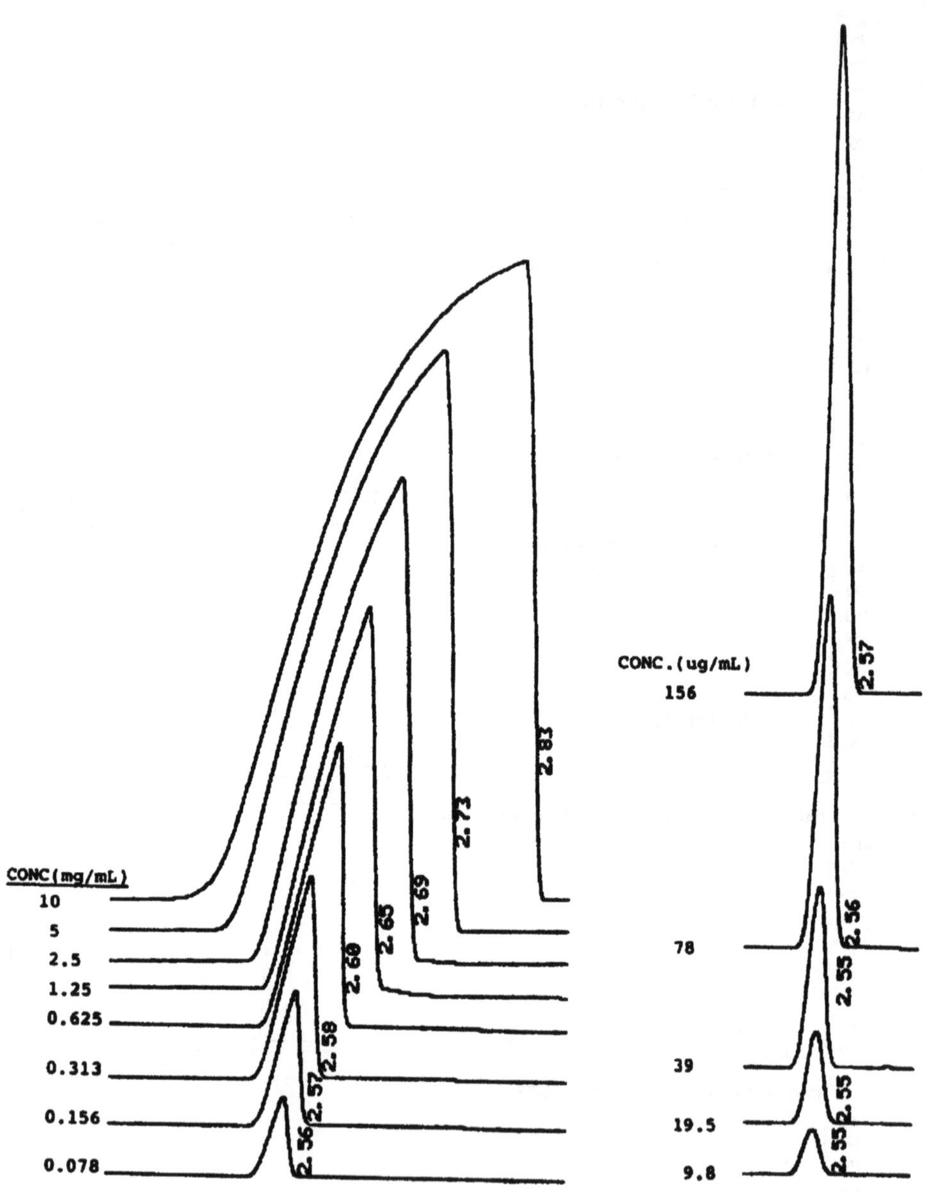

FIGURE 8 Effect of solute concentration on electromigration dispersion. Conditions: capillary, 38 cm to detector × 50 μm; buffer, 20 mM borate, pH 9.2; voltage, 25 kV; temperature, 50°C; injection, 1 s at 0.5 psi vacuum; detection, UV, 230 nm; solute: naproxen. (Reprinted with permission from Ref. 52, copyright Marcel Dekker, Inc.)

with the proper peak area, we have that warm feeling that the separation is under control. Internal standards are not required for impurity determinations by area%. Normalized peak areas (peak area/migration time) should be used when area% is employed to compensate for the impact of peak velocity on area counts [54].

4.7. Have Another Person Run Your Method

The job is not complete until another chemist can independently run your method. I have transferred many methods over the years. Despite the use of comprehensive method documentation, problems often occur during start-up in a new laboratory, particularly when chemists new to CE are involved. These often minor problems are always solved through a dialogue with the operator. Even seemingly minor details can be befuddling. For example, be sure to specify the electrolyte reservoir size in your method. Using too small an electrolyte reservoir can cause precision problems, particularly when multiple runs are performed from a set of electrolytes.

4.8. Method Validation

There have been over 100 papers published in the scientific literature on validation of methods using CE, including a few review articles [55,56]. For those needing an introduction to the field, the following references are best consulted [57,58], as well as an article on validation of LC methods [59]. CE methods validate in the same fashion as LC methods.

5. CONCLUDING COMMENTS

I recently had the task of evaluating a method development study performed by a contract research laboratory and conducted using good laboratory practices (GLPs). Copious amounts of data were generated but the analytical problem was not solved. A tremendous effort was expended studying separation conditions that did not work or could not possibly work.

GLPs are a critical part of all aspects of pharmaceutical testing from clinical trial analyses through quality control of the final drug substance. When developing analytical separation methods, the documentation and discipline required by GLPs may actually interfere with developing the optimal method in the shortest possible time frame. I believe it is preferable to be agile and creative with methods development and quickly abandon dead ends. This is the framework that has been provided in this chapter. Following these guidelines should rapidly lead you to the optimal separation. At this stage, the method is ready for validation and all aspects of GLPs must be applied.

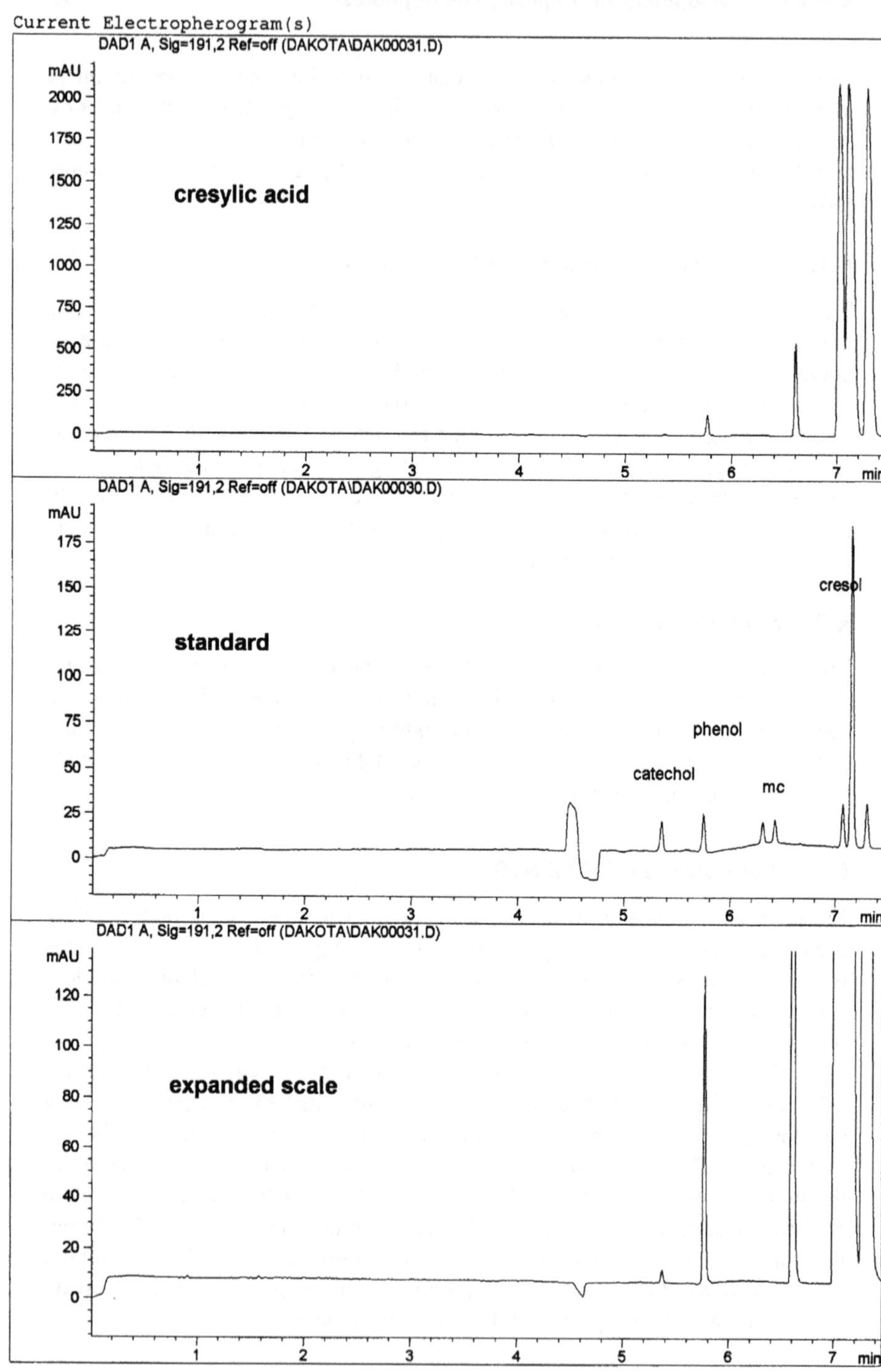
Current Electropherogram(s)
DAD1 A, Sig=191,2 Ref=off (DAKOTA\DAK00031.D)
mAU
2000
1750
1500
1250
1000
750
500
250
0
cresylic acid
1
2
3
4
5
6
7
min
DAD1 A, Sig=191,2 Ref=off (DAKOTA\DAK00030.D)
mAU
175
150
125
100
75
50
25
0
standard
cresol
phenol
catechol
mc
1
2
3
4
5
6
7
min
DAD1 A, Sig=191,2 Ref=off (DAKOTA\DAK00031.D)
mAU
120
100
80
60
40
20
0
expanded scale
1
2
3
4
5
6
7
min

FACING PAGE

FIGURE 9 MECC of cresylic acid and process impurities at pH 7. Conditions: capillary, 72 cm × 50 μm i.d. bubble factor 3; BGE, 20 mM phosphate, 50 mM SDS, pH 7; injection, 50 mbs; detection, UV, 191 nm; temperature, 50°C; voltage, 30 kV (18-s ramp); sample, 1.5% cresylic acid dissolved in water.

REFERENCES

1. Weinberger R. Practical Capillary Electrophoresis. 2nd ed. Boston: Academic Press, 2000:462.
2. Li T, Kennedy RT. Laser-induced fluorescence detection in microcolumn separations. Trends Anal Chem 1998; 17:484.
3. Jandik P, Jones WR, Weston A, Brown PR. Electrophoretic capillary ion analysis: origins, principles, and applications. LC-GC 1991; 9:634.
4. Lauer HH, McManigill D. Capillary zone electrophoresis of proteins in untreated fused silica tubing. Anal Chem 1986; 58:166.
5. Riekkola M-L, Jussila M, Possas SP, Valko IE. Non-aqueous capillary electrophoresis. J Chromatogr A 2000; 892:155.
6. Siouffi AM, Phan-Tan-Luu R. Optimization methods in chromatography and capillary electrophoresis. J Chromatogr A 2000; 892:75.
7. Macka M, Andersson P, Haddad PR. Changes in electrolyte pH due to electrolysis during capillary zone electrophoresis. Anal Chem 1998; 70:743.
8. Nelson RJ, Paulus A, Cohen AS, Guttman A, Karger BL. Use of Peltier thermoelectric devices to control column temperature in high-performance capillary electrophoresis. J Chromatogr 1989; 480:111.
9. Jorgenson JW, Lukacs K. High-resolution separations based on electrophoresis and electroosmosis. J Chromatogr 1981; 218:209.
10. Hudson JC, Golin M, Malcolm M. Capillary zone electrophoresis in a comprehensive screen for basic drugs in whole blood. J Can Soc Forensic Sci 1995; 28:137.
11. Hudson JC, Golin M, Malcome M, Whiting C. F. Capillary zone electrophoresis in a comprehensive screen for drugs of forensic interest in whole blood: an update. J Can Soc Forensic Sci 1998; 31:1.
12. Weinberger R. Separations solutions. Capillary electrophoresis of milk proteins. Am Lab 2001; 33:50.
13. Grune T, Ross GA, Schmidt H, Siems W, Perrett D. Optimized separation of purine bases and nucleosides in human cord plasma by capillary zone electrophoresis. J Chromatogr 1993; 636:105.
14. Emmer A, Roeraade J. Wall deactivation with fluorosurfactants for capillary electrophoresis of biomolecules. Electrophoresis 2001; 22:660.
15. Cordova E, Gao J, Whitesides G. Noncovalent polycationic coatings for capillaries in capillary electrophoresis of proteins. Anal Chem 1997; 69:1370.
16. Terabe S, Otsuka K, Ichikawa K, Tsuchiya A, Ando T. Electrokinetic separations with micellar solutions and open-tubular capillaries. Anal Chem 1984; 56:111.

17. Janini GM, Muschik GM, Issaq HJ. Micellar electrokinetic chromatography in zero-electroosmotic flow environment. J Chromatogr B: Biomed Appl 1996; 683:29.
18. Gorse J, Balchunas AT, Swaile DF, Sepaniak MJ. Effects of organic mobile phase modifiers in micellar electrokinetic capillary chromatography. J High Resolut Chromatogr 1988; 11:554.
19. Terabe S, Miyashita Y, Shibata O, Barnhart ER, Alexander LR, Patterson DG, Karger BL, Hosoyo K, Tanaka N. Separation of highly hydrophobic compounds by cyclodextrin-modified micellar electrokinetic chromatography. J Chromatogr 1990; 516:23.
20. Rasmussen HT, Goebel LK, McNair HM. Micellar electrokinetic chromatography employing sodium alkyl sulfates and Brij 35. J Chromatogr 1990; 517:549.
21. Cugat MJ, Borrull M. Comparative study of capillary zone electrophoresis and micellar electrokinetic capillary chromatography for the separation of naphthalene-disulfonate isomers. Analyst 2000; 125:2236.
22. Luong JHT, Guo Y. Mixed-mode capillary electrokinetic separation of positional explosive isomers using sodium dodecyl sulfate and negative-β-cyclodextrin derivatives. J Chromatogr A 1998; 811:225.
23. Hsiao M-W, Lin S-T. Separation of steroids by micellar electrokinetic capillary chromatography with sodium cholate. J Chromatogr Sci 1997; 35:259.
24. Flurer C. Analysis of macrolide antibiotics by capillary electrophoresis. Electrophoresis 1996; 17:359.
25. Issaq HJ, Horng PLC, Janini GM, Muschik GM. Micellar electrokinetic chromatography using mixed sodium dodecyl sulfate and sodium cholate. J Liq Chromatogr Related Technol 1997; 20:167.
26. Crosby D, El Rassi Z. Micellar electrokinetic capillary chromatography with cationic surfactants. J Liq Chromatogr 1993; 16:2161.
27. Sainthorant C, Morin P, Dreux M, Baudry A, Goetz N. Separation of phenylenediamine, phenol and aminophenol derivatives by micellar electrokinetic chromatography. comparison of the role of anionic and cationic surfactants. J Chromatogr A 1995; 717:167.
28. Mechref Y, Smith JT, El Rassi Z. Micellar electrokinetic capillary chromatography with in situ charged micelles. VII. Expanding the utility of alkylglycoside-borate micelles to acidic and neutral pH for capillary electrophoresis of dansyl amino acids and herbicides. J Liq Chromatogr 1995; 18:3769.
29. Altria KD. Microemulsion electrokinetic chromatography. J Capillary Electrophor 2002; 7:11.
30. Lurie IS, Klein RX, Lebelle M, Brenneisen R, Weinberger R. Chiral resolution of drugs of forensic interest by capillary electrophoresis using cyclodextrins. Anal Chem 1994; 66:4019.
31. Abushoffa AM, Fillet M, Hubert P, Crommen J. Prediction of selectivity for enantiomeric separations of uncharged compounds by capillary electrophoresis involving dual cyclodextrin systems. J Chromatogr A 2002; 948:321.
32. Lurie IS. Separation selectivity in chiral and achiral capillary electrophoresis with mixed cyclodextrins. J Chromatogr A 1997; 792:297.
33. Wright PB, Dorsey JG. Silver(I)-mediated separations by capillary zone electrophoresis and micellar electrokinetic chromatography: argentation electrophoresis. Anal Chem 1996; 68:415.

34. Mosher RA. The use of metal ion-supplemented buffers to enhance the resolution of peptides in capillary zone electrophoresis. Electrophoresis 1990; 11:765.
35. Lloyd DK, Waetzig H. Sodium dodecyl sulfate is an effective between-run rinse for capillary electrophoresis of samples in biological matrixes. J Chromatogr B: Biomed Appl 1995; 663:400.
36. Lurie IS, Bethea MJ, McKibben TD, Hays P, Pellegrini P, Sahai R, Garcia A, Weinberger R. Use of dynamically coated capillaries for the routine analysis of methamphetamine, amphetamine, MDA, MDMA, MDEA and cocaine using capillary electrophoresis. J Forensic Sci 2001; 46:1025.
37. Boone CM, Jonkers EZ, Franke JP, de Zeeuw RA, Ensing K. Dynamically coated capillaries improve the identification power of capillary zone electrophoresis for basic drugs in toxicological analysis. J Chromatogr A 2001; 927:203.
38. Weinberger R. Separations solutions. Sample preparation. Am Lab 2000; 32:48.
39. Clarke MB. Quantitation of nicotine in tobacco products by capillary electrophoresis. J AOAC Int 2002; 85:1.
40. Wey AB, Zhang C-X, Thormann W. Head-column field-amplified sample stacking in binary system capillary electrophoresis. Preparation of extract for determination of opioids in microliter amounts of body fluids. J Chromatogr A 1999; 853:95.
41. Shihabi ZK. Sample stacking by acetonitrile-salt mixtures. J Capillary Electrophor 1995; 2:267.
42. Gebauer P, Thormann W, Bocek P. Sample self-stacking in zone electrophoresis. theoretical description of the zone electrophoretic separation of minor components in the presence of bulk amounts of a sample component with high mobility and like charge. J Chromatogr 1992; 608:47.
43. Gebauer P, Krivankova L, Pantuckova P, Bocek P, Thormann W. Sample self-stacking in capillary zone electrophoresis: behavior of samples containing multiple major coionic components. Electrophoresis 2000; 21:2797.
44. Weinberger R. Separations solutions. Transient isotachophoresis. Am Lab 1997; 29:49.
45. Weinberger R, Sapp E, Moring S. Capillary electrophoresis of urinary porphyrins with absorbance and fluorescence detection. J Chromatogr 1990; 516:271.
46. Burgi D, Chien R-L. Optimization in sample stacking for high-performance capillary electrophoresis. Anal Chem 1991; 63:2042.
47. Aebersold R, Morrison H. Analysis of dilute peptide samples by capillary zone electrophoresis. J Chromatogr 1990; 516:79.
48. Palmer J, Munro NJ, Landers JP. A universal concept for stacking neutral analytes in micellar capillary electrophoresis. Anal Chem 1999; 71:1679.
49. Quirino JP, Terabe S. On-line concentration of neutral analytes for micellar electrokinetic chromatography I. Normal stacking mode. J Chromatogr A 1997; 781: 119.
50. Zhang C-X, Thormann W. Head-column field-amplified sample stacking in binary system capillary electrophoresis: a robust approach providing over 1000-fold sensitivity enhancement. Anal Chem 1996; 68:2523.
51. Wey AB, Thormann W. Head-column field-amplified sample stacking in binary-system capillary electrophoresis. The need for the water plug. Chromatographia 1999; 49(Suppl. 1):S12.

52. Weinberger R, Albin M. Quantitative micellar electrokinetic capillary chromatography: linear dynamic range. J Liq Chromatogr 1991; 14:953.
53. Weinberger R, Sayler K. Separations solutions. Capillary electrophoresis of phenol process streams. Am Lab 2002; 34:68.
54. Huang X, Coleman WF, Zare RN. Analysis of factors causing peak broadening in capillary zone electrophoresis. J Chromatogr 1989; 480:95.
55. Fabre H. Validation for methods using capillary electrophoresis in pharmaceutical compounds. Analysis 1999; 27:155.
56. Waetzig H, Dengenhardt M, Kunkel A. Strategies for capillary electrophoresis. Method development and validation for pharmaceutical and biological applications. Electrophoresis 1998; 19:2695.
57. Huber L. A Primer. Good Laboratory Practice and Current Good Manufacturing Practice. Waldbronn, Germany: Agilent Technologies, 2000:116.
58. Swartz ME, Krull IS. Analytical Method Development and Validation. New York: Marcel Dekker, 1998:92.
59. Huber L. Validation of HPLC methods. Biopharm 1999; 12:64.

3

Capillary Isoelectric Focusing

Ferenc Kilár
University of Pécs, Pécs, Hungary

1. INTRODUCTION

Capillary isoelectric focusing (CIEF) is one of the separation techniques with the highest resolution power. Since the first experiments performed by Hjertén and co-workers in the mid-1980s [1–3], hundreds of papers have appeared about methodological aspects and utilization.

Isoelectric focusing has been commonly performed in flat-bed, slab, or tube gel format [4]. Recently the most frequent application of this technique in biochemical analysis is two-dimensional polyacrylamide gel electrophoresis. The separation of zwitterionic compounds, e.g., proteins, is achieved by a pH gradient which is generated by ampholytes (or other amphoteric buffer components) in an electric field. The sample components focus according to their isoelectric points in the pH gradient between two electrolytes (e.g., sodium hydroxide as catholyte and phosphoric acid as anolyte). Any band broadening caused by thermal diffusion is quickly reduced by the existing pH gradient, because the analytes drifting out from a focused zone always become charged in such a way that they migrate back toward their isoelectric points (into the focused zone). In gel isoelectric focusing the focused zones have to be fixed and stained for visualization.

Since gel methods generally require tedious and time-consuming gel preparation and staining procedures, capillary isoelectric focusing provides a versatile alternative technique for analysing amphoteric compounds. The absence of any matrices reveals several advantages, such as analysis of small molecules or even cellular structures, as well as on-line detection without staining.

In spite of all these clear advantages, capillary isoelectric focusing is still not used as a routine method in bioanalysis, although more and more papers show important applications. Study of the proteome combined with microchip technology may, however, cause sudden progress in this innovative technique. In the past decade several reviews have appeared about CIEF separations [5–17]. This chapter summarizes the theory and the newest innovations, as well as applications of CIEF.

2. CAPILLARY ISOELECTRIC FOCUSING MODES

Capillary isoelectric focusing separations can be discussed from two major viewpoints, the capillary type and the detection techniques used, but other practical questions, e.g., ampholytes, sample pretreatment, etc., should also be considered.

The most distinguishing phenomenon in capillary electrophoresis is whether the experiments are performed in the absence or in the presence of electroosmotic flow (EOF), (see Chapter 6 for details on EOF). Unlike other types of capillary electrophoresis, isoelectric focusing can be performed under both modes. Since the experimental and theoretical principles governing these modes of CIEF are different, they will be discussed separately.

In the absence of EOF (e.g., in non-cross-linked acrylamide-coated capillaries), the focused zones should be mobilized either chemically or with hydrodynamic flow. During this mobilization, maintenance of the electric field serves as a stabilizer to prevent zone broadening. A common name for such a setup is "two-step" capillary isoelectric focusing.

In the presence of electroendosmosis, the mobilization of the focused zones is achieved by the EOF. The so called "single-step" isoelectric focusing is performed, however, under conditions in which EOF is controlled by additives and the dynamic coating of the capillary surface provides reduced bulk flow. This mode of CIEF can also be combined with hydrodynamic elution.

In the following sections, techniques for performing isoelectric focusing in coated or uncoated capillaries in the absence or presence of additives will be discussed. Different electrophoretic and/or hydrodynamic mobilization procedures will also be considered.

2.1. Isoelectric Focusing in Capillaries with Minimized Electroendosmosis

Electroosmotic flow can be reduced and even eliminated through coating of the internal surface of the capillaries. Capillaries coated with, e.g., methylcellulose or non-cross-linked acrylamide have negligible EOF [18]. As a result, the focusing of substances will depend principally on the quality of the ampholytes and other parameters (discussed below), but it is not influenced by electroendosmosis. Isoelectric focusing performed in coated capillaries resembles most the commonly used gel isoelectric focusing methods.

2.1.1. Two-Step Isoelectric Focusing with Electrokinetic Mobilization

The schematic representation of isoelectric focusing in a system introduced by Hjertén and co-workers [1,2,19] is shown in Figure 1.

The substances to be separated are applied to the capillary together with ampholytes. Typically, the entire capillary is filled with the ampholyte/sample mixture and CIEF is accomplished through two consecutive events: (pre)focusing and mobilization. The ampholytes will create the pH gradient within a short time upon application of electric current. At the same time the substances are forced to move toward their isoelectric points in the pH gradient. During this movement the molecules are concentrated in migrating boundaries. Those bound-

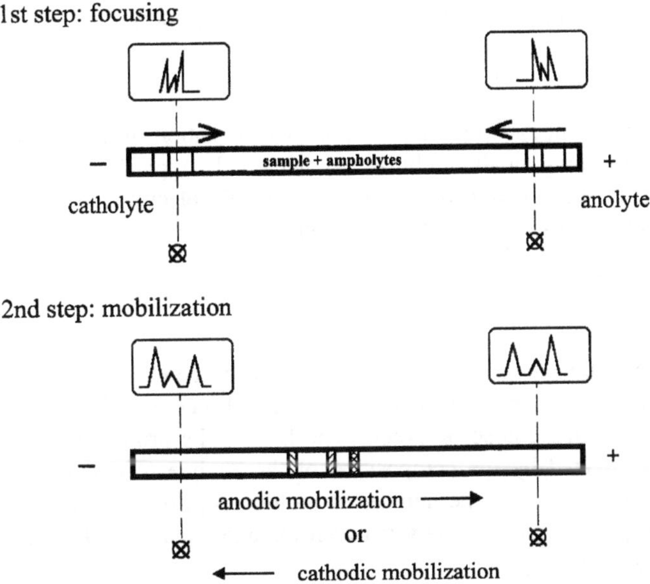

FIGURE 1 Schematic representation of isoelectric focusing in coated capillaries. Both patterns obtained in the focusing and mobilization steps are characteristic of the sample. Mobilization of the focused zones may occur at the anodic or the cathodic end by introducing an appropriate ion in the electrolyte vessel. The order of the appearance of the components in the detector window is opposite in the two steps.

aries may be detected at the detection point and hence "peaks" can be recorded. These "peaks," however, do not feature the common Gaussian distribution pattern, since they are not real electrophoretic zones, although the pattern obtained during this first step is characteristic of the sample. The focused zones are then mobilized electrokinetically by replacing one electrolyte at one end of the tube. For instance, anodic mobilization occurs when the phosphoric acid at the anode is replaced by sodium hydroxide solution or, on the other hand, sodium hydroxide is replaced by sodium chloride containing sodium hydroxide electrolyte at the cathodic end [19]. Commonly a UV light beam is used for on-tube detection.

Hjertén et al. [19] derived expressions which describe the theoretical basis of electrokinetic mobilization. The electroneutrality condition at steady state in the separation tube during focusing is

$$C_{H^+} + \sum C_{NH_3^+} = C_{OH^-} + \sum C_{COO^-} \quad (1)$$

where C_{H^+}, $\Sigma C_{NH_3^+}$, C_{OH^-} and ΣC_{COO^-} are the concentrations in equivalents per liter (or C/cm) of protons, hydroxyl ions, and positive and negative groups in the carrier ampholytes, respectively. Mobilization can be achieved by adding a positive term to the left side of Eq. (1), which then takes the form

$$C_{X^{n+}} + C_{H^+} + \sum C_{NH_3^+} = C_{OH^-} + \sum C_{COO^-} \quad (2)$$

where X^{n+} (n is the valency) represents a cation. This equation illustrates one approach for accomplishing anodic mobilization, namely, by replacing the anolyte used for focusing with a cation which can enter the tube electrophoretically.

The analogous expression for cathodic mobilization is

$$C_{H^+} + \sum C_{NH_3^+} = C_{OH^-} + \sum C_{COO^-} + C_{Y^{m-}} \quad (3)$$

where Y^{m-} is an anion.

The above equations regarding the electroneutrality conditions indicate that the cations (anions) entering the separation tube will cause a pH change, but they do not reveal the course of events. The flux of the protons into the separation tube is affected by the composition of the anolyte (catholyte). At steady state in the focusing step the number of protons, N_{H^+}, from the anolyte electrophoretically passing the boundary between the anolyte and medium per time unit can be expressed by

$$N_{H^+} = \nu_{H^+} + q n_{H^+} \quad (4)$$

where ν_{H^+} is the migration velocity of the protons in the anolyte, q is the cross-sectional area of the tube, and n_{H^+} is the number per volume unit in the anolyte. Since $\nu_{H^+} = E u_{H^+}$ and $E = I/q\kappa$ (where E is the field strength, u_{H^+} is the mobility of the proton in the anolyte, I is the current, and κ is the conductivity in the anolyte), Eq. (4) takes the form

$$N_{H^+} = \frac{I u_{H^+} n_{H^+}}{\kappa} \tag{5}$$

For the mobilization we get a similar expression:

$$N'_{H^+} = \frac{I' u'_{H^+} n'_{H^+}}{\kappa'} \tag{6}$$

where κ' is the conductivity in the anolyte used for mobilization, n'_{H^+} is the number of protons in the same anolyte, and I' is the current in the tube (primed parameters refer to the mobilization step and nonprimed ones to the focusing step). Since in the initial phase of the mobilization I' has about the same value as I in the focusing step, and also because $u_{H^+} \approx u'_{H^+}$, a good approximation is

$$\frac{N_H}{N'_H} = \frac{\kappa' \cdot n_{H^+}}{\kappa \cdot n'_{H^+}} \tag{7}$$

If the number of protons is not changed in the anolyte during focusing and mobilization, then

$$N_{H^+} \kappa = N'_{H^+} \kappa' \tag{8}$$

If we choose conditions where $\kappa < \kappa'$ (or $\kappa << \kappa'$), the ratio of N/N' will be > 1 (or >> 1), and one can therefore state that due to the increase in conductivity, achieved by supplementing the anolyte with a cation, the number of protons entering the tube from the anolyte decreases, which gives rise to a pH increase in the tube. Analogous equations can be derived for the cathodic mobilization, where the introduction of an anion in the catholyte will cause a decrease in pH at the cathodic end of the capillary and, therefore, a mobilization of the pH gradient. Sodium or chloride ions are commonly used [19,20] for anodic or cathodic mobilization, respectively.

When voltage is kept constant during isoelectric focusing, the current decreases in the focusing step due to the increasing resistance of the generated pH gradient (Figure 2). During the electrophoretic mobilization, the change in current is negligible at the beginning, gradually increasing toward the end of the experiment representing the entry of the mobilizing cation or anion in the whole tube [21].

Both patterns obtained by monitoring the moving boundaries during the initial focusing step and recorded in the mobilization step are characteristic of the sample analyses. The order of appearance of the components is opposite in the two steps. The resolution and sensitivity is much lower in the former pattern, although components in relatively large amounts can be identified (see Figure 2). However, the relative amounts of the components cannot be estimated from this pattern.

Electrophoretic mobilization has the advantage that it is also applicable to focusing performed in gel matrices.

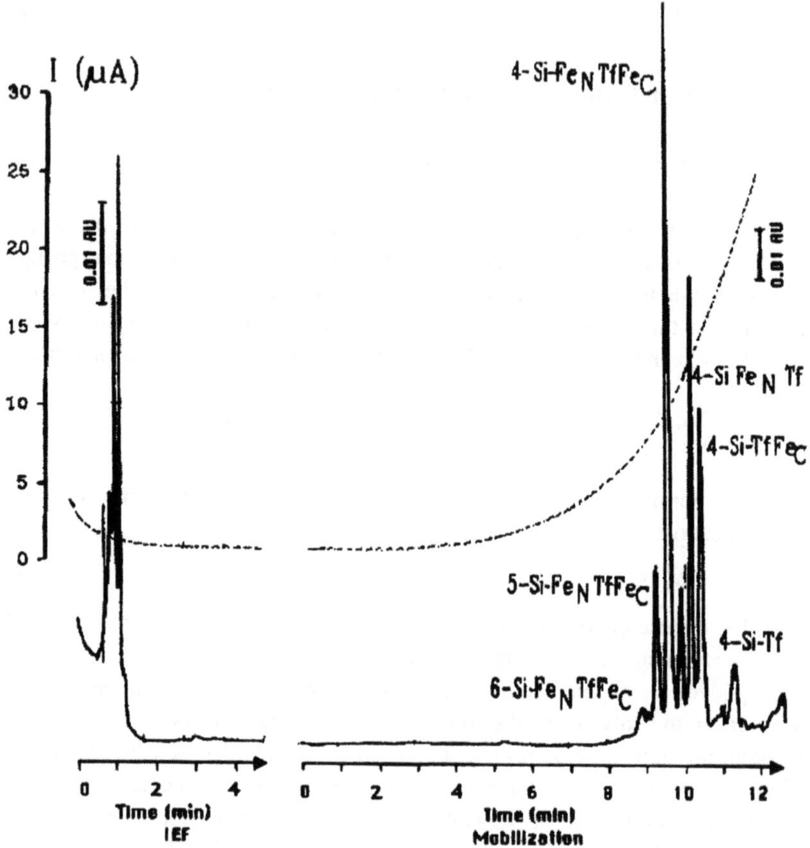

FIGURE 2 High-performance capillary isoelectric focusing experiment of a transferrin sample. Si = sialo; Tf = transferrin; Fe_NTf and TfFe_C = monoferric transferrin forms containing iron at the N- or C-terminal lobe, respectively; Fe_NTfFe_C = diferric transferrin. The current (dotted line) decreases in the focusing step as a result of the immobilization of all the substances in the pH gradient. After a certain time the current reaches a "plateau value," which shows the formation of the steady state in the capillary. Upon replacing the electrolyte at one end of the capillary the current increases, indicating the electrophoretic migration of the mobilizing ion entering the tube. Experimental conditions: tube length, 185 mm; detection point, 155 mm; detection 280 nm; tube diameter, 0.1 mm; voltage, 5000 V; protein concentration, 1 mg/mL; ampholyte, 2% BioLyte 5/7; anolyte, 20 mM H_3PO_4 (focusing)-20 mM NaOH (mobilization); catholyte, 20 mM NaOH; the protein was dissolved in distilled, deionized water. (Reproduced from Kilár, F., *J. Chromatogr.* 545, 1991, 403, with permission of Elsevier Science Publishers.)

2.1.2. Two-Step Isoelectric Focusing with Hydrodynamic Mobilization

As an alternative to electrophoretic mobilization, the focused zones can be mobilized by pressure coupling an appropriate pump to one end of the capillary. In order to avoid distortion of the focused zones due to the hydrodynamic displacement and diffusion, the voltage must be maintained during the displacement. For hydrodynamic displacement of the focused substances the anolyte (or the catholyte) is pumped into the capillary tube at a flow rate of about 60–70 nL/min [2]. The mobilization can be also achieved by applying a precisely regulated vacuum, while maintaining the voltage [22]. In "gravity mobilization" the small difference between the electrolyte solution levels at the two ends of the capillary causes the hydrodynamic mobilization of the zones [6].

2.1.3. Isoelectric Focusing Without Mobilization

The focused pattern can be recorded without mobilizing the pH gradient. The first report of isoelectric focusing in free solution using whole-column absorbance detection [19] was performed in the "free zone electrophoresis apparatus" described by Hjertén [23]. In this case, a 2-mm-i.d. (internal diameter) rotating quartz tube was utilized, and the pattern of focused zones was scanned with high voltage applied by moving the whole tube past a 280-nm UV light beam. A similar system was adapted for smaller capillary tubes [24], where the fused silica capillary was moved through the detector at a speed of 0.30 mm/s. In this latter case, however, the voltage was not maintained during detection, which might result in band broadening. Since inhomogeneities of the capillary wall may cause relatively high noise levels, the method necessitates further improvements. (The "free zone electrophoresis apparatus" has a solution for this problem using a special detection device [23].)

Using whole-column detection with a concentration gradient imaging detection system high-efficiency CIEF separations with subfemtomole detection limits for absolute amounts were obtained, corresponding to 10^2 amol absolute amount of proteins (Figure 3) [10,15, 25–30]. Applying short (e.g., 6-cm) capillaries analyzes can be performed in about 5 min. The detecting UV light beam can be replaced by a concentration gradient detector based on Schlieren optics. Isoelectric focusing of several protein samples performed simultaneously in a capillary array with on-line concentration gradient imaging detection system increases the throughput of the capillary isoelectric focusing technique, and makes the technique comparable to the gel-slab isoelectric focusing technique, with much faster speed of separation and quantitation [26]. The isoelectric point (pI) values of the samples can be directly determined without internal pI markers from their positions inside the capillary after focusing.

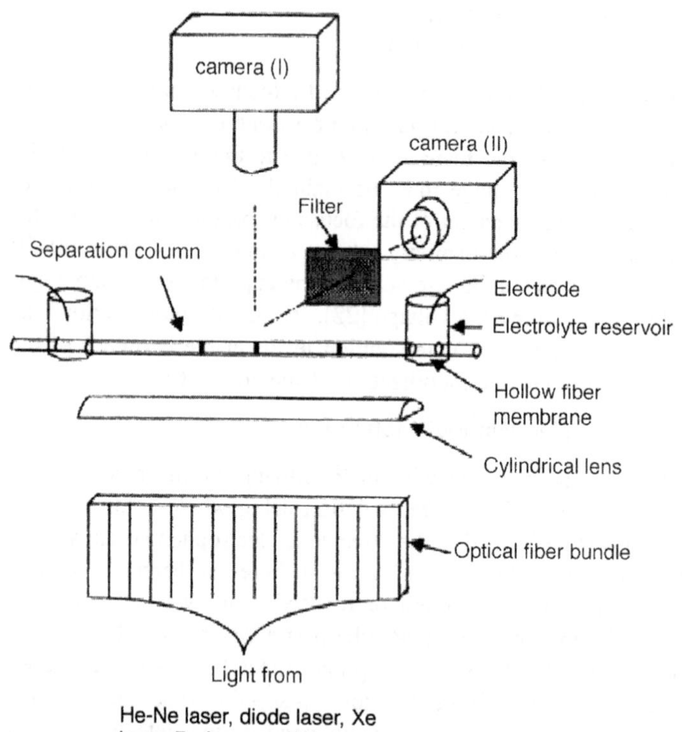

FIGURE 3 Instrument setup of whole-column imaging detection for CIEF. Absorption or refractive index gradient mode with camera (I) placed in the direction of illuminated light; fluorescence mode with camera (II) placed vertically to the direction of illuminated light. (Reproduced from Mao, Q., Pawliszyn, J., *J. Biochem. Biophys. Meth.* 39, 1999, 93–110, with permission of Elsevier Science Publishers.)

2.2. Isoelectric Focusing in Capillaries in the Presence of Electroosmotic Flow

2.2.1. Single-Step Capillary Isoelectric Focusing

Provided that the electroosmotic flow rate is low enough to permit focusing of all substances to occur, isoelectric focusing can be performed in uncoated capillaries. This single-step capillary isoelectric focusing can be achieved by adding a suitable additive to the electrolyte, e.g. hydroxymethyl propylcellulose (HPMC) or methylcellulose (MC), which forms a dynamic coating to reduce EOF and the interaction between analytes and the wall (Figure 4). Thormann et al., and Mazzeo and Krull, reported two different approaches for isoelectric focusing

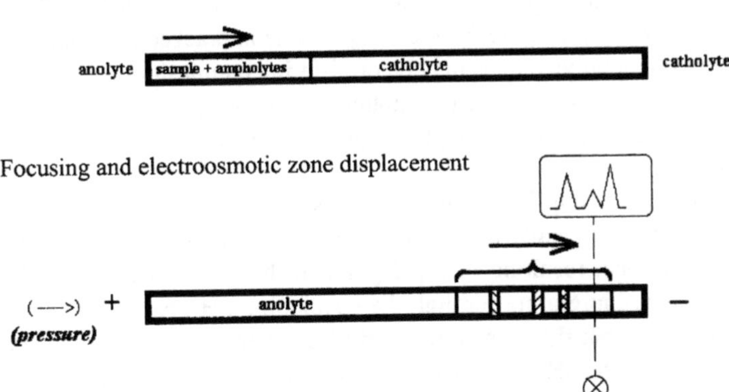

FIGURE 4 Schematic representation of the sample application and the single-step capillary isoelectric focusing with electroosmotic zone displacement. A pressure application may standardize the migration of the focused zones toward the detection point.

experiments in which the EOF served as mobilizer for the focused zones [31–35]. The main difference between the two techniques is that only a small plug of the sample–ampholyte mixture is introduced into a capillary filled with catholyte in one case [31,32] (see Figure 4), as opposed to filling the whole capillary with the sample–ampholyte mixture in the other case [33–35]. With this latter method, it is necessary to add sufficient N,N,N′,N′-tetramethyl ethylenediamine (TEMED) to block the region after the detection point so that substances do not focus in this region. The samples separate in the presence of EOF toward the cathode. Addition of additives reduce both the protein–wall interaction and the EOF, which allows the substances to be focused in the pH gradient, although no steady-state conditions of the analytes can be achieved under these circumstances. However, the separation of the focused zones can be sufficient for analyses.

2.2.2. Single-Step Capillary Isoelectric Focusing with Segmented Injection Protocol

In the previous CIEF setups, sample substances and carrier ampholytes were introduced in mixtures. However, the possible interactions between analytes and ampholytes may disturb the analysis if the sample cannot be applied immediately after the preparation of this mixture [36]. Another limitation is that the pI values of the separands must fall in the same range as the ampholytes. A new setup, in which the separands and the ampolyte solutions are applied in separated zones, prevents the above problems since the automated capillary electrophoretic equipment provides good opportunities to apply several zones in the capillary. In this new setup (Figure 5), the sample zone is bracketed by ampholytes where the two ampholyte zones may be of differing composition or volume [37]. With this technique, separation of amphoteric compounds can be accomplished by ampholyte solutions with pH ranges not including the isoelectric points of the separands.

The isoelectric focusing is a complex process in this system. Ampholyte components positioned originally at the two sides of the sample analytes move through the middle zone, resulting in a pH gradient, but at the same time the sample components are migrating through the developing pH gradient. The experiments demonstrate that the cross-migration of the components separates the analytes irrespective of the pH range covered by the ampholytes (even in experiments with narrow-pH-range ampholytes), although the resolution is dependent on the composition of the ampholyte solutions. The components having pIs outside the actual pH range leave the pH gradient with different migration speeds and separation occurs (Figure 6).

3. METHODOLOGY FOR CAPILLARY ISOELECTRIC FOCUSING

3.1. General Aspects

Isoelectric focusing in capillaries offers several advantages over conventional gel techniques. Some of these are high speed, on-line detection, two-step analy-

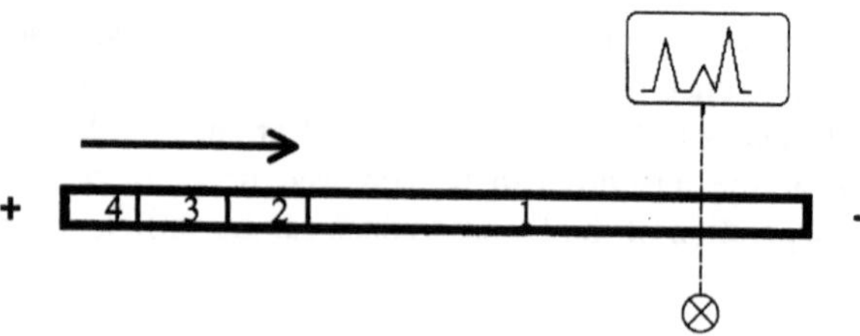

FIGURE 5 Schematic representation of the segmented injection protocol for EOF-driven single-step capillary isoelectric focusing. The following solutions are injected into the capillary: (1) 20 mM NaOH containing 0.015% methylcellulose, (2) ampholyte solution, (3) sample, (4) ampholyte solution. The catholyte at the negative pole is 20 mM NaOH, and the analyte is 10 mM H_3PO_4. The length of the ampholyte–sample–ampholyte "sandwich-zone" is ca. 20–30% of the whole capillary length. The composition of the ampholyte solutions and the relative lengths of the zones can be varied easily in this protocol.

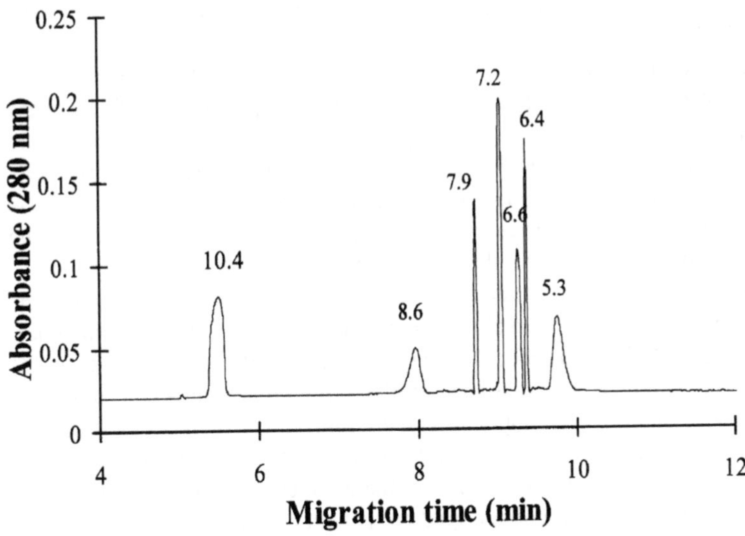

Figure 6 Capillary isoelectric focusing of a mixture of aminomethylated nitrophenols (dyes) with a combination of 2% ampholyte solutions as zone 2 (Bio-Lyte 6/8, 300 mbar-s) and zone 4 (Bio-Lyte 4/6, 500 mbar-s) according to the injection setup in Figure 5. A segmented injection protocol of ampholytes and sample was applied. The sample (750 mbar-s) consisted of dyes having pI: 10.4, 8.6, 7.9, 7.2, 6.6, 6.4, and 5.3; concentration 0.143 mg/mL each. The peaks are assigned in the isoelectropherogram by their pI values. Experimental conditions: applied voltage 20 kV, detection wavelength 280 nm, capillary length 60 cm, effective length 45 cm, i.d. 75 μm, PrinCE modular capillary electrophoresis system. Although the pH range of ampholytes did not cover all the pI values, all dye components have been separated in the complex focusing process.

sis in coated capillaries providing characteristic focusing and mobilization patterns, and selective monitoring using a suitable wavelength for detection. It must be emphasized that isoelectric focusing is a concentrating method, hence the concentrations of the substances in the separated zones are higher than in the original sample solution. Therefore, isoelectric focusing has an advantage over other capillary electrophoretic methods, with perhaps the exception of isotachophoresis, which may also cause sample components to be concentrated.

Sample application can be done either by pressure or vacuum, but not electrophoretically as in other capillary electrophoretic systems. The length of the sample plug in experiments performed in the presence of EOF must be carefully determined. Long sample zones may result a focusing step that will not be completed before the moving pH gradient reaches the detection point

[31]. This may be circumvented by adding a strong base (e.g., TEMED) to the sample [33–35].

The temporal behavior of the current during IEF provides information on the degree of focusing prior to sample detection at a specified location toward the capillary end (see Figure 2). In capillaries without EOF, the end of the focusing step can be estimated from the current, and when the current is below 15–20% of its initial value, mobilization can be started. The progress of the focusing in uncoated capillaries can be similarly followed by monitoring the current while the EOF transfers the pH gradient toward the detection point.

Isoelectric focusing in coated capillaries with minimized EOF can be performed within 15–30 min when single-point detection and electrophoretic mobilization is utilized [22,36,38,39]. The length of the mobilization depends on the applied voltage, length and diameter of the capillary, composition and ionic strength of the sample, ampholyte concentration, length of the focusing step, etc. Faster mobilization of the pH gradient can be obtained by increasing the concentration of, e.g., sodium chloride in the anolyte or catholyte according to Eq. (2) or (3), provided the voltage gradient is kept constant. However, to avoid thermal zone deformation caused by Joule heat, it is advisable to use a concentration of, e.g., NaCl not higher than 0.1 M (but not lower than 0.02 M to avoid long mobilization times). The speed of mobilization can also be increased with hydrodynamic mobilization, but distortion of the peaks by laminar flow band broadening may result and should be avoided by choosing an optimum flow rate.

Applying imaging detection, the experiments can be performed within significantly shorter times [27,40].

In uncoated capillaries, CIEF separations can be carried out in 5–40 min [31–35], depending on several parameters including the length of the capillary, the distance of the detection point from the sample application side, the composition of the electrolytes, voltage applied, etc. Since the dynamic coating of the capillary wall is not easily controlled, the migration times of the analytes may be strongly dependent on the quality of the capillary.

Selective monitoring using a suitable wavelength makes the detection of amphoteric compounds other than high-molecular-weight proteins possible, since no fixation and staining procedures are necessary. Since most ampholytes absorb strongly below 280 nm, low-UV detection is usually not possible in CIEF.

The two-step analysis of samples in IEF with electrophoretic mobilization is especially useful when proteins may become so concentrated during the focusing step that they exceed their solubility in solution and precipitate. Hence, no true focusing pattern can be obtained. In such cases the (pre)focusing pattern is of particular importance (see Figure 2). The resolution in this pattern may be increased by increasing the distance between the detecting UV beam and the end of the capillary tube. Suppressing the tendency for precipitation may be achieved by supplementing the ampholytes with additives such as 10–50% (v/v) ethylene glycol or 0.5–2% (w/v) G 3707 (an efficient, non-UV-absorbing detergent), or other additives [19,41]. The voltage during the mobilization step should be lowered compared to the focusing step, since high current may also induce blocking of the capillary by precipitate formation.

3.2. Practical Considerations

3.2.1. Drift of the pH Gradient in the Absence of Electroendosmosis

In isoelectric focusing, the pH gradient is established using a mixture of synthetic carrier ampholytes. The experiment begins with a transient period (i.e., formation of the pH gradient and of the focused and concentrated zones), followed by a steady state in which the amphoteric substances are in a position where the pH is equal to their isoelectric points. According to the theory, this steady state, once established, should remain invariant as long as the applied current density is constant. In practice, however, this has been shown not to be the case [42]. Both experimental and theoretical approaches have showed that cathodic, anodic, as well as symmetrical drifts of the pH gradient occur after the formation of the pH gradient due to the loss of ampholytes at the acidic and/or cathodic end of the gradient [43]. The extent of drift in ampholyte systems depends strongly on the anolyte and catholyte concentrations. In experiments in which capillaries with minimized EOF are used, an optimum ratio is $2.25[H_3PO_4] = [NaOH]$ [42].

3.2.2. Simulation of Capillary Isoelectric Focusing

The dynamics of the focusing process are efficiently modeled with a 150-component, dynamic electrophoresis simulator that is capable of producing high-resolution focusing data with 140 individual carrier ampholytes (20/pH unit) and at current densities that are used in CIEF. With a focusing capillary of 5-cm length, the predicted focusing dynamics for amphoteric dyes obtained at a constant voltage of 1500 V (300 V/cm) are shown to agree qualitatively with data obtained by whole-column optical imaging (Figure 7). The simulation data provide detailed insight into the dynamics of the focusing process for cases in which the focusing column is sandwiched between 40 mM NaOH (catholyte) and 100 mM phosphoric acid (anolyte) or with the column ends permeable only for OH^- and H^+ at the cathode and anode, respectively. Simulation data reveal that the number of sample boundaries migrating from the two ends of the column to the focusing positions is always equal to the number of sample components. The number of detectable migrating sample boundaries, however, can be lower. The data demonstrate that the model together with imaging monitoring can be used to optimize the CIEF separation conditions [44].

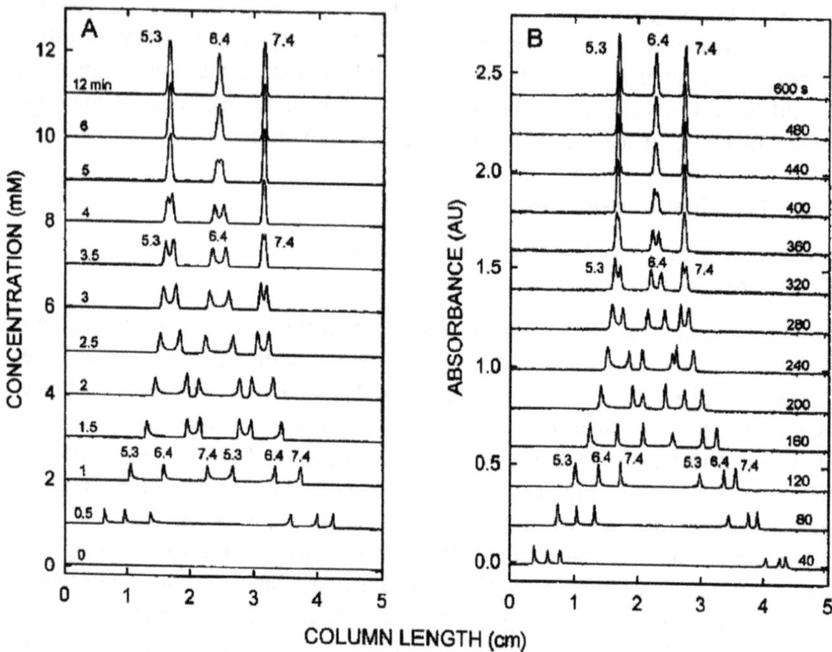

FIGURE 7 Dynamics of (A) the computer-predicted detector response for the pI 5.3, 6.4, and 7.4 dyes during focusing at 1500 V in a gradient produced by 140 carrier ampholytes (0.25 mM each) and (B) the corresponding experimental data with the same three dyes. The cathode is to the right. Graphs at the indicated, successive time points are presented with a *y*-axis offset. (Reproduced from Mao, Q., Pawliszyn, J., Thormann, W., *Anal. Chem.* 72, 2000, 5493–5502, with permission of the American Chemical Society.)

3.2.3 Ampholytes and Additives in the pH Gradient

Experiments with ampholytes from different sources covering the same or similar pH ranges may not provide identical results, indicating differences in the distribution of the components and/or chemical composition [37,39]. A combination of narrow- and broad-range ampholytes provides high efficiency in separation of components with close pI values [37].

One of the most troublesome aspects of isoelectric focusing is how to modulate the slope of the pH gradient to increase resolution (equivalent to pH gradient engineering, as easily available in immobilized pH gradients). A simple solution is offered with addition, to the standard 2-pH-units interval, of separators or spacers, i.e., of amphoteric molecules (either single or in combination) able to locally flatten the pH and increment resolution [8].

The metal complexation of ampholytes may cause unwanted results when iron-containing tools are used for handling the sample (e.g., syringes with metal needles, metal tubing, etc.). This was demonstrated with transferrin, a metal-binding protein, when iron-complexed forms of transferrin appeared in the mobilization pattern of an iron-free sample upon mixing the ampholytes and the protein solution with a Hamilton syringe equipped with metal needle [36].

3.2.4. pI Markers

The determination of the pI of focused substances in isoelectric focusing experiments generally requires internal standards and their "migration time" values (i.e., the time parameter of the peaks in the electropherograms). Such experiments have been done in coated capillaries, and calibration curves obtained with electrophoretic [2] or hydrodynamic mobilization [22] have been constructed. As an alternative, the current measured during electrophoretic mobilization has been used to determine the pI values without having internal standards in every run [37]. Keeping the experimental conditions unchanged, the current parameter characterizes the position of the respective substance in the pH gradient and therefore it can be used for the estimation of the isoelectric point. Experiments with transferrin isoforms show that the pI can be determined with an error of about 0.03 pH unit or less [38].

Several reference substances (pI markers) have been developed (synthesized) for CIEF analyses. The commonly used protein markers can be replaced by aminomethylated nitrophenols (dyes) that are highly soluble in water and have high absorbance at low wavelengths [45,46] or with broad-range (pH 3–10) peptide markers [47,48].

The combination of the highly sensitive capillary isoelectric focusing with (laser-induced) fluorescence detection needs fluorescent pI markers. Commercially available peptides, mostly angiotensin derivatives, have been labeled at their N-terminal amino group with 5-carboxytetramethylrhodamine succinimidyl ester and used for capillary isoelectric focusing with fluorescence detection (He–Ne laser, 543.5 nm) [49]. pH standards (iodoacetylated derivative of tetramethylrhodamine coupled to one cysteine residue in tetra- to tridecapeptides) or other pI markers ranging from pH 3.64 to 10.12 are available for calibration of a wide-range pH gradient formed in a capillary by fluorescence detection [50]. Low-molecular-mass fluorescent compounds excitable in the near-UV region with suitable acidobasic and electrophoretic properties have been tested as isoelectric point (pI) markers (range from 2.1 to 10.3) using both UV photometric and UV excited fluorometric detection [51]. A series of low-molecular-mass fluorescent ampholytes with narrow pI range for argon laser-induced fluorescence (LIF) detection was also developed and tested [52].

3.2.5. Analyte Solubility

A whole family of protein solubilizers, compatible with native structure and maintenance of enzyme activity, is reported for preventing protein precipitation and aggregation at the pI value under conditions of very low ionic strength, as is typical of isoelectric focusing methodologies. In addition to mild solubilizers proposed in the past, such as glycols (glycerol, ethylene, and propylene glycols), nondetergent sulphobetaines, in concentrations up to 1 M, have been found to be quite effective in a number of cases. Other common zwitterions, such as taurine and a few of the Good's buffers (e.g., Bicine, CAPS), are also quite useful in acidic pH gradients and up to pH 8. Addition of sugars, notably saccharose, sorbitol, and, to a lesser extent, sorbose (20% in capillary IEF and in the 30–40% concentration range in gel-slab IPGs), greatly improved protein solubility in the vicinity of the pI. The improvement was dramatic if these sugars were mixed with 0.2 M taurine. In the case of hydrophobic peptide antibiotics, mixtures of 6 M urea and 25% trifluoroethanol were found to markedly improve solubility. All these additives, unlike nonionic or zwitterionic surfactants, have the advantage of remaining monomeric, i.e., of being unable to form micelles, even at concentrations > 1 M. Thus, their elimination from the protein zone can be easily accomplished by gel filtration or by centrifugation through dialysis membranes. Using these additives, capillary IEF of proteins should now be applicable to a number of difficult cases, such as the separation of mildly hydrophobic macroions [41,53].

3.2.6. Desalting

A major problem of capillary isoelectric focusing is that separation performance is usually degraded by the presence of salts within the sample. Normally this requires the removal of these components by some off-line sample clean-up method, prior to analyte separation by CIEF. It is possible, however, to remove high salt levels from samples efficiently by on-tube techniques, e.g., the substitution of the salts with an ampholyte solution in a short focusing step prior to the final analytical isoelectric focusing procedure [54], or on-line voltage ramping of the applied CE voltage [55]. These desalting methods are especially useful for studies of whole human blood and human cerebrospinal fluid which have undergone no manipulation or workup prior to cIEF analysis.

4. ADVANCED DETECTION TECHNIQUES IN CAPILLARY ISOELECTRIC FOCUSING

4.1. On-Tube Whole-Column Imaging

Real-time detectors have been developed, including refractive index gradient, laser-induced fluorescence (LIF), and absorption. Of these, absorption imaging detection is the most practical at the present time, due to its quantitative ability and universal characteristics. Whole-column imaging detection eliminates the mobilization step required for single-point detection after the focusing process (see Sections 2.1.1 and 2.1.3). Therefore, it provides fast analysis speed and avoids the disadvantages associated with the mobilization process, such as distortion of pH gradient and loss in resolution. The light beam (e.g., a laser) is focused into the capillary by a cylindrical lens. An 1024-pixel charge-coupled device (CCD) measures the intensity of light (Figure 3). The on-line imaging detector allows simultaneous separation and detection so that the analysis time for a sample is only 3–5 min [10,15,25–30].

4.2. Off-Tube Detection

4.2.1. Mass Spectrometry Coupling

Two-dimensional separation systems in which the separation selectivity of the two dimensions is orthogonal offer the highest resolving power for complex mixtures. Coupling CIEF to mass spectrometry promises to provide 2-D separations of proteins and their metabolites in an automated format with short analysis time. Successful on-line interfacing of capillary electrophoresis (CE) with electrospray (ES) mass spectrometry (MS) has progressed substantially in recent years [56]. Of particular note also is the development which has occurred in combining the more advanced capillary-based electromigration separation techniques, such as capillary gel electrophoresis (CGE), capillary isoelectric focusing (CIEF), capillary isotachophoresis (CIT), micellar electrokinetic chromatography (MEKC), and capillary electrochromatography (CEC), with ES/MS. The union of these electromigration schemes with MS detection provides a useful and sensitive analytical tool for the separation, quantitation, and identification of biological, therapeutic, environmental, and other important classes of chemical analytes.

In the first on-line CIEF-electrospray mass spectrometry the two-step isoelectric focusing was applied [57]. Following the focusing step, the catholyte reservoir was removed and mobilization of the focused zones into the electrospray source was initiated by infusion of a sheath liquid composed of methanol–water–acetic acid. This method was further modified by incorporating an online microdialysis device between the separation capillary and a transfer capillary connected to the ESI source [58,59]. The microdialysis device (a hollow-fiber dialysis tube sealed into a chamber through which 2% acetic acid was infused) removed sufficient ampholytes to prevent fouling of the ESI source and interference with the electrospray process [60]. An on-line capillary isoelectric focusing-electrospray ionization time-of-flight mass spectrometry (ESI-TOF-MS) as a two-dimensional separation system has been employed for high-resolution analysis of model proteins (myoglobin and beta-lactoglobulin) and human hemoglobin variants C, S, F, and A. The focused proteins in a polyacrylamide-coated capillary are mobilized by replacing the sodium hydroxide catholyte with a sheath liquid solution containing methanol–water–acetic acid. The use of a sheath liquid also establishes the electrical connection at the CIEF capillary terminus, which serves to define the electric field along the capillary and apply an electric voltage for electrospray ionization. At the end of the capillary, the mobilized protein zones are analyzed and identified directly by ESI-TOF-MS [61].

The on-line interfacing of capillary isoelectric focusing with Fourier-transform ion cyclotron resonance-mass spectrometry (FTICR-MS) was shown to be effective for separating minor components of protein mixtures for on-line mass spectral analysis [62–64].

The combination of capillary zone electrophoresis or CIEF with inductively coupled plasma (ICP) MS promises a powerful tool for metal speciation. The on-line hyphenation of these techniques resulted in short separation times (10 min) and a subsequent detection step lasting 100 s. Standard mixtures and body fluids such as human milk and serum have been studied [65].

A stepwise mobilization strategy has been developed for the elution of complex protein mixtures, separated by capillary isoelectric focusing for detection using on-line electrospray ionization mass spectrometry. Carrier polyampholytes are used to establish a pH gradient as well as to control the electroosmotic flow arising from the use of uncoated fused-silica capillaries. Elution of focused protein zones is achieved by controlling the mobilization pressure and voltage, leaving the remaining protein zones focused inside the capillary. Protein zones are stepwise eluted from the capillary by changing the mobilization conditions. Stepwise mobilization improves separation resolution and simplifies coupling with multistage MS (i.e., MSn) analysis, since it allows more effective temporal control of protein elution from the CIEF capillary [66].

To remove carrier ampholytes when electrospray ionisation mass spectrometry is used on-line with capillary isoelectric focusing, a specially designed free-flow electrophoresis device can be coupled to the CIEF system [67].

4.2.2. Chemiluminescence Detection of Proteins

Chemiluminescence detection in capillary electrophoresis (CE) has attracted much attention as a promising way to offer excellent analytical selectivity and sensitivity. Several reagents, such as luminol, acridinium, peroxyoxalate, and tris(2,29-bipyridine)ruthenium(II) complex have been utilized. Since chemiluminescence detection is approximately 10^2–10^6 times more sensitive than spectrophotometric and fluorometric detections, its combination with isoelectric focusing may result in a highly sensitive analytical tool for amphoteric compounds, e.g., proteins and peptides. A detector using luminol–H_2O_2 chemiluminescence has been characterized in a very simple and inexpensive setup, but only pressure-driven mobilization of the zones was effective. [68].

4.3. Off-Line Detection

Capillary isoelectric focusing can be applied as a micropreparative tool for protein analysis by matrix-assisted laser desorption time-of-flight mass spectrometry (MALDI-TOF-MS) [69,70]. The exact timing of the collector steps in the interface is based on determining the velocity of each individual zone measured between two detection points close to the end of the capillary. During the collection a sheath flow fraction collector is used to maintain the permanent electric current.

5. CHIP TECHNOLOGY

The analysis of complex mixtures such as cell extracts is one of the most important tasks of proteome and metabolome studies. Capillary isoelectric focusing has already shown its high resolving power in analysing lysates of microorganisms [71–73], but the analysis time should be significantly reduced. The use of whole-column imaging on chips may provide a solution for this. Although chip technology in capillary electrophoresis was introduced several years ago, the adaptation of capillary isoelectric focusing to microchannels on chips is still in an early stage. It must be noted, however, that a reduction of analysis time in common capillary systems can be simply made with a whole-column detection setup (see Section 2.1.3).

Applying a 200-μm-wide, 10-μm-deep, and 7-cm-long channel etched into planar glass with single-point detection, both chemical and hydrodynamic mobilization were shown to give superior separation efficiency and reproducibility. However, EOF-driven mobilization, which occurs simultaneously with focusing, proved most suitable for miniaturization because of high speed, EOF compatibility, and low instrumentation requirements. A mixture of Cy5-labeled peptides could be focused in less than 30 s, with plate heights of 0.4 μm (410 plates/s) upon optimization [74]. The miniaturization of whole-column imaging capillary isoelectric focusing using a 1.2-cm capillary is shown in Figure 8. The light-emitting diode (LED) helped to simplify and miniaturize the IEF equipment. The development of the absorbance images in coated capillary was obtained within 80–250 s [15].

6. APPLICATIONS OF CAPILLARY ISOELECTRIC FOCUSING

Capillary isoelectric focusing will be applied mostly for proteome studies in the future. Proteome analysis requires fast methods with high separation efficiencies in order to screen the various cell and tissue types for their proteome expression and monitor the effect of environmental conditions and time on this expression. The established two-dimensional gel electrophoresis is by far too slow for a

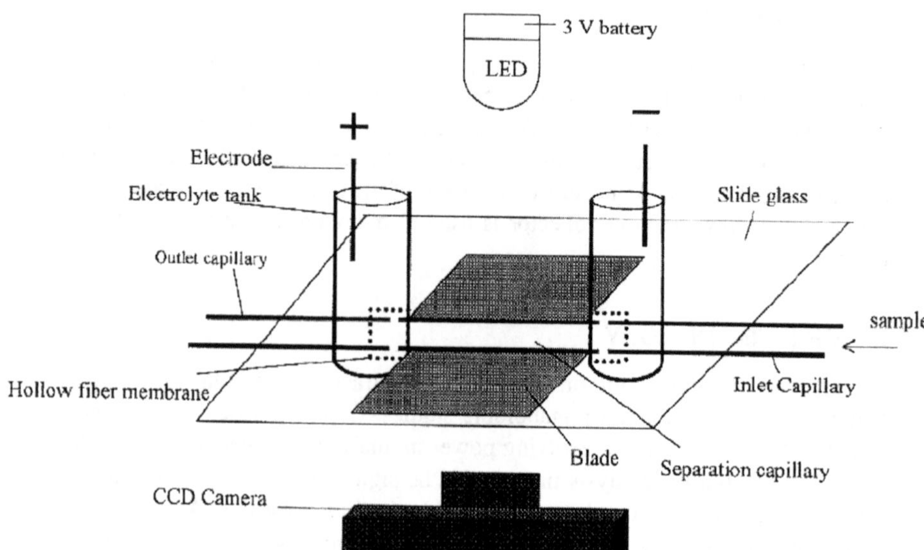

FIGURE 8 Illustration of whole-column imaging miniaturized capillary isoelectric focusing instruments with a LED as light source. A 1.2-cm capillary is used as a separation column for focusing. The outside polyimide coating of the capillary is removed, and the inside surface of capillary is coated with non-cross-linked polyacrylamide to eliminate electroosmotic flow. The two ends of the capillary are connected with inlet and outlet capillaries (with the same i.d. and o.d. as the separation capillary) by two pieces of porous hollow fiber. Two glass tubes are used as electrolyte tanks and glued directly on the glass slide by epoxy glue. The two pieces of hollow fiber are in the electrolyte tanks. The length of the capillary between the two electrolyte tanks is about 0.9 mm.

consequential screening. Moreover, it is not precise enough to observe changes in protein concentrations. High-resolution separations of complex protein mixtures have been made primarily with conventional UV detection until recently. The combination of CIEF with FTICR mass spectrometry has revealed 400–1000 putative proteins in the mass range of 2–100 kDa from total injections of approximately 300 ng protein in single analyses [63,64]. Capillary isoelectric focusing coupled to mass spectrometry (CIEF-MS) and preparative IEF followed by size-exclusion chromatography, hyphenated with MS (PIEF-SEC-MS), promise faster and automated proteome analysis [61,75,76].

Special applications of CIEF have focused on studies of glycoproteins, antibodies, and proteins in serum (e.g., hemoglobin variants [30,37,39,77–90], transferrin forms [21,31,36,80,82,90,91]), and cerebrospinal fluid [92]. The glycoforms of recombinant human tissue-type plasminogen activator [93–95] and various glycoproteins [89,96–102] have been studied. The pI values for scrapie prion protein isolated from sheep and hamster brain were successfully determined. The scrapie-infected sheep sample had peaks with pI values ranging from 5.2 to 3.00 with a major peak at 3.09. The normal sheep brain had pI values that were higher. The hamster adapted scrapie strain had peaks with pI values ranging from 6.47 to 3.8 [103].

Enantioseparation of dansylated as well as AQC-derivatized amino acids by means of capillary isoelectric focusing using various cyclodextrin derivatives is based on the enantio-selective shift of the isoelectric points upon complexation with the chiral selectors. The zwitterionic, diastereomeric analyte-cyclodextrin complexes exhibited differences in the pI values up to more than 0.25 pH units. Hydroxypropyl-beta-cyclodextrin proved to be the best selector for this purpose. The kinetics of complex formation and dissociation is fast enough in most instances to produce single peaks, even with complexation degrees near 0.5 and significant pI shifts [104]. The method offers excellent prospects for preparative applications, as is shown by full-column imaging in continuous free-flow isoelectric focusing separation of enantiomers [105].

Studies on monoclonal antibodies [94,102,106–110] comprise a significant part of CIEF applications. The affinity of an antibody toward its antigen is highly specific to its conformation, in order to have optimal antibody–antigen interaction. The increase of temperature might cause changes in antibody conformations and this change may be reflected in the isoelectric points (pI values), peak shape, and absorbance of the antibody [109].

7. CONCLUSIONS

Capillary isoelectric focusing has proven its separation potential and special advantages in proteome bioanalysis. Several applications have been reported using this unique methodology. Notwithstanding, further improvements are necessary to design fully automated chip-based devices and a more advanced coupling to various detection systems.

ACKNOWLEDGMENT

This work was partially supported by the grants FKFP 0320/2000, Inco-Copernicus ERB IC15 CT98 0322, CEEPUS H-076.

REFERENCES

1. Hjertén S, Zhu M-D. J Chromatogr 1985; 346:265–270.
2. Hjertén S, Kilár F, Liao J-L, Zhu M-D. In: Dunn MJ, ed. Electrophoresis '86. Weinheim, Germany: VCH Verlagsgesellschaft, 1986:451–461.

3. Hjertén S, Elenbring K, Kilár F, Liao JL, Chen AJ, Siebert CJ, Zhu MD. J Chromatogr 1987; 403:47–61.
4. Righetti PG. Isoelectric Focusing: Theory, Methodology and Applications. Amsterdam: Elsevier, 1983.
5. Wehr T, Zhu M, Rodriguez-Diaz R. Meth Enzymol 1996; 270:358–374.
6. Rodriguez-Diaz R, Zhu M, Wehr T. J Chromatogr A 1997; 772:145–160.
7. Rodriguez-Diaz R, Wehr T, Zhu MD. Electrophoresis 1997; 18:2134–2144.
8. Righetti PG, Bossi A, Gelfi C. J Capillary Electrophor 1997; 4:47–59.
9. Righetti PG, Gelfi C, Conti M. J Chromatogr B: Biomed Sci Appl 1997; 699, 91–104.
10. Fang X, Tragas C, Wu J, Mao Q, Pawliszyn J. Electrophoresis 1998; 19:2290–2295.
11. Manabe T. Electrophoresis 1999; 20:3116–3121.
12. Issaq HJ. J Liq Chromatogr Rel Technol 2002; 25:1153–1170.
13. Dolnik V, Hutterer KM. Electrophoresis 2001; 22:4163–4178.
14. von Brocke A, Nicholson G, Bayer E. Electrophoresis 2001; 22:1251–1266.
15. Wu XZ, Sze SK, Pawliszyn J. Electrophoresis 2001; 22:3968–3971.
16. Wehr T, Rodriguez-Diaz R, Zhu M. Chromatography 2001; 53:S45–S58.
17. Issaq HJ. Electrophoresis 2000; 21:1921–1939.
18. Hjertén S. J Chromatogr 1985; 347:191–198.
19. Hjertén S, Liao JL, Yao KQ. J Chromatogr 1987; 387:127–138.
20. Manabe T, Miyamoto H, Iwasaki A. Electrophoresis 1997; 18:92–97.
21. Kilár F. J Chromatogr A 1991; 545:403–406.
22. Chen SM. Wiktorowicz JE. Anal Biochem 1992; 206:84–90.
23. Hjertén S. Chromatogr Rev 1967; 9:122–219.
24. Wang TS, Hartwick RA. Anal Chem 1992; 64:1745–1747.
25. Wu J, Pawliszyn J. J Chromatography A 1992; 608:121–130.
26. Wu J, Pawliszyn J. Electrophoresis 1993; 14:469–474.
27. Wu J, Pawliszyn J. Anal Chem 1992; 64:224–227.
28. Mao Q, Pawliszyn J. J Biochem Biophysl Meth 1999; 39:93–110.
29. Huang TM, Pawliszyn J. Analyst 2000; 125:1231–1233.
30. Tragas C, Pawliszyn J. Electrophoresis 2000; 21:227–237.
31. Thormann W, Caslavska J, Molteni S, Chmelik J. J Chromatogr 1992; 589:321–327.
32. Molteni S, Thormann W. J Chromatogr A 1993; 638:187–193.
33. Mazzeo JR, Krull IS. Anal Chem 1991; 63:2852–2857.
34. Mazzeo JR, Krull IS. J Chromatogr A 1992; 606:291–296.
35. Mazzeo JR, Krull IS. J Microcolumn Sep 1992; 4:29–33.
36. Kilár F, Hjertén S. Electrophoresis 1989; 10:23–29.
37. Kilár F, Végvári Á, Mód A. J Chromatogr A 1998; 813:349–360.
38. Kilár F, Hjertén S. J Chromatogr 1989; 480:351–357.
39. Zhu M, Rodriguez-Diaz R, Wehr T, Siebert C. J Chromatogr A 1992; 608:225–237.
40. Wu XZ, Wu J, Pawliszyn J. LC GC N Am 2001; 19:526– + .
41. Conti M, Galassi M, Bossi A, Righetti PG. J Chromatogr A 1997; 757:237–245.

42. Mosher RA, Saville DA, Thormann W. The Dynamics of Electrophoresis. Weinheim, Germany: VCH Verlagsgesellschaft, 1992.
43. Mosher RA, Thormann W. Electrophoresis 1990; 11:717–723.
44. Mao Q, Pawliszyn J, Thormann W. Anal Chem 2000; 72:5493–5502.
45. Slais K, Friedl Z. J Chromatogr A 1994; 661:249–256.
46. Caslavska J, Molteni S, Chmelik J, Slais K, Matulik F, Thormann W. J Chromatogr A 1994; 680:549–559.
47. Shimura K, Wang Z, Matsumoto H, Kasai K. Electrophoresis 2000; 21:603–610.
48. Shimura K, Zhi W, Matsumoto H, Kasai K. Anal Chem 2000; 72:4747–4757.
49. Shimura K, Kasai K. Electrophoresis 1995; 16:1479–1484.
50. Shimura K, Kamiya KI, Matsumoto H, Kasai KI. Anal Chem 2002; 74:1046–1053.
51. Horka M, Willimann T, Blum M, Nording P, Friedl Z, Slais K. J Chromatogr A 2001; 916:65–71.
52. Slais K, Horka M, Novackova J, Friedl Z. Electrophoresis 2002; 23:1682–1688.
53. Righetti PG, Bossi A. Anal Chim Acta 1998; 372:1–19.
54. Liao JL, Zhang R. J Chromatogr A 1994; 684:143–148.
55. Clarke NJ, Tomlinson AJ, Naylor S. J Am Soc Mass Spectrom 1997; 8:743–748.
56. Banks JF. Electrophoresis 1997; 18:2255–2266.
57. Tang Q, Harrata AK, Lee CS. Anal Chem 1995; 67:3515–3519.
58. Lamoree MH, Tjaden UR, van der Greef J. J Chromatogr A 1997; 777:31–39.
59. Lamoree MH, van der Hoeven RAM, Tjaden UR, van der Greef J. J Mass Spectrom 1998; 33:453–460.
60. Yang LY, Lee CS, Hofstadler SA, Smith RD. Anal Chem 1998; 70:4945–4950.
61. Wei J, Lee CS, Lazar IM, Lee ML. J Microcolumn Sep 1999; 11:193–197.
62. Severs JC, Hofstadler SA, Zhao Z, Senh RT, Smith RD. Electrophoresis 1996; 17:1808–1817.
63. Jensen PK, Pasa-Tolic L, Anderson GA, Horner JA, Lipton MS, Bruce JE, Smith RD. Anal Chem 1999; 71:2076–2084.
64. Jensen PK, Pasa-Tolic L, Peden KK, Martinovic S, Lipton MS, Anderson GA, Tolic N, Wong KK, Smith RD. Electrophoresis 2000; 21:1372–1380.
65. Michalke B, Schramel P. J Chromatogr A 1998; 807:71–80.
66. Zhang CX, Xiang F, Pasa-Tolic L, Anderson GA, Veenstra TD, Smith RD. Anal Chem 2000; 72:1462–1468.
67. Chartogne A, Tjaden UR, van der Greef J. Rapid Commun Mass Spectrom 2000; 14:1269–1274.
68. Hashimoto M, Tsukagoshi K, Nakajima R, Kondo K. J Chromatogr A 1999; 852: 597–601.
69. Foret F, Muller O, Thorne J, Gotzinger W, Karger BL. J Chromatogr A 1995; 716:157–166.
70. Minarik M, Foret F, Karger BL. Electrophoresis 2000; 21:247–254.
71. Shen YF, Berger SJ, Smith RD. Anal Chem 2000; 72:4603–4607.
72. Shen YF, Berger SJ, Smith RD. J Chromatogr A 2001; 914:257–264.
73. Shen YF, Xiang F, Veenstra TD, Fung EN, Smith RD. Anal Chem 1999; 71: 5348–5353.

74. Hofmann O, Che DP, Cruickshank KA, Muller UR. Anal Chem 1999; 71:678–686.
75. Hille JM, Freed AL, Watzig H. Electrophoresis 2001; 22:4035–4052.
76. Wei J, Yang L, Harrata AK, Lee CS. Electrophoresis 1998; 19:2356–2360.
77. Righetti PG, Gianazza E, Bianchi-Bosisio A, Wajcman H, Cossu G. Electrophoresis 1989; 10:595–599.
78. Zhu M, Wehr T, Levi V, Rodriguez-Diaz R, Shiffer K, Cao ZA. J Chromatogr A 1993; 652:119–129.
79. Hempe JM. FASEB J 1994; 8:A639–A639.
80. Wu J, Pawliszyn J. J Chromatogr B: Biomed Sci Appl 1994; 657:327–332.
81. Conti M, Gelfi C, Righetti PG. Electrophoresis 1995; 16:1485–1491.
82. Pritchett TJ. Electrophoresis 1996; 17:1195–1201.
83. Conti M, Gelfi C, Bosisio AB, Righetti PG. Electrophoresis 1996; 17:1590–1596.
84. Mohammad AA, Okorodudu AO, Bissell MG, Dow P, Reger G, Meier A, Guodagno P, Petersen JR. Clin Chem 1997; 43:1798–1799.
85. Hempe JM, Granger JN, Craver RD. Electrophoresis 1997; 18:1785–1795.
86. Mario N, Baudin B, Giboudeau J. J Chromatogr B: Biomed Sci Appl 1998; 706: 123–129.
87. Warrier RP, Nocerino A, Hempe J, Craver R. Pediatr Res 1999; 45:895.
88. Jenkins MA, Ratnaike S. Clin Chim Acta 1999; 289:121–132.
89. Lopez-Soto-Yarritu P, Diez-Masa JC, Cifuentes A, de Frutos M. J Chromatogr A 2002; 968:221–228.
90. Liu T, Shao XX, Zeng R, Xia QC. Acta Biochim Biophys Sinica 2002; 34:423–432.
91. Richards MP, Huang TL. J Chromatogr B 1997; 690:43–54.
92. Manabe T, Miyamoto H, Inoue K, Nakatsu M, Arai M. Electrophoresis 1999; 20: 3677–3683.
93. Tran NT, Taverna M, Chevalier M, Ferrier D. J Chromatogr A 2000; 866:121–135.
94. Santora LC, Krull IS, Grant K. Anal Biochem 1999; 275:98–108.
95. Thorne JM, Goetzinger WK, Chen AB, Moorhouse KG, Karger BL. J Chromatogr A 1996; 744:155–165.
96. Tran NT, Daali Y, Cherkaoui S, Taverna M, Neeser JR, Veuthey JL. J Chromatogr A 2001; 929:151–163.
97. Hui JPM, Lanthier P, White TC, McHugh SG, Yaguchi M, Roy R, Thibault P. J Chromatogr B 2001; 752:349–368.
98. Pantazaki A, Taverna M, Vidal-Madjar C. Anal Chim Acta 1999; 383:137–156.
99. Taverna M, Tran NT, Merry T, Horvath E, Ferrier D. Electrophoresis 1998; 19: 2572–2594.
100. Wu J, Li SC, Watson A. J Chromatogr A 1998; 817:163–171.
101. Mulders JWM, Derksen M, Swolfs A, Maris F. Biologicals 1997; 25:269–281.
102. Kubach J, Grimm R. J Chromatogr A 1996; 737:281–289.
103. Schmerr MJ, Cutlip RC, Jenny A. J Chromatogr A 1998; 802:135–141.
104. Rizzi AM, Kremser L. Electrophoresis 1999; 20:3410–3416.
105. Spanik I, Lim P, Vigh G. J Chromatogr A 2002; 960:241–246.

106. Kundu S, Fenters C. J Capillary Electrophor 1995; 2:273–277.
107. Liu X, Sosic Z, Krull IS. J Chromatogr A 1996; 735:165–190.
108. Hagmann ML, Kionka C, Schreiner M, Schwer C. J Chromatogr A 1998; 816: 49–58.
109. Dai HJ, Krull IS. J Chromatogr A 1998; 807:121–128.
110. Janini G, Saptharishi N, Waselus M, Soman G. Electrophoresis 2002; 23:1605–1611.

4

Capillary Gel Electrophoresis and Related Microseparation Techniques

András Guttman*
Torrey Mesa Research Institute, San Diego, California, U.S.A.

1. INTRODUCTION

At the beginning of the new millennium, moving from the age of genomics to the age of functional genomics and proteomics, we expect to see high-resolution separation techniques used in an integrated and automated fashion to solve formidable separation problems and provide the means for large-scale biomedical applications. Capillary gel electrophoresis greatly enhances the productivity of the analysis of biologically important molecules including nucleic acids, proteins/peptides, complex carbohydrates, and small molecules by automating current manual procedures and reducing both the analysis time and human intervention from sample loading to data analysis. As the advent of capillary gel electrophoresis has already made it possible to sequence the human genome [1], we anticipate that this technique will reveal global changes in the genome and proteome level, bringing about revolutionary transition in our views of living systems on the molecular basis.

Electrophoresis experiments in glass tubes were reported as early as in the nineteenth century, but the first real breakthrough occurred in the first half of the twentieth century when the Swedish chemist Arne Tiselius applied free-solution electrophoresis—i.e., moving boundary—to serum protein analysis, for which he later received the 1937 Nobel Prize [2]. In less than two decades, just after the striking scientific discovery of the double-helical structure of DNA by Watson and Crick in 1953 [3] and the following unveiling of the genetic code, electrophoresis became a standard and indispensable tool in the field of modern

**Current affiliation*: Diversa Co., San Diego, California, U.S.A.

biochemistry and molecular biology. Since the 1960s, electrophoresis methods employing anticonvective sieving media, such as starch, agarose, and polyacrylamide gels, have become the norm for nucleic acid and protein analysis. In the last decade of the twentieth century, an automated and high-performance electric field-mediated differential migration technique, capillary electrophoresis, was introduced in almost every aspect of basic and applied biomedical and clinical research. Using narrow-bore fused silica capillaries, filled with cross-linked gels or non-cross-linked linear polymer networks, unprecedented high resolving power was achieved in separation of biologically important macromolecules. As an instrumental approach to electrophoresis, capillary electrophoresis offers on-line detection and full automation. The method is ideally suited for handling microliter amounts of sample material, and the throughput is superior to that of conventional approaches. Capillary gel electrophoresis separates complex mixtures in just minutes, even in multicapillary format, with excellent reproducibility, generating in this way a large amount of data. The availability of high-capacity computer systems capable of rigorous qualitative and quantitative analysis of the separation profiles enables us to establish, store, and operate with large databases. This is why capillary gel electrophoresis and related microseparation techniques (e.g., electrophoresis microchips) are quickly becoming important separation and characterization tools in analytical biochemistry and molecular biology.

The aim of this chapter is to cover the theoretical and practical aspects of capillary gel electrophoresis. It also provides an overview of the key application areas of nucleic acid, protein, and complex carbohydrate analysis, affinity-based methodologies, as well as related microseparation methods such as ultra-thin-layer gel electrophoresis and electric field-mediated separations on microchips. It also gives the reader a better understanding of how to utilize this technology, and determine which actual method will provide appropriate technical solutions to problems that may have be perceived as more fundamental. Micropreparative aspects and applications are discussed in Chapter 12.

2. BASIC PRINCIPLES AND THEORY OF CAPILLARY GEL ELECTROPHORESIS OF BIOPOLYMERS

2.1. Electrophoretic Migration

By the application of a uniform electric field (E) to a polyion with a net charge of Q, the electrical force (F_e) is defined as

$$F_e = QE \tag{1}$$

In gel or polymer network solution, the applied electric field strength results in electrophoretic migration, but a frictional force (F_f) acts in the opposite direction:

$$F_f = f\left(\frac{dx}{dt}\right) \tag{2}$$

where f is the translational friction coefficient and dx and dt are the distance and time increments, respectively. Differences in shape, size, and overall charge of the solute molecules result in variances in electrophoretic mobilities, providing the basis of the electrophoretic separation. Under steady-state conditions, F_e and F_f are counterbalanced, thus the solute migrates with a steady-state velocity of v:

$$v = \frac{dx}{dt} = \frac{EQ}{f} \tag{3}$$

The electrophoretic mobility (μ) is defined as the velocity per unit field strength:

$$\mu = \frac{v}{E} \tag{4}$$

Retardation of the solute molecules in gel or polymer filled capillaries is a function of the separation matrix concentration (P) and its physical interactions with electrophoresed molecules defined by the retardation coefficient (K_R):

$$\mu = \mu_0 \exp(-K_R P) \tag{5}$$

where μ is the apparent electrophoretic mobility and μ_0 is the free solution mobility (i.e., with no sieving matrix) of the analyte.[4]

When the average pore size of the matrix is in the same size range as the hydrodynamic radius of the migrating analyte molecule, classical sieving exists (Ogston regime, Figure 1 [5]). In this case the retardation coefficient (K_R) is a function of molecular weight (MW) of the analyte at constant polymer concentration [6]:

$$\mu \sim \exp(-\mathrm{MW}) \tag{6}$$

The mobility of the solute molecules is an apparent logarithmic function of the molecular weight, usually resulting in linear so-called Ferguson plots [7] crossing each other at zero gel concentration.

The Ogston theory assumes that the migrating solute behaves as an unperturbed spherical object with comparable size to the pores of the gel. However, large biopolymers (DNA, proteins, complex carbohydrate molecules) can migrate through polymer networks pores significantly smaller than their size [8]. This phenomenon is referred to as reptation (Figure 1, reptation regime), which describes an electric field-mediated migration model for large biopolymers as "snakelike" motion through the much smaller gel pores [9–11]. The reptation model suggests an inverse relationship between the size (molecular weight) and the mobility of the analyte molecules (slope value close or equal to −1):

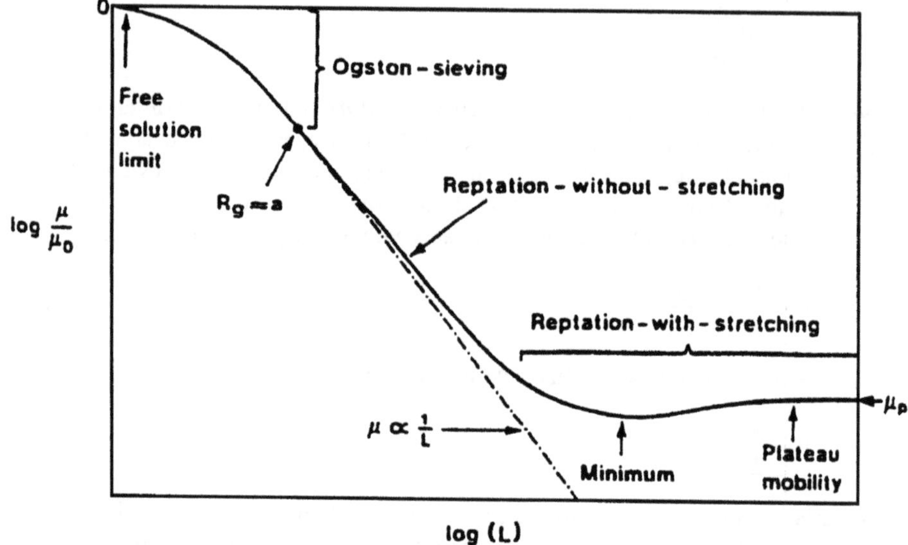

FIGURE 1 Schematic representation of the relationship between the logarithmic normalized electrophoretic mobility (μ/μ_o) and the molecular size (L). R_g, radius of gyration. (Reproduced with permission from Ref. 161.)

$$\mu \sim \frac{1}{\text{MW}} \tag{7}$$

At extremely high electric field strengths, reptation turns into biased reptation mode (Figure 1, biased reptation regime) and the resulting mobility of the analyte is described by

$$\mu \sim \left(\frac{1}{\text{MW}} + bE^a\right) \tag{8}$$

where b is a function of the mesh size of the sieving matrix, the charge, and the segment length of the migrating polyions, and $1 < a < 2$. Figure 1 depicts the double-logarithmic plots of solute mobility as a function of molecular weight, to identify the Ogston, reptation, and biased reptation regime.

2.2. Secondary Equilibrium

If there is any special additive in the separation gel and/or buffer system, such as a complexing ligand (L), the negatively charged polyion (P^{n-}) distributes between the complex ($PL_m^{(n \pm m)-}$) and the electrolyte. Depending on the charge of the ligand, it may increase (Eqs. (9a) and (9b), e.g., sodium dodecyl sulfate (SDS)–protein complex–staining dye [12]) or decrease (Eqs. (10a) and (10b), e.g., DNA–intercalator dyes [13]) the charge of the resulting complex:

$$P^{n-} + mL^{-} \rightleftarrows PL_m^{(n+m)-} \tag{9a}$$

$$K = \frac{[PL_m^{(n+m)-}]}{[P^{n-}] \cdot [L^{-}]^m} \tag{9b}$$

$$P^{n-} + mL^{+} \rightleftarrows PL_m^{(n-m)-} \qquad \text{where } m < n \tag{10a}$$

$$K = \frac{[PL_m^{(n-m)-}]}{[P^{n-}] \cdot [L^{+}]^m} \tag{10b}$$

where K is the complex formation constant, m is the number of the ligand molecules in the complex, and n is the total number of charges on the polyion. As a first approximation, n can be considered the same as the base number in the case of DNA or the number of SDS molecules in the case of SDS–protein complexes.

The electrophoretic velocity (v) of the polyion–ligand complex is described as

$$v = \frac{\ell}{t_M} = \mu_P \cdot E \cdot R_P \tag{11}$$

where ℓ is the effective separation length (from the injection to the detection point), t_M is the migration time of the polyion–ligand complex, μ_P is the electrophoretic mobility of the polyion, and E is the applied electric field strength. The molar ratio of the complexed ligand, R_P, is given by Eq. (12) in the case of similarly charged ligand and analyte molecules,

$$R_P = \frac{[PL_m^{(n+m)-}]}{c_P} = \frac{K \cdot [P^{n-}] \cdot [L^{-}]^m}{c_P} \tag{12}$$

and by Eq. (13) for oppositely charged solute and ligand molecules,

$$R_P = \frac{[P^{n-}]}{c_P} = \frac{1}{1 + K[L^{+}]^m} \tag{13}$$

where c_P is the total polyion concentration. Since the usually limited amount of ligand binds only a fraction of the complex, combining Eqs. (11) with (12) and (13) results in Eq. (14) in the case of similarly charged ligand and analyte molecules ($P^n \sim c_P$),

$$v = \mu_P \cdot E \cdot K \cdot [L^{-}]^m \tag{14}$$

and Eq. (15) for oppositely charged polyion and ligand molecules,

$$v = \mu_P \cdot E \cdot \frac{1}{1 + K[L^+]^m} \tag{15}$$

Equations (14) and (15) suggest the extent of the increase (e.g., SDS–protein complex–staining dye [12]) or decrease (e.g., DNA–intercalator dyes [13]) in electrophoretic velocity when the negatively or positively charged ligand binds the analyte, respectively. Please note that one should also take into consideration the mass change of the polyion–ligand complex, i.e., the additional mass of the ligand molecules that increases the resulting mass of the complex in both instances.

2.3. Efficiency and Resolution

In capillary gel electrophoresis, one of the major contributors to band broadening, besides the injection and detection extra-column effects, is the longitudinal diffusion of the solute molecules in the capillary tube [14]. The theoretical plate number (N) is characteristic of column efficiency:

$$N = \mu \frac{E \cdot \ell}{2D} \tag{16}$$

where μ is the electrophoretic mobility, D is the diffusion coefficient of the solute in the separation gel-buffer system, and ℓ is the effective column length.

Resolution (R_s) between two peaks can be calculated from the differences of their electrophoretic mobilities ($\Delta\mu$) [15]:

$$R_s = 0.18 \cdot \Delta\mu \sqrt{\frac{E \cdot \ell}{D \cdot \bar{\mu}}} \tag{17}$$

where $\bar{\mu}$ is the mean mobility of the sample components of interest. As one can see, Eqs. (16) and (17) suggest that higher applied electric field and lower solute diffusion coefficient will result in higher separation efficiency (N) and concomitantly higher resolution (R_s). One of the limiting factors is the so-called Joule heat (Q_j) generated by the applied power ($P = V \times I$) [16]:

$$Q_j = \frac{P}{r^2 \cdot I \cdot L} \tag{18}$$

where I is the current, L is the total length (electrode to electrode), and r is the radius of the capillary. Due to the dependence of the electrophoretic mobility and complex formation constant on the temperature, efficient temperature control during capillary electrophoresis separation is important in order to attain good reproducibility. Modern, automated capillary electrophoresis instruments are equipped with effective liquid or air cooling systems to address temperature change-related problems.

3. CAPILLARY COATINGS

Due to the sensitivity to solute–wall interactions and the distortive effects of electroosmotic flow (EOF), mostly coated (covalent or dynamic) tubes are employed in capillary gel electrophoresis. Uncoated columns possess strong electroosmotic character (EOF) at higher than neutral pH, and may also exhibit strong absorption to the walls of various sample types. In addition, at high pH, the EOF can be so strong that even the negatively charged solute molecules migrate toward the anode and thus are not detected in the regularly used normal polarity mode (anode at the injection side). Apparently, gel or polymer solution-filled bare fused silica capillaries last for only a few runs with real biological samples before deteriorating. Appropriate capillary coatings modify (usually reduce) electroosmotic flow and also prevent sample interactions with the inner surface of the capillary. With a proper coating in place, the capillary can be used at high pH for longer time periods and extensively washed (even with strong acids). Coated capillaries also feature good run-to-run migration time reproducibility and provide high peak efficiency. For this reason, most capillary gel electrophoresis separations are conducted in coated capillaries. Both dynamic and covalently coated capillaries have been proved useful in capillary electrophoresis with gels or polymer solutions, as summarized in two recent reviews discussing the current state of the art of coating [17,18].

3.1. Covalent Coatings

In the mid 1980s, Hjerten introduced a special silanization technique using γ-methacryloxypropyl-trimethoxysilane and subsequent cross-linking the surface-bound methylacryl groups with linear polyacrylamide to eliminate EOF and minimize wall adsorption of the analyte molecules [19]. Later, Cohen et al. used similar polyacrylamide-coated tubes for SDS-capillary gel electrophoresis separation of various protein mixtures [20]. The coated capillary they used exhibited no apparent interaction with the SDS–protein complexes in the analyte mixture and resulted in good separation with sharp peaks. Poppe and co-workers covalently coated fused silica capillaries with polyethylene glycol (600) through the same bifunctional reagent [21]. To decrease hydrolysis-mediated deterioration of the coating at higher pH, Novotny's group introduced a polyacrylamide coating similar to that described earlier, but the coating was bound to the silica wall through Si–C, rather than Si–O–Si bonds [22]. The Si–C bond is more hydrolytically stable, resulting in improved stability over the wide pH range of 2–10.5. Various existing liquid chromatography (LC) and gas chromatography (GC) type coatings were also attempted with gel- or polymer network-filled capillaries [23,24]. Ganzler et al. reported a dextran coating and filling the column with dextran polymer for protein analysis by SDS-capillary gel electrophoresis [25]. Their coated capillary was quite stable and featured good run-to-run migration-time reproducibility. Recently, Shieh et al. [26] developed a novel approach, combining the advantages of both dynamic and covalent coatings, i.e., a thin layer of polyacrylamide is dynamically coated to the inner surface of the capillary, followed by allylamine treatment [27]. The capillary coated in this way showed stable performance with biological samples for more than 200 injections (acid rinses between runs). Covalently coated capillaries also offer high-speed separations with little or no requirement for preseparation capillary equilibration.

3.2. Physical (Noncovalent) Coatings

Due to the extra work involved in producing covalently coated capillary columns, various kinds of noncovalent coatings have been reported, such as anionic, polymeric, and nonionic/zwitterionic (Figure 2) [17]. During the dynamic

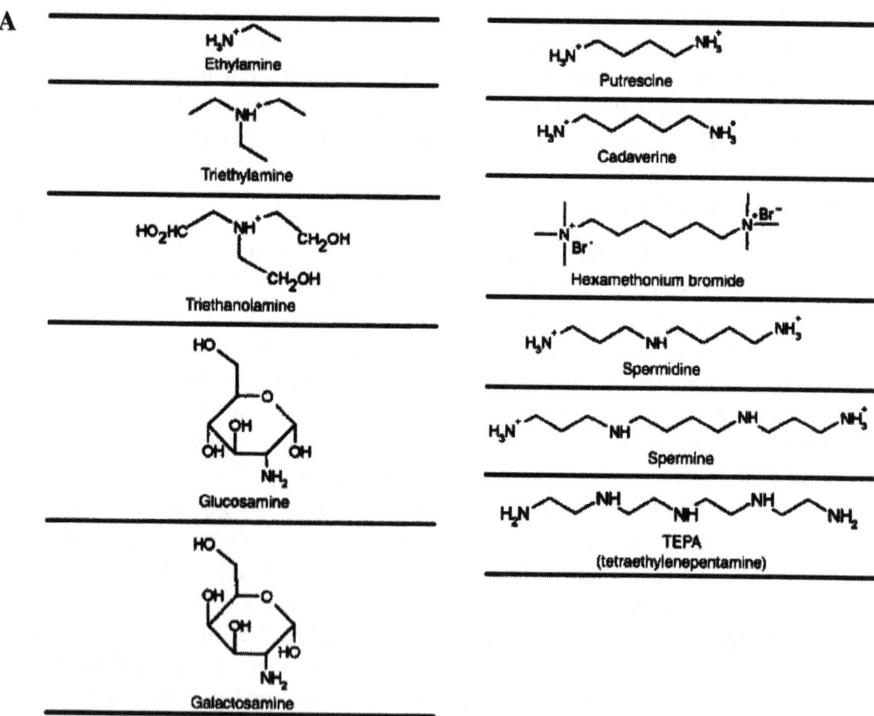

FIGURE 2a Formulas of the various dynamic surface-coating materials: (A) amines; (B) polymers; (C) nonionic and zwitterionic surfactants. (Reproduced with permission from Ref. 17.)

B

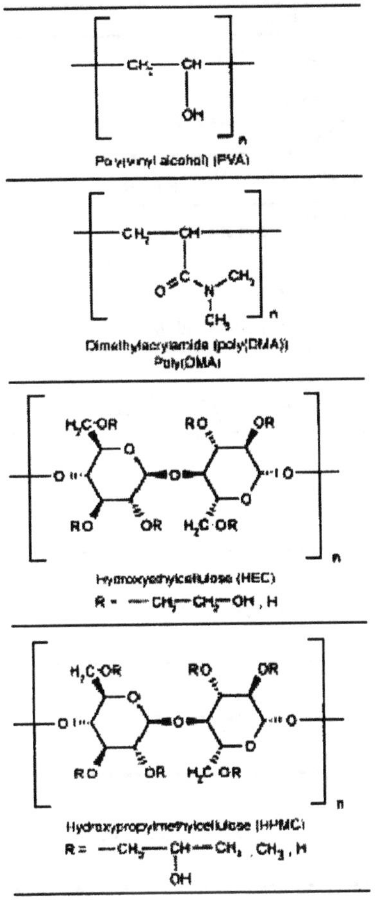

Poly(vinyl alcohol) (PVA)

Dimethylacrylamide (poly(DMA))
Poly(DMA)

Hydroxyethylcellulose (HEC)
R = $-CH_2-CH_2-OH$, H

Hydroxypropylmethylcellulose (HPMC)

FIGURE 2b Continued.

coating procedure, a thin layer of coating material is adsorbed onto the inner capillary surface prior to or during the separation process. Dynamic coatings are easily applied to any bare fused silica capillary and readily regenerated whenever is necessary, even before or after each run. In the early 1990s, rapid protein separations with good resolution were demonstrated with dynamically coated capillaries [26,28], but they required longer equilibration times prior to use and sometimes were not so efficient and compatible with real biological samples.

Landers and co-workers compared various surface derivatization agents in conjunction with coating polymers for DNA heteroduplex analysis of the breast cancer susceptibility gene (BRCA1) [29]. The most effective coating combina-

C

Detergent	Formula	CMC (mM)	M_r
Triton X-100 (polyethylene glycol *tert*-octylphenyl ether)	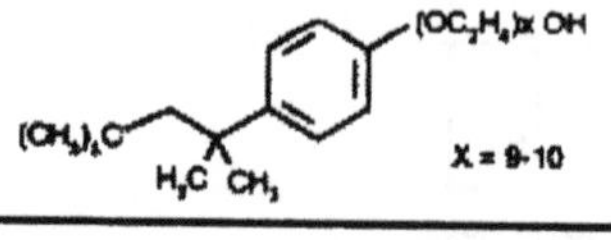	0.02–0.09	625
Brij 35 (polyoxyethylene lauryl ether)	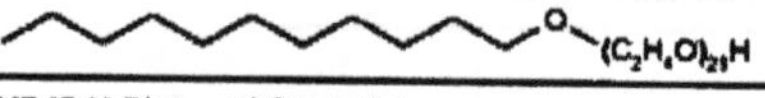	0.05–0.1	1225
Tween 20 (polyoxyethylene sorbitanmonolaurate)	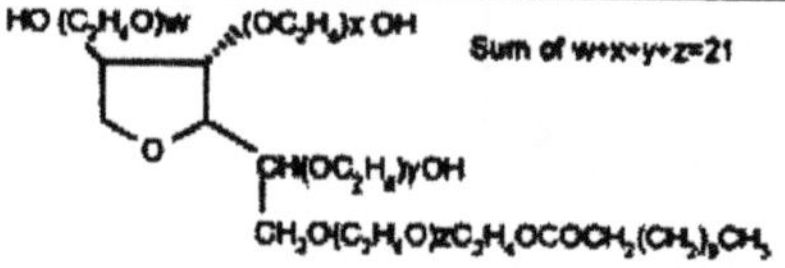	0.06	1228
CHAPS (3-[(3-cholamidopropyl) dimethylammonio] -1-propanesulfonate)	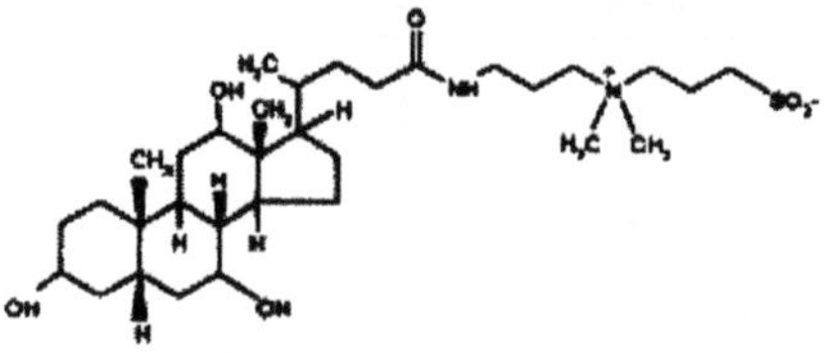	6–10	614.9
Caprylyl sulfobetaine SB-10 (*N*-decyl-*N*,*N*-dimethyl-3-ammonio-1-propanesulfonate)	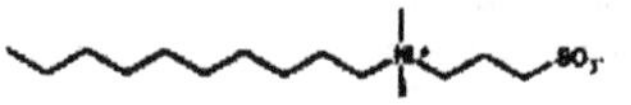	25–40	307.6
Lauryl sulfobetaine SB-12 (*N*-dodecyl-*N*,*N*-dimethyl-3-ammonio-1-propanesulfonate)	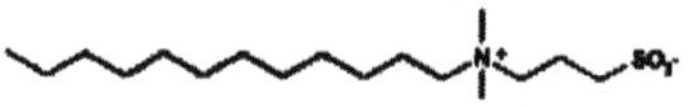	2–4	335.6
Palmityl sulfobetaine SB-16 (*N*-hexadecyl-*N*,*N*-dimethyl-3-ammonio-1-propanesulfonate	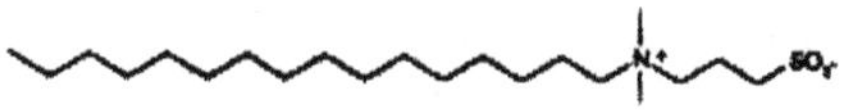	0.01–0.06	391.6

FIGURE 2c Continued.

tion was the chlorodimethyloctylsilane (OCT) with polyvinylpyrrolidone polymer that was used to separate the relevant DNA fragments in hydroxyethylcellulose matrix within 10 min. Applications of short-chain polydimethylacrylamide as sieving and wall coating medium was also beneficial for the electrophoretic separation and mutation analysis of DNA fragments in uncoated capillaries [30]. Another approach applied absorbed polydimethylacrylamide-co-allyl glycidyl ether coating to capillary electrophoresis analysis of DNA fragments. This coating was found to be stable even under harsh conditions of highly alkaline pHs, elevated temperatures, and denaturing conditions that usually rapidly deteriorate most other coatings [31].

4. SEPARATION MATRICES

The primary difference between classical polyacrylamide or agarose gel electrophoresis and capillary electrophoresis with gels or polymer networks is the use of narrow-bore fused silica capillary columns containing a sieving medium. Capillaries were filled first with cross-linked polyacrylamide similar to that used in traditional gel electrophoresis (PAGE), and have proven to be particularly useful in the analysis of DNA and protein molecules. While gel-filled capillaries were originally prepared by the users, recently more and more manufacturers have offered prefilled, ready-to-use columns. In classical polyacrylamide gel electrophoresis, mostly cross-linked gels are used. The three-dimensional structure of the gel creates a molecular sieve enabling size separation of biopolymers when they migrate through the gel media. However, unlike slab PAGE, in capillary gel electrophoresis, both cross-linked gels and high- or low-viscosity non-cross-linked polymer solutions are applied. In traditional slab-gel electrophoresis, the separation matrix is filled into the separation chamber (between two glass or plastic plates) and connected to the buffer tanks. The gel medium is polymerized and filled into the chambers, usually by the user prior to use. In capillary gel electrophoresis, narrow-bore capillaries are used that contain the sieving matrix. The separation is carried out at high voltages, resulting in rapid separations with good efficiency. Depending on the preferred separation medium, the matrix can be polymerized in situ inside the capillary or filled and replaced by applying pressure on the polymer medium-filled container, which is connected to the separation capillary. As gels may vary from viscous fluids to solids [32], the two types of gels used in capillary electrophoresis are the high-viscosity chemical gels (cross-linked) and low-viscosity physical gels (non-cross-linked). It is important to note that physical gels offer several advantages over their chemical counterparts, such as longer shelf life, lower viscosity, and ease of manufacturing. These gels are less sensitive to changes in temperature, and can be used with pressure injection. Today, both types of gels are extensively employed for the separation of various biopolymers. For more details, see the comprehensive review on electrophoresis separation media in Ref. 33.

4.1. Cross-Linked (Chemical) Gels, Structure and Properties

Chemical gels are covalently cross-linked polymer networks, featuring very high viscosity and well-defined pore structure. Polyacrylamide is the most widely used chemical gel material, usually cross-linked with N,N-methylene-bisacrylamide (BIS). The pore size of the gel is determined by the relative concentration of monomer and cross-linker used during polymerization ($\%T$, total monomer concentration; and $\%C$, cross-linker concentration as a percent of the total monomer and cross-linker concentration [34]). Highly cross-linked ($\sim 5\%C$) polyacrylamide gels are very rigid and cannot be removed from the capillary after polymerization. For stabilization, most chemical gels are covalently attached to the inner capillary surface via a bifunctional reagent. Samples can only be introduced into cross-linked gel-filled capillaries by the electrokinetic injection method. In electrokinetic injection the sample migrates into the capillary by simply starting the electrophoresis process from the sample compartment. While this approach usually results in sharp peaks with some sample preconcentration (if the separation and sample buffers are chosen appropriately [35]), the method suffers from poor peak-area (injection amount) reproducibility and therefore does not readily support quantitative applications. Chemical gels are sensitive to changes in temperature, pH, and high voltage, e.g., small bubbles can be formed due to phase-transition phenomena (gel shrinking) [32] during polymerization or sometimes even during separation. On the other hand, chemical gels provide extremely high resolving power, especially for the separation of low-molecular-weight oligomers.

The pioneering work of Karger'group [20] revealed the possibility of filling cross-linked polyacrylamide gels into narrow-bore capillaries and demonstrated the excellent separation power of this system by resolving the two chains of insuline in less than 8 min, using a 7.5%*T*, 3.3%*C* polyacrylamide gel containing 8 M urea and 0.1% SDS. The first report on single-nucleotide resolution separation of the pdA_{40-60} DNA ladder (Figure 3) by Guttman et al. [36] opened up new horizons in modern capillary gel electrophoresis-based DNA sequencing

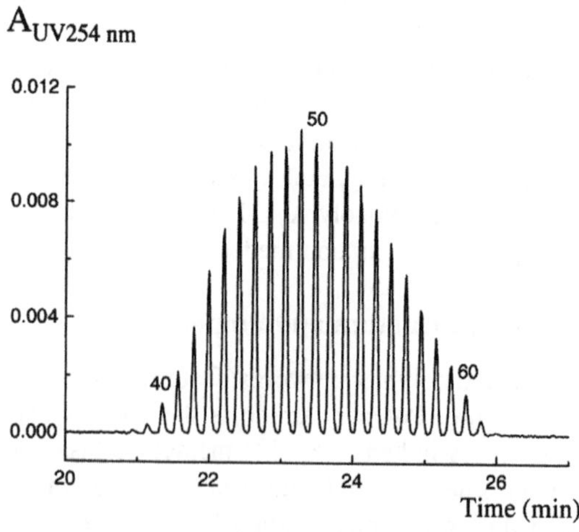

FIGURE 3 The first capillary gel electrophoresis-based single-stranded DNA separation (pdA_{40-60}). Conditions: polyacrylamide gel (7.5%*T*/3.3%*C*) in 7 M urea, 100 mM Tris, and 250 mM borate (pH 8.3), $E = 350$ V/cm; detection, 254 nm; capillary, $\ell = 30$ cm (effective), 75 μm i.d. (Reproduced with permission from Ref. 36.)

analysis, which actually made possible the faster-than-anticipated completion of the Human Genome Project [1]. In continuation of the successful introduction of gel-filled capillaries for high-resolution separation of DNA molecules, Swerdlow and Gesteland reported in 1990 the development of the first capillary gel electrophoresis-based DNA sequencer [37]. This novel approach revealed the enormous separation power of capillary gel electrophoresis, capable of resolving DNA molecules differing by only one nucleotide, exactly what was required for DNA sequencing. Chemical gels have been used since as sieving media mainly for separating shorter single-stranded DNA molecules and also for separation of proteins by capillary SDS-gel electrophoresis [38]. Vegvari and Hjerten have recently introduced a new polyacrylamide gel formulation, cross-linked with allyl-beta-cyclodextrin, for the separation of DNA fragments [39]. Interestingly, the resolving power of their gel was almost independent of the concentration of the cross-linker, in contrast to regularly cross-linked polyacrylamide gels with N,N′-methylene-bisacrylamide.

4.2. Non-Cross-Linked Linear Polymer Solutions (Physical Gels)

The most promising advances in capillary electrophoresis separation of biopolymers came from the exploration of novel separation matrices. Linear polymers, so-called physical gels or polymer networks, have become very popular and widely used these days in capillary electrophoresis of biologically important polymers [28,35,40]. Non-cross-linked linear polymers are not attached to the inside wall of the capillary and feature very flexible dynamic pore structure. Actually, the pore size of these gels is defined by dynamic interactions between the polymer chains, and can be varied at any time by changing such variables as capillary temperature, separation voltage, salt concentration, or pH. Physical gels are not heat-sensitive, and even if a thin layer is attached to the capillary wall (see dynamic coating), the separation matrix can usually be simply replaced in the capillary by applying pressure to the reservoir containing the separation matrix. Thus fresh separation medium can be used for each analysis, preventing any contamination from previous samples. Physical gels support both electrokinetic and pressure injection methods. In pressure injection, introduction of the sample occurs by applying constant pressure for a short period of time on the sample vial connected to the capillary. This method also enables sample stacking and offers excellent run-to-run peak-area reproducibility supporting routine quantitative analysis.

4.2.1. Non-Cross-Linked Polyacrylamide and Agarose

In the first report of Heiger et al., non-cross-linked polyacrylamide was introduced into coated capillary columns and applied to the separation of double-stranded DNA fragments ranging up to several thousand base pairs [41] (Figure

4). The same group studied the sequence-dependent migration behavior of ds-DNA fragments in capillary electrophoresis using a linear polyacrylamide (LPA) matrix [42] and found significant conformational effects under high electric field strengths. Later, Kenndler and co-workers demonstrated the usefulness of linear polyacrylamide gels for the separation of SDS–protein complexes ranging from 17.8 to 77 kDa in 60 min [43]. Regnier and co-workers [44] and others

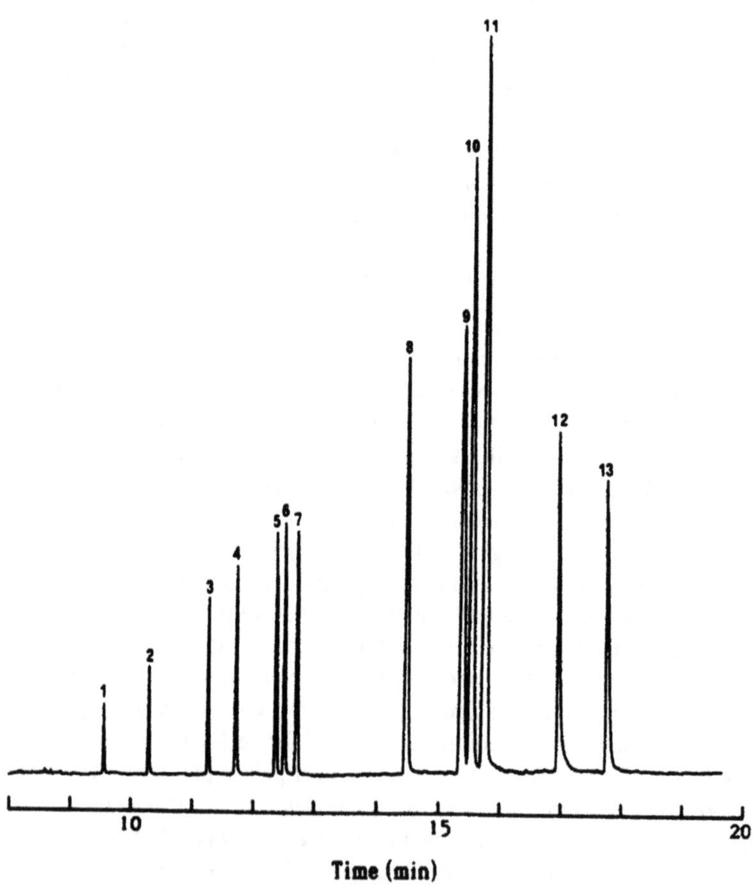

FIGURE 4 Separation of Hae III digest of ϕX 174 DNA, *Eco*R I digest of pBR322, and *Eco*R I digest of M13mp18 on 3%T/0.5%C linear polyacrylamide separation matrix. Peaks 1–13: 72, 118, 194, 234, 271, 281, 310, 603, 872, 1078, 1353, 4353, 7253 bp. Conditions: $\ell = 30$ cm($L = 40$ cm), $E = 250$ V/cm; buffer, 100 mM TBE; detection, 254 nm; injection, 10 kV/0.5 s. (Reproduced with permission from Ref. 41.)

[45,46] successfully applied linear polyacrylamide gels for the separation of standard proteins and biological samples as well. Recent advances in capillary electrophoresis of DNA fragments in linear poly(*n*-substituted acrylamides) are thoroughly reviewed in Ref. 47.

Agarose gel-filled capillaries were studied extensively by Bocek and co-workers [48] in the separation of dsDNA molecules. Chemically modified agarose gels or composite agarose–non-cross-linked polymer gels, capable of resolving several-base-pair differences in DNA fragments of several hundreds of base pairs in length, have also been developed [49].

4.2.2. Derivatized Celluloses

Derivatized celluloses were introduced in the early 1990s by Brownlee and co-workers [50] as efficient sieving matrices to separate biologically important polymers in capillary columns. More recently, low-concentration entangled linear polymers, such as dextran, polyethylene oxide, hydroxyethylcellulose, pollulane, and polyvinyl alcohol, have also been employed as physical gels for separation of biopolmers. Standard proteins ranging from 14 to 97 kDa were separated on hydroxypropylcellulose in less than 30 min with no apparent influence of various buffer compositions on the sieving performance [51]. Morris and co-workers have investigated the migration of double-stranded (ds) DNA molecules in semidilute hydroxyethylcellulose (HEC) solution and demonstrated segmental DNA motion allowing quantitative description of the changing shape of DNA as it interacts with the sieving polymer [52]. With a 3-D view of the migrating DNA molecules, they observed U-shape conformations oriented at an angle to the microscope plane, as well as ambiguities and artifacts resulting from loss of information from DNA segments not in focus. A polymer coil shrinking theory was compared with the existing entanglement solution theory and showed promising potential for semidilute solutions in DNA fragment analysis [53]. Gibson and Sepaniak reported that axial diffusion, even when adjusted for kinetic conditions using the Einstein relationship, did not account for the total observed band variance [54]. Barron's group studied the impact of polymer hydrophobicity on the properties and performance of DNA sequencing matrices for capillary electrophoresis [55]. Their study highlighted the importance of polymer hydrophobicity for high-performance DNA sequencing matrices to form robust, highly entangled polymer networks and to minimize the hydrophobic interactions between the polymers and the fluorophore-labeled sequencing fragments. The same group also developed novel DNA sequencing matrices with a thermally controllable "viscosity switch" [56].

4.2.3. Polyethylene Oxide and Polyvinylpyrrolidone

Polyethylene oxide (PEO) and polyvinylpyrrolidone (PVP) solutions both proved to be good separation matrices in capillary electrophoresis-based DNA sequencing, also featuring self-coating properties [57–61]. Dye-induced mobility shifts were also characterized as affecting DNA sequencing in a polyethylene oxide sieving matrix [62]. A versatile low-viscosity polyethylene oxide-based sieving matrix was developed for nondenaturing DNA separations in capillary array electrophoresis [63]. Rapid molecular diagnostics of 21-hydroxylase deficiency was investigated by Barta et al. [64] detecting the most common mutations in the 21-hydroxylase gene using primer extension technique and capillary electrophoresis with polyvinylpyrrolidone sieving and a wall coating matrix. The Cy5-labeled primers and the two possible primer extension products (mutant and wild type) were completely separated in 90 s on a 10-cm-effective-length capillary (Figure 5).

4.3. Copolymers, Composite Gels, and Other Alternative Matrices

Besides the regularly used cross-linked and linear polyacryamide and their derivatives, polydimethylacryamide, polyethyleneoxide, polyvinylpirrolidone, polyethyleneglycol (also with fluorocarbon tails), hydroxyethylcellulose, and various cellulose detivatives, other polysaccarides, etc., have proved useful for size separation of biopolymers in narrow-bore capillaries (see review in Ref. 65), and some alternative matrices have also been adapted. Recently introduced novel termoresponsive copolymers, comprising hydrophobic and hydrophilic blocks, such as pluronics [66–68] and polyisopropylacrylamide grafted with polyethylene oxide chains [69], have also exhibited promising results. A grafted copolymer of poly (N-isopropylacrylamide)-γ-polyethyleneoxide with self-coating ability and slightly adjustable viscosity properties was applied for rapid and high-resolution separation of dsDNA fragments. These matrices have pronounced temperature-dependent viscosity transition point, implying promising implementations. In particular, thermoresponsive polymers can offer some practical advantages for capillary electrophoresis, such as easier handling and loading of the viscous polymer solutions without the requirement for a high-pressure manifold. For example, Barron's group has constructed an interesting "viscosity switch" material responding to changes of temperature, pH, or ionic strength [56]. These matrices are based on copolymers of acrylamide derivatives with variable hydrophobicity and feature reversible, temperature-controlled viscosity switch from high-viscosity solutions at room temperature to low-viscosity colloid dispersions at elevated temperatures. High resolving power and good DNA sequencing performance (up to 463 bases in 78 min) was achieved with these sieving media. Viovy and co-workers introduced a novel block copolymer thermoassociating matrix for DNA sequencing (Figure 6) [70]. These comb polymers are made of hydrophilic polyacrylamide backbone, grafted with poly-N-isopropylacrylamide side chains and characterized by lower critical solution temperature. These matrices combined easy loading with high sieving perfor-

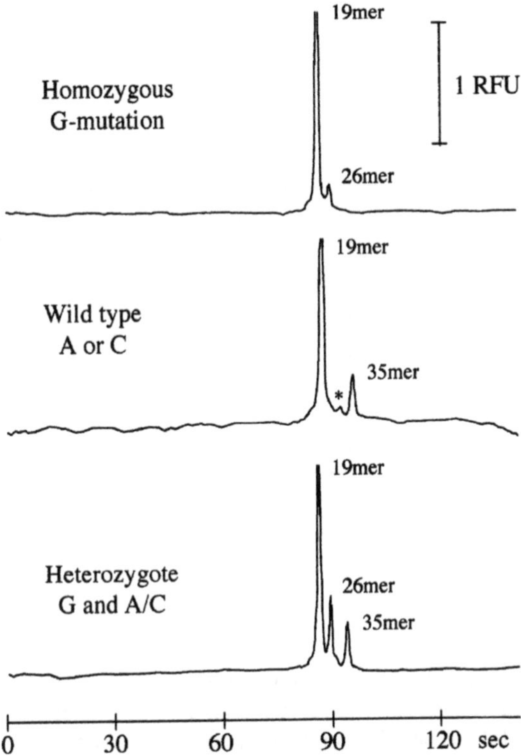

FIGURE 5 Ultrafast SNP analysis by primer extension and capillary electrophoresis. Upper panel: electropherogram of the mutant homozygote (G-mutation) with the 19-mer primer peak and the 26-mer product; Middle panel: electropherogram of the wild type with the primer and the 35-mer product; Lower panel: electropherogram of the heterozygote with all three peaks (19-, 26-, and 35-mers). Conditions: capillary, $\ell = 10$ cm (effective) ($L = 30$ cm), i.d. 75 μm; separation matrix and running buffer, 10% PVP (MW 1,300,000) in 1 × TBE; applied voltage, 20 kV; injection, 30 s/10 kV; temperature, 30°C. (Reproduced with permission from Ref. 64.)

mance due to switching between a low- and a high-viscosity state by temperature change. Thermothickening properties of these polymers, due to formation of transient intermolecular cross-links at higher temperatures, offer clear advantages for DNA sequencing, e.g., easy handling of low-viscosity solutions at low temperatures and excellent sieving characteristics of high-viscosity solutions at elevated temperatures. The rheological behavior and separation performance were in correlation with their mirostructure. Sequencing read lengths as high as 800 bases were attained in 1 h using these polymers.

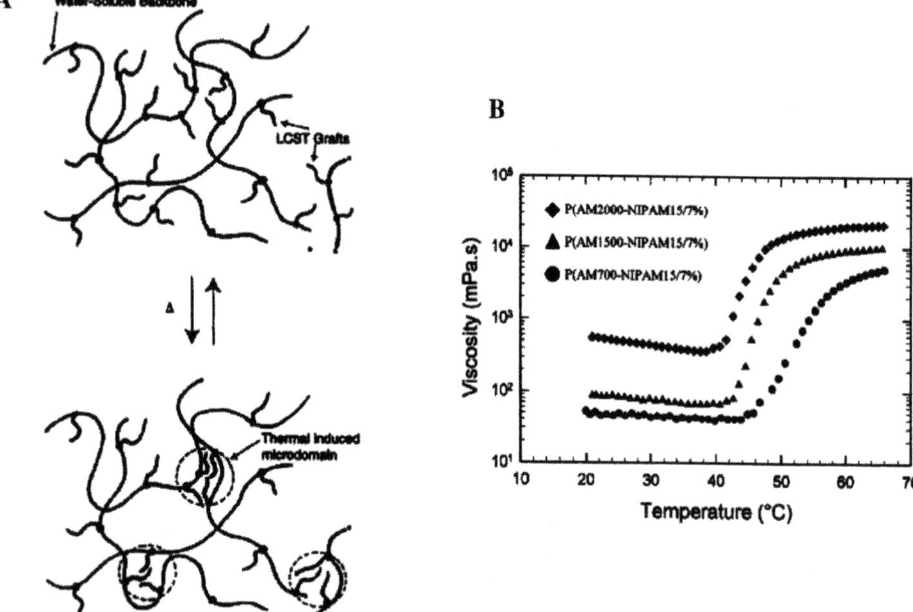

FIGURE 6 (A) Simplified view of a mechanism for thermothickening by associating block copolymers. By heating, the low critical solution temperature grafts undergo a microphase separation and create micelle-like aggregates, which act as transient cross-links. (B) Viscosity versus temperature for different synthetized polymers, dependence on main-chain molecular weight. (Reproduced with permission from Ref. 70.)

A low-concentration (0.25%) solution of the natural polysaccharide glucomannan was applied as a sieving additive for the separation of a range of DNA fragments up to 1400 bp in capillary electrophoresis [71]. The effect of temperature and viscosity of pullulan sieving medium on electrophoretic behavior of SDS–proteins in capillary electrophoresis was studied by Nakatani et al. [72]. A low-viscosity polysaccharide matrix, TreviSol, has been characterized in capillary electrophoresis for the separation of DNA fragments and exhibited good separation ability for larger DNA fragments [73]. The same group evaluated composite agarose/hydroxyethylcellulose matrices for the separation of DNA fragments by capillary electrophoresis [74]. Relative to homogenous gels, these composite matrices provided enhanced separation selectivity, especially for DNA fragments larger than 100 bp. In another approach, a viscosity-adjustable PEO99/PEO69/PEO99 block copolymer [67] and a triblock polymer of polyethyleneoxide–polypropyleneoxide–polyethyleneoxide [66] was successfully applied to obtain high-speed separation of DNA fragments by capillary electrophoresis.

Recently, Chiari et al. synthetized sugar-bearing polyacrylamide copolymers as sieving matrices and capillary coatings for DNA electrophoresis [75] and also copolymerized poly(N,N-dimethylacrylamide) with hydrophilic monomers to improve separation performance [76]. A copolymer of acrylamide and β-D-glucopyranoside was used as a low-viscosity and high-capacity sieving matrix for the separation of dsDNA molecules. The growth of the polymer chains was controlled by the different reactivity of the two monomers. The chain length was inversely proportional to the number of glucose residues incorporated into the copolymer [77]. Purified galactomannans from guaran, tara gum, and locust bean gum were attempted as sieving media for DNA sequencing in capillary electrophoresis. Separation efficiency exceeded 1 million theoretical plates for DNA fragments less than 600 bases long [78]. Pluronic copolymer liquid crystals were successfully used as unique, replaceable media in capillary gel electrophoresis [68]. Separation of model mixtures of peptides/proteins was attempted in surface-modified capillaries filled with pluronic liquid crystals acting as secondary partition mechanism [79,80].

Solutions of monomeric nonionic surfactants, *n*-alkyl polyoxyethylene ethers, behave as dynamic polymer structures and therefore can be used as sieving matrices for DNA fragment analysis in capillary columns [60]. Surfactant solutions offer several advantages over regularly used linear polymers. Some of the most important ones are the ease of matrix preparation, solution homogeneity, stable structure, low viscosity, and self-coating properties to reduce EOF. Good and rapid separation of both dsDNA fragments and single-stranded sequencing ladders were reported. Separation of dsDNA fragments was also successfully attempted in a transient interpenetrating network of polyvinylpyrrolidone with polyacrylamide [81] and poly (N,N-dimethylacrylamide) [82]. Viscosity measurements revealed that this interpenetrating network had significantly higher viscosity that of a simple mixture containing the same amount of components. The particularly good sieving ability of the system was attributed to the increase in the number of entanglements by the more extended polymer chains.

Other additives, such as mannitol, were attempted to enhance the sieving ability of hydroxy-propylmethylcellulose (HPMC) for dsDNA fragment analysis [83,84]. The authors suggested that a mannitol chain is formed through hydrogen bonding among mannitol, HPMC, and borate, reshaping the network and decreasing the pore size. Addition of glycerol to an entangled solution of HPMC enhanced separation performance of DNA fragments in capillary electrophoresis, probably due to the formation of dimeric 1:2 borate:didiol complexes with both glycerol and HPMC [85].

5. MAJOR APPLICATIONS

5.1. Separation of DNA

Denaturing capillary gel electrophoresis is utilized mainly in size separation of relatively short single-stranded oligonucleotides and in DNA sequencing. The most commonly used denaturing agents are urea and formamide. Nondenaturing gels or polymer networks are used when separation should be based on the size and shape of nucleic acids, as well as to reveal secondary structure differences. A comprehensive review on capillary electrophoresis of DNA in sieving polymers for molecular diagnostics applications, such as screening for inherited human genetic defects, quantitative gene dosage, microbiology/virology, forensic analysis, and therapeutic antisense DNA separation, is given in Ref. 86.

5.1.1. DNA Sequencing

One of the most important applications of denaturing capillary gel electrophoresis is DNA sequencing. During the last decade, the lack of adequate stability of cross-linked polyacrylamide gels within microbore columns initiated a rapid development to find novel, more capillary-friendly sieving matrices. The first breakthrough demonstrated the usefulness of non-cross-linked polymeric solutions to attain rapid separation of single-stranded DNA molecules, also enabling the application of high temperature during electrophoresis [87]. Choosing the appropriate electrophoresis parameters (e.g., temperature) for the analysis of DNA sequencing fragments also plays an important role in obtaining high-speed and high-read-length separations. Temperatures as high as 80°C were successfully applied to reach a sequencing speed of more than 1000 bases per hour (Figure 7) [88]. Another important issue to overcome was the high salt content of the DNA sequencing reaction mixture. Sample preparation and cleanup turned out to be a significant part of capillary gel electrophoresis-based DNA sequencing using electrokinetic injection. Conventionally, DNA sequencing samples are pro-

---

FACING PAGE

FIGURE 7 Electrophoretic separation of DNA sequencing fragments generated on ssM13mp18 with BigDye-labeled universal (−21) primer and AmpliTaq FS at the optimum experimental conditions: 2.0% (w/w) LPA 9 MDa and 0.5% (w/w) 50 kDa LPA, 200 V/cm, and 60°C. Conditions: effective length $\ell = 30$ cm ($L = 45$ cm), 75μm i.d., 365 μm o.d.; polyvinyl alcohol-coated capillary; running buffer (both cathode and anode), 50 mM Tris/50 mM TAPS/2 mM EDTA. Cathode running buffer also contained 7 M urea, the same as in the separation matrix. The samples were injected at a constant electric field of 25 V/cm (0.7 μA) for 10 s and electrophoresed at 200 V/cm (10.2 μA) at 60°C. (Reproduced with permission from Ref. 88.)

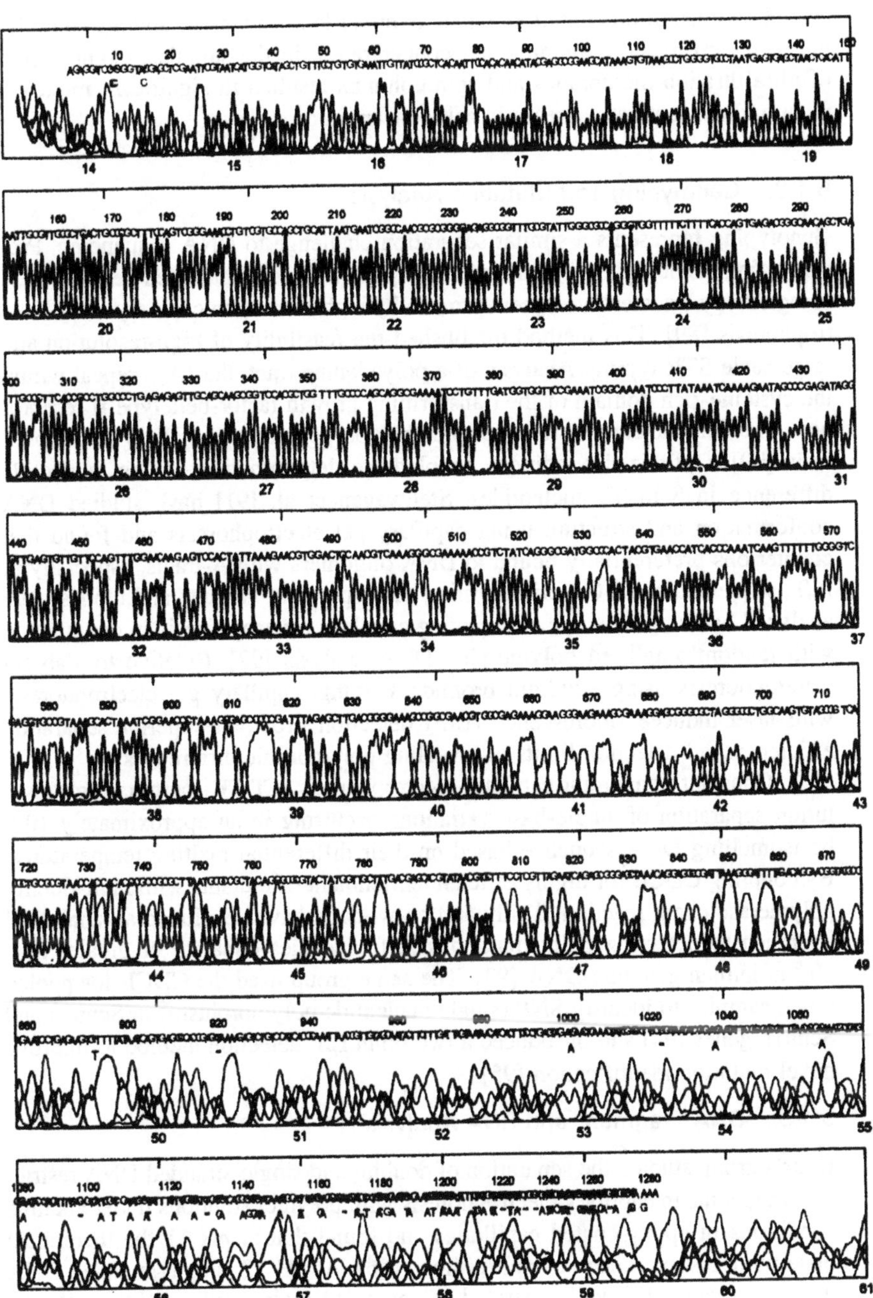
migration time (min)

cessed by ethanol precipitation, which is rather labor-intensive and also unreliable in regard to the remaining salt concentration in the samples. Combination of ultrafiltration membranes and spin columns resulted in significant improvement and consequent decrease in salt concentration.

5.1.2. Genotyping and Mutation Analysis

Genotyping represents a similar separation challenge to DNA sequencing. Precise sizing of multiplexed short tandem repeat (STR) loci has been demonstrated using energy-transfer fluorescent primers by high-resolution capillary array electrophoresis [89]. This method established the feasibility of high-resolution and large-scale STR typing. Analysis of a polyadenine tract, the $(A)_{10}$ repeat within the cysteine-rich domain of the transforming growth factor-beta type II receptor gene in colorectal cancer was attempted by non-gel-sieving capillary electrophoresis [90]. Optimized conditions enabled the determination of one nucleotide difference in 8 to 32 nucleotides. Stellwagen et al. [91] have studied DNA conformation and structure using capillary gel electrophoresis and found that counterions preferentially bound to DNA oligomers with A-tracts, especially in A_nT_n sequence motif. Genetic profiling of grape plant variants and clones were analyzed by using dynamic size-sieving capillary electrophoresis in conjunction with random amplified polymorphic DNA analysis [92]. Relative to slab-gel electrophoresis using ethidium bromide staining, capillary gel electrophoresis with laser-induced fluorescence (LIF) detection provided superior separation efficiency and detection limits in revealing polymorphic differences.

Constant denaturant capillary electrophoresis (CDCE) permits high-resolution separation of single-base variations occurring in an approximately 100-bp isomelting DNA sequence based on their differential melting temperatures. By coupling CDCE for highly efficient enrichment of mutants with high-fidelity polymerase chain reaction, Thilly and co-workers have developed an analytical approach to detecting point mutations at frequencies equal to or greater than 10^{-6} in human genomic DNA [93]. The same group used the CDCE for pooled blood samples to identify SNP (single nucleotide polymorphism) in Scnn1a and Scnn1b genes [94] and introduced a two point LIF detection method to improve novel mutation identification [95].

5.1.3. DNA Fragment and RNA Analysis

Kuhr's group studied the separation of double- and single-stranded DNA restriction fragments in capillary electrophoresis with polymer solutions under alkaline conditions in epoxy-coated capillaries and found that at pH 11 the theoretical plate numbers exceeded several millions [96]. At pH 12, single-stranded DNA molecules were still well separated in entangled hydroxyethylcellulose (HEC) solutions, but the resolution decreased significantly in dilute polymer solutions.

Heller thoroughly studied the separation of double-stranded and single-stranded DNA in a linear poly-N,N-dimethylacrylamide matrix and found significant differences between the experimental data and predicted scaling laws [97,98]. Analysis of gamma-radiation-induced damage to plasmid DNA was evaluated by using dynamic size-sieving capillary electrophoresis [99]. On-column sample preconcentration and separation of DNA samples was demonstrated in an open tubular capillary system utilizing electroosmotic flow with a polyethylene oxide sieving matrix [100]. This polymer also featured wall coating capability, especially in low-ionic-strength buffer systems. Capillary electrophoresis of RNA in dilute and semidilute polymer solutions of HEC and the dependence of solute mobility on its chain length was consistent with separation by transient entanglement mechanism in dilute solutions [101]. Another recently reported application employed 1% PVP in 1 × TBE (Tris-boric acid–EDTA) buffer with 4 M urea and 0.5 μM ethidium bromide separation medium for automated and quantitative RNA screening by capillary electrophoresis, using commercially available instrumentation and reagents, enabling high-throughput and large-scale analysis of RNA samples (Figure 8) [102].

Other major applications on size sieving capillary electrophoresis of DNA are discussed in detail for diagnostics [103], DNA typing [104], antisense DNA analysis [105–108], linear and supercoiled DNA separation [109], single-stranded conformation polymorphism study [110], DNA sequencing [87,111–113], and DNA restriction fragments [114]. A comprehensive review of the specific application of molecular diagnostics is given in Ref. 115.

5.2. Analysis of Proteins and Complex Carbohydrates from Glycoproteins

5.2.1. Capillary SDS-Gel Electrophoresis

Capillary SDS-gel electrophoresis is a rapid automated separation and characterization technique for protein molecules and is contemplated as a modern instrumental approach to sodium dodecylsulfate–polyacrylamide slab-gel electrophoresis (SDS-PAGE). Size separation of SDS–protein complexes can be readily attained in coated capillaries filled with cross-linked gels or non-cross-linked polymer networks. Figure 9 depicts one of the early applications of the technique for the analysis of a standard protein test mixture ranging in size from 14.2 to 205 kDa.

Capillary SDS-gel electrophoresis has proved to be a very important separation tool for rapid molecular-weight estimation and purity check of recombinant proteins in the modern biotechnology industry. A comprehensive review of capillary gel electrophoresis of proteins was published by Guttman in 1996 [40]. Since then, the field has moved toward biomedical and biotechnology applica-

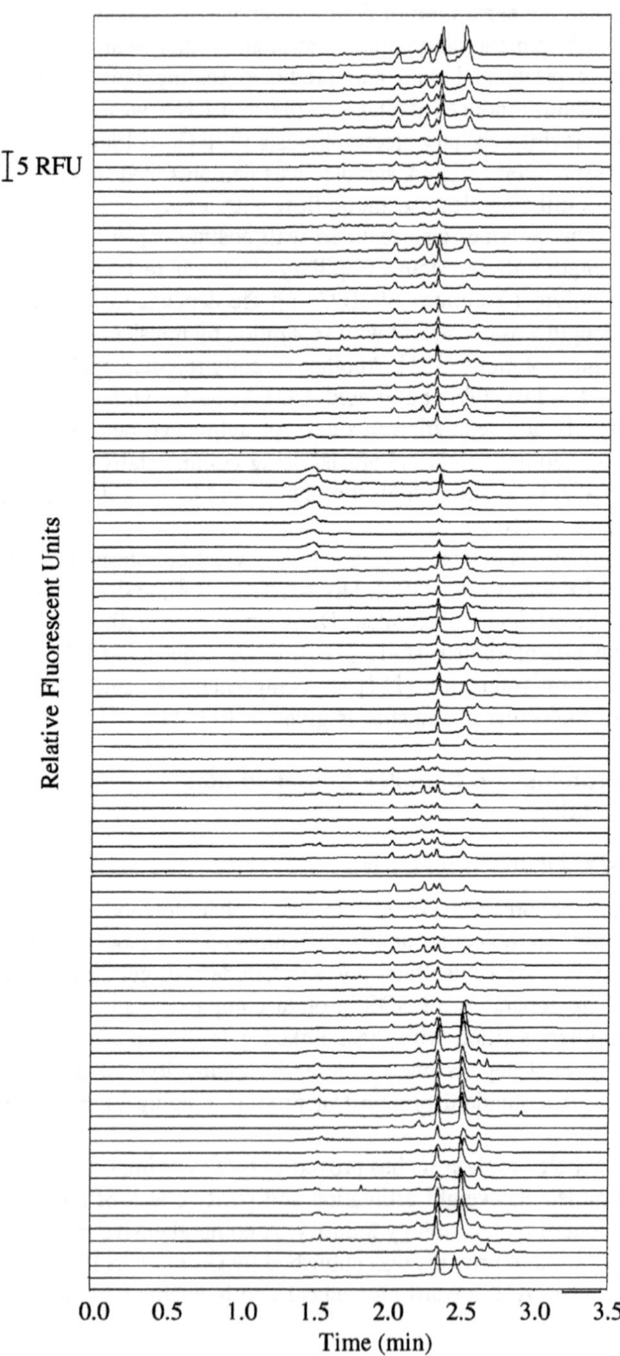
5 RFU
Relative Fluorescent Units
0.0
0.5
1.0
1.5
2.0
2.5
3.0
3.5
Time (min)

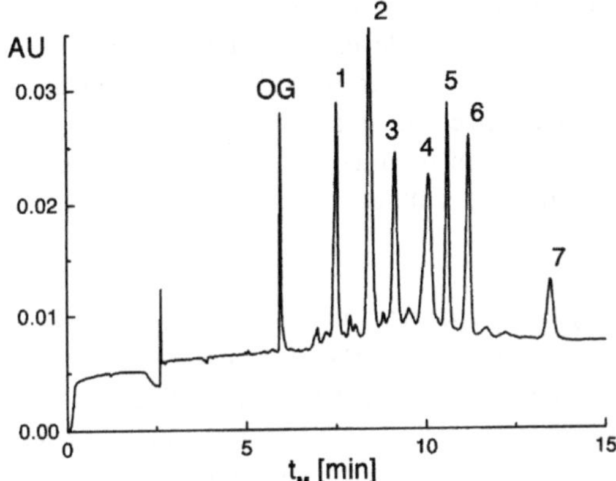

Figure 9 Capillary SDS-gel electrophoresis trace of a protein testmixture: (1) β-lactalbumin, MW 14,200; (2) carbonic anhydrase, MW 29,000; (3) ovalbumin, MW 45,000; (4) bovine serum albumin, MW 66,000; (5) phosphorylase B, MW 97,400; (6) β-galactosidase, MW 116,000; (7) myosin, MW 205,000; (OG) tracking dye Orange-G. Conditions: E = 300 V/cm; Injection, 100-ng protein mix; detection, 214 nm. (Reproduced with permission from Ref. 162.)

tions. Hunt and Nasabeh explored a biotechnology perspective of the use of capillary electrophoresis SDS nongel sieving of a therapeutic recombinant monoclonal antibody [116]. They used precolumn fluorophore labeling of rMAb to obtain low-nanomolar detection limits. Their assay illustrated the advantages of enhanced precision and robustness, speed, ease of use, and on-line detection in monitoring bulk manufacture of protein pharmaceuticals.

Kilar and co-workers successfully studied outer-membrane proteins, lipopolysaccharides, hemolysin, and the in-vivo and in-vitro virulence of wild-type *Proteus penneri* 357 and its two isogenic mutant variants (a transposon and a

---

FACING PAGE

Figure 8 Automated high-throughput RNA analysis by capillary electrophoresis. Typical batch processing profiles of a 96-well sample plate. Total RNA sample preparations from rice (traces 1–76 from top), arabidopsis (traces 77–95), and yeast (trace 96); 6 μL each in 96-well plate. Conditions: 50-μm-i.d. capillary, $\ell = 10$ cm ($L = 30$ cm); sieving medium, 1% PVP (polyvinylpirrolidone, MW = 1.3 MDa), 4 M urea, 1 × TBE, 0.5 μM ethidium bromide; $E = 500$ V/cm; 25°C. RNA samples were diluted in deionized water and denatured at 65°C for 5 min prior to analysis. Sample tray was stored at 4°C in the CE instrument during processing. Injection: vacuum (5 s at 3.44 kPa). Separation matrix was replaced after each run, 2 min at 551 kPa. (Reproduced with permission from Ref. 102.)

spontaneous mutant) using capillary electrophoresis with dynamic sieving [117]. Capillary electrophoresis was found to be suitable for comparative analysis of bacterial protein patterns of genetic variants and also provided valuable insights in connection with bacteriological virulence. Other interesting applications of capillary SDS-gel electrophoresis have been reported for protein characterization of bacterial lysates [118] and profiling of human serum proteins [119].

5.2.2. Fluorophore Labeling

A sodium dodecyl sulfate-capillary electrophoresis technique of proteins in a sieving matrix using laser-induced fluorescence (LIF) detection was reported by Craig et al. [120] utilizing a two-spectral channel detector that resolved fluorescence from the samples and standards covalently labeled by two different dyes. Others found picomolar detection limits with noncovalent fluorogenic labeling of proteins using Sypro Red dye [121]. Dovichi and co-workers reported rapid and efficient capillary SDS-gel electrophoresis separation of proteins by characterizing HT29 human colon cancer adenocarcinoma cells [122]. The cells were lysed inside a capillary, followed by protein denaturation with SDS and fluorophore labeling with 3-(2-furoyl)-quinoline-2-carboxyaldehide and separated using 8% pollulan sieving solution. Typical resolution was found to be around 30 protein components of a single HT29 cell, similar to the peak capacity of SDS-PAGE. Fluorescent detection provided high sensitivity, ranging from 10^{-10} to 10^{-11} M. Single-cell-level analysis was completed in 45 min. Other research groups reported the separation and comparative analysis of apolipoproteins by capillary zone and capillary SDS-gel electrophoresis [123,124].

Complex carbohydrates released from glycoproteins were readily profiled by capillary gel electrophoresis with LIF-based detection of 1-aminopyrene-3,6,8-trisulfonic acid (APTS)-labeled sugar molecules (Figure 10) [125]. High-mannose-type oligosaccharides of ribonuclease B were derivatized by APTS and separated by capillary electrophoresis using polyethylene oxide separation medium [126].

5.3. Capillary Affinity Gel Electrophoresis

5.3.1. Soluble Ligand Method

Affinity-type ligand molecules may be either soluble or immobilized in the gel–buffer system [13]. With the use of soluble ligands (Figure 11A), analyte complexes having a broad range of physical or chemical properties can be formed. If the ligand is very small compared to the analyte molecules, the change in mobility of the complex will be relatively small or negligible, especially if the ligand has little or zero charge. A substantial effect on the electrophoretic mobility is expected if the ligand and the complex are comparable in size or if the

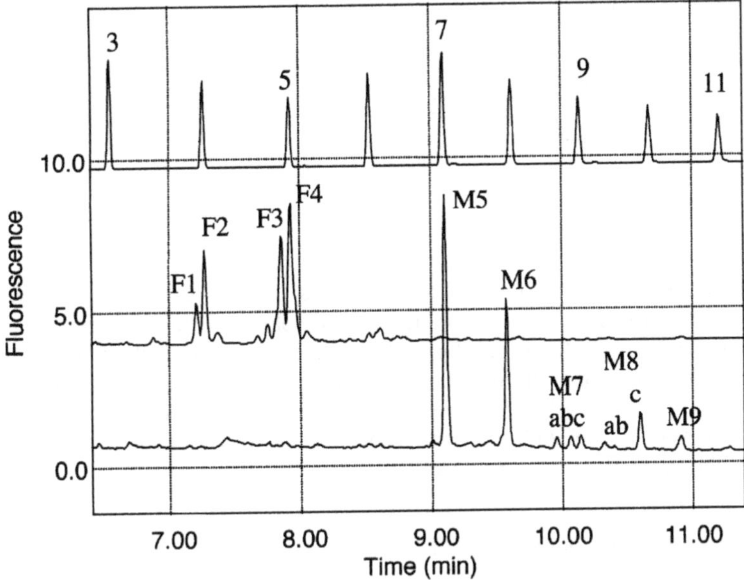

FIGURE 10 Electropherograms of APTS-labeled glycans from bovine fetuin (middle trace) and bovine ribonuclease B (lower trace) compared to the maltooligosaccharide ladder standard (upper trace). Numbers on the upper trace correspond to the degree of polymerization of the glucose oligomers. Peaks: F1 = tetrasialo-triantennary-2xα2,6; F2 = tetrasialo- triantennary-2xα2,3; F3 = trisialo-triantennary-2xα2,6; F4 = trisialo-triantennary-2xα2,3; M5-M9: Mannose 5-Mannose 9. Conditions: Capillary, $\ell = 40$ cm (effective) neutrally coated capillary (eCAP™ N-CHO) with 50-μm i.d.; buffer, 25 mM acetate, pH 4.75; detection, laser-induced fluorescence, excitation 488 nm, emission 520 nm; applied field strength, 500 V/cm; temperature, 20°C. (Reproduced with permission from Ref. 125.)

ligand is small but highly charged. In this case the interaction may result in a strong decrease or increase in the electrophoretic mobility of the complex. An example of selectivity manipulation by soluble affinity ligand interactions was evaluated by the addition of intercalating agents to the separation gel–buffer system in dsDNA separations [13]. In this instance, the ligand, ethidium bromide, intercalated between the two strands of the DNA double helix during the separation process. As it is positively charged at the separation pH, the complexation reduces the migration times of all the DNA fragments [Eq. (15)] and therefore increases the separation time window (enhanced resolution).

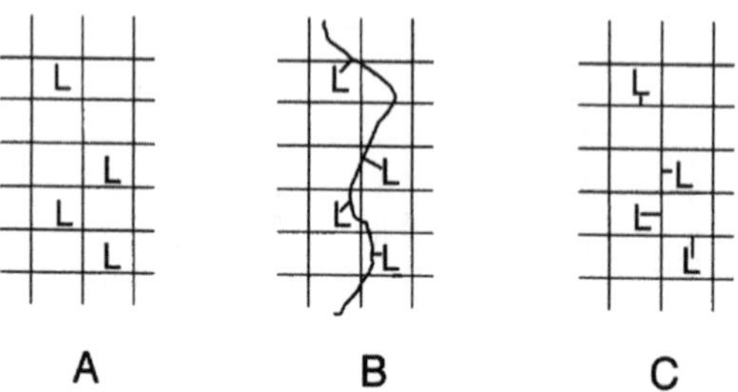

FIGURE 11 Schematic representation of techniques used for solubilized and immobilized affinity ligands in polyacrylamide gels. (A) Solubilized ligand method: the ligand can move freely in the gel–buffer system. (B) Macroligand method: the solution of acrylamide contains the macroligand that becomes entrapped within the gel matrix after polymerization. (C) Chemically bound ligand: direct copolymerization of polyacrylamide gel with the copolymerizable derivative of the ligand. (Reproduced with permission from Ref. 13.)

5.3.2. Immobilized Ligand Method

The ligand can be immobilized by physical entrapping into (Figure 11B) or chemical bonding (Figure 11C) to the polymer sieving matrix. The entrapping method was demonstrated for the separation of chiral compounds by incorporating cyclodextrins into polyacrylamide gel-filled capillary column [127]. Cyclodextrins are nonionic cyclic polysaccharides of glucose with the shape of a toroid or hollow truncated cone. The cavity is relatively hydrophobic, whereas the external faces are hydrophilic. The torus of the larger circumference holds chiral secondary hydroxyl groups. By incorporation of cyclodextrins into a small-pore cross-linked polyacrylamide gel matrix, the complexing agent was practically immobilized. Cyclodextrins were used for chiral separations, since chiral selectivity arose from the entrance of the cavity with the chiral glucose moiety. A homogenous polyacrylamide-allyl-β-cyclodextrin copolymer gel was used for the separation of drug enantiomers [128]. The large pore size defined by the composition of the gel ensured the necessary electroosmotic flow for proper operation. A review article on capillary affinity electrophoresis was recently published in Ref. 129.

6. RELATED MICROSEPARATION TECHNIQUES USING GELS OR ENTANGLED POLYMER SOLUTIONS

6.1. Ultra-Thin-Layer Gel Electrophoresis

Ultra-thin-layer gel electrophoresis is a combination of slab-gel electrophoresis and capillary gel electrophoresis [130], providing a multilane separation platform (a plurality of virtual channels) with excellent heat-dissipation characteristics allowing the application of high voltages necessary to obtain rapid and efficient analysis of biopolymers. Detection of the separated bands is usually accomplished in real time by continuous LIF scanning of the separation lanes. While gel and buffer compositions used in ultra-thin-layer gel electrophoresis are practically the same as in conventional procedures, the efficiency is increased due to its reduced cross-sectional area.

6.1.1. DNA Analysis

Vertical thin-layer gels were first introduced in high throughput automated DNA sequencing [131]. In general, the thickness of ultra-thin-layer separation platforms ranges from 0.050 to 0.25 mm. Horizontal, ultra-thin-layer polyacrylamide gel electrophoresis was implemented by van den Berg [132] in a multizonal format for large-scale analysis of polymerase chain reaction (PCR) products in order to reveal short-sequence-repeat (SSR) polymorphisms. Detection of ds-DNA fragments in the range of 200–3000 bp was accomplished by silver staining, and up to 400 PCR samples were analyzed in 2 h. Later, Ewing's group introduced a novel approach for DNA fragment analysis, which combined the parallel processing capabilities of slab gels with the advantages of sample introduction employing a single capillary column [133]. Gels were fabricated by using 57-µm spacers between two quartz plates, and a single capillary was used to introduce dsDNA fragments into the gel-filled ultra-thin-layer separation platform. Their capillary sample introduction approach allowed multiple samples to be rapidly deposited along the edge of the ultra-thin slab gels for consequent separation. The method was applied for large-scale analysis of PCR-amplified short tandem repeats (STR) using nondenaturing polyacrylamide gel and ethidium bromide labeling with LIF detection of the amplified fragments. This technique demonstrated promising results for increasing the throughput of STR analysis [134]. Recent publications of Guttman and co-workers reported the development and implementation of an automated, horizontal ultra-thin-layer agarose gel electrophoresis system, equipped with integrated scanning laser-induced fluorescence detection, for large-scale DNA fragment analysis [135]. The automated ultra-thin-layer agarose gel electrophoresis setup consisted of a high-voltage power supply, ultra-thin-layer separation cassette with built in buffer reservoirs, and a fiber-optic bundle-based scanning detection system. A lens set, connected to the illumination/detection block via a fiber-optic bundle, scanned across the ultra-thin-layer separation gel by means a high-speed translation stage. A laser excitation source (532-nm frequency-doubled Nd–YAG laser) and an avalanche photodiode were connected to the central excitation fiber and the surrounding collecting fibers of the fiber-optic bundle, respectively [136].

In addition to the use of conventional preseparation labeling, ultra-thin-layer gel electrophoresis systems readily accommodate fluorophore labeling during the separation process ("in migratio"), such as by complexation with novel, high-sensitivity staining dyes. Sample loading onto the ultra-thin separation platform is easily accomplished by membrane-mediated loading technology, which also enables robotic spotting of multiple samples [137]. The samples are injected manually or automatically (robots) onto the surface of the loading membrane tabs, outside of the separation/detection platform. The sample spotted membrane is then placed into the injection (cathode) side of the ultra-thin-layer separation platform, in intimate contact with the straight gel edge. Under the influence of an electric field, the sample components migrate into the gel. This novel sample injection method is readily automated and can be applied to most high-throughput thin-layer slab-gel electrophoresis-based DNA analysis applications (e.g., automated DNA sequencing, genotyping, etc.).

Figure 12 shows simultaneous separation of various dsDNA fragment mixtures over a broad size range from 20 bp up to 23,130 bp, using a single agarose gel composition. The migrating bands were visualized by "in migratio" ethidium bromide staining (50 nM). The first five lanes show the dilution series of the 100-bp DNA ladder representing 25, 10, 5, 2.5, and 1.25 ng total amount of DNA injected (lanes 1–5), corresponding to 863, 300, 170, 80, and 40 pg per band, respectively. Based on these results, the limit of detection (LOD) of the automated high-performance ultra-thin-layer agarose gel electrophoresis system using laser-induced fluorescence/avalanche photodiode detection was 40 pg DNA per band. Lane 6 shows the separation of the φX174 DNA Hae-III restriction digest mixture ranging from 72 to 1353 bp. Lane 7 depicts a rapid (< 17-min) and high-resolution separation of the pBR322 DNA Msp-I restriction digest mixture. The high separation efficiency of the system enabled the authors to obtain baseline resolution of the 4-bp difference between the 242- and 238-bp fragments (see arrow). Lane 8 shows the separation of the lambda DNA Hind-III restriction digest fragments on the same gel composition, ranging up to 23,130 bp in size. Lanes 9–13 show the separations of the 10-, 50-, 100-, 200-, and 500-bp DNA sizing ladders, respectively. The high resolving power of the system was also demonstrated by the nice separation of the 10-bp ladder (lane 9), ranging from 20 to 320 bp. Similar high-resolution separations were observed for the other ladders.

6.1.2. Protein Analysis

Isoelectric focusing and electrophoresis were used for protein mapping and to study protein extraction in horizontal ultra-thin-layer format [138]. The 0.12–0.36-mm-thick polyacrylamide gel layer was deposited onto tiny glass plates (e.g., microscope slides). The method enabled the analysis of 1-ng tissue culture specimens. Ultra-thin-layer polyacrylamide isoelectric focusing gel was also employed in two-dimensional analysis of plant and fungal proteins. Marlow et al. [139] reported on the use of 0.2-mm semirigid backing (polyester)-supported

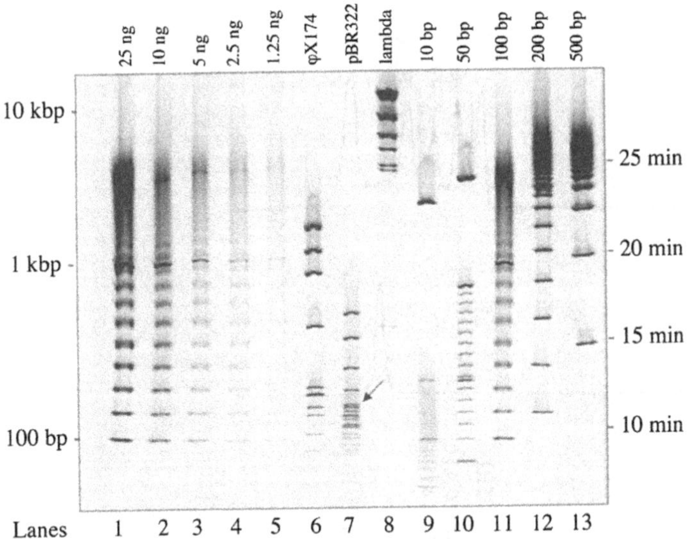

FIGURE 12 Separation of DNA sizing ladders and restriction digest fragment mixtures by automated ultra-thin-layer agarose gel electrophoresis. Lanes 1–5, dilution series of a 100-bp ladder (25, 10, 5, 2.5, and 1.25 ng total DNA injected); lane 6, φX174 DNA Hae-III restriction digest mixture; lane 7, pBR322 DNA Msp-I restriction digest mixture; lane 8, lambda DNA Hind-III restriction digest mixture; lanes 9–13, 10-bp, 50-bp, 100-bp, 200-bp, and 500-bp DNA sizing ladders. Conditions: effective separation length, 6 cm; separation matrix, 2% agarose gel in 0.5 × TBE buffer containing 25 nM ethidium bromide; running buffer, 0.5 × TBE; applied voltage, 750 V; temperature, 25°C; Injection, membrane-mediated, 0.5 μL sample per tab. (Reproduced with permission from Ref. 135.)

IEF gels as the first dimension. This ultra-thin-layer IEF gel quickly dried on the backing after the focusing step. Then narrow strips were cut from the gel and applied to the second dimension. Yakhyayev et al. [140] used cellophane foil to support ultra-thin-layer isoelectric focusing gels. As the polyacrylamide gel was firmly attached to the cellophane foil, it provided good protection from mechanical damage and enabled easy handling. Since cellophane is permeable to ions, the use of this combination alleviated difficulties of the removal of thin gels from the support. Using a similar approach, proteins were also separated under nondenaturing conditions and transferred onto a nitrocellulose membrane followed by enzyme assay-based detection.

Automated ultra-thin-layer gel electrophoresis was applied for large-scale analysis of SDS–protein complexes and for molecular-mass estimation of the proteins in the sample [12]. Figure 13 compares the separations of two differ-

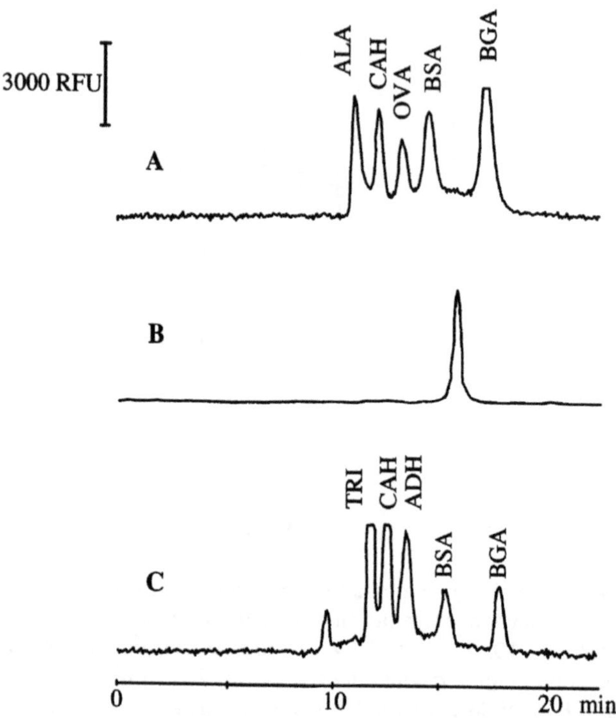

FIGURE 13 Molecular-mass analysis of phosphorylase B (panel B) and the separation of the noncovalently (Sypro Red) labeled protein markers (panel A; ALA, α-lactalbumin; CBA, carbonic anhydrase; OVA, ovalbumin; BSA, bovine serum albumin; BGA, β-galactosidase) and the covalently (FITC) labeled protein markers (panel C; TRI, tripsin inhibitor; CAH, carbonic anhydrase; ADH, alcohol dehydrogenase; BSA, bovine serum albumin; BGA, β-galactosidase). Separation conditions: gel, 1% agarose, 2% linear polyacrylamide (LPA, MW 700,000–1,000,000) in 50 mM Tris, 50 mM TAPS, 0.05% SDS (pH 8.4); separation buffer, 50 mM Tris, 50 mM TAPS, 0.05% SDS (pH 8.4); separation voltage, 420 V, current, 5 mA; gel thickness, 190 μm; effective separation length, 3.5 cm; temperature, 25°C; sample loading, 0.2 μL into 2.5 × 4 × 0.19-mm injection wells. Sample buffer contained 0.05% SDS and 1 × Sypro Red. (Reproduced with permission from Ref. 141.)

ently labeled protein standard mixtures for molecular-mass estimation of phosphorylase B, using ultra-thin-layer SDS-gel electrophoresis. Panel A depicts the separation of the noncovalently (Sypro Red) stained five-protein test mixture of α-lactalbumin (ALA), carbonic anhydrase (CBA), ovalbumin (OVA), bovine serum albumin (BSA), and β-galactosidase (BGA). The trace in Panel B corresponds to the analysis of noncovalently (SR) labeled phosphorylase B. Panel C shows the separation of the covalently (FITC) labeled five-protein test mixture of tripsin inhibitor (TRI), carbonic anhydrase (CBA), alcohol dehydrogenase (ADH), bovine serum albumin (BSA), and β-galactosidase (BGA). Note that noncovalent labeling of the sample and standard proteins took place immediately prior to loading. Detection of the migrating SDS–protein complexes was accomplished in real time during the electrophoresis separation process. The actual separation distance was only 3.5 cm, in order to obtain high resolution but still rapid analysis time [141]. A standard curve was constructed for molecular-mass estimation of the unknown proteins by plotting the logarithmic molecular masses of the five test proteins (lane B) against their electrophoretic mobilities. A second-order polynomial function provided the best fitting of the standard curve ($r^2 = 0.9999$). This finding was in contrast to previous reports describing a linear relationship between the logarithmic molecular mass and electrophoretic mobility, but was probably caused by the noncovalent attachment of the negatively charged staining dye, which increased the charge and, concomitantly, the overall electrophoretic velocity of the complex [Eq. (14)]. Based on the calibration curve obtained, the molecular mass of the phosphorylase B peak in Panel B was estimated to be 97.250, representing only a 0.25% error compared to the literature value of MW 97,400.

6.2. Gel Electrophoresis in Microfabricated Devices

Recently emerging microfluidics-based analytical techniques have brought the promise to further increase the speed and throughput of electric field mediated separations [142]. Early feasibility experiments proving the usefulness of doing electrophoresis in microfabricated devices was done more than a decade ago [143]. Electrophoresis microchips were developed using the techniques elaborated by the semiconductor industry, suggesting that channels and other functional elements can be fabricated in glass substrates by microlithography. With the advent of the so-called simple cross and double-T injector structure, well-defined amounts of samples can be readily analyzed on electrophoresis microchips [144]. Samples are typically loaded electrokinetically into the injector cross region, then the analyte molecules are separated by applying the electric field not only along the separation channel but to a smaller extent to the sample and waste reservoirs as well, to prevent bleeding of the sample into the separation channel [145]. The separated solute molecules are then most frequently visualized by confocal microscopy with laser-induced fluorescence detection. Short injection plugs, high electric field strengths, and short effective channel lengths result in separations in seconds with extremely high efficiencies because of minimized extra-column broadening effects [146]. Electrophoresis microchips were applied to the analysis of amino acids [147], DNA restriction fragments [148–150], PCR products [151], DNA sequencing [152,153], and rapid PCR, preconcentration, and DNA analysis [154]. Microfabrication also offers a

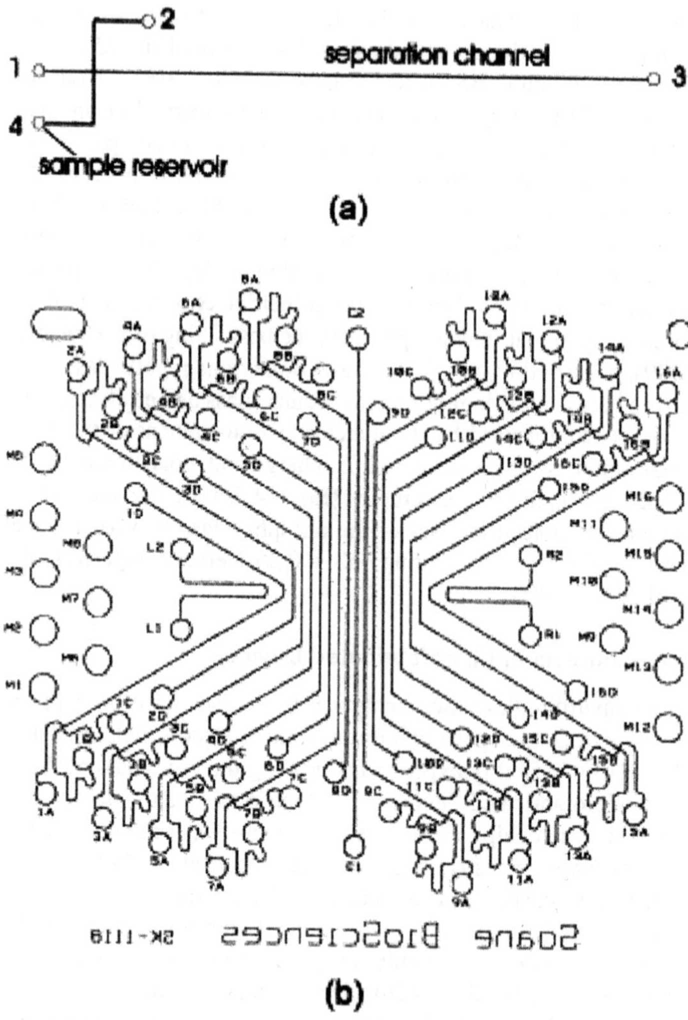

FIGURE 14 (a) Channel layout of a single-channel chip. (b) Layout of channels for a multichannel plastic chip. The leftmost, center, and rightmost channels are reference channels which, when filled with fluorescein, are used for proper alignment of the card under the beam. The remaining 16 channels are used for separations. The topology of each channel pattern is identical to that illustrated in (a). The chip is approximately 8.5 cm square. (Reproduced with permission from Ref. 157.)

novel approach to high-performance micro- and nanovolume liquid chromatography through the application of micromachined chromatographic phase supports [155]. Ronai et al. used polyvinylpyrrolidone-filled glass microchips to evaluate the influence of operational variables on the separation of dsDNA molecules [156]. Effects of sieving matrix concentration (Ferguson plot), migration characteristics (reptation plot), separation temperature (Arrhenius plot), electric field strength, and intercalator dye concentration were all thoroughly examined. Introducing parallelization, rapid separations of alleles of the D1S80 locus was performed in a 16-channel plastic multichannel microdevice in less than 10 min, and the chip was replicated from a microfabricated master and laminated with a plastic film (Figure 14) [157]. Shi et al. [158] introduced a pressurized capillary array system to simultaneously load 96 samples into 96 sample wells of a radial microchannel array electrophoresis microplate for high-throughput DNA sizing. As a result, 96 samples were analyzed in less than 90 s per microplate, demonstrating the power of microfabricated devices for large-scale and high-performance nucleic acid characterization.

Integrated microfabricated device technology has opened up new horizons in bioseparations for the biotechnology industry. Entering the era of genomics and proteomics, we expect to see a paradigm shift toward miniaturized, high-resolution separation techniques used in integrated and automated fashion to solve formidable separation problems and provide means for ultra-high-throughput analysis. Most current separation protocols for DNA and protein analysis, already in use in molecular biology and biotechnology labs, can be readily transferred to microfabricated devices. Another real strength of miniaturization is the possibility of integrating existing methods/functionalities in a way that allows sample preparation, reactions, analysis, and even fraction collections to be carried out on a single microchip (lab-on-a-chip) [159,160]. Microfabricated devices are intrinsically acquiescent to full automation, enabling large-scale analyses with considerably less human intervention than conventional techniques, resulting in significant savings in time, labor, and expense.

ACKNOWLEDGMENT

The kind support of Syngenta Research and Technology is greatly appreciated.

REFERENCES

1. Venter JC, Adams MD, Myers EW, et al. Science 2001; 291:1304–1351.
2. Tiselius A. Trans Faraday Soc 1937; 33:524.
3. Watson JD, Crick FHC. Nature 1953; 171:964–967.
4. Andrews AT. Electrophoresis. 2nd ed. Oxford, UK: Claredon Press, 1986.
5. Ogston AG. Trans Faraday Soc 1958; 54:1754–1757.

6. Grossman PD, Menchen S, Hershey D. GATA 1992; 9:9–16.
7. Ferguson KA. Metab Clin Exp 1964; 13:985–1002.
8. Lumpkin OJ, Dejardin P, Zimm BH. Biopolymers 1985; 24:1573–1593.
9. De Gennes PG. Scaling Concept in Polymer Physics. Ithaca, NY: Cornell University Press, 1979.
10. Slater GW, Noolandi J. Biopolymers 1989; 28:1781–1791.
11. Viovy JL, Duke T. Electrophoresis 1993; 14:322–329.
12. Csapo Z, Gerstner A, Sasvari-Szekely M, Guttman A. Anal Chem 2000; 72:2519–2525.
13. Guttman A, Cooke N. Anal Chem 1991; 63:2038–2042.
14. Terabe S, Otsuka K, Ando T. Anal Chem 1989; 61:251–260.
15. Karger BL, Cohen AS, Guttman A. J Chromatogr 1989;492:585–614.
16. Nelson RJ, Paulus A, Cohen AS, Guttman A, Karger BL. J Chromatogr 1989; 480:111–127.
17. Righetti PG, Gelfi C, Verzola B, Castelletti L. Electrophoresis 2001; 22:603–611.
18. Horvath J, Dolnik V. Electrophoresis 2001; 22:644–655.
19. Hjerten S. J Chromatogr 1985; 347:191–197.
20. Cohen AS, Karger BL. J Chromatogr 1987; 397:409–517.
21. Bruin G, Chang J, Kuhlman R, Zegers K, Kraak J, Poppe H. J Chromatogr 1989; 471:429 – 439.
22. Cobb KA, Dolnik V, Novotny M. Anal Chem 1990; 62:2478–2483.
23. Li SFY. Capillary Electrophoresis. Amsterdam: Elsevier, 1993.
24. Barta C, Sasvari-Szekely M, Guttman A. J Chromatogr A 1998; 817:281–286.
25. Ganzler K, Greve KS, Cohen AS, Karger BL, Guttman A, Cooke N. Anal Chem 1992; 64:2665–2671.
26. Shieh P, Hoang D, Guttman A, Cooke N. J Chromatogr A 1994; 676:219–226.
27. Shieh P. US Patent 5,462,646 1995.
28. Guttman A. US Patent 5,332,481 1994.
29. Tian H, Brody LC, Mao D, Landers JP. Anal Chem 2000; 72:5483–5492.
30. Ren J, Ulvik A, Refsum H, Ueland PM. Anal Biochem 1999; 276:188–194.
31. Chiari M, Cretich M, Horvath J. Electrophoresis 2000; 21:1521–1526.
32. Tanaka T. Sci Am 1981; 244:124–138.
33. Righetti PG, Gelfi C. J Chromatogr B 1997; 699:63–75.
34. Hjerten S. Arch Biochem Biophys 1962; 1:147–151.
35. Karger BL, Chu YH, Foret F. Annu Rev Biophys Biomol Struct 1995; 24:579–610.
36. Guttman A, Paulus A, Cohen AS, Karger BL, Rodriguez H, Hancock WS. Proceedings. In: Schafer-Nielsen C, ed. Electrophoresis 1988. Weinheim, Germany: VCH, 151–159.
37. Swerdlow H, Gesteland R. Nucleic Acids Res 1990; 18:1415–1419.
38. Tsuji K. J Chromatogr 1991; 550:823–830.
39. Vegvari A, Hjerten S. J Chromatogr A 2002; 960:221–227.
40. Guttman A. Electrophoresis 1996; 17:1333–1341.
41. Heiger DN, Cohen AS, Karger BL. J Chromatogr 1990; 516:33–48.
42. Berka J, Pariat YF, Muller O, Hebenbrock K, Heiger DN, Foret F, Karger BL. Electrophoresis 1995; 16:377–388.
43. Widhalm A, Schwer C, Blass D, Kenndler E. J Chromatogr 1991; 546:446–451.

44. Wu D, Regnier F. J Chromatogr 1992; 608:349–356.
45. Werner W, Demorest D, Wictorowicz JE. Electrophoresis 1993; 14:759–763.
46. Habenbrock K, Schugerl K, Freitag R. Electrophoresis 1993; 14:753–758.
47. Righetti PG, Gelfi C. Anal Biochem 1997; 244:195–207.
48. Bocek P, Chrambach A. Electrophoresis 1991; 12:1059–1061.
49. Soto D, Sukumar S. PCR Meth Appl 1992; 2:96–98.
50. Schwartz HE, Ulfelder K, Sunzeri FJ, Busch MP, Brownlee RG. J Chromatogr 1991; 559:267–283.
51. Hu S, Zhang Z, Cook LM, Carpenter EJ, Dovichi NJ. J Chromatogr A 2000; 894: 291–296.
52. de Carmejane O, Yamaguchi Y, Todorov TI, Morris MD. Electrophoresis 2001; 22:2433–2441.
53. Jin Y, Lin B, Fung YS. Electrophoresis 2001; 22:2150–2158.
54. Gibson TJ, Sepaniak MJ. J Chromatogr B 1997; 695:103–111.
55. Albarghouthi MN, Buchholz BA, Doherty EA, Bogdan FM, Zhou H, Barron AE. Electrophoresis 2001; 22:737–747.
56. Buchholz BA, Doherty EA, Albarghouthi MN, Bogdan FM, Zhou H, Barron AE. Anal Chem 2001; 73:157–164.
57. Kim Y, Yeung ES. J Chromatogr A 1997; 781:315–325.
58. Gao Q, Yeung ES. Anal Chem 1998; 70:1382–1388.
59. Wei W, Yeung ES. J Chromatogr B 2000; 745:221–230.
60. Wei W, Yeung ES. Anal Chem 2001; 73:1776–1783.
61. Song JM, Yeung ES. Electrophoresis 2001; 22:748–754.
62. Tan H, Yeung ES. Electrophoresis 1997; 18:2893–2900.
63. Madabhushi RS, Vainer M, Dolnik V, Enad S, Barker DL, Harris DW, Mansfield ES. Electrophoresis 1997; 18:104–111.
64. Barta C, Ronai Z, Sasvari-Szekely M, Guttman A. Electrophoresis 2001; 22:779–782.
65. Albarghouthi MN, Barron AE. Electrophoresis 2001; 21:4946–4111.
66. Liang D, Chu B. Electrophoresis 1998; 19:2447–2453.
67. Wu C, Liu T, Chu B. Electrophoresis 1998; 19:231–241.
68. Rill RL, Liu Y, Van Winkle DH, Locke BR. J Chromatogr A 1998; 817:287–295.
69. Liang D, Song L, Zhou S, Zaitsev VS, Chu B. Electrophoresis 1999; 20:2856–2863.
70. Sudor J, Barbier V, Thirot S, Godfrin D, Hourdet D, Millequant M, Blanchard J, Viovy JL. Electrophoresis 2001; 22:720–728.
71. Izumi T, Yamaguchi M, Yoneda K, Isobe T, Okuyama T, Shinoda T. J Chromatogr A 1993; 652:41–46.
72. Nakatani M, Shibukawa A, Nakagawa T. Electrophoresis 1996; 17:1210–1213.
73. Siles BA, Collier GB. J Capillary Electrophor 1996; 3:313–321.
74. Siles BA, Anderson DE, Buchanan NS, Warder MF. Electrophoresis 1997; 18: 1980–1989.
75. Chiari M, Cretich M, Riva S, Casali M. Electrophoresis 2001; 22:699–706.
76. Chiari M, Cretich M, Consonni R. Electrophoresis 2002; 23:536–541.
77. Chiari M, Damin F, Melis A, Consonni R. Electrophoresis 1998; 19:3154–3159.
78. Dolnik V, Gurske WA, Padua A. Electrophoresis 2001; 22:707–719.

79. Miksik I, Deyl Z. J Chromatogr B 2000; 739:109–116.
80. Miksik I, Deyl Z, Kasicka V. J Chromatogr B 2000; 741:37–42.
81. Song L, Liu T, Liang D, Fang D, Chu B. Electrophoresis 2001; 22:3688–3698.
82. Wang Y, Liang D, Hao J, Fang D, Chu B. Electrophoresis 2002; 23:1460–1466.
83. Shen Y, Xu Q, Han F, Ding K, Song F, Fan Y, Zhu N, Wu G, Lin B. Electrophoresis 1999; 20:1822–1828.
84. Han F, Lin B. Se Pu 1998; 16:489–491.
85. Cheng J, Mitchelson KR. Anal Chem 1994; 66:4210–4214.
86. Righetti PG, Gelfi C. J Capillary Electrophor 1999; 6:119–124.
87. Ruiz-Martinez MC, Berka J, Belenkii A, Foret F, Miller AW, Karger BL. Anal Chem 1993; 65:2851–2858.
88. Salas-Solano O, Carrilho E, Kotler L, Miller AW, Goetzinger W, Sosic Z, Karger BL. Anal Chem 1998; 70:3996–4003.
89. Wang Y, Wallin JM, Ju J, Sensabaugh GF, Mathies RA. Electrophoresis 1996; 17:1485–1490.
90. Oto M, Koguchi A, Yuasa Y. Clin Chem 1997; 43:759–763.
91. Stellwagen N, Gelfi C, Righetti PG. Electrophoresis 2002; 23:167–175.
92. Siles BA, O'Neil KA, Fox MA, Anderson DE, Kuntz AF, Ranganath SC, Morris AC. J Agric Food Chem 2000; 48:5903–5912.
93. Li-Sucholeiki XC, Khrapko K, Andre PC, Marcelino LA, Karger BL, Thilly WG. Electrophoresis 1999; 20:1224–1232.
94. Xue MZ, Bonny O, Morgenthaler S, Bochud M, Mooser V, Thilly WG, Schild L, Leong-Morgenthaler PM. Clin Chem 2002; 48:718–728.
95. Ekstrom PO, Wasserkort R, Minarik M, Foret F, Thilly WG. Biotechniques 2000; 29:582–589.
96. Liu Y, Kuhr WG. Anal Chem 1999; 71:1668–1673.
97. Heller C. Electrophoresis 1999; 20:1962–1977.
98. Heller C. Electrophoresis 1999; 20:1978–1986.
99. Nevins SA, Siles BA, Nackerdien ZE. J Chromatogr B 2000; 741:243–255.
100. Hsieh MM, Tseng WL, Chang HT. Electrophoresis 2000; 21:2904–2910.
101. Todorov TI, de Carmejane O, Walter NG, Morris MD. Electrophoresis 2001; 22: 2442–2447.
102. Khandurina J, Chang HS, Wanders B, Guttman A. BioTechniques 2002; 32:1226–1228.
103. Siles BA, O'Neil KA, Tung DL, Bazar L, Collier GB, Lovelace CI. J Capillary Electrophor 1998; 5:51–58.
104. Isenberg AR, McCord BR, Koons BW, Budowle B, Allen RO. Electrophoresis 1996; 17:1505–1511.
105. Gelfi C, Perego M, Righetti PG. Electrophoresis 1996; 17:1470–1475.
106. Barme I, Bruin GJ, Paulus A, Ehrat M. Electrophoresis 1998; 19:1445–1451.
107. Khan K, Van Schepdael A, Saison-Behmoaras T, Van Aerschot A, Hoogmartens J. Electrophoresis 1998; 19:2163–2168.
108. Chen SH, Gallo JM. Electrophoresis 1998; 19:2861–2869.
109. Oana H, Hammond RW, Schwinefus JJ, Wang SC, Doi M, Morris MD. Anal Chem 1998; 70:574–579.
110. Ren J, Ulvik A, Ueland PM, Refsum H. Anal Biochem 1997; 245:79–84.

111. Zhang J, Fang Y, Hou JY, Ren HJ, Jiang R, Roos P, Dovichi NJ. Anal Chem 1995; 67:4589–4593.
112. Bashkin J, Marsh M, Barker D, Johnston R. Appl Theor Electrophor 1996; 6: 23–28.
113. Quesada MA. Curr Opin Biotechnol 1997; 8:82–93.
114. Kleemiss MH, Gilges M, Schomburg G. Electrophoresis 1993; 14:515–522.
115. Righetti PG, Gelfi C. Electrophoresis 1997; 18:1709–1714.
116. Hunt G, Nashabeh W. Anal Chem 1999; 71:2390–2397.
117. Kustos I, Toth V, Kocsis B, Kerepesi I, Emody L, Kilar F. Electrophoresis 2000; 21:3020–3007.
118. Kustos I, Kocsis B, Kerepesi I, Kilar F. Electrophoresis 1998; 19:2317–2323.
119. Manabe T, Oota H, Mukai J. Electrophoresis 1998; 19:2308–2316.
120. Craig DB, Polakowski RM, Arriaga E, Wong JC, Ahmadzadeh H, Stathakis C, Dovichi NJ. Electrophoresis 1998; 19:2175–2178.
121. Harvey MD, Bandilla D, Banks PR. Electrophoresis 1998; 19:2169–2174.
122. Hu S, Zhang L, Cook LM, Dovichi NJ. Electrophoresis 2001; 22:3677–3682.
123. Proctor SD, Mamo JC. J Lipid Res 1997; 38:410–414.
124. Stocks J, Nanjee MN, Miller NE. J Lipid Res 1998; 39:218–227.
125. Guttman A. Nature 1996; 380:461–462.
126. Guttman A, Pritchett T. Electrophoresis 1995; 16:1906–1911.
127. Guttman A, Paulus A, Cohen AS, Grinberg N, Karger BL. J Chromatogr 1988; 448:41–53.
128. Vegvari A, Foldesi A, Hetenyi C, Kocnegarova O, Schmid MG, Kudirkaite V, Hjerten S. Electrophoresis 2000; 21:3116–3125.
129. Baba Y. J Biochem Biophys Meth 1999; 41:91–101.
130. Guttman A, Ronai Z. Electrophoresis 2000; 21:3952–3964.
131. Smith LM, Sanders JZ, Kaiser RJ, Hughes P, Dodd C, Connell CR, Heiner C, Kent SB, Hood LE. Nature 1986; 321:674–679.
132. van den Berg BM. Electrophoresis 1997; 18:2861–2864.
133. Ewing AG, Gavin PF, Hietpas PB, Bullard KM. Nature Medicine 1997; 3:97–99.
134. Bullard KM, Hietpas PB, Ewing AG. Electrophoresis 1998; 19:71–75.
135. Guttman A. LC GC Magazine 1999; 17:1020–1026.
136. Trost P, Guttman A. Anal Chem 1998; 70:3930–3935.
137. Guttman A. Anal Chem 1999; 71:3598–3602.
138. Inczedy-Marcsek M, Lindner E, Hassler R, Zwack-Megele G, Roisen F, Yorke G. Acta Hystochem Suppl 1988; 36:377–394.
139. Marlow GC, Wurst DE, Loschke DC. Electrophoresis 1988; 9:693–704.
140. Yakhyayev AV, Voronkova IM, Sukhanov VA. Electrophoresis 1991; 12:680–682.
141. Guttman A, Ronai Z, Csapo Z, Gerstner A, Sasvari-Szekely M. J Chromatogr A 2000; 894:329–336.
142. Manz A, Becker H, eds. Microsystem Technology in Chemistry and Life Sciences. Berlin, Germany: Springer-Verlag, 1999.
143. Manz A, Graber N, Widmer HM. Sensors Actuators B 1990; 1:244–248.
144. Effenhauser CS, Bruin GJ M, Paulus A. Electrophoresis 1997; 18:2203–2213.
145. Jacobson SC, Ramsey JM. Electrophoresis 1995; 16:481–486.

146. Effenhauser CS. Integrated chip-based microcolumn separation systems. In: Manz A, Becker H, eds. Microsystem Technology in Chemistry and Life Sciences. Heidelberg, Germany: Springer, 1999; pp. 51–82.
147. Harrison DJ, Fan ZH, Seiler K, Manz A, Widmer HM. Anal Chim Acta 1993; 283:361–366.
148. Jacobson SC, Hergenroder R, Koutny LB, Ramsey JM. Anal Chem 1994; 66: 1114–1118.
149. Wooley AT, Mathies RA. Proc Natl Acad Sci USA 1994; 91:11348–11352.
150. Wooley AT, Sensabaugh GF, Mathies RA. Anal Chem 1997; 69:2181–2186.
151. Effenhauser CS, Paulus A, Manz A, Widmer HM. Anal Chem 1994; 66:2949–2953.
152. Paulus A. Am Lab 1998; 30:59–62.
153. Schmalzing D, Adourian A, Koutny L, Ziaugra L, Matsudaira P, Ehrlich D. Anal Chem 1998; 70:2303–2310.
154. Khandurina, J, McKnight TE, Jacobson SC, Waters LC, Foote RS, Ramsey JM. Anal Chem 2000; 72:2995–3000.
155. He B, Tait N, Regnier F. Anal Chem 1998; 70:3790–3797.
156. Ronai Z, Barta C, Sasvari-Szekely M, Guttman A. Electrophoresis 2001; 22:294–299.
157. Sassi AP, Paulus A, Cruzado ID, Bjornson T, Hooper HH. J Chromatogr A 2000; 894:203–217.
158. Shi Y, Simpson PC, Schere JR, Wexler D, Skibola C, Smith MT, Mathies RA. Anal Chem 1999; 71:5354–5361.
159. Khandurina J, Guttman A. J Chromatogr 2002; 943:159–183.
160. Khandurina J, Chovan T, Guttman A. Anal Chem 2002; 74:1737–1740.
161. Noolandi J. Annu Rev Phys Chem 1992; 43:237–256.
162. Guttman A, Shieh P, Lindahl J, Cooke N. J Chromatogr 1994; 676:227–231.

5

Affinity Capillary Electrophoresis

Vadim M. Okun and Ernst Kenndler
University of Vienna, Vienna, Austria

1. INTRODUCTION

Electrophoretic methods, based on the differential migration of ionic analytes under the influence of an electric field, are of utmost importance in the biochemical field, as many analytes are charged or chargeable, e.g., proteins or nucleic acids. Classical electrophoresis in slabs is one of the mostly applied techniques in bioanalysis. However, for several decades electrophoresis in the capillary format, in the last years on chips, has become more and more common due to its advantages compared to the classical format: it needs lower sample amount, it is compatible for direct coupling with mass spectrometry (MS), it is well suited for automation, and it exhibits outstanding separation efficiency. Therefore many techniques which are well established in slab-gel electrophoresis have been transformed to capillary electrophoresis (CE). This includes those for size-based separations of nucleic acids or protein–sodium dodecyl sulfate (SDS) complexes, which are carried out in sieving media replaceably filled into the capillary. It is thus no surprise that affinity methods have been introduced into CE with success and, by the analogy with slab-gel electrophoresis, named affinity CE (ACE).

It should be mentioned that originally the term ACE was used only for a certain CE technique in which a migration shift of a protein was under study with its ligand dissolved in a background electrolyte (BGE). However, in the past decade the term ACE became more general and now includes a collection of techniques in which affinity binding (either on- or off-column) is used in conjunction with CE separation. It is therefore sometimes difficult to clearly distinguish ACE methods from a number of others in which separation is facilitated by interactions of an analyte with buffer additives such as detergents, chiral selectors, etc. To clarify the definition of ACE, we propose to use this term in the cases dealing with a *highly specific* bioaffinity interaction, on- or off-column, used either to derive parameters of such interaction or to determine an analyte quantitatively. In this regard, for example, enantioseparation cannot be considered as ACE, since an interaction between a chiral selector and enantiomers is not highly specific. Nevertheless, even with this restriction, there is still a number of techniques which are not considered in this review simply because of the space limit. Among them, ligand immobilization on a capillary wall to study weak affinity interactions, enzyme-mediated microanalysis (EMMA), and electroinjection analysis must be mentioned.

This review is structured based on two different approaches in performing ACE—equilibrium case for the study of weak to moderate affinities and non-equilibrium case for moderate to high affinities. Special attention is focused on the evaluation of different practical advantages and limitations, which are characteristic for both modes of ACE.

2. ELECTROPHORESIS AND ITS COMBINATION WITH IMMUNOREACTIONS

Combination of electrophoresis in different formats with affinity reactions started in the early 1950s, when Tiselius cells were used for the first time to characterise antigen–antibody interactions. A real milestone report in the field was the determination of the equilibrium constants for the binding of Ca^{2+} and Zn^{2+} to serum albumin by gel electrophoresis in 1960 [1]. Finally, the term "affinity electrophoresis" was proposed [2,3]. Its modification or, better said, its particular case, named immunoelectrophoresis, has been used for many years in reports relying on agarose gel separation in conjunction with immunoprecipitation. Appearance of precipitated zones on a gel is indicative for the antigen–antibody reaction and can be used for the identification of an analyte and also for its quantification.

The first report on the use of affinity interactions in conjunction with CE appeared in 1989 [4]. The authors were able to separate an immunoaffinity complex from unbound antigen and antibody. This report can be considered as a starting point in the general development of a number of powerful techniques, known nowadays as ACE.

3. ACE: TWO APPROACHES

ACE is carried out experimentally in different arrangements according to the binding constant between the analyte and the ligand, that is, according to the "off" and "on" rates of affinity reaction. This is obvious, as the complex should not deteriorate significantly during the CE run. The equilibrium constant for the reversible reaction

$$A + B \Leftrightarrow AB \tag{1}$$

with on-rate constant k_{on} and off-rate constant k_{off} is expressed as the ratio of the two rate constants according to

$$K_A = K_D^{-1} = \frac{k_{on}}{k_{off}} = \frac{[AB]}{[A][B]} \tag{2}$$

where the square brackets indicate the equilibrium concentrations; K_A is the affinity (binding, association) constant, and K_D is the dissociation constant.

Therefore two main arrangements are applied. For a slow rate of dissociation related to the CE run-time scale the sample can be preincubated prior to electrophoresis and the reaction set to equilibrium. After equilibrium is obtained, an aliquot of the reaction mixture is analyzed by CE, which is used simply as a method to separate and quantify the analyte, ligand, and complex in the pre-equilibrated mixture. During the CE run, which is carried out in a buffer without reactants, no equilibrium is established. However, as the rate of dissociation is comparably slow, the reactants and the complex present in the mixture can be separated and their concentrations (which are those of the incubation equilibrium) determined quantitatively. Change of the mutual concentrations of analyte and ligand in the mixture allows conclusion about the binding constant and the stoichiometry of the complex formed.

The second main approach is applicable when the rates of formation and dissociation are fast compared to the CE run. In this case one reactant is added to the buffer, and the change in mobility of the analyte injected as usual in capillary zone electrophoresis (CZE) is measured. This method is not different from those in which other fast chemical equilibria are used, e.g., acid–base equilibria for the adjustment of the degree of protonation via a constant pH of the buffer or complexation equilibria with additives such as cyclodextrins. In all cases the position of the equilibrium is influenced by the reactant concentration, which is reflected by the according change of the mobility. The obtained titration curve (mobility versus ligand concentration) again allows the derivation of the equilibrium constant according to the mass action law. The underlying theoretical principles and application areas of both approaches will be described further.

3.1. Nonequilibrium Case: Off-Column ACE

3.1.1. Determination of Binding Constants

As ACE can in principle differentiate between free and bound antigen (Ag) (or free and bound antibody, Ab), the method can be used for affinity-constant determination. In contrary to a "classical" ACE, the basis for K_D determination in the nonequilibrium mode is quantification of the peak of unbound reagent. Usually, a series of mixtures—keeping one of the reagents at a constant concentration and varying the amount of the other—is prepared, incubated until the equilibrium is reached, and analyzed by CE. Peaks of unbound reagent and/or affinity complex are quantified for each sample and a binding curve reflecting the dependence of a bound versus free reagent can be created. The binding constants are then calculated either by direct fitting of the binding curve or by linear transformation of equations obtained from the law of mass action. For a simple, fully reversible binding reaction

$$\mathrm{A} + \mathrm{B} \Leftrightarrow \mathrm{AB} \tag{3}$$

the equilibrium (or binding/affinity/association) constant, K_A, is defined by

$$K_A = \frac{k_a}{k_d} = \frac{[\mathrm{AB}]}{[\mathrm{A}][\mathrm{B}]} \tag{4}$$

where k_a and k_d are the rate constants of association and dissociation, respectively, and [AB], [A], and [B] are concentrations of affinity complex and reagents in equilibrium. If we consider

$$[\mathrm{A}] = (\mathrm{A}) - [\mathrm{AB}] \tag{5}$$

where (A) is initial (or total) concentration of a reagent A (notice, this is true only for a 1:1 reaction), then the equilibrium concentration of the complex [AB] can be expressed as

$$[\mathrm{AB}] = \frac{K_A[\mathrm{B}]}{1 + K_A[\mathrm{B}]} \cdot (\mathrm{A}) \tag{6}$$

This equations expresses the dependence of a bound reagent [AB] (once again, in the case of 1:1 binding it is equal to the amount of complex in equilibrium) versus free reagent [B]. Plotting [AB] versus [B] results in a typical binding isotherm with hyperbolic shape. The double-reciprocal form of the binding isotherm,

$$\frac{1}{[\mathrm{AB}]} = \frac{1}{K_A(\mathrm{A})[\mathrm{B}]} + \frac{1}{(\mathrm{A})} \tag{7}$$

is known as the Langmuir equation and gives a straight line, enabling determination of K_A as a negative intercept with the x axis.

Another type of linear transformation, which also can be easily derived from the law of mass action [Eq. (4)] is the so-called Scatchard plot. According to it, the ratio of a bound reagent [AB] and free reagent [B] versus bound reagent [AB] gives a straight line with a slope equal to negative K_A (or negative reciprocal K_D), and the intercept with the x axis is equal to the amount of available binding sites:

$$\frac{[\mathrm{AB}]}{[\mathrm{B}]} = K_A \cdot ((\mathrm{A}) - [\mathrm{AB}]) \tag{8}$$

As for the Langmuir transformation, this type of linearization is valid only for 1:1 binding. Both types of analysis have been widely used in conventional immunoassay for at least 50 years and recently were transferred to CE without modification.

When the stoichiometry of binding exceeds the simplest 1:1 case (e.g., hapten–antibody, which is usually 2:1), we can formulate the equilibrium for the hapten with the divalent (uniform) receptor with invariant identical affinities (an "ideal" system) as

$$K_A = \frac{k_a}{k_d} = \frac{[\mathrm{AB}]}{[\mathrm{A}][\mathrm{B}]} \tag{4}$$

where $[\mathrm{AB}]_1$ and $[\mathrm{AB}]_2$ are the concentrations of bound A at sites 1 and 2, and $F1$ and $F2$ are the fractions of bound A to total B at sites 1 and 2, respectively. The total concentration of A at both sites is the sum of the concentrations at the individual sites. If we suppose noncooperative binding (both sites have the same affinity), then $[\mathrm{AB}]_1 = [\mathrm{AB}]_2$ and the binding constant at any individual binding site K_A equals $K_{A,1} = K_{A,2}$. We can formulate these equations for n sites as

$$F = \frac{[\mathrm{AB}]}{(\mathrm{B})} = n\,\frac{K_A[\mathrm{A}]}{1 + K_A[\mathrm{A}]} \tag{10}$$

with $n = 2$ for the present case.

The direct plot of F versus the free hapten concentration [A] gives a hyperbolic curve with a plateau due to saturation where F reaches n (Figure1). The concentration of [A] at $y = n/2$ equals the value of $1/K_A$.

We can linearize Eq. (10) by rearrangement to

$$F \cdot (1 + K_A[\mathrm{A}]) = nK_A[\mathrm{A}] \tag{11}$$

which gives

$$F = nK_A[\mathrm{A}] - FK_A[\mathrm{A}] \tag{12}$$

or

$$\frac{F}{[\mathrm{A}]} = nK_A - FK_A \tag{13}$$

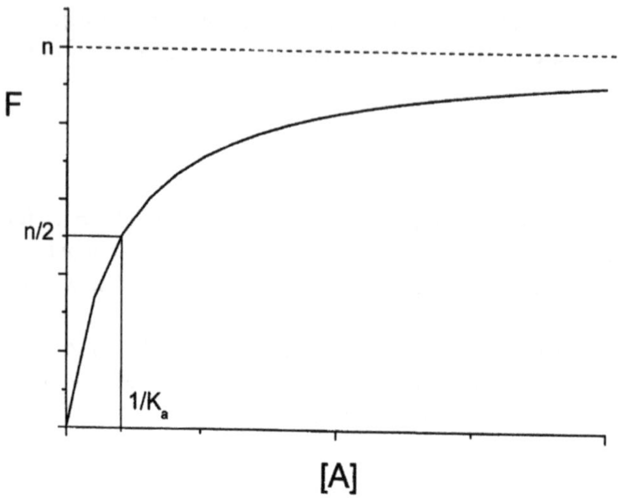

FIGURE 1 Plot of the ratio of bound reagent A to total reagent B as a function of free reagent A.

which is a form of the Scatchard plot. It gives a straight line when $F/[A]$ is plotted versus F with a slope corresponding to $-K_A$, the intercept with the y axis equal to nK_A, and the intercept with the x axis equal to n. Thus, if a studied system can be considered as an interaction between monovalent and multivalent reagents, void of cooperativity, its affinity can be correctly determined with this method provided the amount of monovalent reagent was monitored. The resulted K_A will reflect an average affinity between a monovalent reagent and a single binding site.

Affinity determination by CE is a rather simple method which has, however, certain drawbacks and restrictions, limiting its wider application. General prerequisites for reliable analysis include those common for all types of CE-based immunoassay on the one hand and also some specific features on the other. It is obvious that this approach can be used only for complexes with slow "off" rates to ensure that the dissociation which occurs during separation does not have a significant effect on the resulted binding constant. It was pointed out that as a rule of thumb the complex dissociation rate constant should be less than $0.105/t$, where t is the separation time [5]. Fronting or tailing of the CE peaks can indicate that the dissociation rate is too high to get a correct K_A value.

Another important issue is the time of reaction. It must be long enough to ensure that equilibrium is established. Although in homogeneous solutions equilibrium can be reached faster than in solid-phase systems, this must be substantially studied especially for very large macromolecular assemblies.

Regardless from the type of linear transformation, the determination of bound and free reagent is conducted by a comparison with a preestablished calibration line. Therefore special attention must be paid to the proper selection of the CE conditions. Protein sorption or any loss of reagent during electrophoresis should be avoided. Freshly performed calibration is strongly recommended, particularly when fluorescence detection is used. The necessity to conduct calibration is an obvious drawback of the method, since in many cases it is a very time-consuming procedure.

A certain bias in correct K_D determination can be introduced by stacking of the sample zone, occurring when the injected sample has an ionic strength less than BGE. Stacking leads to an uncontrolled increase in concentration of the interacting reagents which may disturb equilibrium [6].

One of the main prerequisites known from conventional immunoanalysis is a proper adjustment of reagents concentration. Their range must be near K_D of the complex. In both extreme cases, namely, when the concentration is much higher or much lower than K_D, the binding isotherm cannot be created. In the first case it will be an increasing straight line without hyperbolic shape; in the latter no binding will be seen at all. Since K_D values of many affinity systems are less than 10^{-7} M, the limit of detection of UV-based CE detectors (which is normally about 10^{-6} M) is a strong restricting factor. In the case of higher-affinity systems the use of fluorescence or laser-induced fluorescence (LIF) detection should be considered. This needs pre- or postcolumn labeling, which might, however, change the conformation integrity of the reagent and hence its affinity.

It is not always necessary to linearize the binding curve, as direct fitting with computer software can be used instead to determine K_A. However, one should keep in mind that the deviation from the straight line which might be seen on either a Scatchard or Langmuir plot can be very informative. It can indicate the presence of multiple binding sites with different affinities or cooperativity effect between the binding sites. Two distinct parts of the linearized curve with different slopes sometimes reflects a system with 2:1 binding stoichiometry (e.g., IgG–monovalent antigen). In some cases, when differences in slope are especially noticeable, each binding constant can be calculated from each slope. However, the latter must be used with great care, since it is often hard to distinguish between negative cooperativity and multiple binding.

Above we described only rather simple cases, for which the stoichiometry of interaction is 1:1 or *n*:1, and cooperativity effect is absent. In strict terms, neither of the mentioned linear transformations is suited for more complex systems such as, e.g., bivalent antibodies and multivalent antigens. Only rough estimation of the total affinity is possible. Separate determinations of the affinity of all interactions occurring in a studied multivalent system (see, e.g., Ref. 7) leads to more complicated calculation; its consideration is, however, outside the scope of the present discussion.

Applications. From the first successful reports on protein analysis by CE, it became obvious that the unique separation power of the method can be also applied for ligand–receptor analysis to separate bound from free analyte. This is the main prerequisite for evaluating biomolecular interaction by most of the conventional methods.

Heegaard et al. [8] seem to be the first who used this approach to separate free synthetic peptide derived from the heparin-binding region of human serum amyloid P component from its complex with anionic carbohydrates. Quantification of the peptide in a series of affinity mixtures was accomplished by a comparison of the relevant peak areas with the calibration curve. Using UV detection, the authors obtained a dissociation constant of about 10^{-5} M, which is in the range of the sensitivity such detection can provide.

One year later these authors demonstrated how ACE can be used to quantify the binding of Ca^{2+} and phosphorylcholine to human C-reactive protein (CRP) [9]. The authors emphasized that CE conditions provided physiological ionic strength and pH for the interaction, which, in turn, enabled correct determination of the binding constant. For the studied system, K_D was in the range of 10^{-6} M, in good agreement with previous analyses by gel filtration and equilibrium dialysis.

Lin et al. [10] applied CE with UV detection to determine the binding constant for slow phosphoserine–anti-phosphoserine interaction kinetics. The sensitivity of the detector was sufficiently high to measure K_D in the 10^{-5} M range. The Langmuir equation was used to linearize the binding curve.

A similar approach with Scatchard analysis was used by Li et al. [11] for the measurement of the binding constants between purified calf thymus DNA and a library of designed tetrapeptides.

Tao et al. [12] applied much more sensitive LIF detection to measure an antigen–antibody dissociation constant which was in the picomolar range. The studied system included fluorescein isothiocyanate (FITC)-labeled insulin and monoclonal anti-insulin C-terminal pentapeptide. Scatchard linear transformation was used to determine the binding constant. These authors were also the first to apply a competitive analysis for the determination of competitive binding constant between unlabeled antigen and its antibody. This approach is well known from conventional immunoassay and was transferred directly to CE. The competitive mode of analysis utilized a competition between FITC–insulin, whose concentration was kept constant, and unlabeled insulin for a limited number of antibodies (Figure 2). A plot was derived showing the bound-over-free ratio of FITC–insulin as a function of insulin concentration. Applying the usual equations for both binding constants (FITC–insulin/antibody and insulin/antibody) and rearranging using expressions for the initial amounts of both antigens

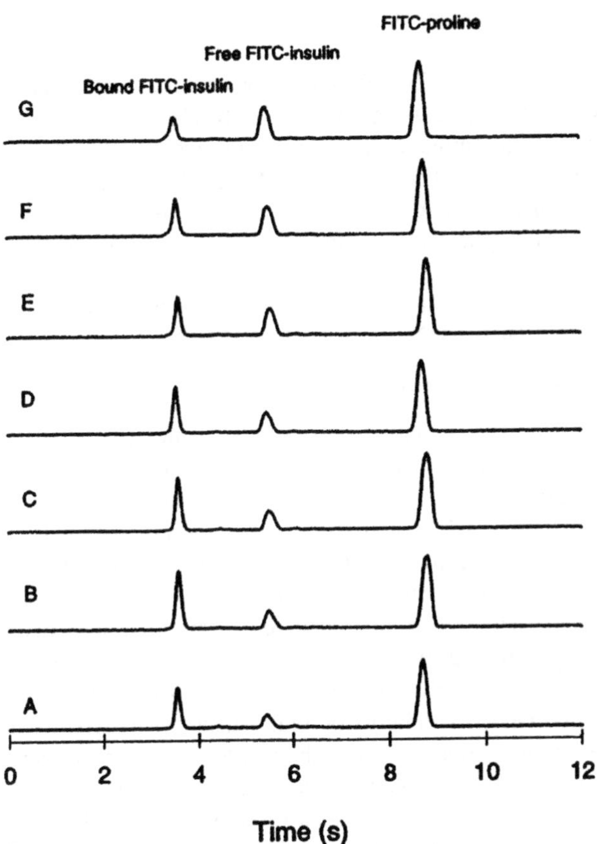

FIGURE 2 Affinity CE for the determination of competitive binding constant between unlabeled bovine insulin and the antibody. All samples contained fixed amount of FITC–insulin and antibody and increasing amount of bovine insulin. (Reprinted from Ref. 12.)

and antibody, the authors obtained a formula which permitted them to determine K_D from the plot by a single- or double-parameter curve-fit. It is obvious that for the competitive mode of K_D determination one of the two binding constants (here it is for FITC–insulin–antibody complex) must be known in advance. This approach was slightly modified by Zhou et al. [13], who used CE with LIF detection to evaluate the binding between the potential anti-HIV drug phosphorothioate oligodeoxynucleotide and viral envelope glycoprotein gp120. The Scatchard equation was used to derive the binding constant, which in turn permitted them further to obtain a competition constant for the unlabeled oligodeoxynucleotide and glycoprotein.

The binding constant between FITC-labeled bovine serum albumin and its monoclonal antibody was determined using ACE with LIF [14]. Interaction of the enzyme cyclophilin with the immunosuppressive drug cyclosporine A was quantitatively assessed by CE with UV detection [15]. In the latter case the authors showed in particular how the difference between the sensitivity of the detector and K_D of the studied system can severely hamper the correct determination of binding constants.

Recently, several reviews have been published dealing with an estimation of binding constants by off-column ACE analysis [5,16,17]

3.1.2. Binding Stoichiometry Determination

Binding stoichiometry is one of the quantitative parameters which characterizes certain biospecific binding. Generally, it reflects a maximum amount of a compound which can bind to another compound (so-called theoretical stoichiometry; e.g., hapten–IgG has stoichiometry 2:1, biotin–biotin-binding protein streptavidin has 4:1, etc.) and sometimes can be predicted from structural considerations or microscopy findings. Theoretical stoichiometry might, however, differ from what is occurring in a real system. Steric hindrance, low affinity, partially damaged binding sites, improper folding, or other constrains might lead to lack of complete occupancy of all theoretically available sites. Determination of this parameter is of high importance especially for multivalent systems, since it allows prediction of the binding topology and might help to localize the binding sites. In addition, knowledge of the stoichiometry can indicate whether the number of ligand molecules attached to the multivalent receptor is compatible with the theoretical value or whether attachment occurs in an unexpected way. CE analysis is especially favorable for stoichiometry determination since it involves homogeneous interaction in which all binding sites are accessible, in comparison to solid-phase assays, in which a certain number of binding sites are always masked due to immobilization.

The first attempt to determine the amount of binding sites in capillary gel electrophoresis was reported by Rose [18]. Later Chu et al. developed a general approach for the CE-related determination of binding stoichiometry for both weak and tight affinity interactions [19]. For tight affinity interactions a series of affinity mixtures with a fixed protein concentration and increasing ligand concentration was prepared and analyzed by CE. The peak area of unbound ligand was plotted versus protein:ligand ratio and a break in the slope of the straight line was indicative of complete saturation and, hence, stoichiometry. The authors reported stoichiometry determination for interaction of anti-HSA monoclonal antibody to its antigen HSA and for biotin–streptavidin interaction.

This approach, though well suited for systems with high affinity, has certain drawbacks when analyzing moderate to high-affinity systems. If the concentration range of the reagents is comparable with the dissociation constant, than unbound ligand will appear before (sometimes long before) saturation. This leads to an error in estimation of a break in the slope, a point which is crucial for this approach.

Later Okun et al. [20] demonstrated that another approach can be used. These authors prepared a series of mixtures with a fixed concentration of a ligand and increasing concentrations of a protein. Upon CE analysis a ligand peak was quantified and plotted against total protein concentration. The intercept with the x axis indicated stoichiometry. The theory behind this approach was developed later in our laboratory and reported in a number of publications, relying on affinity interactions of human rhinoviruses to their neutralizing agents [21,22]. Here we discuss the theory and also emphasize certain limitations of this mode of assay.

For 1:1 stoichiometry as expressed in Eq. (1), the functional dependence of a reagent [A] (which is measured by CE, e.g., from the peak heights or areas) can be derived by substituting the relations between the initial and the equilibrium concentrations in Eq. (2) according to

$$\begin{aligned}[A] &= (A) - [AB] \\ [B] &= (B) - [AB]\end{aligned} \tag{14}$$

which gives the equilibrium concentration of A as a function of the initial concentration of the reactants:

$$[A] = (A) - \frac{(A) + (B) + K_D}{2} \pm \sqrt{\left\{\frac{(A) + (B) + K_D}{2}\right\}^2 - (A)(B)} \tag{15}$$

or, expressed as the relative change in concentration (or peak area or height),

$$\frac{[A]}{(A)} = 1 - \frac{(A) + (B) + K_D}{2(A)} \pm \sqrt{\left\{\frac{(A) + (B) + K_D}{2(A)}\right\}^2 - \frac{(B)}{(A)}} \tag{16}$$

It can be seen that for given initial concentrations (A) and (B) the decrease of the equilibrium concentration of A is only a function of K_D, and the dissociation constant can be derived by fitting the experimental CE data of the decrease of the peak area of A to Eq. (16).

The shape of this function depends on the value of the dissociation constant, K_D, related to the concentrations of the reactants, and is depicted for different cases in Figure 3. One extreme is given by the situation in which the dissociation constant (note its dimensions are mol/L) is 5 orders of magnitude smaller than the initial concentration of A (curve indicated by 10^{-5} in the figure). Here a stable complex is formed and no significant redissociation is observed. There-

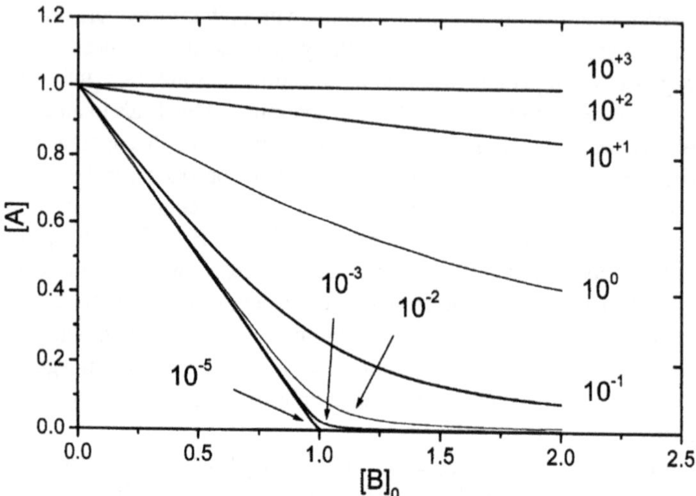

FIGURE 3 Dependence of the equilibrium concentration [A] of the analyte on the initial concentration $[B_0]$ (indicated by (B) in the text) of ligand added to the reaction mixture before CZE analysis. The curves are calculated for different values of the dissociation constants (between 10^{-5}- and 10^{3}-fold of the initial concentration of A). Reaction between analyte and ligand is according to A + B = AB. The initial concentration (A) is kept constant, and increasing initial concentrations (B) are added to the reaction mixture. All concentrations are given in arbitrary units (including the dimension of K_D). (Reprinted from Ref. 22.)

fore A is continuously consumed from the incubation mixture when increasing amounts of B are added, and a linear decrease of [A] is measured. When an equimolar amount of B is added, the entire amount of A is consumed and transformed into the complex; consequently the concentration of A in the incubation mixtures falls to zero. Note that at this point stoichiometric amounts of A and B are present in the mixture. All systems with dissociation constants 10^{-5} or smaller (related to the initial reactant concentrations) give the same picture.

The other extreme is found when the dissociation constant is several orders of magnitude larger that the initial analyte concentration, indicated by 10^{+3} [K_D exceeds (A) by factor 1000]. Here the equilibrium is nearly quantitatively at the side of the dissociated forms, so A remains constant, independent of the amount of B initially added [at least in the concentration range of (B) depicted]. Also here the equilibrium constant cannot be derived.

In the range of reactant concentration relative to a dissociation constant between these two cases, the dependence of [A] on (B) is visibly nonlinear. [It is obvious that all curves are nonlinear, according to Eq. (15); however, in the extreme ranges discussed above they have a quasi-linear shape.] Its shape depends strongly on the magnitude of the dissociation constant. Note that the discussion is based so far on monovalent complexes with 1:1 stoichiometry. However, other stoichiometries can be derived for the case of strong binding, as the concentration of B when [A] reaches zero (the intercept of the line on the x axis) gives the stoichiometric composition of the complex.

An example of the application of the method is the determination of the stoichiometry of binding between human rhinovirus (HRV) and recombinant soluble fragments of the low-density lipoprotein receptor MBP-VLDLR$_{2\text{-}3}$ as shown in Figure 4 [22]. The electropherograms are obtained from a mixture with a constant initial amount of MBP-VLDLR$_{2\text{-}3}$ (1.8 μM) with increasing amount of HRV2 added. The decrease of the receptor peak with the concomitant appearance of the complex peak can be observed. If the peak height of the receptor is plotted versus the initially added concentration of HRV2, the curve shown in Figure 5 results. It has a similar decay as the curves in Figure 4, with intermediate concentration in the K_D range. In order to estimate the stoichiomet-

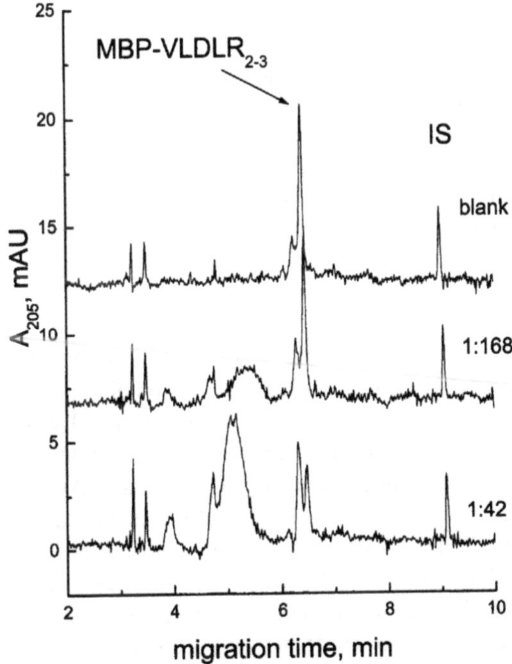

FIGURE 4 A set of ACE separations demonstrating the decrease in a peak of MBP-VLDLR$_{2\text{-}3}$ upon attachment to HRV2. (Reprinted from Ref. 22.)

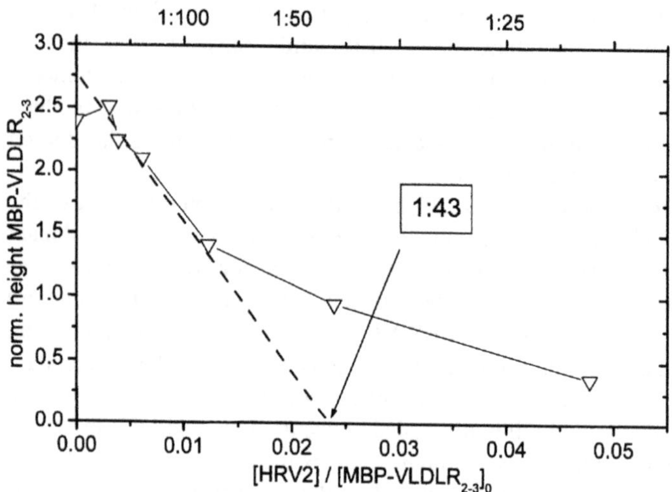

FIGURE 5 Stoichiometry determination for the system MBP-VLDLR$_{2-3}$/HRV2. The plot was derived from the CE separations shown in Figure 4. (Reprinted from Ref. 22.)

ric concentration, we should extrapolate the linear part of the curve at low HRV2 concentrations to a peak height of zero. In the present case, a composition of 43 receptor fragments per virion is obtained. It is obvious from Figure 5 that this extrapolation will lead to an underestimation of the stoichiometric composition. This is in reasonable agreement with the next larger possible value of 60, the most likely stoichiometric composition of the complex according to the topology of the icosahedral structure of the virus capsid.

3.1.3. CE-Based Immunoassays

Immunoaffinity interactions, as a major part of all kinds of ligand–receptor binding, are involved in a wide variety of clinical, biochemical, and pharmaceutical analyses due to their unique selectivity and high affinity. Generally, an antigen (Ag), which can be either a small molecule or a large macromolecular assembly, reacts specifically with its antibody (Ab), forming a stable Ag–Ab complex. Usually, high affinity of interaction ($K_D = 10^{-8}$–10^{-12} M) enables determination of various analytes even in trace levels, whereas its unique specificity (especially in the case of monoclonal antibodies) permits measurements in the presence of structurally or chemically similar compounds and in very complex biological matrixes such as urine, blood, tissues, fermentation liquid, etc.

The combination of immunoassays with capillary electrophoresis, known as CE-based immunoassay or CE-IA, is a relatively new approach which has gained an increasing popularity in the last years. This method combines all features of immunoaffinity interaction with the unique separation potential of CE. In CE-IA a precolumn incubation of an Ag–Ab mixture is followed by CE separation of the complex formed from unbound reagent. Generally, with this method the formation of an affinity complex and/or the consumption of Ag (Ab) can be monitored upon their separation, which in turn enables quantitative determination of any of the reagent and binding parameters. The prerequisites for this mode of analysis are generally the same as for any type of precolumn ACE: (a) reagents must be preincubated for a time sufficient to reach equilibrium; (b) the complex formed must be stable enough to reach the detector point; and (c) electrophoretic migration of at least one of the reagents must change upon complexation. In principle, a wide variety of Ag–Ab systems, such as those in which Ag is a relatively large or highly charged molecule, meet these requirements.

CE-IA offers several advantages over conventional immunoassays. First, unlike solid-phase assays, in which one of the reactants is immobilized on a solid support, reagents in CE-IA are free in solution (so-called homogeneous mode of analysis), which may sometimes give more correct results, especially when K_D is going to be measured. This is because immobilization might alter the native conformation of the reagent and, hence, the affinity of binding. For multivalent systems such as, e.g., viruses and mAb, this approach enables determination of the total number of binding sites, whereas immobilization always masks a certain number of them. The absence of a solid phase in CE-IA also permits a decrease in the incubation time, since diffusion in solution is much faster than on the surface of a solid support. Nonspecific bindings to the surface, which are the reason for sometimes high background and narrow dynamic range of calibration in ELISA-like techniques, are also eliminated. Second, the method permits simultaneous determination of multiple analytes and assessment of their specific activities. Third, it allows direct visualization of complex formation and its stability. Rapidity of the method, its flexibility, and the potential for on-line quantification are also among the advantages of the method. However, the low throughput of CE-IA is its obvious limitation. The new instrumental design with a capillary arrow or an application of microchips might overcome this drawback.

Like conventional immunoassays, CE-IA can be performed in either noncompetitive or competitive format. Sometimes different approaches for direct and indirect modes of analysis are also considered.

Noncompetitive Format. Noncompetitive assay is based on migration differentiation between the Ag–Ab complex and unbound reagent(s) provided by CE. An excess amount of a reagent is added to the sample, which contains a certain amount of an analyte. The affinity reaction which occurs in this case can be described by the well-known equations

$$\text{Ab} + \text{Ag (excess)} \rightarrow \text{Ab} - \text{Ag} + \text{Ag (excess)} \tag{17}$$

$$\text{Ag} + \text{Ab (excess)} \rightarrow \text{Ag} - \text{Ab} + \text{Ab (excess)} \tag{18}$$

Depending on the nature of the target, either Ag or Ab is added in excess. The excess is needed to ensure that all the analyte of interest is being complexed. A large excess facilitates a shift of equilibrium toward complex formation and in some cases can compensate for reduced affinity of the system. After incubation, the mixture is analyzed by CE. A peak of an affinity complex or a peak(s) of unbound reagent(s), resolved from each other and compared with that of calibration standards, is then used for analyte quantification.

However, UV detection, which is commonly used for CZE analysis, is lacking in sensitivity and selectivity. The usual limit of detection for proteins is higher than 10^{-6} M. Moreover, many constituents stemming from biological matrixes may severely hamper the quantification by co-eluting with the peak of interest. To circumvent the latter, a highly charged reactant must be used to shift its peak far from the elution range of biological constituents, which might not be possible in all cases. Therefore, in practice, in most noncompetitive CE-IA, the reactant which is added in excess needs to be labeled with a fluorescent tag (indicated as Ag* or Ab*), permitting the use of laser-induced fluorescence (LIF) detection This provides the necessary selectivity and greatly enhances the sensitivity of analysis (in best cases down to 10^{-12} M of an analyte). Direct and indirect modes of analysis can be used if the system fulfils the general requirements for a noncompetitive assay. The first is based on quantification of the complex peak (Ag–Ab* or Ab–Ag*). The higher the amount of an analyte in the sample, the higher will be the peak of the complex. On the other hand, an indirect method is based on quantification of an added reactant (Ag* or Ab*) which is consumed due to the complex formation. In this case the higher amount of an analyte in the sample, the lower will be the peak of the added reactant. For many systems the first approach seems preferable, since a small decrease in the peak of the added reactant might be not noticeable if it has been taken in a large excess. However, in several cases this peak is more suited for quantification due to CE conditions which can be more favorable for this reactant than for Ag–Ab* or Ab–Ag* complexes. If an added reactant is a small organic compound or peptide (Ag*), its peak can be quantified more correctly, and thus an indirect mode of analysis is preferable.

Applications. The general feasibility of CE to separate immunocomplex from unbound reagent was first demonstrated by Grossman et al. [4] and Nielsen et al. [23], whose reports set the stage for further development of quantification techniques in CE-IA. The first quantitative noncompetitive CE assay of insulin was reported by Schultz and Kennedy [24]. To increase the sensitivity of analysis, the authors used FITC-labeled insulin and LIF detection. Due to the separation power of CE, multiple products of fluorescence labeling of insulin did not hamper analysis (see Figure 6). Two types of calibration curves were con-

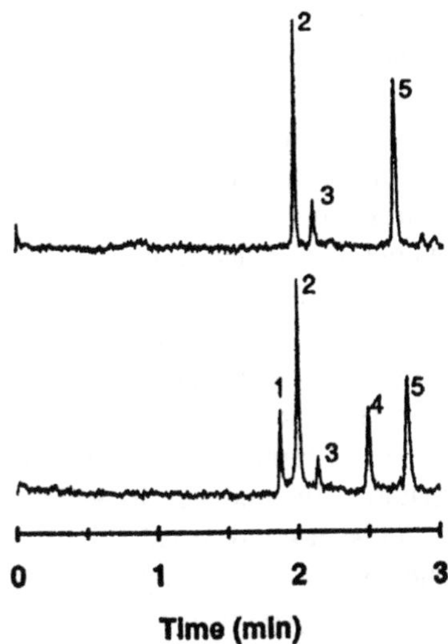

FIGURE 6 Representative electropherograms showing the principle of noncompetitive CE-IA. The top trace shows the blank sample of FITC-labeled insulin (note the presence of multiple labeled products); the bottom trace shows CE separation of a mixture of FITC–insulin and Fab. (Reprinted from Ref. 24.)

structed, reflecting the formation of the complex and the consumption of FITC–insulin. Another approach utilizing fluorescence labeling of antibody instead of antigen was proposed by Shimura and Karger [25]. An interesting feature of their study is that they proposed to use capillary isoelectric focusing (CIEF) to separate bound from free analyte (for details on affinity CIEF see Chapter 4). To avoid multiple site labeling the authors used a previously developed labeling technique which utilizes covalent attachment of tetramethylrhodamine to the Fab′ fragment through the exposed thiol groups in the hinge region.

One of the most important issues of noncompetitive CE-IA is separation of bound from free antibody, which is not always straightforward (see discussions above). This probably explains why most effort was put forth to develop a methodology facilitating their separation rather than to expand the application field of the assay. One approach to overcome the obstacle was first proposed by Chen et al. [26], who formulated the strategy of charge modifying. Their idea was to introduce charges into antibody, thus altering its electrophoretic mobility. The authors proposed to modify the intrinsic charge of an antibody by random succinylation of the free amines of the lysine residues. However, succinylation altered the antibody and complex charges equally, which limited the separation efficiency. Other reports exploited the charge modification introduced by highly charged ligands, e.g., biotin-labeled oligonucleotide [19,20]. This approach appeared to be very useful in analyses of complex biological matrixes such as fermentation liquid. Highly charged oligonucleotide residue shifted the ligand far from the elution range of most of the fermentation products, thus permitting its quantification and hence an indirect determination of a biotin-binding protein present in the fermentation liquid.

A very interesting attempt to facilitate separation of bound from free antibody was reported by Fuchs et al. [27]. To avoid equal charge modification, authors suggested using a matched pair of antibodies to perform the labeling reaction and the mobility shifting independently. Both antibodies are raised against the analyte, but the first contains an appropriate label whereas the second is highly charged. In equilibrium a tertiary complex is formed, which contains both antibodies and the analyte. Upon CE separation the complex is well resolved from the unbound labeled antibody and can be easily detected.

The report of Tim et al. [28] was a further successful development of a strategy of charge modifying. The authors added a charged competitive ligand in a background electrolyte (BGE) to facilitate separation of preformed affinity complex from unbound reagent. A noncompetitive direct CE-IA was developed for the analysis of the drug dorzolamid with fluorescently labeled carbonic anhydrase. *Para*-carboxybenzenesulfonamide, as a charged competitive ligand dissolved in a BGE, shifted a peak of unbound carbonic anhydrase to a shorter migration time, enabling correct quantification of dorzolamid–carbonic anhydrase complex. An obvious requirement of such an assay is that the competitive ligand exhibits much less affinity toward the labeled receptor, to avoid dissociation of the preformed complex during CE separation.

Electrophoretic heterogeneity of whole antibody or even of active Fab′ proteolytic fragments is still an issue in the development of noncompetitive assay. To get a uniform peak in CE, Hafner et al. [29] proposed using single-chain antibody variable-region fragments (scFv), which can be produced by recombinant technology.

Instead of using antibody as a label receptor, other kinds of affinity reagents can be applied. Aptamers are single-stranded DNA or RNA oligonucleotides which can bind specifically to a target molecule with a high affinity. The first report on aptamers as ligands in affinity CE was done by German et al. [30]. Fluorescently labeled aptamer (A*) reacted with IgE to form a complex which was well resolved from unbound excess of A*. Both direct and indirect quantification was possible based on two types of calibration curves, one which reflected complex formation and another which indicated the consumption of aptamer. Later, a similar mode of assay was applied for the analysis of reverse transcriptase from human immunodeficiency virus with fluorescently labeled aptamers [31].

Competitive Format. Although UV detection can be used in the competitive format, its sensitivity is not high enough to provide a necessary limit of detection (LOD). Therefore, in almost all cases reported in the literature so far, LIF detection is used and, consequently, one of the reagents is fluorescently labeled.

As in the competitive mode of conventional immunoassays, at least one of the reagents must be limited to provide competition. If the analyte of interest is Ag, then a known amount of fluorescently labeled Ag* (so-called fluorescence tracer) is added to the sample together with a limited amount of antibody Ab. Analyte, present in a sample, competes with the tracer for a limited number of antibody-binding sites. The affinity reaction which occurs in such a mode is described by

$$\text{Ag(analyte)} + \text{Ag}^* + \text{Ab(limited)} \rightarrow \text{Ag}^* - \text{Ab} + \text{Ag} - \text{Ab} + \text{Ag}^* \qquad (19)$$

As soon as equilibrium is established, the sample is analyzed by CE. Ideally, CE separation should reveal two distinct peaks, free Ag* (if it still exists in solution) and bound Ag*–Ab. Both peaks and even their ratio can be used for quantification of the analyte present in the sample. In such a mode, the higher the amount of the analyte, the more Ag* is displaced from the complex and, consequently, the higher the peak of Ag* and the lower the peak of Ag*–Ab.

If the analyte of interest is Ab, a known amount of the tracer, fluorescently labeled Ab*, and a limited amount of Ag are added to the sample. Analyte Ab competes with tracer Ab* for a limited amount of Ag. The affinity reaction is described by

$$\text{Ab(analyte)} + \text{Ab}^* + \text{Ag(limited)} \rightarrow \text{Ab}^* - \text{Ag} + \text{Ab} - \text{Ag} + \text{Ab}^* \qquad (20)$$

CE with LIF of this mixture should reveal two resolved peaks corresponding to bound and free Ab* (if it still exists in solution). As in the above case, both peaks and their ratio can be used for analyte quantification. Here the higher the amount of Ab in the sample, the more Ab* is displaced from the complex, and hence the higher the peak of the tracer and the lower of the complex. Besides CE-related requirements which the analyzed system should meet (see above), special attention must be paid to proper adjustment of the amount of labeled Ag* and Ab added to the sample. In both extreme cases of incorrect adjustment, namely, when the amount of either Ag* or Ab is too high, the results might be outside the dynamic range of the calibration curve and hence incorrect. These requirements are very typical for any mode of competitive immunoassay, including ELISA-like methods, and are discussed in detail in related books (see, e.g., Ref. 32).

The majority of CE-IA developed so far utilizes the competitive mode. This is despite the fact that theoretically, noncompetitive assays possess several advantages over competitive assays, which holds also for conventional immunoassays. First, there is a lower limit of detection, since the sensitivity is based on the specific activity of the label (fluorimetric, radioactive, enzymatic, etc.) rather than on the binding constant between Ag and Ab. In contrast, in competitive mode the sensitivity is limited by the affinity of the antibody used. Of course, this does not mean that antibody with very low affinity can be used in the former case. However, the lack of affinity can be compensated (at least partially) by using a very large excess of a reagent, although in this case nonspecific binding and/or system overloading will be a limiting factor. Second, noncompetitive assay provides broader linear dynamic range, since it measures the number of analyte-bound sites, contrary to competitive assay, which measures the number of sites remaining free after binding with the analyte. This advantage, however, cannot be considered if an indirect mode of noncompetitive assay is used. Finally, a methodology for noncompetitive assay can be developed more easily, since it avoids the very tedious adjustment of the amounts of immunoreagents (ratio between tracer and antibody).

Despite these advantages, there are certain practical problems which significantly restrict the development of noncompetitive analyses. If Ag is an analyte to be measured, then antibody should be fluorescently labeled, which very often results in heterogeneous products. Thiol-specific labeling at a single cystein residue can be considered [33], and antibody fragments (Fab′or scFv) can be used instead of the whole IgG to minimize heterogeneity. However, antibody fragments are known to exhibit reduced affinity compared with IgG, which might be another issue. In addition, if analyte Ag is a small, not highly charged compound (like most drugs), its complexing with Ab* will not change noticeably the migration of the complex compared with free Ab*, and hence the separation of Ab* and Ag–Ab* becomes almost impossible. The case situation when the analyte is Ab and labeled Ag* should be added does not suffer from this technical issue, since CE conditions can be easily developed which facilitate separation of Ab or Ab–Ag* complex from unbound Ag*. However, in the majority of tasks the analyte to be measured is Ag.

It should be mentioned that a mode of competitive assay when Ab is the analyte [see (Eq. (29)], encounters the same problem. As a conclusion, a careful consideration of the nature of the analyte to be determined is strongly recommended when constructing an appropriate mode of CE-IA.

Applications. It is outside the scope of this review to examine all applications dealing with competitive-mode CE-IA, since more than 50 reports have been published so far. Instead we refer the reader to several comprehensive reviews [5,34–36] covering almost all applications in the field. In this chapter we consider only some representative contributions.

Competitive CE-IA was reported for the first time by the same group of authors who first developed noncompetitive CE-IA [24]. Insulin as an analyte and FITC–insulin as a tracer compete for the limiting number of binding sites at Fab′ fragments of anti-insulin antibody. CE was capable of resolving all three forms of a tracer, stemming from insulin targeting at different locations, from FITC–insulin/Fab′ complex, the latter emerging as two distinct peaks due to differential migration of complexes with differently tagged insulin. Two types of calibration curve were constructed, reflecting the decreasing of the tracer peak and increasing of the complex peak, respectively, with increasing amounts of analyte (insulin) added. Both curves were not linear at high concentrations of insulin, due to saturation of the Fab′ with insulin. The reported detection limit was 3 nM of insulin, and the authors discussed the possibility further improving the LOD. Later these findings were extended to a more complicated real system for the determination of insulin content of single islets of Langerhans [37]. Further development of the method allowed coupling the CE system with on-line determination of insulin, in which the entire separation after each injection was achieved in less than 25 s [38].

It was mentioned that the unique separation power of CE enables simultaneous determination of different analytes. An attempt to analyze two drugs of abuse in one run was reported by Chen et al. [39]. Morphine and phencyclidine were fluorescently derivatized with different labels and electrophoretically resolved from each other, thus enabling their simultaneous quantification using two different antibodies.

Biomedical analysis, in particular, analysis of drugs, is a major area of competitive CE-IA. A significant contribution to this field was made by the group of Thormann [40–43]. All immunoreagents were taken from standard fluorescence polarization immunoassay kits. An interesting feature of almost all reports of this group is that CE separation of affinity complexes was conducted in the presence of SDS, which seemingly does not deteriorate, at least significantly, the complex stability. These authors also presented a simultaneous urinary analysis of four drugs of abuse: methadone (M), opiates (O), cocaine metabolite benzoylecgonine (C), and amphetamine/methamphetamine (A). Multianalyte CE-IA is shown in Figure 7. The limit of detection was 30 ng/mL.

A group of scientists from Peking University [44,45] proposed conducting CE-IA in a thermally reversible hydrogel added to the buffer as a replaceable packing material. Antibodies were immobilized in the gel and a mixture of tracer and analyte was injected. The presence of gel enhanced the peak shapes and resolution, whereas immobilized antibody suppressed dissociation during the CE run. The latter makes this assay a kind of a mixture between on-line and off-line analysis modes.

3.1.4. Affinity Mode of Capillary Isoelectric Focusing

Capillary isoelectric focusing (CIEF) is one of the modes of capillary electrophoresis which relies on a separation of amphoteric solutes with different iso-

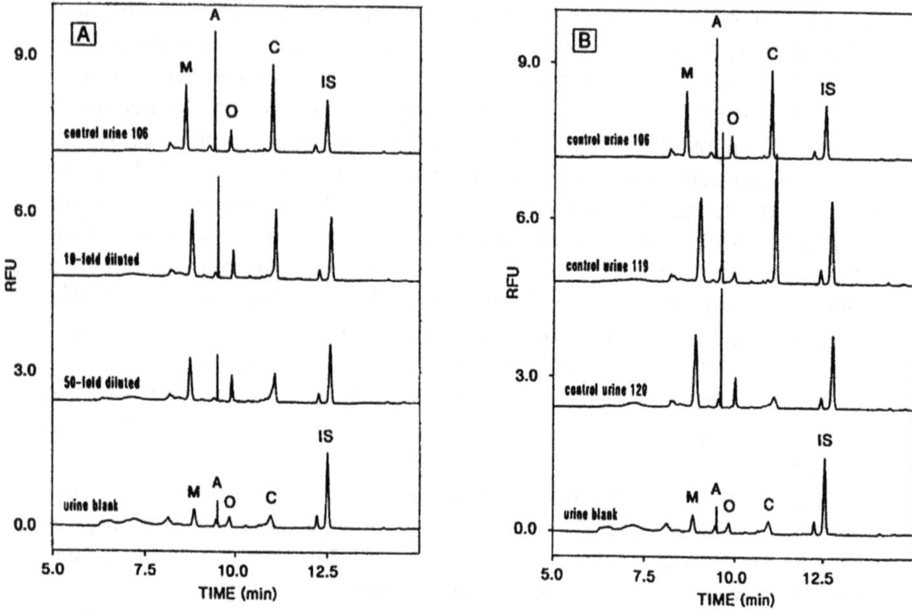

FIGURE 7 Multianalyte CE-IA of four drugs of abuse. Left panel (A) shows electropherograms of samples with different dilutions; right panel (B) shows control analysis in real urine samples. Peak identification: M, methadone; C, cocaine metabolite benzoylecgonine; A, amphetamine/methamphetamine; IS, internal standard. (Reprinted from Ref. 42.)

electric points (pI) in a formed pH gradient [46]). Since most reagents involved in biospecific interactions are amphoteric (polypeptides, proteins), their separation is possible in CIEF provided the difference in their pIs is sufficiently large to ensure the separation in a given pH gradient. Moreover, affinity interaction of a large biomolecule with even a small ligand can in some cases lead to a drastic change in conformation of the former and hence to noticeable changes in its pI. In addition, CIEF includes all features of the homogeneous mode of analysis, which is especially favorable for the correct assessment of affinity interactions.

The methodology of affinity CIEF is very similar to that of ACE with preequilibrated mixtures and is obviously suited only for complexes with a relatively high affinity and slow "off" rates. A series of mixtures with a fixed amount of one reagent and variable amounts of the other must be prepared in ampholyte solution, incubated to reach equilibrium, and analyzed by CIEF. Analysis must be capable of resolving bound from free reagent, setting the stage for quantification of the interaction. General prerequisites for this mode of analysis are the same as for other modes of ACE. However, a certain restriction limits its wide application: to be focused and detected, at least one of the reagents must be amphoteric. This means that, e.g., oligonucleotides or relative aptamers cannot be analyzed by this mode. In additional, the methodology of CIEF requires that EOF to be eliminated or sufficiently suppressed. This can be achieved either by polymer coating of the inner capillary wall, which almost completely eliminates EOF (so-called two-step CIEF when focusing and mobilization occur in series), or by adding an appropriate polymer to ampholyte solutions, thus suppressing EOF (one-step CIEF when focusing and mobilization occur simultaneously). In some cases EOF can also be reduced by application of a counterpressure [47].

The idea of using CIEF to separate bound from free antibody was realized for the first time by Shimura et al. [25]. These authors quantified human growth hormone (hGH) using tetramethylrodamine/iodoacetamide-labeled anti-hGH Fab′ fragment. Excess fluorescently labeled Fab′ was added to solutions with variable amounts of hGH and, upon incubation, Fab′–Ag complex was separated from the excess Fab′ by CIEF. LIF detection was used, which provided a very low detection limit of 5×10^{-12} M of methionyl hGH. In this report different forms of Ag were detected simultaneously in one run, since complexes of labeled Fab′ with non-, mono-, and di-deaminated variants of met-hGH exhibited different isoelectric points and hence could be resolved by CIEF.

The feasibility of studying simultaneously the specific activities of different protein isoforms is a clear advantage of affinity CE over conventional immunoassays such as ELISA. This is even more pronounced in the case of affinity CIEF, since protein isoforms may have very similar structure and are not always resolved by CZE. In contrast, protein analysis by CIEF has a higher resolving power and isoforms with only slight differences in their pI can be resolved and quantified. This was demonstrated by Okun [48] in his report on the analysis of a new biotin-binding protein actinavidin. According to CZE and HPLC analysis, actinavidin was 99% pure, whereas CIEF in nondenaturing conditions with UV detection revealed about 14 peaks associated with different protein conformations. To assess whether all conformations exhibited affinity toward biotin, a series of affinity mixtures with increasing amounts of biotin added were prepared and analyzed by cIEF (Figure 8). Analysis resulted in only two isoforms remaining intact, while all others disappeared. A new single peak appeared, reflecting the formation of actinavidin–biotin complex with pI different from that of actinavidin. It became clear that the pI of most affinity complexes is not just an average value between pI's of interacting reagents, but rather reflects a new conformation of complex molecule. These findings also support the idea of significant stabilization of a protein upon its complexation.

Recently another report appeared which deals with imunoassay of serum α_1–antitripsin by affinity CIEF [33]. Fluorescently labeled Fab′ fragments were

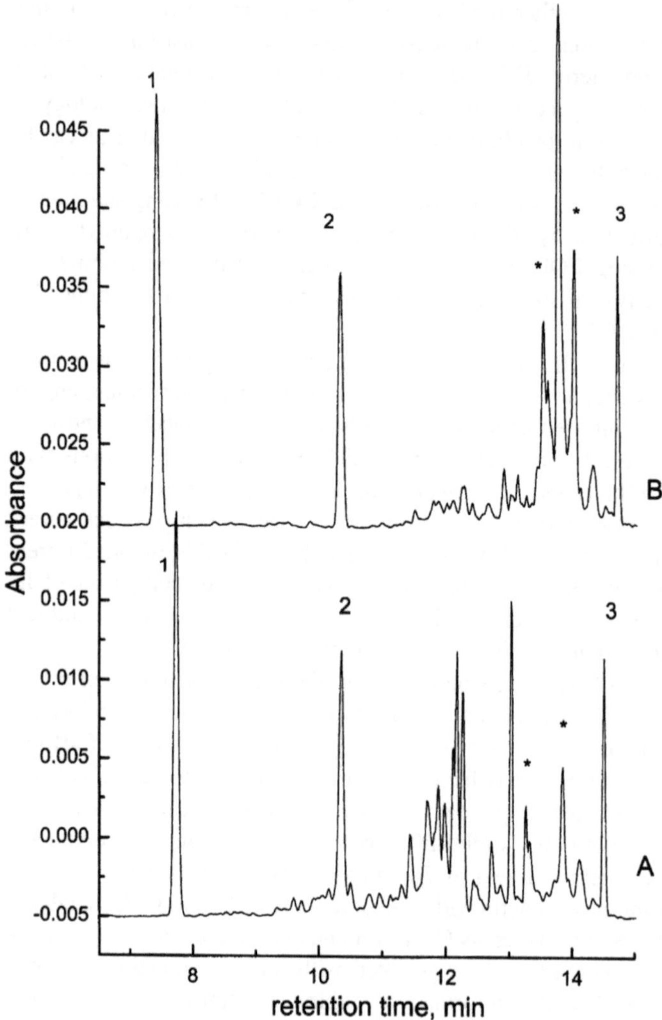

Figure 8 Affinity CIEF of actinavidin in native conditions (bottom trace) and a mixture of actinavidin and its ligand biotin (top traces). Arabic numerals represent pI standards, asterisk shows protein isoforms, which remained intact upon attachment of biotin. (Reprinted from Ref. 48.)

used in conjunction with LIF detection. As usual, the method was based on the separation of bound from free labeled antibody fragments. To reduce multiple labeling, Fab′ was labeled with tetramethylrodamine on the single cystein residue at the hinge region. Detection limit was around 2 ng/mL with a linear calibration curve, covering about three orders of magnitude.

To the best of our knowledge, these are all the reports so far relying on affinity investigations by CIEF. Probably, the methodology of CIEF itself needs further development in terms of stability and reproducibility. Obviously, the potential of this method has not yet been fully appreciated. We believe that the affinity approach in CIEF is a very promising technique, especially for the assessment of specific activities of different but closely related protein isoforms, where the separation power of CIEF is clearly superior to that of CZE.

3.2. Equilibrium Case: One or Both Reactants Added to BGE

Although affinity capillary electrophoresis (ACE) in its "classical" mode (one of the reagent is dissolved in a BGE, another is injected) is the most widely used technique in the literature, other capillary electrophoretic methods exist which are even more favorable concerning the information about binding parameters obtainable: the Hummel-Dreyer (HD) method, frontal analysis (FA), the vacancy peak (VP) method, and vacancy affinity capillary electrophoresis (VACE) (see, e.g., Refs. 49–57). All the methods need as a precondition that the equilibrium between the reactants (say, protein P, drug D, and complex formed PD) is established rapidly compared to the dislocation of the electrophoretically migrating zones. The experimental setup of the HD and the ACE methods is identical, and so is the setup for the VP and the VACE methods. FA differs from all the other techniques.

In HD and ACE the capillary is filled with the buffering background electrolyte to which one of the reactants is added in varying concentrations. The sample consists of the second reactant, which is injected into the capillary and run by electrophoresis. In the VC and the VACE techniques, the buffering BGE contains both the protein and the drug. The concentration of one reactant is constant, whereas that of the second reactant is varied. Plain buffer is injected and the peaks emerging in the electropherogram are recorded. ACE and VACE are similar in the way that the mobility shift of a reactant is measured, whereas HD and VC are based on the quantitation of the electrophoretic peaks (either that of the analytes or the corresponding vacancies) from areas or heights.

3.2.1. Affinity Capillary Electrophoresis (ACE)

In ACE the shift of mobility is used for the determination of the binding constant. The experimental setup is the same as for the HD method described below, but the parameter that is related to the extent of binding is the electrophoretic mobility (and not the peak area as for HD), i.e., its change upon varying the concentration of the reactant added to the BGE. The mobility of the injected reactant will vary between its mobility in the unbound state and the mobility of the complex. It is a prerequisite for the applicability of ACE that the mobility of the free reactant and that of the complex differ, otherwise no change will be observed upon increasing the concentration of the additive. It should be noted that in this case of equal mobilities the HD method still could be used. The ACE method has severe limitations when there is multiple binding between the analyte and the ligand, as the values for both the number of binding sites and the successive binding constants are uncertain.

As the kinetics of complex formation and dissociation are fast enough, the ligand B can be added to the BGE as mentioned above, and the analyte is injected as usual in CZE in a narrow zone. When the complex has a different mobility than the free analyte, the mobility of A changes successively with ligand concentration. If we consider again the equilibrium given in Eq. (1), the mobility of the analyte, μ_A, during migration is composed from its mobility, μ_A^f, as free species A, weighted by the fraction of free analyte, ϕ, and the mobility of the complex, μ_{AB}, also weighted by the complex fraction, which is $(1 - \phi)$.

Fraction ϕ is given by

$$\phi = \frac{[A]}{[A] + [AB]} \tag{21}$$

which upon application of the mass action law [Eq. (2)] gives

$$\phi = \frac{1}{1 + K_A[B]} \tag{22}$$

and accordingly,

$$\mu_A = \mu_A^f \phi + \mu_{AB}(1 - \phi) = \mu_A^f \frac{1}{1 + K_A[B]} + \mu_{AB} \frac{K_A[B]}{1 + K_A[B]} = \frac{\mu_A^f + \mu_{AB} K_A[B]}{1 + K_A[B]} \tag{23}$$

Note that the measured mobility should be corrected for changes of physical properties of the BGE solution upon addition of ligands. Normally the viscosity increases when significant amounts of additives are dissolved. In these cases a correction must be made which takes such changes into account, e.g., by using marker compounds that do not interact with the ligands.

We can express the experimentally measured (corrected) mobility in relation to the mobility of the free species A as a function of the varying concentration of the ligand, B, and obtain

$$(\mu_A - \mu_A^f) = \frac{(\mu_{AB} - \mu_A^f)K_A[B]}{1 + K_A[B]} \tag{24}$$

from which we can derive K_A.

In practice, we add B at a concentration significantly higher than the concentration of the analyte, A, and can therefore neglect the change in [B], which can thus be considered constant. When we plot the difference of the mobility $(\mu_A - \mu_A^f)$ versus [B], we obtain a nonlinear curve whose shape depends on the magnitude of K_A. In this case we obtain K_A by fitting Eq. (24) to the experimental data.

For completeness, the different linearized forms of Eq. (24) are given, which have had great practical importance in the past and still have. Double-reciprocal transformation leads to

$$\frac{1}{(\mu_A - \mu_A^f)} = \frac{1}{(\mu_{AB} - \mu_A^f)K_a}\frac{1}{[B]} + \frac{1}{(\mu_{AB} - \mu_A^f)} \tag{25}$$

which gives a straight line for the reciprocal of the difference $(\mu_A - \mu_A^f)$ plotted against 1/[B]. The y-reciprocal transformation gives

$$\frac{[B]}{(\mu_A - \mu_A^f)} = \frac{[B]}{(\mu_{AB} - \mu_A^f)} + \frac{1}{(\mu_{AB} - \mu_A^f)K_a} \tag{26}$$

and x-reciprocal transformation gives curves known as Scatchard plots:

$$\frac{1}{(\mu_A - \mu_A^f)} = \frac{1}{(\mu_{AB} - \mu_A^f)K_a}\frac{1}{[B]} + \frac{1}{(\mu_{AB} - \mu_A^f)} \tag{27}$$

However, linearization of Eq. (24) is of less importance today, as computers are easily available, and nonlinear curve-fitting is routine.

3.2.2. Hummel-Dreyer (HD) Method

In the HD method, after injection of one reactant dissolved in plain buffer, a peak appears that is indicative for the complex formed, and a negative peak emerges at the migration time of the reactant added to the BGE. The area of this second peak is related to the amount of reactant present in the BGE and consumed upon complex formation during migration of the injected compound. This amount can be derived from the peak area by calibration. From the electropherograms obtained at varying concentrations of reactant added to the BGE, the amount of drug bound to the protein can be determined: it is obtained from that concentration where the negative peak vanishes. It should be noted that with this method, information about the stoichiometry of the complex can be obtained, in addition to the binding constants.

3.2.3. Vacancy Peak (VP) Method

In the vacancy peak method the capillary is filled with a buffer that contains both reactants. The concentration of one additive is constant, that of the second is varied. Plain buffer is injected, and two peaks can be recorded in the electropherogram: one negative peak indicates the vacancy of the complex and the one free reactant, the second negative peak is due to the vacancy in the free drug. The peak area of the second peak depends on the amount of free drug in the buffer, and can be quantified as in the HD method by either internal or external calibration. Then the drug is dissolved in the plain buffer and injected. For a large excess of drug, the negative second peak vanishes, and a positive peak appears upon further increase of concentration of drug in the sample. The concentration where the peak vanishes gives information about the stoichiometry of the complex.

3.2.4. Vacancy Affinity Electrophoresis (VACE)

The setup is identical for VP and VACE. In VACE, instead of measuring the peak area, the shift in mobility of the vacancy peak emerging after injection of plain buffer is used to determine the binding constants. The concentration of one reactant (say, the drug) in the buffer is varied, and that of the second, the protein, is kept constant. When the concentration of the drug is low, only a small fraction of the protein is complexed, and the vacancy peak due to the protein migrates with about the mobility of the free protein. If the concentration of the drug in the BGE is high enough, the complex is formed nearly quantitatively, and the vacancy peak related to the protein reaches its other extreme value, namely, the mobility of the complex. As usual, the situation in between these extremes leads to an average mobility of the vacancy peak from the weighted contributions of both forms. Similar considerations can be undertaken for the second, the drug peak. From the shifts in mobilities, the number of binding sites and the successive complex constants can be derived.

3.2.5. Frontal Analysis

The experimental setup for frontal analysis differs from those described above. In FA the capillary is first filled with run buffer. Then a large sample plug is inserted, which consists of an equilibrium mixture of drug and protein. We consider two preconditions: (a) the complex and the protein have the same mobilities; (b) the mobility of the drug is significantly different from that of the complex. Upon application of an electrical field, the species move electrophoretically. Due to the different mobilities of the drug on the one hand, and the complex and protein on the other, the drug will migrate out of the initially mixed sample zone. Two zones with plateaus will be formed: that of the drug which migrates out of the initial zone, and that formed by the remaining mixture of complex and protein (note that in this zone equilibrium is established). The concentration in the zone formed by the drug is equal to the free drug concentration in the initial sample mixture. From the height of the plateau of the drug zone, the free drug concentration can be obtained by calibration. As the total concentration of the drug is known, and its free concentration determined from the plateau, the binding parameters can be derived.

An example of frontal analysis is given in Figure 9 for the complex formed between warfarin and bovine serum albumin (BSA) [49]. The formation of the free warfarin zone with increasing height upon addition of varying amounts of this ligand is clearly seen. If we assume for the present reaction two classes of identical binding sites, the ratio, F, of bound drug to total protein has the form

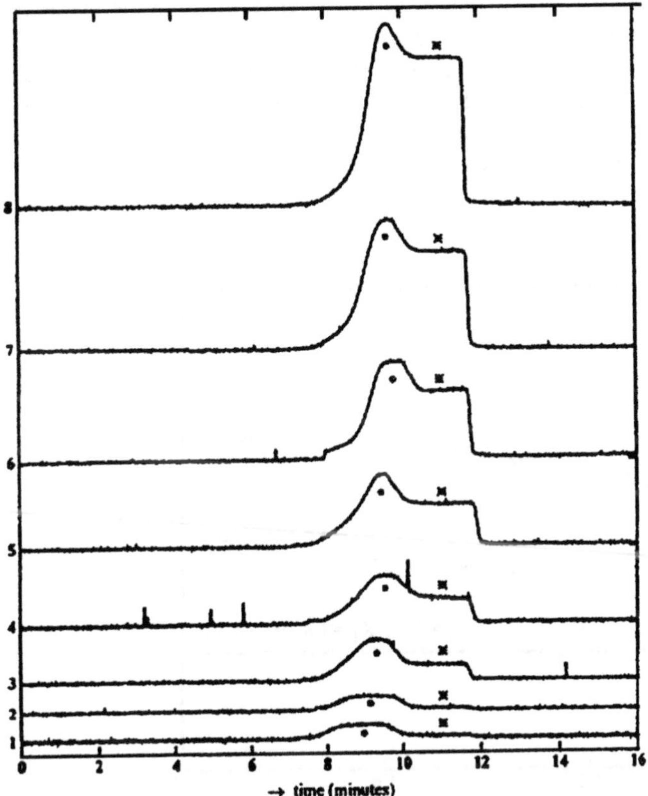

Figure 9 Frontal analysis applied for the study of warfarin–BSA affinity. Peak identification: (•) BSA–warfarin complex and free BSA; (*) free warfarin. (Reprinted from Ref. 49.)

$$F = \frac{[\text{bound A}]}{[\text{total B}]} = n_1 \frac{K_1[\text{A}]}{1 + K_1[\text{A}]} + n_2 \frac{K_2[\text{A}]}{1 + K_2[\text{A}]} \tag{28}$$

and the results of the FA analysis (with the height of the plateau representing the concentration of the free drug) leads to the binding curve shown in Figure 10. By fitting the experimentally obtained data to Eq. (28), both, the number of sites, n_1 and n_2 of the two classes, and the corresponding association constants, K_1 and K_2, can be derived. They are inserted in the figure.

4. CONCLUSIONS

After a decade of intensive development, ACE is widely recognized nowadays as a powerful tool in the study of different kinds of biomolecular interactions. More than 400 scientific reports related to ACE can be found in the literature, covering almost all fields of bioanalytical chemistry. Unique features of homogeneous analysis coupled with the separation power of CE makes ACE especially favorable for precise determination of affinity parameters, such as binding constants and binding stoichiometries. Automation, multicapillary arrows, and chip technology increase throughput of ACE analysis, a factor which still limits

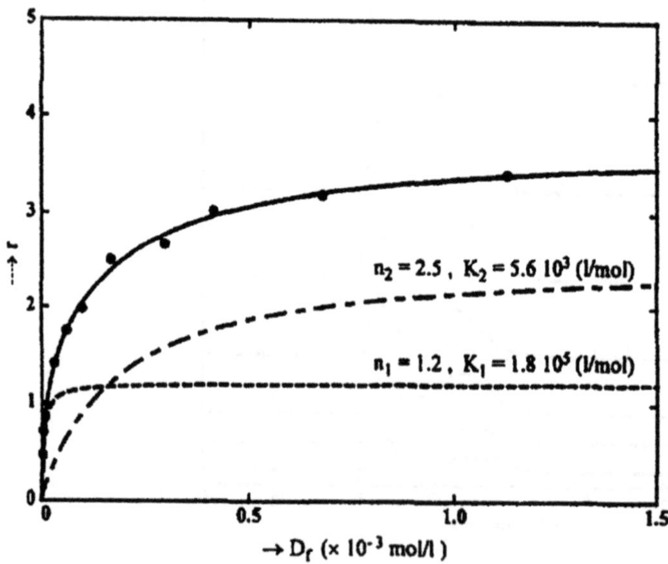

FIGURE 10 Binding curves for the system warfarin–BSA, derived from frontal analysis. (Reprinted from Ref. 49.)

its routine application in clinical, forensic, or environmental fields. However, it is unlikely that ACE in, e.g., immunoanalysis will become as well recognized and routinely used as conventional immunoanalysis such as ELISA. The LOD provided by CE-IA is still worse than in solid-phase analysis, even when LIF detection is used. There is also no evidence that CE-IA will be able to deal every day with thousands of routine analyses, since its reproducibility and robustness are still not sufficient. It is therefore no surprise that the majority of ACE assays have been conducted so far in a research environment, where all the features of ACE make it clearly superior to conventional methods of analysis. We thus believe that in various research studies, dealing with affinity investigation of new compounds (in particular, with affinity properties of new recombinant proteins, for which proper folding can be very crucial), this method will definitely become a major choice.

REFERENCES

1. Waldmann-Meyer H. J Biol Chem 1960; 235:3337–3345.
2. Horejsi V, Kocourek J. Biochim Biophys Acta 1974; 336:338–343.
3. Bog-Hansen IC. Anal Biochem 1973; 56:480–488.
4. Grossman PD, Colburn JC, Lauer HH, Nielsen RG, Riggin RM, Sittampalam GS, Rickard EC. Anal Chem 1989; 61:1186–1194.
5. Heegaard NH, Kennedy RT. J Chromatogr B Anal Technol Biomed Life Sci 2002; 768:93–103.
6. Heegaard NH, Kennedy RT. Electrophoresis 1999; 20:3122–3133.
7. Mammen M, Gomez FA, Whitesides GM. Anal Chem 1995; 67:3526–3535.
8. Heegaard NH, Robey FA. Anal Chem 1992; 64:2479–2482.
9. Heegaard NH, Robey FA. J Immunol Meth 1993; 166:103–110.
10. Lin S, Hsu SM. Electrophoresis 1997; 18:2042–2046.
11. Li C, Martin LM. Anal Biochem 1998; 263:72–78.
12. Tao L, Kennedy RT. Electrophoresis 1997; 18:112–117.
13. Zhou W, Tomer KB, Khaledi MG. Anal Biochem 2000; 284:334–341.
14. Wang QG, Luo GA, Yeung WSB. Chem J Chin Universities Chin 1999; 20:1551–1553.
15. Kiessig S, Bang H, Thunecke F. J Chromatogr A 1999; 853:469–477.
16. Tseng WL, Chang HT, Hsu SM, Chen RJ, Lin S. Electrophoresis 2002; 23:836–846.
17. Tanaka Y, Terabe S. J Chromatogr B Anal Technol Biomed Life Sci 2002; 768: 81–92.
18. Rose DJ. Anal Chem 1993; 65:3545–3549.
19. Chu YH, Lees WJ, Stassinopoulos A, Walsh CT. Biochemistry 1994; 33:10616–10621.
20. Okun VM, Bilitewski U. Electrophoresis 1996; 17:1627–1632.
21. Okun VM, Moser R, Ronacher B, Kenndler E, Blaas D. J Biol Chem 2001; 276: 1057–1062.

22. Okun VM, Moser R, Blaas D, Kenndler E. Anal Chem 2001; 73:3900–3906.
23. Nielsen RG, Rickard EC, Santa PF, Sharknas DA, Sittampalam GS. J Chromatogr 1991; 539:177–185.
24. Schultz NM, Kennedy RT. Anal Chem 1993; 65:3161–3165.
25. Shimura K, Karger BL. Anal Chem 1994; 66:9–15.
26. Chen FT, Pentoney SL Jr. J Chromatogr A 1994; 680:425–430.
27. Fuchs M, Nashabeh WA, Schmalzing DR. 1997 U.S. Patent 5,630,924.
28. Tim RC, Kautz RA, Karger BL. Electrophoresis 2000; 21:220–226.
29. Hafner FT, Kautz RA, Iverson BL, Tim RC, Karger BL. Anal Chem 2000; 72: 5779–5786.
30. German I, Buchanan DD, Kennedy RT. Anal Chem 1998; 70:4540–4545.
31. Pavski V, Le XC. Anal Chem 2001; 73:6070–6076.
32. Connors KA. Binding Constants, the Measurement of Molecular Complex Stability. New York: John Wiley, 1987.
33. Shimura K, Hoshino M, Kamiya K, Katoh K, Hisada S, Matsumoto H, Kasai K. Electrophoresis 2002; 23:909–917.
34. Bao JJ. J Chromatogr B Biomed Sci Appl 1997; 699:463–480.
35. Schmalzing D, Nashabeh W. Electrophoresis 1997; 18:2184–2193.
36. Schmalzing D, Buonocore S, Piggee C. Electrophoresis 2000; 21:3919–3930.
37. Schultz NM, Huang L, Kennedy RT. Anal Chem 1995; 67:924–929.
38. Tao L, Kennedy RT. Anal Chem 1996; 68:3899–3906.
39. Chen FT, Evangelista RA. Clin Chem 1994; 40:1819–1822.
40. Steinmann L, Caslavska J, Thormann W. Electrophoresis 1995; 16:1912–1916.
41. Steinmann L, Thormann W. Electrophoresis 1996; 17:1348–1356.
42. Caslavska J, Allemann D, Thormann W. J Chromatogr A 1999; 838:197–211.
43. Thormann W, Lanz M, Caslavska J, Siegenthaler P, Portmann R. Electrophoresis 1998; 19:57–65.
44. Zhang XX, Li J, Gao J, Sun L, Chang WB. J Chromatogr A 2000; 895:1–7.
45. Wang YC, Su P, Zhang XX, Chang WB. Anal Chem 2001; 73:5616–5619.
46. Wehr T, Rodriguez-Diaz R, Zhu M. Capillary electrophoresis of proteins. In: Cazes J, ed. Chromatographic Science Series. Vol. 80. New York: Marcel Dekker, 1998.
47. Schnabel U, Groiss F, Blaas D, Kenndler E. Anal Chem 1996; 68:4300–4303.
48. Okun VM. Electrophoresis 1998; 19:427–432.
49. Busch MH, Carels LB, Boelens HF, Kraak JC, Poppe H. J Chromatogr A 1997; 777:311–328.
50. Busch MH, Kraak JC, Poppe H. J Chromatogr A 1997; 777:329–353.
51. Chu YH, Lees WJ, Stassinopoulos A, Walsh CT. Biochemistry 1994; 33 :10616–10621.
52. Colton IJ, Carbeck JD, Rao J, Whitesides GM. Electrophoresis 1998; 19:367–382.
53. Heegaard NH. J Mol Recognition 1998; 11:141–148.
54. Heegaard NH, Nilsson S, Guzman NA. J Chromatogr B Biomed Sci Appl 1998; 715:29–54.
55. Heegaard NH, Kennedy RT. Electrophoresis 1999; 20:3122–3133.
56. Rundlett KL, Armstrong DW. Electrophoresis 1997; 18:2194–2202.
57. Rundlett KL, Armstrong DW. Electrophoresis 2001; 22:1419–1427.

6

Electroosmotic Mobility and Conductivity in Microchannels

Emily Wen
Merck Research Laboratories, West Point, Pennsylvania, U.S.A.

Anurag S. Rathore
Pharmacia Corporation, North Chesterfield, Missouri, U.S.A.

Csaba Horváth
Yale University, New Haven, Connecticut, U.S.A.

1. INTRODUCTION

The trend in analytical chemistry is clearly toward nanotechnology. With increasing research interest in combinatory chemistry, genomics, and proteomics, the need for analytical techniques that can handle reduced sample sizes has become more than ever. One such technique is capillary electrochromatography (CEC), which combines features of both capillary zone electrophoresis (CZE) and microscale high-performance liquid chromatography (μ-HPLC). In CEC the liquid mobile phase is an electrolyte in contact with the electrostatically charged surface of the stationary phase. As a result, electroosmotic flow (EOF) is generated in the column due to the presence of fixed charges on the wetted surfaces in the column. The separation of ionized sample components is brought about by differences in their retention by the stationary phase and by the difference of their electrophoretic mobility [1–5].

The use of electroosmotic flow of the mobile phase in HPLC was first suggested in 1974 [6] to replace pressure-driven flow. However, it became a practical proposition only after the introduction of fused silica capillary columns in gas chromatography. The growing need in life sciences and biotechnology for high-performance analytical separation methods [7,8] led to the capillarization of electrophoresis and opened the door for the development of CEC, which received attention as a promising and versatile high-performance analytical separation technique. It also attracted separation scientists who found intriguing the merger of chromatography and electrophoresis, two major high-performance separation methods that had not been successful in earlier times [9–14]. The results of recent work indicate that CEC has the potential to become a powerful analytical tool in chromatography of substances of biological interest [15–17].

In order to make CEC a versatile separation technique and to exploit its full potential, control and optimization of electrochromatographic conditions to generate and sustain high-EOF velocities are extremely important. Speed enhancement can be achieved through better column engineering. In the present review, the theoretical basis of the generation and control of electroosmotic flow in CEC will be presented, and it will be followed by discussion of various factors that can influence the speed of separation. Finally, the chapter concludes with novel column designs. This chapter focuses on developments that have been made in the last 5 years.

2. THEORY OF CEC

2.1. Electrosmotic Flow

Various forms of equations describing electroosmotic flow in CEC are all based on the work of von Smoluchowski in the early 1900s [18]. The theory of von Smoluchowski specifies the electroosmotic mobility μ_{eo} of an electrolyte solution along a flat charge surface under the influence of an electric field, E, as

$$\mu_{eo} = \frac{\varepsilon_o \varepsilon_r \zeta}{\eta} \tag{1}$$

where ε_o, ε_r, ζ, η are the permittivity of vacuum, the dielectric constant, the zeta potential of the surface, and viscosity of the electrolyte solution, respectively. Under conditions where the double-layer thickness is small, Eq. (1) was proven to work well in open capillary systems [19].

The system was further extended [20] to open cylindrical capillaries and to cases where the double-layer thickness is not negligible,

$$\mu_{eo} = \frac{\varepsilon_o \varepsilon_r \zeta}{\eta} \left[1 - \frac{I_0(\kappa r)}{I_0(\kappa a)} \right] \tag{2}$$

where κ, r, and a are the reciprocal of the electrical double-layer thickness, the distance of the solute from the center of the capillary, and the radius of the capillary, respectively. I_0 is the zeroth-order modified Bessel function of the first kind.

Equation (2) is valid for surface potential of up to 50 mV. For packed capillary columns, the dependence of electroosmotic flow velocity on operating parameters, based on Eq. (2) was examined and it was found that electroosmotic flow velocity increases with column porosity, particle diameter, and concentration of bulk electrolyte [21].

A detailed analysis of the theories of electroosmotic flow in porous media was presented earlier [22] of the theories by Overbeek [23–25] and Dukhin and his co-workers [26–30]. Overbeek extended von Smoluchowski's work to packed capillaries under conditions of low electric field strength. The model can be applied to porous or nonporous packing particles of any shape, and the particles can be assumed to be nonconducting, have uniform zeta potential, and a thin double layer. The average EOF velocity in a column for CEC can be expressed as

$$<u_r> = <u_p> \left[1 + \left(\frac{d_p}{a}\right)\left(\frac{2}{\beta}\right)\left(\frac{\zeta_w}{\zeta_p} - 1\right)\right] \tag{3}$$

where $<u_r>$ and $<u_p>$ are the average local velocity and average velocity at the particle surface, respectively. These two parameters provide more accuracy in determining u_r and u_p. d_p is the particle diameter, R is the inner radius of the capillary, ζ_w and ζ_p are the zeta potentials at the capillary wall and particle surface, respectively, and β is a dimensionless constant given by

$$\beta = 3\sqrt{\frac{\alpha(1 - \varepsilon_c)}{2}} \tag{4}$$

where ε_c is the total porosity of the column, and α is a dimensionless packing parameter which depends on the structure of the packing and the shape of the particles.

Both Dukhin's and Mischuk's groups [26–30] had proposed a model for "super high" EOF velocity obtained when the packing particles have higher conductivity than the surrounding medium, and the phenomenon was termed "electroosmosis of the second kind." It has been claimed that EOF velocities 10-fold higher than what is currently attained are possible. This unexpectedly high EOF velocity was assumed to be the result of the tangential and normal components of the electric field at the surface of a curved ion exchange particle as shown in Figure 1. Part of this enhancement in EOF was attributed to the "electroosmotic whirlwind" around a conductive spherical particle. In an electrolyte solution under the influence of a strong electric field, the conductive particles will have a polarized double layer which is responsible for the high EOF as shown schematically in Figure 2. According to the predictions, the enhancement of EOF enhancement increases as both the particle size and the conductivity of the particles increased, as shown in Figure 3. The authors pointed out that in

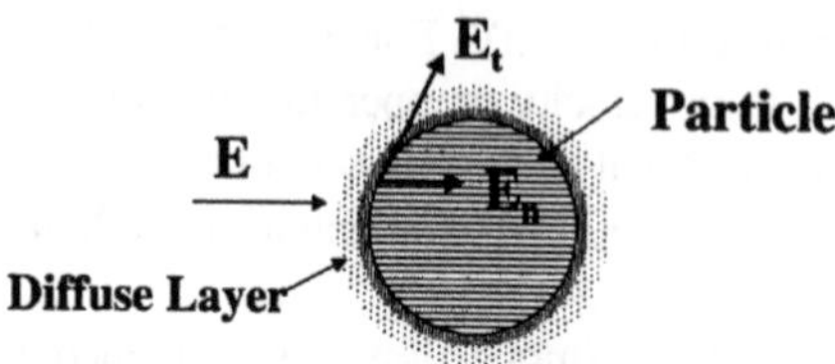

FIGURE 1 Schematic illustration of normal and tangential components of the electric field acting on the diffuse layer at a charged curved surface. (Reprinted with permission from Ref. 22, copyright 1997, with permission from Elsevier Science.)

order for the enhancement in EOF to take place, the conductivity of the column packing material has to be higher than that currently available.

Another method to evaluate the EOF mobility in a packed column has also been proposed in the literature [31]. The EOF mobility was expressed in terms similar to Eq. (1), and we obtain

$$\mu_{eo,packed} = \frac{\varepsilon_o \varepsilon_r \zeta_p}{\eta} \tag{5}$$

Since the mobility in a packed capillary is dependent on both the zeta potential and the column structure, it is important to take into account of the column architecture when EOF mobility is evaluated. An expression for calculating the "actual" interstitial mobility for a solute in a packed column is [31,32]

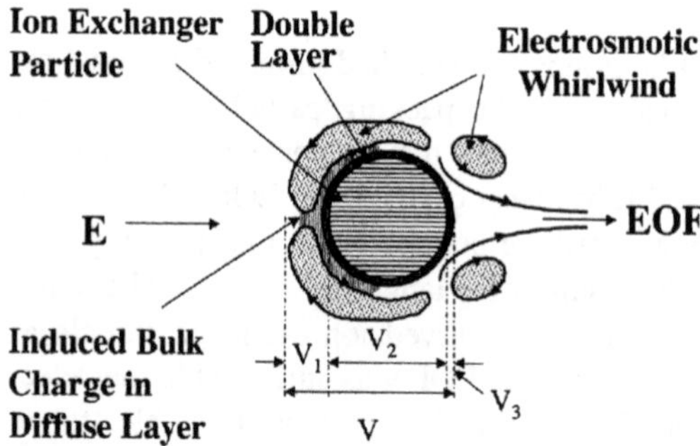

FIGURE 2 Illustration of the loci of the induced bulk charge layer and electroosmotic whirlwind around a highly conductive spherical ion-exchanger particle immersed in an electrolyte solution of relatively low conductivity under the influence of high electric field. (Reprinted with permission from Ref. 22, copyright 1997, with persmission from Elsevier Science.)

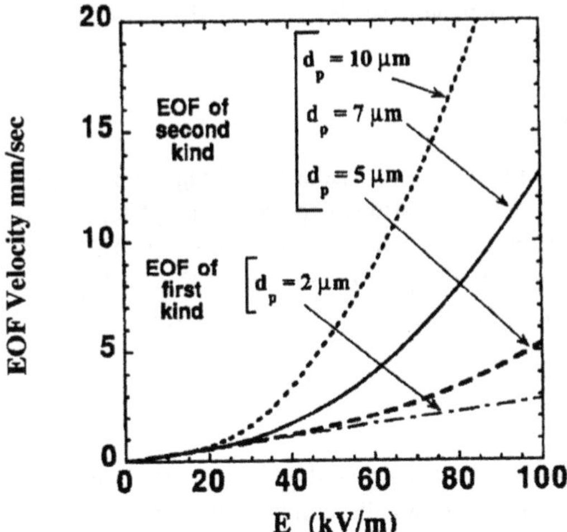

FIGURE 3 Plots of the average EOF velocity versus the electrical field strength according to Eq. (34) with the particle diameter as the parameter. Conditions: $\sigma_p/\sigma_b = 5 \times 102$, $\delta = 500$ Å; $\zeta_w = \zeta_p = 100$ mV, $\varepsilon = 80$; $\varepsilon_o = 8.85 \times 10^{-12}$ C V^{-1} m^{-1} and $\eta = 10^{-3}$ kg s^{-1} m^{-1}. (Reprinted with permission from Ref. 22, copyright 1997, with permission from Elsevier Science.)

$$\mu_{eo,\text{packed}} = \frac{L_e^2}{t_{o,\text{packed}} V_{\text{packed}}} = \frac{\tau^2 L_{\text{packed}}^2}{t_{o,\text{packed}} V_{\text{packed}}} \tag{6}$$

where μ_{eo}, packed is electroosmotic mobility of the packed segment, L_e is the actual or "equivalent" flow path through the packed segment, $t_{o,\text{packed}}$ is the migration time of a tracer in the packed segment of the column, V_{packed} is the voltage across the packed segment, and $\tau = L_e/L$ (L is the length of the column) is the tortuosity of the packing. The actual EOF mobility is evaluated after accounting for the tortuosity and depends only on the zeta potential of the packing surface, and it is a unique property of the stationary phase. The "equivalent" length, L_e, can be evaluated as

$$L_e = L\sqrt{\frac{i_{\text{open}}}{i_{\text{packed}}}} - L_{\text{open}} \tag{7}$$

where i_{open} and i_{packed} are the currents measured with the capillary tubing in the absence and presence of the packing, respectively, and L_{open} is the length of the packed segment. Another way to estimate the zeta potential at the surface is by calculating the conductivity ratio, which will be described in detail in the next section.

2.2. Electrical Conductivity

Electrical conductivity measurements in porous media have long been used for evaluation of porosity and permeability of various geological samples [33–36]. In CEC, it is important to characterize the column permeability in the appropriate manner. One way to characterize the permeability in CEC of the column packing is by the conductivity ratio, ϕ, which is defined as [36,37]

$$\phi = \frac{\sigma_{\text{packed}}}{\sigma_{\text{open}}} \tag{8}$$

where σ_{packed} is the conductivity of the packed segment and σ_{open} is the conductivity of the open segment. Equation (8) shows the conductivity ratio, ϕ, as the ratio of the conductivities of the packed and the open segments of a CEC column. In other words, it represents the ratio of electrical conductivity of the packing saturated with an electrolyte solution to the conductivity of the same electrolyte in bulk solution.

For the estimation of ϕ, empirical relationships can be used. These include the Tobias equation [38], in which the conductivity ratio is related to the porosity by the relationship

$$\phi = \varepsilon_c^{1.5} \tag{9}$$

and the Slawinski equation [39],

$$\phi = \frac{\varepsilon_c}{(1.3219 - 0.3219\varepsilon_c)^2} \tag{10}$$

It should be noted that in both equations, the conductivity ratio of a porous packing is independent of particle size, column length, or electric field strength.

Experimentally, by preparing columns with different lengths of the packed segment, the following relationships can then be used to obtain the conductivity of each segment [32]:

$$\sigma_{\text{open}} = \frac{Li_{\text{open}}}{VA_{\text{open}}} \tag{11}$$

$$\sigma_{\text{packed}} = \frac{Li_{\text{open}} i_{\text{packed}} L_{\text{packed}}}{V[i_{\text{open}} L - i_{\text{packed}} L_{\text{open}}] A_{\text{open}}} \tag{12}$$

where i_{open} and i_{packed} are the currents measured with a capillary tubing of length L and cross-sectional area A_{open}, in the absence and the presence of the packing, when a potential drop, V, is applied across it. L_{packed} is the length of the packed segment.

3. FACTORS AFFECTING FLOW AND CONDUCTIVITY

3.1. Effect of Particle Size and Porosity

The EOF velocity depends on the column packing structure and pore size of the packing material. Particle size and porosity both can greatly influence the electroosmotic flow and conductivity, and the two factors are interrelated. In a packed capillary, the porosity of the column is based on the pore size, the particle size, and the capillary diameter, and can be estimated by using the conductivity ratio. Several groups have found good correlation between the porosity of a packed column and the conductivity ratio [9,32,37,40]. In general, the conductivity ratio for most packing materials was found to be between 0.3 to 0.6, with conductivity ratio increases as the nominal pore size increases.

A CEC column can be considered as a collection of capillary tubes with the average channel size between the particles corresponds to the diameter of each tube. Using such a capillary bundle model, the following relationship between the mean channel diameter and the particle size was derived [37]:

$$d = \frac{0.42 d_p \varepsilon_c}{1 - \varepsilon_c} \tag{13}$$

where d and d_p are the mean channel diameter and mean particle diameter, respectively. This offers another way to calculate the porosity of a packed column.

Since electroosmotic flow can exist in both the interparticle and intraparticle spaces, numerous studies have focused on the existence of intraparticle flow in CEC. Several groups have investigated the existence of electroosmotic flow in wide-pore materials [41–44]. A model was developed to estimate the extent of perfusive flow in CEC packed with macroporous particles [41] by employing the Rice and Whitehead relationship. Results showed the presence of intraparticle EOF in large-pore packings (> 1000 Å) at buffer concentrations as low as 1.0 mM. Additional parameters had been investigated [43,44] to control intraparticle flow by the application of pressure to electro-driven flow. Enhancement in mass transfer processes was obtained at low pore flow velocities under the application of pressure. The authors pointed out that macroporous particles could be used as an alternative to very small particles, as smaller particles were difficult to pack uniformly into capillary columns.

3.2. Effect of Buffer Concentration

The ionic strength of the mobile phase can influence both the zeta potential of the column and the viscosity of the medium. Theory predicts that increasing the ionic strength would reduce the electrical double-layer thickness, and hence decrease the electroosmotic flow velocity. Experimental data has confirmed these predictions [9,21,42,45–47]. For example, it was observed that for different phosphate concentrations (0.04, 0.2, and 1 mM), the EOF mobility in an open tubular column increased linearly with the electric field strength as the ionic strength was decreased, as shown in Figure 4 [21]. However, in a packed column in CEC, the EOF mobility attained a maximum value at some intermediate concentration. The results were attributed to heat generation inside the column and different packing densities. The same was observed [42] when the buffer concentration was extended up to 100 mM, as shown in Figure 5. The EOF velocity increased in a nonlinear fashion at applied voltages greater than 15 kV, which was also attributed to Joule heating at high buffer concentration and electric field.

The use of wide-pore packings in CEC was examined [46] and it was found that for normal pore size packings (120 Å), there is a gradual decrease in

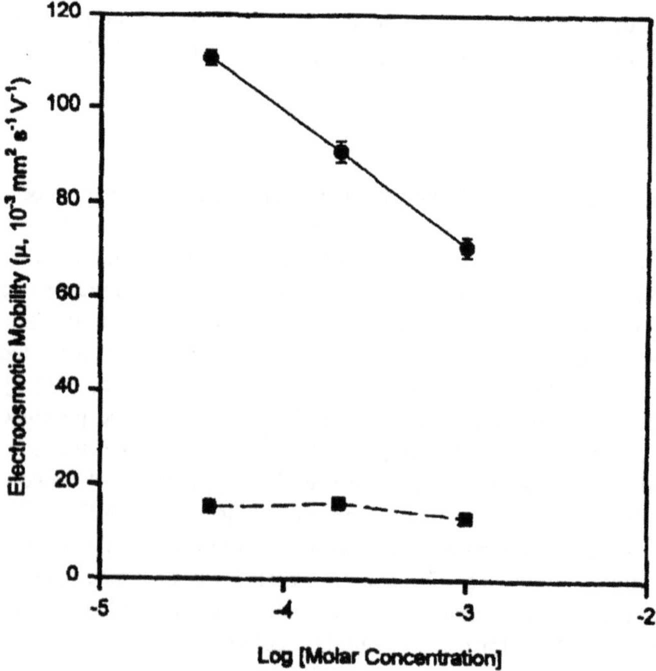

FIGURE 4 Effect of electrolyte concentration on electroosmotic mobility in open tube (upper) and in packed column (lower). (Reprinted with permission from Ref. 21, copyright 1997, with permission from Elsevier Science.)

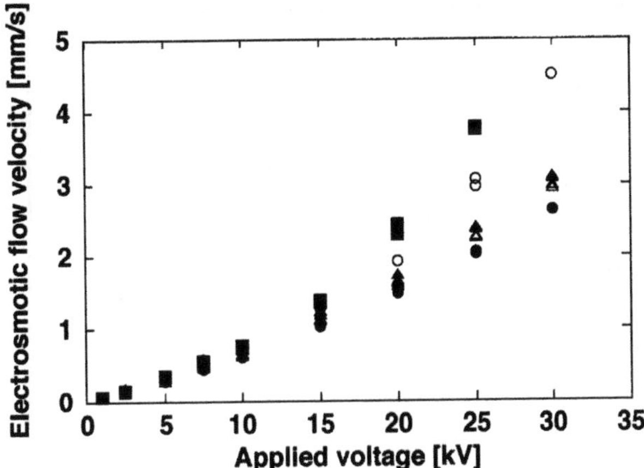

FIGURE 5 Plot of the electroosmotic flow velocity against the applied voltage with the buffer concentration as the parameter. Eluent: (△) 5, (●) 10, (▲) 20, (○) 50, (■) 100 mM Na-phosphate, pH 7.0. (Reprinted with permission from Ref. 42, copyright 1999, with permission from Elsevier Science.)

EOF with increasing ionic strength. However, for larger pore size packings (>500 Å), the EOF increases with increasing ionic strength. The results are illustrated in Figure 6. The phenomenon was attributed to the existence of pore flow, with a larger portion of the total flow existing inside the macroporous particles, resulting in a reduction in the double-layer thickness and faster EOF at higher buffer concentrations.

High salt concentrations have been used in the separation of peptides and proteins by isocratic CEC in columns packed with different particles [48,49]. Salt concentrations as high as 200 mM were employed. As shown in Figure 7, the EOF is constant between 50 and 200 mM NaCl, but the migration velocity of protein increases between 100 and 200 mM NaCl. A relationship that correlates the salt concentration with the EOF mobility has been developed as shown in the following relationship [49]:

$$\log \mu_{eo} = \log\left(\frac{\sigma_s}{e\eta(2000e^2 N_A/\varepsilon_o \varepsilon_r k_B T)^{1.2}}\right) - 0.5 \log m \qquad (14)$$

where σ_s is the surface charge density, e is the elementary charge, N_A is Avogadro's number, k_B is the Boltzmann constant, T is the temperature in kelvin, and m is the molal concentration. This equations shows that in an ideal case there is a linear correlation between the logarithmic mobility and the logarithmic concentration.

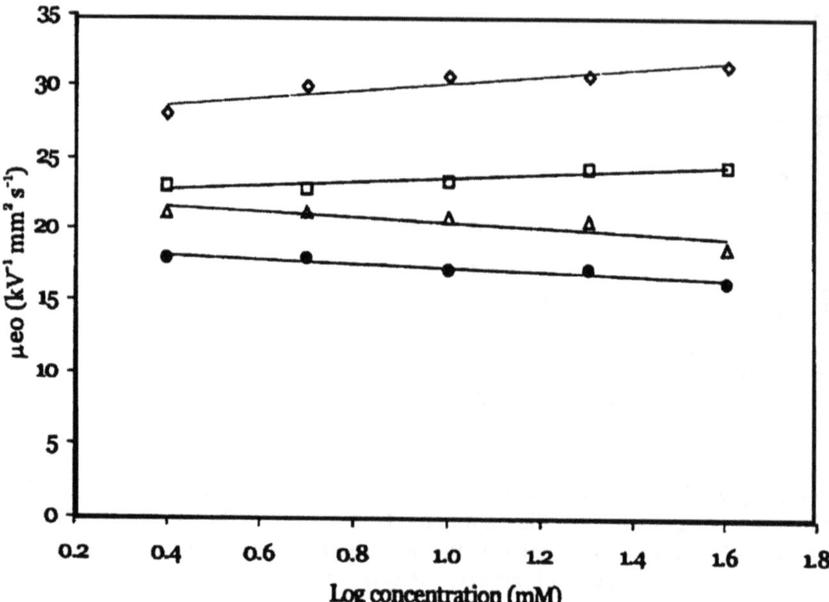

FIGURE 6 The effect of increasing the ionic strength on the EOF mobility of Nucleosil C-18 wide-pore octadecylated silica stationary phases. Conditions: column: 55 cm (40 cm packed length) × 100 μm i.d. Eluent: actonitrile-Tris (pH 9.0) (60:40, v/v) ionic strength variable. Temperature: 25°C. Dectection: 214 nm. Voltage: 30 kV. (◇) 7-μm Nucleosil 4000 Å, (□) 7-μm Nucleosil 500 Å, (Δ) 5-μm Nucleosil 120 Å, and (●) 3-μm Spherisorb ODS-1, 80 Å. (Reprinted with permission from Ref. 46, copyright 2001, with permission from Elsevier Science.)

3.3. EFFECT OF TEMPERATURE

It is well known in most cases that the speed of chromatographic separations can be increased by increasing the temperature due to enhancement of the transport properties, such as diffusivity and fluidity. The temperature dependence of viscosity of the mobile phase is usually described by the Andrade equation [50] as

$$\eta = A \exp\left(\frac{B}{T}\right) \tag{15}$$

where A and B are empirical constants for a given liquid and T is the absolute temperature. Moreover, because zeta potential is dependent on the dielectric constant, which is dependent on the temperature, a change in temperature will

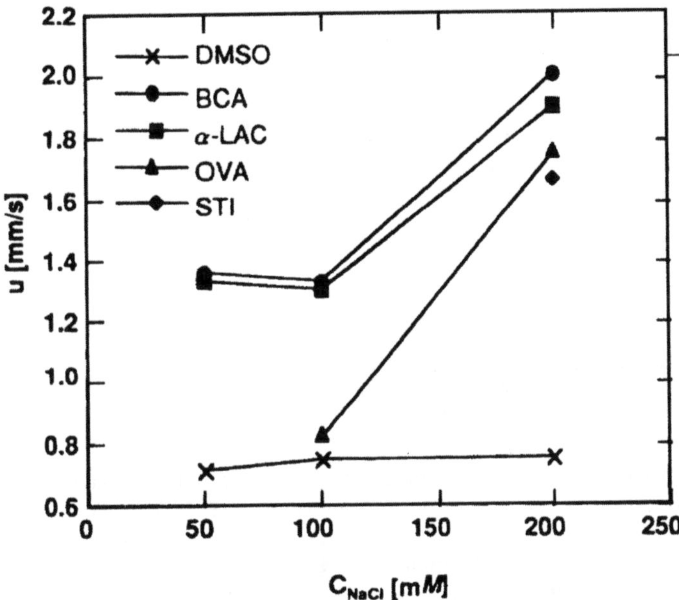

FIGURE 7 Effect of salt concentration on migration velocity of DMSO, bovine carbonic anhydrase (BCA), α-lactalbumin (α-LAC), soybean trypsin inhibitor (STI), and ovalbumin (OVA). Conditions: column: 50 μm i.d. × 340/260 mm, packed with strong anion exchanger; mobile phase, 50, 100, and 200 mM NaCl in 5 mM phosphate buffer, pH 7.0; applied voltage, −15 kV; injection, 6 s at −8 kV; UV detection, 200 nm. (Reprinted with permission from Ref. 48, copyright 2000, American Chemical Society.)

also affect the zeta potential at the surface. The dependence of the EOF mobility can then be expressed as [51]

$$\mu_{eo} = A_{eo} \exp\left(\frac{-B_{eo}}{T}\right) \tag{16}$$

The preexponential factor, A_{eo}, depends on the dielectric properties of the buffer and the zeta potential at the surface of the packing. B_{eo} represents the activation energy associated with the migration process.

At temperatures ranging from 20 to 60°C, it was shown [42] that the EOF velocity increases linearly with temperature as seen in Figure 8. On average, the increase is about 2% per 1°C. Moreover, the plate height did not change significantly over the whole temperature range, as shown in Figure 9. The effect of ionic strength on the EOF mobility at 20, 30, and 40°C was also studied [47],

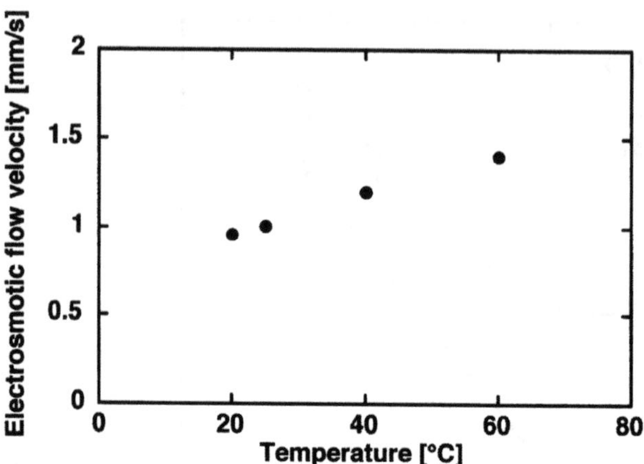

FIGURE 8 Plot of electroosmotic flow velocity against temperature. Column, 50 µm × 21/29-cm capillary packed with 5-µm Spherisorb ODS 300 Å; eluent, 10 mM Na-phosphate in 2:3 water:ACN mixture (v/v); samples, (□) benzaldehyde, (▲) naphthalene, (Δ) biphenyl, (○) fluorene, and (■) *m*-terphenyl. (Reprinted from Ref. 42, copyright 1999, with permission from Elsevier Science.)

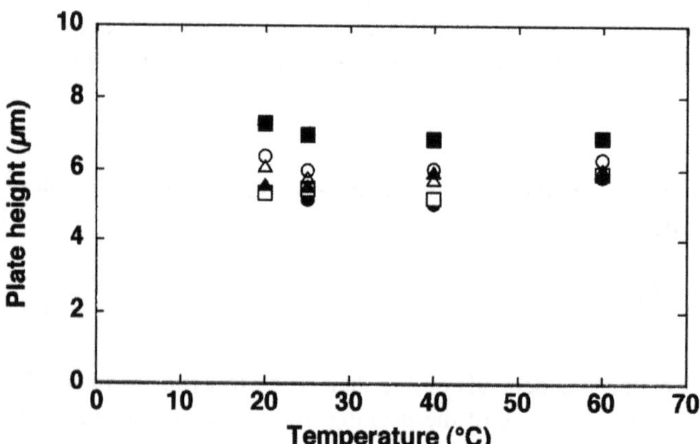

FIGURE 9 Plot of the plate height against temperature. Column, 50 µm × 21/29-cm capillary packed with 5-µm Spherisorb ODS 300 Å; eluent, 10 mM Na-phosphate in 2:3 water:ACN mixture (v/v); samples, (□) benzaldehyde, (▲) naphthalene, (Δ) biphenyl, (○) fluorene, and (■) *m*-terphenyl. (Reprinted from Ref. 42, copyright 1999, with permission from Elsevier Science.)

and the results are shown in Figure 10. As expected, at each individual buffer concentration tested, the EOF mobility was higher for the higher temperatures. In order to achieve optimal EOF mobility, high electrolyte concentration and high temperature are preferred.

Recently, rapid separation of peptide and proteins by CEC was obtained by increasing the separation temperature [52]. A tryptic digest of cytochrome *c* was obtained in less than 5 min at 55°C. Arrhenius plots of logarithmic μ against the reciprocal absolute temperature for both peptides and proteins also yielded straight lines.

3.4. Effect of Organic Solvent

Changes in the organic solvent will affect the viscosity, dielectric constant, and zeta potential of the mobile phase by changing of the double-layer thickness. The influence of selected organic solvents on the EOF mobility in open capillaries has been measured [53]. It was found that at high pH, EOF mobility is generally reduced with increasing concentration of the organic solvent. In CEC, the most commonly used columns are packed with octadecylated silica (ODS,

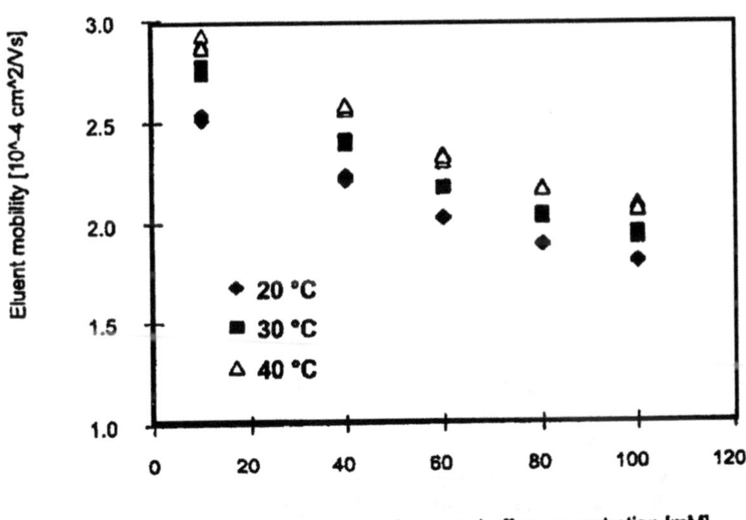

FIGURE 10 Effect of buffer concentration and temperature on EOF in CEC. Column CEC-Hypersil C18, 3 μm, 250 (350) × 0.1 mm; mobile phase 80% acetonitrile/Tris-HCl, pH 8, buffer concentration and temperature given in the figure; voltage, 20 kV; dead-time marker, thiourea. (Reprinted with permission from Ref. 47, copyright 1997, Wiley.)

C-18) particles, which requires the use of organic modifier in the mobile phase, usually acetonitrile, and thus the EOF velocity can be influenced by the type of organic modifier [54]. The EOF mobility seems to correlate well with the ratio of dielectric constant to viscosity, as also shown in Figure 11. With these columns, which were either packed with ODS particles, or coated with C-18 moieties on the capillary wall (as in open tubular or OT-CEC), it was observed that as the percentage of acetonitrile was increased, the EOF velocity increased [9,46,55,56]. For columns packed with ion exchangers or coated with ionic functionalities, the EOF velocity actually decreased with increasing percentage of acetonitrile [15,57–59]. The observed change in EOF with acetonitrile concentration is probably not the sole result of differences in the dielectric constant-to-viscosity ratio, but also variations of the charge surface may play an important role [56,60]. In columns packed with ODS particles, an increase in organic modifier concentration results not only in an increase of the ϵ/η ratio or the mobile phase, but also a change in the column structure and structure of the double layer, and an increase in the charge density on the surface of the capillary wall, which could contribute to the EOF mobility. In another work, the effects of different types of organic modifier on the EOF mobilities was examined [61],

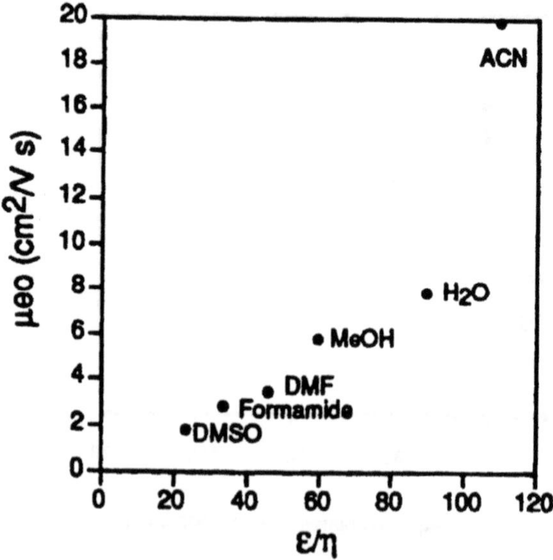

FIGURE 11 Relationship of electroosmotic mobility (μ_{eo}) to dielectric/viscosity ratio (ε/η). (Reprinted with permission from Ref. 54, copyright 1997, American Chemical Society.)

and the results are shown in Figure 12, which compares the EOF mobility to the ratio ε/η for solutions containing both organic modifiers and water. The highest EOF mobility can be obtained with acetonitrile–water mixtures, which is likely due to the relatively high dielectric constant and low viscosity of the acetonitrile–water mixtures. However, a direct correlation of ε/η to EOF mobilities was not obtained, further suggesting that zeta potential of the surface probably plays an important role in controlling the EOF mobility of the system.

4. COLUMN ENGINEERING

Separation columns are the heart of any CEC techniques. In order for CEC to become a versatile analytical technique, it is important to have columns with good stability, high EOF velocity generation, and affinity for different analytes. To this end, several different types of capillary columns have been developed for the separation of biomolecules. In this section, a summary of recent advances in column technology, including open tubular columns, duplex columns, and monolithic columns, will be presented. A comparison of different column configurations is shown in Figure 13.

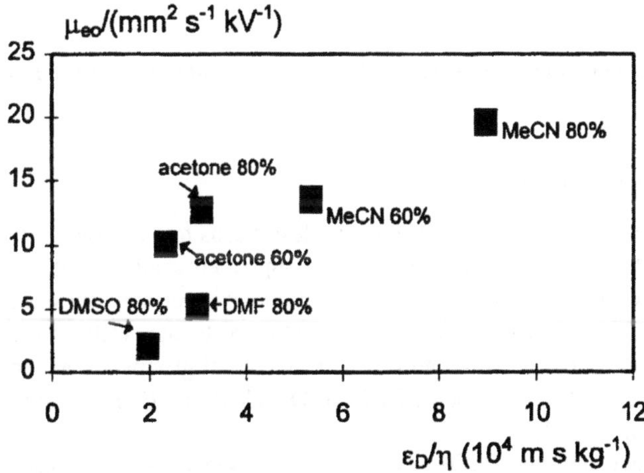

FIGURE 12 Schematic illustration of the voltage drop, *V*, and electric field strength, *E*, across (a) a 30-cm-long open capillary column; (b) a CEC column with a packed segment of 10 cm and an open segment of 20 cm; and (c) a CEC column with a packed segment of 20 cm and an open segment of 10 cm. (Reprinted with permission from Ref. 9, copyright 1997, with permission from Elsevier Science.)

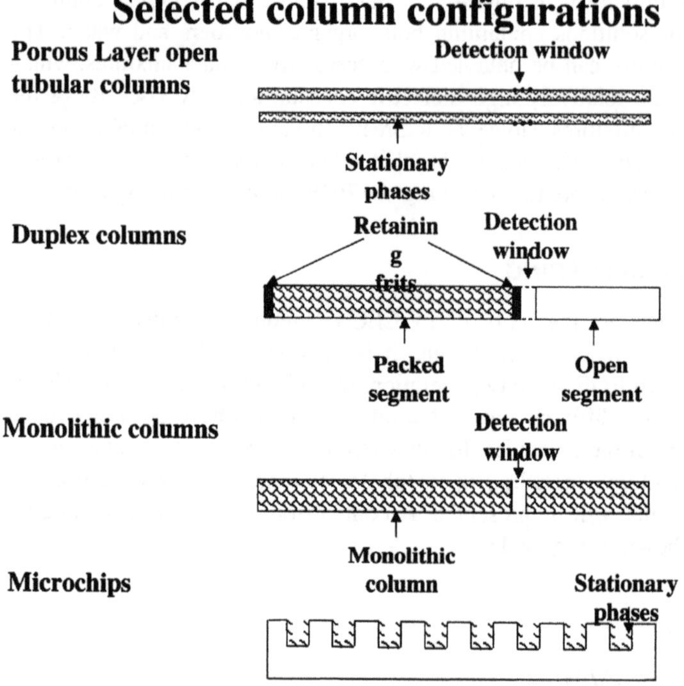

FIGURE 13 Schematic illustration of selected column configurations discussed in Section 4.

4.1. Open Tubular Columns

Open tubular capillary columns (OT) offer the simplest means to carry out CEC experiments. They are capillaries bonded with a wall-supported stationary phase, which can be a coated polymer, bonded molecular monolayer, or a synthesized porous layer network. The stationary phase that is attached to the walls of the capillary is typically less than 25 μm, and charged functional groups are added such that the walls are charged to support EOF. In OT-CEC, the EOF mobility can be calculated using Eq. (1).

Porous layer open tubular (PLOT) columns were used to separate basic proteins and peptides [15]. The use of these types of columns was prompted by their high permeability and by the relatively high loading capacity due to an increased surface area by the porous layer. The authors showed that under conditions of reversed-phase chromatography at acidic pH, the EOF mobility was over 8-fold higher than that in raw fused-silica capillary, which is an indication of the high surface charge present in the porous layer. As expected, the EOF mobility for the PLOT column decreases with increasing salt concentration, and the decrease in mobility is greater than that obtained with packed capillary columns of comparable chromatographic surface. It is believed that this is due to the "openness" of the column. Similar results with increasing salt concentration were obtained [56] with an OT column with a porous silica layer chemically modified with C18. The authors have found that the EOF mobility has an inverse relationship with the square root of the buffer concentration in the range of 1–50 mM phosphate buffer.

Extensive studies in OT-CEC have been carried out [11,62–70] for different applications. OT capillaries modified with different functional groups were used to separate different protein and peptide mixtures. By varying the pH of the buffer, the authors were able to characterized the columns through shifts in the EOF velocity with changes in pH [67,69]. They had prepared OT capillaries with *n*-octadecyl-, chlolesteryl-, and diol bonded phases. The effects of applied field strength and solvent were studied on both the octadecyl- and cholesterol-modified capillaries [64]. As expected, the EOF velocity increased with increasing field strength for both types of columns at two different pH values. The use of three different organic modifiers was examined: methanol, trifluorothylene (TFE), and acetonitrile, and it was found that in all cases, methanol offers the highest EOF velocity. However, some nonlinear behavior of the EOF velocity was observed under certain solvent conditions, which the authors attributed to changes in the bonded-phase morphology as the solvent and field strength differ.

Another way to control the EOF is to vary the composition of the surface-bonded sol-gel stationary phase [60]. A sol-gel ODS-coated column was prepared with significantly stronger EOF mobility than the uncoated column. It was found that the magnitude of EOF increases (shifts toward more negative values) when the percentage of organic modifier in the mobile phase was increased.

In other published studies with OT-CEC [68,71–74], the EOF was generally suppressed by the chosen coating. Separation was carried out by a combination of the electrophoretic mobility of the solute and retention on the stationary phase on the capillary wall. For example, affinity OT-CEC was performed to separate isomeric toluidines using calixarene-coated capillaries [74], isomeric aminophenols, isomeric dihydroxybenzenes, and isomeric nitrophenols using β-cyclodextrin-coated capillaries [73], and β-lactoglobulin mixtures using an aptameric-coated OT-CEC [72]. In all these cases, the authors showed that the EOF mobility generated by such columns was greatly reduced and the chromatography mechanism is the main driving force in the separation of various biomolecules.

4.2. Duplex Columns and Auxiliary Columns

The most commonly used CEC column has a packed and an open segment with a detection window in the middle. The packed segment is usually packed with commercially available stationary phases such as octadecylated silica particles, ion-exchange particles, or mixed-mode column materials, and the packing particles are kept in place by means of retaining frits. Varying the fraction of the packed section versus the open section has shown a dependence on the overall EOF velocity [9,75]. Changing the length of the two segments also leads to different electric field strengths in the two segments, since each segment has different conductance. In one work, the effect of varying packed length on the voltage drop and electric field strength in three different column was examined [9]. It was shown that there is an abrupt change in the field strength at the interface of the packed and open segments, as shown in Figure 14. In another

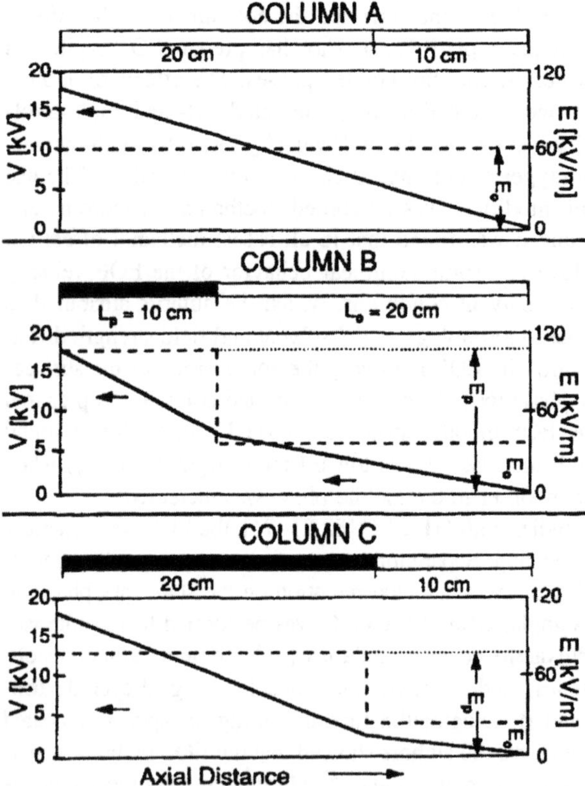

FIGURE 14 Dependence of the flow velocity in the interstitial space of the packed segment and the velocities in the open segment on 1 for $\sigma_{open}/\sigma_{packed} = 3.1$. The inequality of the two velocities in the open segment, $u_{eo,open}$ and $u^{a}_{eo,open}$, illustrates the effect of the pressure differential $P_i - P_o$. The schematic of the CEC column at the top illustrates the discontinuity in flow velocity at the interface of the two column segments. (Reprinted with permission from Ref. 31, copyright 1998, American Chemical Society.)

work, the dependence of the EOF velocity in the packed and open segments with varying dimensionless packed length was evaluated [75]. The dimensionless packed length, $\lambda = L_p/L$, is the ratio of packed length to the total length of the column, and the correlation with EOF is shown in Figure 15. The EOF velocity decreases about three-fold in both the packed and the open segments when λ increases from 0 to 1.

A new approach to speeding up EOF was recently suggested [76] and evaluated [77] by different groups of researchers. The work entailed the use of a tandem arrangement of a separating column and an auxiliary column, which is used to boost EOF velocity in the separating column. The authors found that the average EOF velocity in the separating column increased linearly with

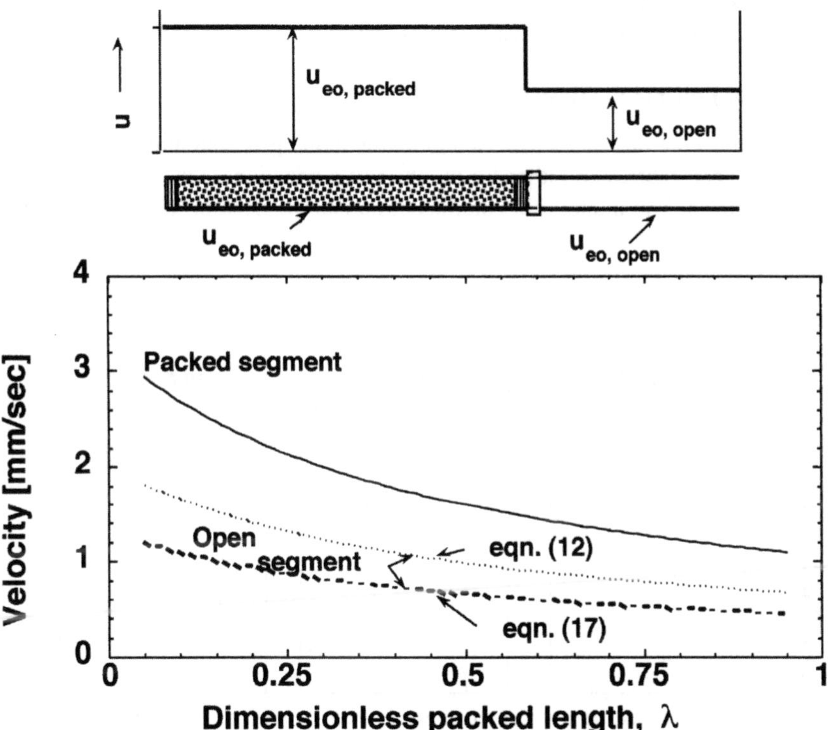

FIGURE 15 Plots of the migration time of the EOF marker against the dimensionless length of the separating column, λ. The conditions were 8 cm of separating column with varying lengths of (solid) open tubular or (dash) auxiliary columns packed with silica particles, pore size 300 Å; applied voltage, 10 kV. Eluent, (a) 2 mM, (b) 5 mM, (c) 10 mM Na-phosphate buffer, pH 7.0, in 7:3 ACN:water. (Reprinted with permission from Ref. 77, copyright 2001, Wiley.)

increasing fractional length of the auxiliary column. The results were later confirmed and extended theoretically [77]. By examining auxiliary columns packed with different particles of varying pores sizes and of different lengths, it was shown that the EOF mobility increases the pore size and with the length of the auxiliary columns. However, as seen from Figure 16, an equidiameter open tubular auxiliary column offers greater enhancement of EOF velocity than columns packed with currently available stationary phases. Also, the ionic strength of the mobile phase was shown not to have a large effect on the conductivity ratio, which was expected because the ratio is a property of the packings only.

One of the disadvantages of silica-based particles is the dependence of the EOF on the pH of the mobile phase. Ion-exchange particles have been used to increase the surface charge, and they provide stronger and more stable EOF over a wide pH range. However, no advantage was found in using columns packed with strong cation-exchange-type particles as compared to one packed with traditional octadecyl-silica particles [59].

4.3. Monolithic Columns

Monolithic columns are packed columns containing a continuous porous bed that has been fabricated either by in-situ polymerization or by sintering of porous particles. The monolithic columns eliminate the need to fabricate retaining

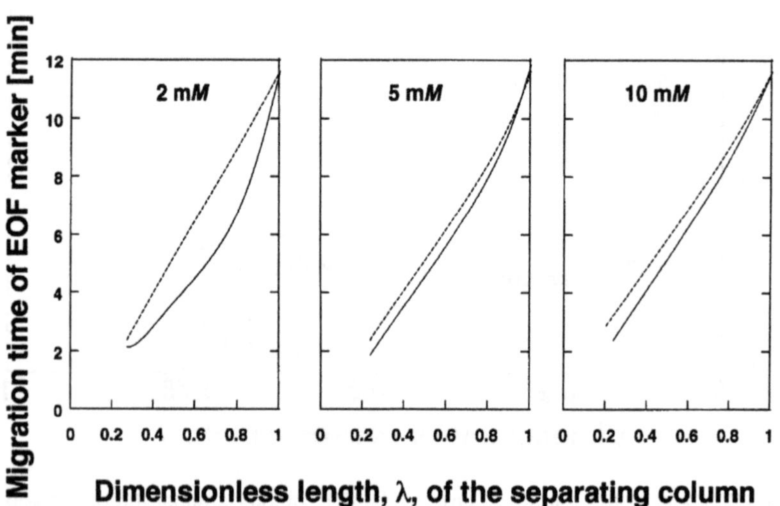

FIGURE 16 Electrosmotic velocity (μ_{eo}) plotted versus ratio of dielectric constant (ε) to viscosity (η) of the corresponding solvent mixture. (Reprinted with permission from Ref. 61, copyright 2000, with permission from Elsevier Science.)

frits and since most monolithic columns do not have an open segment, this eliminates the possibility of having discontinuities between different segments. With its numerous advantages, the monolithic column has attracted a lot of attention, and numerous papers have demonstrated the success of using monoliths in CEC-based separations [55,57,78–81].

Early work in the 1970s extended the use of monolithic columns from liquid chromatography to CEC [82–84]. Monolithic columns were prepared by in-situ polymerization of mixtures of butyl methacrylate, ethylene dimethacrylate, and 2-acrylamido-2-methyl-1-propane-sulfonic acid (AMPS) in the presence of a porogenic solvent. The authors found that by increasing the levels of AMPS used, and thus increasing the charge functionalities on the surface of the monolith, they could increase the EOF flow rate of the columns. Monolithic columns with different average pore diameters were also prepared. A two-fold increase in EOF velocity was observed as the mean pore size increased from 250 to 1300 nm [83]. In a separate part of the study [84], the effect of varying the applied voltage, length of the capillaries, percent organic solvent, and pH was examined. As expected, the EOF velocity is a linear function of the applied voltage. However, the slopes of plots of applied voltage against EOF velocity for monolithic columns of different lengths were not the same. The difference can be attributed to the number and type of charged groups present, and the permeability of the monolithic columns. As for the effect of organic solvent, the EOF velocity reached a minimum between acetonitrile concentration of 40–50%, which is consistent with changes in the dielectric constant/viscosity ratio of the solvent. The monolithic columns were shown to be stable at a wide range of pH, and the EOF velocity increases with the pH, reaching a maximum at a pH of 11. Increasing pH results in an increase in the ionization level of the sulfonic acid functionalities on the monoliths, and hence an increase in EOF velocity.

In a separate study, monolithic columns were prepared with different polymers, which were functionalized with weak or strong anion-exchange groups. For these columns, the EOF velocity was greatly affected by the percentage of cross-linking monomer in the polymerization mixture [58]. Increasing the percentage of cross-linking monomer (ethylene dimethacrylate in this case) from 20% to 40% increased the EOF velocity from 1.56 to 2.55 mm/s. The decrease was caused by a reduction of the column permeability due to the higher cross-linking monolith.

For columns with sintered ODS monolith packing, the EOF velocity was found to increase linearly with the field strength but decreased with the acetonitrile concentration [55]. In addition, mixed-mode reversed-phase and ion-exchange monolithic columns have been prepared [85]. Their EOF mobility was almost constant in the pH range from 2.5 to 6.5 but decreased with increasing phosphate concentration. This could be explained by a decrease in the double-layer thickness at higher phosphate concentrations. On the other hand, the EOF mobility decreases with the acetonitrile concentration, as the viscosity of the mobile phase decreases.

Recently, rapid separation of enantiomers by CEC had been carried out in less than 1 min [86], and the high reproducibility of the separation was consistent in over 30 runs. Batch-to-batch reproducibility was reported to be within 5% relative standard deviation as measured by the migration time of an unretained solute using nine different monolithic columns, and run-to-run reproducibility of within 1% relative standard deviation [87].

4.4. Chips and Other Kinds of Microchannels

With the advent of microfabrication technology, chip-based microdevices are becoming more popular. Reducing the size of analytical separation devices will enhance performance in terms of increased separation speed, higher sample throughput, and the possibility for parallel processing. Controlling the EOF is very important in microfabricated fluid networks, and numerous models [88–91] have been proposed for the flow field in rectangular channels. In general, channel geometry has a major effect on column efficiency [91]. Also, it was found that the volumetric flow rate in a rectangular channel varied linearly with channel height (h) for electrically driven flow, in contrast to cubic (h^3) for pressure-driven flow [89].

Also, EOF velocity in plastic microchannels had been imaged using video imaging of caged fluorescent dye [92]. Dispersion of the uncaged dye for microchannels composed of poly-(dimethylsiloxane) (PDMS) was similar to that found in fused silica capillaries. The calculated EOF mobilities in different mobile-phase systems for various channels (acrylic, PDMS, acrylic/PDMS hybrid) were similar. This work shows that EOF profile is similar in microfabricated devices and in capillary columns.

CEC on microchips can be carried out in several different formats, for instance, as coated open channels, and by using polymeric porous monolith and with a stationary phase constructed by micromachining [93–96]. The solvent used to cast the polymer usually contains charged functionalities in order to generate and sustain EOF and allow the separation channel to be conditioned without the need for high-pressure pumps. In open-channel microchip CEC, several different channels have been coated with different silanizing agents, and it was found that the EOF flow velocity was diminished by 10–25% after the coating procedure [93]. However, there was no apparent correlation between the geometric properties of the chips (coated with different silanizing agents) and the corresponding reduction in EOF.

5. CONCLUSIONS

CEC has fast become a popular technique in separation science, as it combines advantages of both HPLC and CZE. This work has examined the recent progress in CEC, covering various aspects that affect the flow in the microchannels currently employed in CEC applications. Both the theoretical and experimental aspects of column architectures for CEC were discussed. Thus far, there have been several attempts to describe the flow fields of EOF in porus media and in packed capillary columns. Starting with the work of von Smoulochouski and Overbeek, to Rathore and his co-workers recently. Numerous factors, such as temperature and ionic strength, have been shown to have a large effect on the EOF, and the first part of this chapter examined some of the recent observations in the field. The second part of the chapter focused on the various column technologies and the fabrication of separation media and capillaries specifically for CEC.

Although all the fundamental work on the generation of EOF is leading to a better understanding of CEC, more research is still required. For example, if "electroosmosis of the second kind" could be brought about in CEC systems of practical significance, it would greatly facilitate rapid separations and the use of long columns that allow the generation of a large number of theoretical plates. Monolithic columns have emerged as an attractive alternative to packed capillary column, due to the absence of retaining frits and the choice of chemistries they offer, promising the preparation of reproducible, well-defined, and easily prepared capillary columns. However, much remains to be done in this area.

ACKNOWLEDGMENTS

This work was supported by Grant GM20993 from the National Institutes of Health, U.S. Public Health Service.

REFERENCES

1. Rathore AS, Horváth CS. In: Deyl Z, Svec F, eds. Capillary Electrochromatography. Amsterdam: Elsevier, 2001, Chap. 1.
2. Knox JH. J Chromatogr A 1994; 680:3.
3. Dittmann MM, Rozing GP. J Chromatogr A 1996; 744:63.
4. Pyell U. Adv Chromatogr 2001; 41:1.
5. Bartle KD, Myers P. J Chromatogr A 2001; 916:3.
6. Pretorius V, Hopkins BJ, Schieke JD. J Chromatogr A 1974; 99:23.
7. Jorgenson JW, Lukacs KD. J Chromatogr 1981; 218:209.
8. Boughtflower RJ, Underwood T, Paterson CJ. Chromatographia 1995; 40:329.
9. Choudhary G, Horváth Cs. J Chromatogr A 1997; 781:161.
10. Dittmann MM, Rozing GP. Biomed Chromatogr 1998; 12:136.
11. Pesek JJ, Matyska MT. Electrophoresis 1997; 18:2228.
12. Smith NW, Evans MB. Chromatographia 1994; 38:649.
13. Ståhlberg J. J Chromatogr A 2000; 887:187.
14. Yang C, El Rassi Z. Electrophoresis 1998; 19:2061.
15. Huang X, Zhang J, Horváth Cs. J Chromatogr A 1999; 858:91.

16. Zhang M, El Rassi Z. Electrophoresis 2000; 21:3126.
17. Zhang S, Huang X, Yao N, Horváth Cs. J Chromatogr A 2002; 948:193.
18. von Smolouchowski M. Int Acad Sci Cracovie 1903; 1:184.
19. Paul PH, Garguilo MG, Rakestraw DJ. Anal Chem 1998; 70:2459.
20. Rice CL, Whitehead R. J Phys Chem 1965; 69:4017.
21. Wan Q-H. J Chromatogr A 1997; 782:181.
22. Rathore AS, Horváth Cs. J Chromatogr A 1997; 781:185.
23. Overbeek JTG, Wijga PWO. Rec Trav Chim 1946; 65:556.
24. Overbeek JTG. In: Kruyt HR, ed. Colloid Science. New York: New York, 1952: 115.
25. Overbeek JTG. In: Kruyt HR, ed. Colloid Science. New York: Elsevier, 1952:194.
26. Baran AA, Babich YA, Tarovsky AA, Mischuk NA. Colloids Surf 1992; 68:141.
27. Dukhin SS. Adv Colloid Interface Sci 1991; 35:173.
28. Dukhin SS, Mishchuk NA. J Membr Sci 1993; 79:199.
29. Mishchuk NA, Takhistov PV. Colloids Surf A: Physicochem Eng Aspects 1995; 95:119.
30. Mishchuk NA, Barany S, Tarovsky AA, Madai F. Colloids Surf A: Physicochem Eng Aspects 1998; 140:43.
31. Rathore AS, Horváth Cs. Anal Chem 1998; 70:3069.
32. Rathore AS, Wen E, Horváth Cs. Anal Chem 1999; 71:2633.
33. Wong P, Koplik J, Tomanic JP. Phys Rev B 1984; 30:6606.
34. Masliyah JH. Electrokinetic Transport Phenomena. Alberta, Canada: Aostra, 1994.
35. Liu SJ, Masliyah JH. Susp Fund Appl Petroleum Ind 1996; 251:227.
36. Archie GE. Trans AIME 1942; 146:54.
37. Wan Q-H. J Phys Chem B 1997; 101:4860.
38. de la Rue RE, Tobias CW. J Electrochem Soc 1959; 106:827.
39. Slawinski AJ. J Chim Phys 1926; 23:710.
40. Vallano PT, Remcho VT. J Phys Chem B 2001; 105:3223.
41. Vallano PT, Remcho VT. J Chromatogr A 2000; 887:125.
42. Wen E, Asiaie R, Horváth Cs. J Chromatogr A 1999; 855:349.
43. Stol R, Poppe J, Kok WT. J Chromatogr A 2000; 887:199.
44. Stol R, Poppe H, Kok WT. Anal Chem 2001; 73:3332.
45. Dittmann MM, Rozing GP, Ross G, Adam T, Unger KK. J Capillary Electrophor 1997; 4:201.
46. Dearie HS, Spikmans V, Smit NW, Moffatt F, Wren SA, Evans KP. J Chromatogr A 2001; 929:123.
47. Dittmann MM, Rozing GP. J Micro 1997; Sept 9:399.
48. Zhang J, Huang X, Zhang S, Horváth Cs. Anal Chem 2000; 72:3022.
49. Zhang J, Zhang S, Horváth Cs. J Chromatogr A 2002; 953:239.
50. Andrade ENdC. Nature 1930; 125:309.
51. Ma S, Horváth Cs. J Chromatogr A 1998; 825:55.
52. Zhang S, Zhang J, Horváth Cs. J Chromatogr A 2001; 914:189.
53. Schwer C, Kenndler E. Anal Chem 1991; 63:1801.
54. Wright PB, Lister AS, Dorsey JG. Anal Chem 1997; 69:3251.
55. Asiaie R, Huang X, Farnan D, Horváth Cs. J Chromatogr A 1998; 806:251.
56. Crego AL, Martinez J, Marina ML. J Chromatogr A 20002; 869:329.

57. Zhang S, Huang X, Zhang J, Horváth Cs. J Chromatogr A 2000; 887:465.
58. Lämmerhofer M, Svec F, Frechet JM, Lindner W. J Chromatogr 2001; A 925:265.
59. Cikalo MG, Bartle KD, Myers P. Anal Chem 1999; 71:1820.
60. Hayes JD, Malik A. Anal Chem 2001; 73:987.
61. Banholczer A, Pyell U. J Chromatogr A 2000; 869:363.
62. Pesek JJ, Matyska MT. J Chromatogr A 2000; 887:31.
63. Matyska MT, Pesek JJ, Boysen RI, Heam MT. Anal Chem 2001; 73:5116.
64. Matyska MT, Pesek JJ, Boysen I, Hearn TW. Electrophoresis 2001; 22:2620.
65. Pesek JJ, Matyska MT, Mauskar L. J Chromatogr A 1997; 763:307.
66. Pesek JJ, Matyska MT. J Capillary Electrophor 1997; 4:213.
67. Pesek JJ, Matyska MT, Cho S. J Chromatogr A 1999; 845:237.
68. Pesek JJ, Matyska MT, Swedberg S, Udivar S. Electrophoresis 1999; 20:2343.
69. Matyska MT, Pesek JJ, Yang L. J Chromatogr A 2000; 887:497.
70. Catabay AP, Sawada H, Jinno K, Pesek JJ, Matyska MT. J Capillary Electrophor 1998; 5:89.
71. Matyska MT, Pesek JJ, Boysen RI, Hearn MTW. J Chromatogr A 2001; 924:211.
72. Rehder MA, McGown LB. Electrophoresis 2001; 22:3759.
73. Wang Y, Zeng Z, Guan N, Cheng J. Electrophoresis 2001; 22:2167.
74. Zeng Z, Xie C, Li H, Han H, Chen Y. Electrophoresis 2002; 23:1272.
75. Rathore AS, Horváth Cs. Anal Chem 1998; 70:3271.
76. Yang C, El Rassi Z. Electrophoresis 1999; 20:18.
77. Wen E, Rathore AS, Horváth Cs. Electrophoresis 2001; 22:3720.
78. Zhang M, El Rassi Z. Electrophoresis 2001; 22:2593.
79. Gusev I, Huang X, Horváth Cs. J Chromatogr A 1999; 855:273.
80. Svec F, Peters EC, Sykora D, Frechet JM. J Chromatogr A 2000; 887:3.
81. Zou H, Ye M. Electrophoresis 2000; 21:4073.
82. Peters EC, Petro M, Svec F, Frechet JM. Anal Chem 1997; 69:3646.
83. Peters EC, Petro M, Svec F, Frechet JM. Anal Chem 1998; 70:2288.
84. Peters EC, Petro M, Svec F, Frechet JM. Anal Chem 1998; 70:2296.
85. Wu R, Zou H, Fu H, Jin W, Ye M. Electrophoresis 2002; 23:1239.
86. Schweitz L, Andersson LI, Nilsson S. Anal Chim Acta 2001; 435:43.
87. Hoegger D, Freitag R. J Chromatogr A 2001; 914:211.
88. Mala GM, Chun Y, Dongqing L. Colloids Surf A: Physicochem Eng Aspects 1998; 135:109.
89. Conlisk AT, McFerran J, Zheng Z, Hansford D. Anal Chem 2002; 74:2139.
90. Poppe H. J Chromatogr A 2002; 948:3.
91. Zhang X, Regnier FE. J Chromatogr A 2000; 869:319.
92. Ross D, Johnson TJ, Locascio LO. Anal Chem 2001; 73:2509.
93. Kutter JP, Jacobson SC, Matsubara N, Ramsey JM. Anal Chem 1998; 70:3291.
94. Ericson C, Holm J, Ericson T, Hjertén S. Anal Chem 2000; 72:81.
95. Throckmorton DJ, Shepodd TJ, Singh AK. Anal Chem 2002; 74:784.
96. He B, Ji J, Regnier FE. J Chromatogr A 1999; 853:257.

7

Factors Influencing Performance in Capillary Electrochromatography on Silica Columns

Keith D. Bartle and Peter Myers
University of Leeds, Leeds, United Kingdom

1. INTRODUCTION

A wide variety of approaches are currently being used in the fabrication and technology of columns for capillary electrochromatography (CEC). Continuous polymer bed, or "monolithic" columns (see Section 3.4), manufactured by in-situ polymerization within the columns, have been used in numerous application areas and have been shown to be highly efficient. In a second approach, a sol-gel process is employed to form a silica xerogel within the capillary, followed by bonding of the stationary-phase group; alternatively, the separation medium itself may be polymerized in situ.

Open-tubular columns, in which the wall is coated with stationary phase, either physically or by bonding through chemical derivatisation, also have considerable potential. Dittmann and Rozing have recently summarized [1] theoretical treatments of CEC in open-tubular columns (OTCEC) and compared these with experimental results. The plug-flow velocity profile reduces theoretical plate heights (H) for non- or slightly retained solutes, but an increase in retention leads to a rapid increase in H. However, column diameters in OTCEC may be larger than in open-tubular liquid chromatography (OTLC), with consequent improved ease of practical use.

Clearly, the above approaches offer significant advances in CEC and its applications, and may eventually subvert the problems inherent in the use of silica particles. Many workers, however, have chosen to use columns packed with bonded silica; indeed, it has been estimated that up to 90% of reported CEC separations have been carried out on octadecylsilyl (ODS)-bonded silica. In the early stages of progress of a new technique [2–5], in which understanding of many of the basic processes is still sought, research has centered largely on columns analogous to the silica columns which have proved so successful in high-performance liquid chromatography (HPLC). A wide range of silica particles is available, because of intense activity in HPLC development; the chemistry of the HPLC separation process is increasingly understood; the direct transfer of HPLC methods to CEC versions with improved resolution is, perforce, most likely to employ silica-based columns; and, if "customized" stationary phases for CEC are required, these may be synthesized through well-established procedures, in which, most usually, either chloro- or alkoxysilane derivatives are reacted with surface silanol groups.

The purpose of this chapter is to review progress made toward understanding some of the factors which influence the performance of silica-based columns in CEC.

2. PREPARATION OF SILICA COLUMNS FOR CEC

The most commonly used columns for CEC have so far consisted of a small-diameter (usually 50–100-μm internal diameter) fused silica tubing with protective polyimide coating packed with, most often, 3-μm-diameter silica particles with bonded organic groups. Much attention has been paid to the factors which have a major bearing on the properties of the resulting column: the means of packing the bed and the procedure used to hold the packed bed in the column.

2.1. Packing CEC Columns with Silica Particles

In column chromatography with both electrodrive (CEC) and pressure drive (HPLC), a number of processes [6] bring about the broadening of solute bands [7]: (a) eddy diffusion, originating from the variation of flow paths through the packed bed; (b) axial molecular diffusion; (c) resistance to mass transfer in the mobile and stationary phases; and (d) effects arising from "dead" volumes in the system. The smaller values of H and greater plate numbers N ($=L/H$) in CEC in comparison with HPLC arise from reduced contributions to H from factors (a) and (c) above, and it is clearly of great importance to pack columns for CEC for which (a) is minimized. Plug flow in CEC substantially reduces the eddy diffusion term in comparison with parabolic flow in HPLC. This term is also proportional to column particle diameter (d_p), so the use of smaller particles should further reduce the contribution to H. Also, the contribution to H from slow mass transfer in the mobile phase, $C_M\bar{u}$, where $\bar{u}$ is average linear mobile-phase velocity, is proportional to d_p^2, again favoring small particles.

The use of electroosmotic flow (EOF) to drive the mobile-phase flow results in plug flow in the channels between particles, thus reducing $C_M\bar{u}$ by a factor which can be estimated by considering such a channel as an open tube diameter d_c, for which C_M can be related to d_c and retention factor, k, by the Golay equation [8]

$$C_m = \frac{f(k)d_c^2}{D_m} \tag{1}$$

where D_m is the diffusion coefficient of the solute in the mobile phase. Because of the differing expressions [9] for $f(k)$ for plug and parabolic flow, it turns out that the contribution to H from this source in CEC is about half that in micro-HPLC, and regular packing of the column bed is desirable.

Horvath and Lin [10,11] showed how the different contributions to peak dispersion in a packed bed could be defined in an expression for H, which was applied [12] to CEC by Dittman et al.; it was demonstrated that the major contribution of eddy diffusion to H in HPLC was considerably reduced in CEC. Minimal reduced plate heights $h = (H/d_p)$, generally found to be near 2 in HPLC, should be reduced to near 1 or less, giving rise to plate numbers over 10^5 for a 40-cm-long CEC column with 3-μm particles. Horvath and Lin showed [11] how their expression for H could be simplified to

$$h = \frac{B}{\nu} + A\nu^{1/3} + C \tag{2}$$

where ν is mobile-phase reduced velocity; Eq. (2) is identical to the empirical Knox equation, commonly used to monitor performance in HPLC. That the second (A) term is important in defining the advantages of CEC has been addressed [13] by Tallarek et al., who used pulsed magnetic field gradient nuclear magnetic resonance to study flow dynamics, and found an A term smaller by a factor of 2.5 in CEC than in micro-HPLC on the same column; Wen et al. reached [14] similar conclusions from chromatographic measurements.

The dependence of the A term on what Knox has described [15] as the "goodness" of packing of the column, taken with the wide divergence of plate numbers reported by different workers for columns packed with ODS-bonded silica with the same nominal d_p, has led to the fabrication of columns for CEC being regarded [16] as being a "skill requiring experience." Not surprisingly, a variety of procedures have been proposed for the packing of capillaries for CEC with silica particles.

Electrokinetic, centrifugal, and gravitational packing have all been employed, although the pumping of liquid slurries has been most common. A supercritical carbon dioxide medium was also found to give rise to highly efficient columns. Colon et al. have reviewed [16] these methods. Packing columns by means of electrically driven flow [17] is an alternative to pressure packing. The (charged) silica particles are driven by the EOF into the capillary, which is simultaneously vibrated. Centrifugal packing uses [18] the forces generated when (supported) capillaries are rapidly rotated at up to 300 rpm around a central reservoir. Particles can also be introduced into the capillary by gravitational forces [19]. Both the latter two procedures are thought [16] to be most effective if the silica particles are slurried in a low-viscosity solvent such as acetone, which allows more rapid particle velocity.

The most commonly used procedure for CEC column packing, filling columns using a liquid slurry pressurized by an HPLC pump [20], was investigated [21] in depth by Angus et al., for numerous bonded silica packings. For ODS-bonded silica, packing solvents of low viscosity, especially acetonitrile, acetone, and methanol, were found to be most effective, in agreement with the results of Robson et al., who attributed [22] the high efficiency of the CEC columns they packed using supercritical carbon dioxide slurries to the low viscosity of the packing solvent. Angus et al. emphasized [21] the importance of preventing particle aggregation during packing, which could be achieved by sonication of the slurry before packing. Extension of investigations to other hydrophobic silica-bonded phases led to specific recommendations of packing conditions for, e.g., C_8 and phenyl-bonded silica.

In a comparative study [23] of silica CEC column packing methods, a large number of 50-μm-i.d. fused-silica capillaries were packed with 3-μm ODS-bonded silica from the same batch, by four methods: supercritical fluid or liquid (acetonitrile) carrier, each with and without ultrasonic vibration of the column during packing. The resulting columns were evaluated by measuring the EOF velocity, migration time, efficiency, and retention factors for constituents of a neutral test mixture with identical conditions. The results of this careful comparison revealed that there were no significant differences between the properties of columns packed by different methods or in the success rate of separation; mean values of N of $\geq 2 \times 10^5$ plates/m were observed for columns packed by all four methods. Differences in column performance were found to originate in random variations between replicate columns, and not between packing methods.

In another study in which packing methods were compared by preparing columns with the same bonded silica material, Colon et al. determined [16] retention factors and efficiencies for duplicate columns packed electrokinetically, centripetally, and using pressure (liquid and supercritical carbon dioxide slurries). While electrokinetic packing was found to be the simplest, centripetal and supercritical CO_2 packing were also found to give good efficiencies. Small-particle-size materials may, however, be handled more easily by electrokinetic, centripetal, and gravity packing.

2.2. Column Frits and Alternatives

Unless confined, silica particles in a CEC column are likely to migrate under the influence of the applied electric field. Retaining frits are therefore usually manufactured at the beginning and end of the packed bed by sintering the silica-column packing under hydrothermal conditions after prolonged washing with acetonitrile/water and then water; frits may be also generated by UV radiation from a plug of sol-gel material containing hydrochloric acid, water, toluene, and a photoinitiator.

The more usual sintering procedure is commonly carried out with the aid of a heated coil [24] (although a portable fiber-optic splicer has also been recommended [21]), but can give rise to a series of problems which make the frit the "Achilles' heel" of column manufacture [16]. Because frit fabrication by heating removes the protective polyimide coating from the fused-silica tubing column body, frit manufacture may not be reproducible, and heating the packing material pyrolytically removes the bonded groups bonded to the silica particles. Changes in zeta potential are therefore likely at the frit, leading to changes in EOF velocity and hence bubble formation, which may severely impede analytical CEC.

A number of approaches have been suggested to minimize bubble formation, including pressurization, degassing, decreasing buffer concentration, and adding surfactant to the mobile phase; among these, pressurization has, in our experience, been found especially effective, although specialized equipment is required. However, the problem can be tackled at the source by two procedures [25]: (a) minimizing the length of the frit to approximately 3 mm, which can be achieved by cutting back the frit and butt-connecting with a PTFE sleeve a length of empty capillary with a detector window formed by removing the polyimide coating—a procedure which also improves significantly the durability of the column; and (b) rebonding stationary phase organic groups onto the frit surface, to overcome the effect of loss of carbon-containing groups during thermal frit formation. The bonded groups are easily restored to the silica frit by treatment with, for example, chlorodimethlyloctadecylsilane, thus minimizing the change in zeta potential.

Attempts have been reported to avoid the necessity for a sintered frit by the use of tapers [26–28] at the capillary end which can retain particles with diameter smaller than the orifice by a so-called [27] keystone effect, thus avoiding the need for a frit (Figure 1). Column durability is then again increased by butt-connecting a length of unpacked capillary on which the detector window is located. The positioning of a single controlled-size bead in the end of the column which is then butt-connected to a piece of empty smaller i.d. has also been found effective [28] (Figure 2).

a) **Preparing the tapers**
A e.g. 60 cm length of a fused silica capillary is sealed at the middle with a microflame torch.
The seal is cut to yield two tapered capillaries.
Non-tapered ends to be coupled are ground plane and smooth with P4000 (wet).
Tapers are ground to form the desired orifice (i.d. 10-50 μm).

b) **Coupling the segments dead volume free**
The ground ends are carefully alligned and pushed together.
The dual PTFE/FEP-connector is shrunk onto the junction.

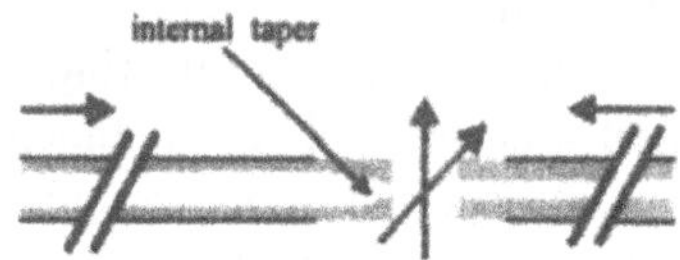

c) **Slurry preparation**
10 -20 mg beads are ultrasonicated for 20 minutes in 70 - 150 μl acetone (or methanol).

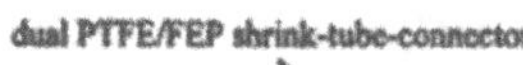

d) **Packing the capillary**
The slurry is flushed in.

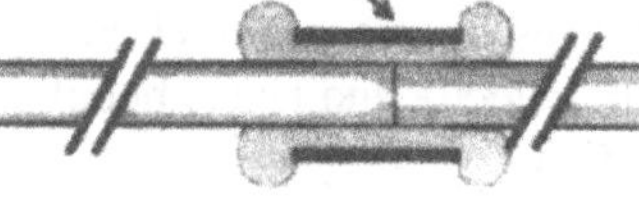

The stationary phase beads are allowed to settle under ultrasonication.
Pressure drop to zero over a period of 20 min.
The capillary is flushed with water

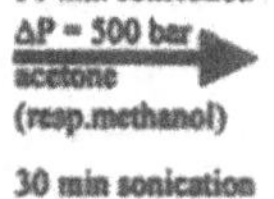

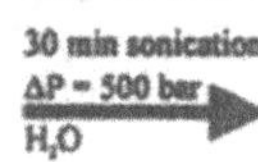

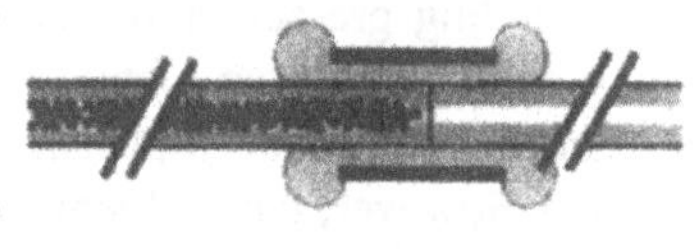

e) **Frit sintering (T=500°C)**

f) **Burning the detection window**

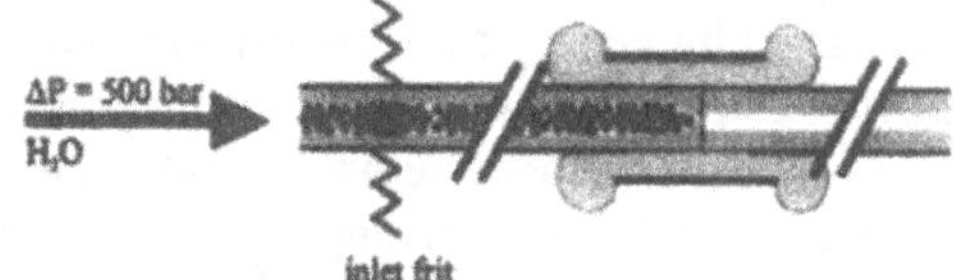

g) **Conditioning**
Column is flushed for 20 minutes with mobile phase (Δ P =150 bar) followed by elektrokinetic conditioning: 45 min. at 10 kV with a 25 min. voltage - ramp
45 min. at 15 kV with a 5 min. voltlage - ramp

h) **Storage**
Column is flushed for 30 min with iso-propanol (Δ P =150 bar)
Capillary is stored with both ends immersed in iso-propanol filled vials

i) **Single - Frit Column**

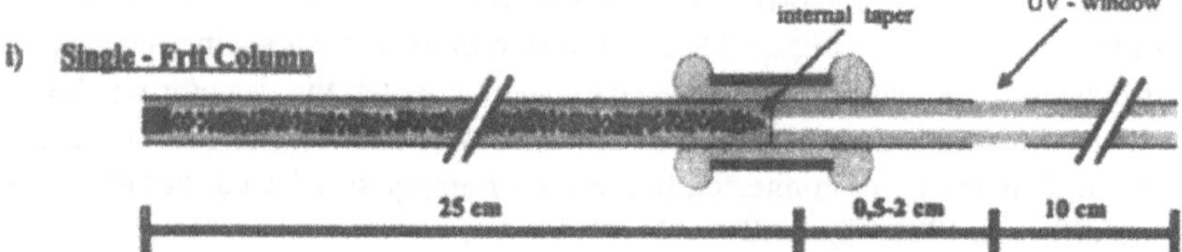

FIGURE 1 An alternative to frit formation in CEC columns: use of the "keystone" effect.

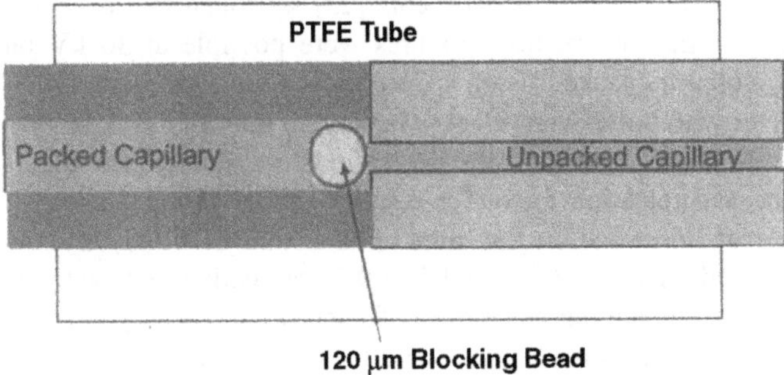

FIGURE 2 Use of a single controlled size bead as an alternative to frit formation.

2.3. Column Repeatability and Reproducibility

Extensive experience of the preparation of bonded-silica columns for CEC has suggested to us that all packing methods have similar problems associated with the fragility of the fused-silica tubing material, especially at the frit and detection window, where the polyimide coating had been removed. A less commonly met effect is the production of loosely packed silica particle beds with voids, giving rise to unstable currents. Approximately 70% of capillaries could be successfully packed by an experienced operator, with approximately 80% of packed and tested columns giving satisfactory CEC [23].

Reproducibility during column preparation is a significant problem in CEC. Preparation methods involving pumped slurries were all found to produce generally highly efficient (> 200,000 plates/m for 3-μm ODS particles), but within a batch of columns packed by the same method, the relative standard deviation (RSD) of EOF, and migration time and retention factor of a standard were [23], respectively, 7–14%, 5–22%, and 9–30%. These values are particularly relevant to considerations of the transfer of HPLC methods to CEC.

If silica CEC columns are subject to some variability during preparation, a given column, once installed, allows highly repeatable separation in the short term. Over a series of 4–10 repeat injections on to an ODS-silica column, test mixture retention times varied [22] by only 0.1–0.2% RSD at voltages between 10 and 30 kV. In the same tests, detected peak heights and areas varied only by approximately 1% RSD. Excellent linearity of peak areas and heights as a function of concentration was also observed.

2.4. Column Lifetime and Robustness

Bonded-silica CEC capillary lifetimes of >1000 consecutive analyses have been reported [29]. In our preliminary work [22] we found that columns could be used over periods of months. In later experiments [23] we found that over 300 consecutive analyses of the neutral test mixtures were possible at 30 kV on 25-cm × 50-μm-i.d. columns packed using supercritical CO_2 with 3-μm ODS-silica particle diameter with buffer vials changed every 10 runs and sample vials every 50 runs. Over 30 consecutive runs, the percent relative standard deviations of the retention time and retention factor for benzophenone were 1.7 and 2.5. Over a longer series of injections on the same column, the EOF velocity and retention factors showed small (~2% and 1.5–3.0%, respectively) systematic increases, with small accompanying changes in efficiency and peak symmetry.

Column lifetimes were, however, much shorter in tests in which a "real" sample of natural-product origin was investigated [28]. A supercritical carbon dioxide extract of red peppers was analyzed by CEC for its content of capsacain and its derivatives. Approximately 30 consecutive injections were made into a column similar to that above, and operated under the same conditions. Beyond this number there was substantial degradation of column performance. Partial restoration of EOF velocity and column efficiency was achieved by flushing the column with buffer for up to 2 h, but it is clear that the small mass of stationary phase in 50–100-μm-i.d. CEC columns imposes stringent sample clean-up requirements if analytical capability is to be maintained in natural-product chemistry. Similarly, if CEC is to play a major role in bioanalysis [29], e.g., in the separation of complex mixtures of peptides from combinatorial synthesis or derived enzymatically from proteins, the problems of sample clean-up must be addressed [5], along with the reproducibility and precision of high-throughput analysis.

3. INFLUENCE OF COLUMN PACKING MATERIAL ON EOF AND COLUMN EFFICIENCY

3.1. Particle Size and Distribution

The majority of CEC separations reported to date have been carried out on silica stationary phases that were originally developed for HPLC. Large differences have been observed [30] in using these HPLC phases in CEC. The differences seen in the separations of polycyclic aromatic hydrocarbon mixtures are far greater in CEC than would be expected for the corresponding HPLC separation. Two main reasons for this have been put forward: (a) the differences in the packing of the capillary columns for CEC; and (b) the particle size distribution of the materials, although all were nominally defined as 3 μm.

The particle size distribution of packing materials has been shown to vary not only from one manufacturer to another but also with the measurement method used. For HPLC specifications, it is normal for manufacturers to give a particle size distribution based on an area distribution. However, this distribution has been pointed out [30] not to give a true distribution of the fine material below 1.5 μm, and it is this material that can cause difficulties in the packing of capillary columns. The particle size distribution problem of the silica particles was also pointed out by Knox [15], who argued that better particle fractionation or the use of monodisperse particles is required for the production of reproducible capillary columns.

In early theoretical work, Knox and Grant [31] showed how very high efficiencies could be obtained in CEC if submicrometer particles were used. To date, however, there are still conflicting reports on the applications of small monodisperse solid silicas and small nonmonodisperse <2-μm porous particles in CEC. In our laboratory we have been unsuccessful in the packing of capillary columns using such small particles, and problems in the agglomeration of packing materials and discontinuities in the packed bed have always been evident. These problems lead to poor column durability and performance. However, monodisperse stationary phases in the range of 0.2–0.3 μm diameter have been successfully packed by Adam et al. [32]. Regardless, we are still left with the very interesting observation [33] that better efficiencies were observed for CEC on very small internal-diameter columns when 3-μm ODS-bonded silica particles were packed into 30-μm-i.d. fused silica and gave reduced plate heights below 1.

3.2. Particle Porosity

The pores within the HPLC particles used in CEC have a distribution in the range of 8–10 nm. This will not in general support EOF, owing to the double-layer overlap that occurs in these narrow pores, and in CEC using these particles flow is assumed to occur in the interstitial region of the space between particles. However, if the packing material contains fine particles of the order of <2 μm, then these may pack into the interstitial space and so in turn cause double-layer overlap and so prevent EOF.

Furthermore, this again turns on how well the small-diameter capillary columns can be packed and relates to the observations outlined in the previous section. Intraparticulate EOF has advantages over interparticle flow from the standpoint of efficiency, as it would be expected that the plate height contributions from eddy diffusion and slow mass transfer in the mobile phase will be less when the pores are large enough to support EOF. If EOF occurs in the pores, then mass transfer effects are also minimized as the pores support flow. To investigate this, the use of wide-pore material has been studied by Li and Remcho [34], who evaluated 7-μm-diameter silica particles with pore diameters up to 400 nm. They found that for such material, substantial reductions in the plate height, H, were observed when using buffer concentrations from 10 to 50 mM, but no plate height reductions were observed when using buffer concentrations from 1 to 10 mM. These observations are consistent with the theory of

Rice and Whitehead [35] and show that as the double-layer thickness is reduced, the perfusive character of the particles is increased and hence the reduction in *H*. However, these results have often been ignored, and to the authors' knowledge no new CEC phases have been developed using larger particles that afford better and more straightforward packing with large pores to give the advantage of lower plate heights.

3.3. Bonded Phases for CEC

For use in reverse-phase-type CEC separations it is important to have both charged groups, in the case of silica, silanols, and traditional surface functionality such as C_4–C_{18}, but results from the HPLC-type supports used in CEC can show poor separation efficiencies due mainly to the flow-rate dependence on pH and the instability of silica at high pH, and the lack of a charged surface at low pH. The use of "total endcapping" or polymer encapsulation as employed for HPLC supports cannot be used for CEC, as there is a need for a charged surface. Work has been published [36,37] on combining ion-exchange groups with C_4–C_{18} ligands to form mixed-mode types of materials. With the presence of strong ion-exchange groups on the surface of the silica, the EOF was reported [28] to remain constant over a wide pH range, but problems of silica dissolution and analyte adsorption due to the presence of the ion-exchange sites still remain and do not allow these materials to have general applicability. A novel approach has been developed [38] that deactivates the charged silanols but yet retains a strong and stable EOF over a wide pH range by utilizing a transition-metal complex into a polysiloxane backbone and coating this onto a silica support. The positively charged transition metal is surrounded by a cyclic ligand that has been designed to overcome analyte adsorption associated with the normal ion-exchange materials. Using the principle of this phase could lead to the development of a new range of bondings that in turn could extend the scope and reproducibility of packed-column CEC.

3.4. Porous Monoliths

Apart from the problems outlined in the previous sections on the packing of small particles into fused silica tubes suitable for CEC columns, there is the added problem of frit formation (Section 2.2), and detection window formation. A way to circumvent these problems is to use a continuous polymer bed or monolith. The EOF, and the bulk flow, is generated in the network of pores, the size, and to some extent the shape of which can be controlled by the reaction conditions of the polymerization. The advantage is therefore that no frits are required and so the problems of heating and destabilization of the fused silica at the frit are removed. For CEC, three types of continuous polymer monoliths have been reported: (a) organic; (b) inorganic; and (c) hybrid. An extensive review of monolithic stationary phases for liquid and capillary electrochromatography has recently been published by Zou et al. [39]. The most commonly reported monoliths for CEC are the organic type, and these in general are prepared by polymerizing organic monomers in the presence of a porogen [40,41]. Three types of polymers have been used in the CEC monoliths: acrylamides, methacrylate esters and styrenes, of which the acrylamides seem to be the most common. However, in all cases the methods of manufacture are very similar in that first the fused silica column is pretreated, either by acid/alkali washing or by silanization; then the porous monolith is prepared in the fused silica capillary; and then finally any additional functionality is bonded onto the matrix. It is at this later stage where the differences in the manufacturing of monoliths for HPLC and CEC use occur. However, in none of the reviewed papers do any of the authors discuss how the individual column monoliths are quality-controlled with respect to the total surface area, pore size, and volume. Claessens et al. [42] discussed the reproducibility and stability of their monoliths manufactured from acrylates. Although they argued that the reproducibility of both the EOF and the retention factor is satisfactory, they stated that for routine analysis improvements must be made to decrease the variation in plate numbers from one monolith to another. They did not measure the individual surface areas or the pore sizes on the individual monoliths.

Figure 3 shows a styrene monolith made in our laboratory, with the corresponding pore distribution.

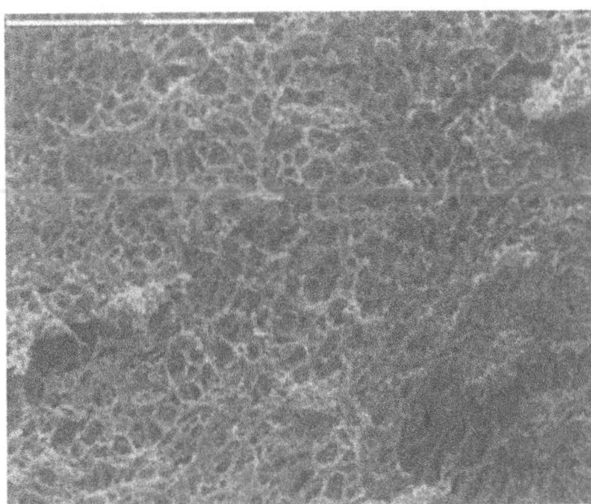

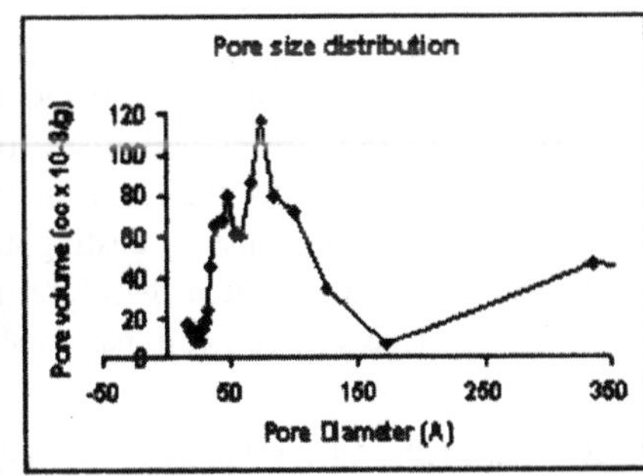

FIGURE 3 (a) Micrograph of a styrene monolithic packing with (b) the corresponding pore distribution.

Inorganic monoliths have typically been manufactured from sol-gel-type reactions involving the polycondensation of alkoxysilanes in the presence of an organic polymer. The porosity and pore size are controlled by the percentage of organic polymer in the mixture. The monolith matrix is then usually derivatized by common reversed-phase silanes and in some cases the coupling of charged groups to enhance the EOF [43]. Another method of manufacturing the inorganic monoliths has been the manufacture of particle-fixed monoliths. These are prepared by fixing conventional spherical packing materials in the fused silica capillary by sintering the particles, once packed into a bed [44], by a thermal treatment. Particle-entrapping [45] monoliths are prepared by introducing an entrapping solution into the column after it has been packed with silica particles. The work of Lee [46] showed how particle-entrapped columns could be made by packing the columns with supercritical CO_2 and entrapping by the hydrolysis and polycondensation of a methoxy silane. All these workers have shown how the entrapping changes the retention characteristics of the entrapped column compared to the packed column. Particle-loaded monolithic columns [47] are very similar to particle-entrapped columns except that the packing of particles and entrapping occurs in a single step. However, the results from this type of monolith have been disappointing [48].

Monolithic columns have different properties and perhaps some advantages over packed columns for CEC, but to date the disadvantages outweigh the advantages. For the most common organic polymers, most of the derived monoliths are known to swell in organic solvents, and this can lead to instability. The derived monoliths have a low inherent surface area compared to packed columns, which of course leads to a low column capacity. However, the main problem with all these monoliths is in the column-to-column reproducibility. In packed columns the particles used can be well characterized before being packed into a column; i.e., the particle size, surface area, pore size, and pore volume are controlled in large batches. Hence, from a 1-kg batch of 3-μm silica, thousands of CEC columns can be made, all with the same physical particle characteristics. The only difference arises from the packing procedure, and this variation can easily be checked from the efficiency of the column. With monolithic columns a batch is made by casting a 1-m column, and this is then cut into 4 × 25-cm lengths. One batch therefore provides four columns, and for controlled work the physical specifications of these must be known. These data are lacking.

4. INFLUENCE OF MOBILE-PHASE PROPERTIES

The dependence of EOF mobility, μ_{EOF}, on mobile-phase properties has been explored theoretically, and confirmed in a number of experiments on silica and bonded silica columns [49]. Comparisons have also been made with the results of corresponding experiments in open-tubular capillary electrophoresis (CE) columns under similar conditions [49].

4.1. pH

The well-known fall-off in CE of μ_{EOF} with decreasing pH consequent on the decrease in silanol group ionization was also observed in CEC (Figure 4), although the EOF was now greater. The zeta potential, ζ, is proportional to the surface charge (σ) and to the thickness, δ, of the layer of counterions near the surface of the silica particle via [50]

$$\zeta = \frac{\sigma}{\varepsilon_o} \frac{\delta}{\varepsilon_r} \tag{3}$$

where ε_o represents the permittivity of a vacuum, and ε_r is the dielectric constant. The EOF mobility depends [51] on ζ according to

$$\mu_{EOF} = \frac{\varepsilon_o \varepsilon_r \zeta}{\eta} \tag{4}$$

where η is electrolyte viscosity. It follows that a reduction in σ as Si–OH ionization is suppressed leads to a reduction in μ_{EOF}. This effect has been recorded many times for ODS-bonded silica, but is also observed [33] for *t*-butyl and cyano-bonded silica packings (Figure 4). For both the latter phases, μ_{EOF} was generally greater than for ODS, a possible consequence of lower concentrations of bonded groups.

If, however, the group bonded to silica carries a negative charge, as does the sulphonic acid group of strong-cation exchange (SCX) phases, the effect of

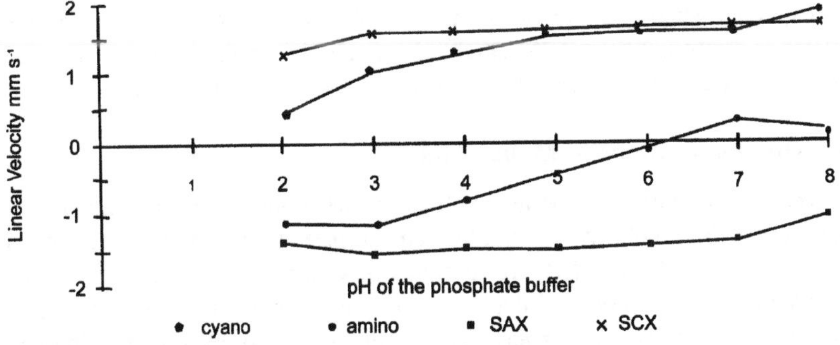

FIGURE 4 Variation of EOF velocity with mobile phase pH for silica packings with different bonded groups.

pH on EOF velocity is much reduced, as the SO_3 groups remain charged even at high hydrogen-ion concentrations. Correspondingly, a strong anion-exchange (SAX) silica packing has reversed EOF which is virtually independent of pH. On aminopropyl-bonded silica, the EOF is reversed with changing pH, offering the possibility, by changing pH, of augmenting a separation based on partition by an electromigration either with or against the EOF.

4.2. Ionic Strength

The effect of ionic strength, I, on the magnitude of the EOF can be predicted from the influence of I on δ, given [52] by

$$\delta = \left(\frac{\varepsilon_o \varepsilon_r RT}{2F^2 I}\right)^{0.5} \tag{5}$$

where F, R, and T are the Faraday constant, universal gas constant, and absolute temperature, respectively. Because of the dependence of zeta potential and hence EOF mobility on δ, it follows that a graph of μ_{EOF} against $I^{-0.5}$ should be linear. This has been confirmed in experiments [49] over the range 1–20 mM phosphate in acetonitrile/water solution for both open-tubular (CE) and CEC columns packed with ODS-bonded silica. The increase of μ_{EOF} with $I^{-0.5}$ was similar for both CE and CEC at higher ionic strengths, but below 5 mM phosphate concentration the CEC mobility leveled off (Figure 5) and began to drop for silica packings with pore size 8 nm, while continuing to rise in CE. This difference is in accordance with theories of double-layer overlap which predict that, for plug flow in a capillary channel, the diameter d would need to far exceed δ. Flow profiles calculated [35] by Rice and Whitehead suggest a decline in EOF when $d/\delta \leq 50$ and particularly when $d/\delta \leq 10$. For a 40% reduction in EOF velocity over that in a tube with infinite d, considered [31] to be an acceptable limit, d must exceed 10δ. In a packed tube a significant loss in EOF velocity is expected when $d_p \leq 40\delta$, since the ratio of mean channel diameter/particle diameter is approximately 0.25. Double-layer overlap within pores of a silica packing material should occur at $I \leq 2.5$ mM, where δ is about 4 nm.

4.3. Organic Content of Mobile Phase

The type and proportion of organic solvent in the mobile phase is predicted to influence the EOF mobility through the ratio of permittivity to viscosity ε_r/η in Eq. (3). Typical values of ε_r/η for mixtures of water with a variety of organic solvents are listed in Table 1, calculated with the assumption of no contribution from the buffer component. Experimental findings [49] for experiments in open tubes are in good agreement, with acetone, acetonitrile, and methanol all showing an EOF minimum around 50–70% organic, although the increase observed

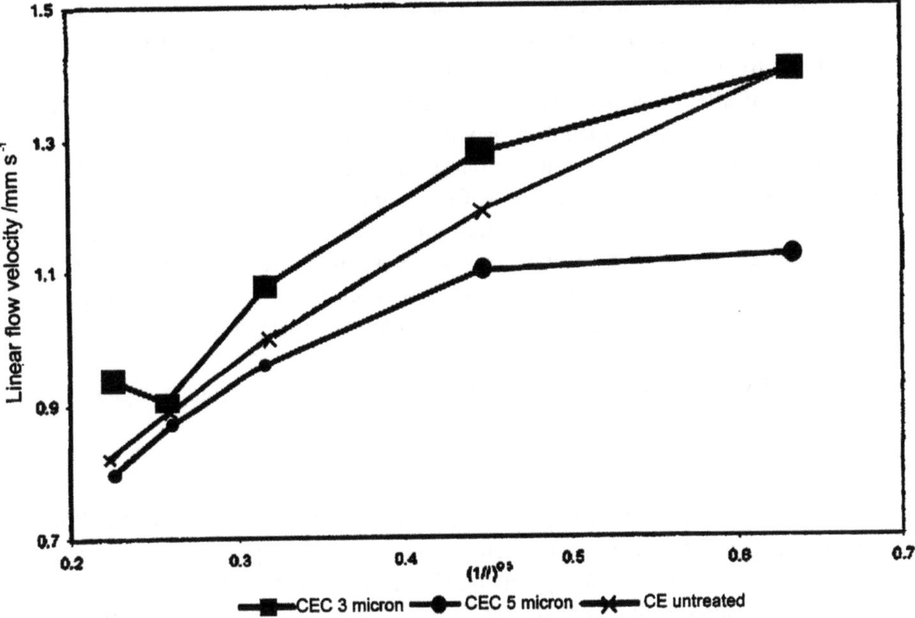

FIGURE 5 Variation of EOF velocity with ionic strength in CEC and CE.

when using a higher acetonitrile content was far greater than that anticipated. Wright et al. also witnessed [53] this behavior for acetonitrile/water systems without supporting electrolyte, and explained it by changes in solvent polarity and hydrogen-bond donor ability. Acetonitrile/buffer is generally selected as the mobile phase in CEC. However, many recommended reversed-phase HPLC

TABLE 1 Approximate ε_r/η Ratios for Binary Mixtures with Water at 25°C

	ε_r/η (cP^{-1}) for varying % organic content		
Solvent	0%	50%	100%
Acetone	88	23	68
Acetonitrile	88	75	105
Methanol	88	37	60
2-Propanol	88	15	7

Source: From Ref. 49.

methods employ methanol/water as a mobile phase, and the reduced EOF velocity consequent on the much smaller values of ε_r/η for this solvent system means that direct transfer of HPLC to CEC may not be possible.

Use of alternative solvents such as acetonitrile/buffer may, however, be possible. In HPLC the principle of isoeluotropy is well known—the reproduction of retention behavior for mobile phases of similar eluotropic strengths. Isoeluotropy has been demonstrated [54] for neutral solutes in CEC on ODS-bonded silica. As discussed above, retention times are much longer for a methanol/buffer (80/20) mobile phase than for an acetonitrile/buffer (70/30) which has similar eluotropic strength, but retention factors are much more similar (Figure 6). Deviations from unit gradient may arise from the choice of thiourea as retention marker.

For HPLC the retention factor, k, is related to the percentage organic in the aqueous mobile phase, p, via

$$\ln k = \ln k_o - ap \tag{6}$$

where k_o is the retention factor for water eluent, and a is a constant. In CEC experiments, as the percentage of acetonitrile in the mobile phase was increased, there was a linear fall in ln k, suggesting [54] that well-established theories used in HPLC method development should be equally applicable to the separation of neutral molecules in CEC.

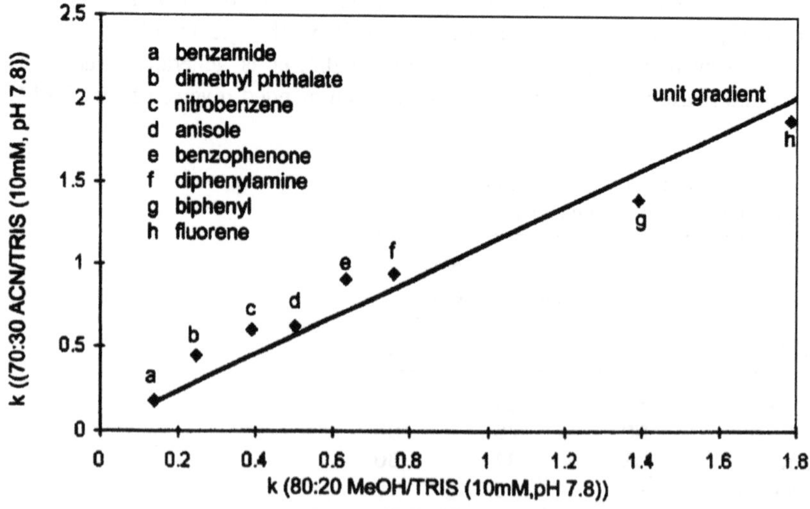

FIGURE 6 Relation of retention factors (k) in CEC of model compounds on ODS bonded silica with mobile phases of the same eluotropic strength.

4.4. Column Temperature

The small heat capacity of silica CEC columns means that column temperature is easily changed. Temperature influences EOF velocity through its effect on zeta potential and mobile-phase viscosity [Eq. (4)]. Increased temperature reduces η via an exponential relation:

$$\eta \propto \exp\left(\frac{\text{constant}}{\text{RT}}\right) \tag{7}$$

ζ depends directly on temperature [Eqs. (3) and (5)], and also indirectly through the temperature dependence of relative permittivity:

$$\varepsilon_r = a + bT + cT^2 \tag{8}$$

where a, b, and c are constants.

These complex and interrelated factors result in an increase of approximately 60% in EOF velocity between 10 and 60°C on a C_{18} silica column with acetonitrile/water mobile phase, with the temperature dependence of viscosity probably dominant. Over this limited temperature range, and perhaps surprisingly, μ_{EOF} was observed [55] to be related to T via linear graphs of μ_{EOF} against $T^{1/2}$, T^{-1}, and even T, as well as ln μ_{EOF} against T^{-1}. The practical result is shorter analysis times at higher temperatures as analyte retention times fall linearly [22] with increasing T.

Retention factors in CEC are also reduced by increasing column temperature because of increased partition into the mobile phase; van't Hoff plots of ln k versus T^{-1} are generally [7,55] linear, and the slopes of such plots may differ sufficiently for column selectivity to be changed by temperature variation. For example, in the CEC of a number of diuretic drugs on ODS-bonded silica at temperatures between 15 and 60°C, the resolution of chlorothiazide and hydrochlorothiazide increases [7] with decreasing temperature, and the relative retention of chlorothalidone and hydroflumethazide is reversed with increasing temperature variations of k with temperature, which may make [56] temperature programming a useful technique in CEC.

Van't Hoff plots may be used to compare the entropies of solute transfer, ΔS, between mobile and stationary phases in CEC and (pressure-driven) micro-HPLC on the same column. Djordjevic et al. found [57] that ΔS was more negative for electrodrive, a difference attributed to Joule heating, which was thought to bring about differences between set and actual column temperatures.

Jiskra et al. preferred [55] the alternative explanation that in CEC, (a) EOF on a silica support increases the ordering of the hydrocarbon chains of the stationary phase and thus increases ordering of the solutes during interaction with the stationary phase, and (b) the mobile phase becomes more ordered under the influence of the applied electric field, and is thus more disturbed by the cavity containing a solute molecule. They pointed out that the latter effect would also explain the higher ΔS values in CEC for polar compounds, as such molecules penetrate into the mobile phase through, for example, hydrogen bonding. Increased organization of stationary-phase hydrocarbon chains also confirms more negative values of the enthalpy change for transfer between the phases in CEC and thus better contact of the solutes with the stationary phase.

5. CEC APPLICATIONS ON SILICA COLUMNS

CEC suffers from the classical "Catch 22" situation in that, to develop applications outside of a university environment, industry requires purpose-designed CEC instruments that are capable of performing isocratic and gradient CEC. However, instrument manufacturers will not develop such instrumentation until they see clear advantages in the technique or that there is a clear demand for such instruments. So, do we have a unique technique that can offer some orthogonality to HPLC, can give rapid separations with increased peak capacity over HPLC, but is going nowhere? A review of the literature shows that most of the reported applications copy those of HPLC but for a very much reduced subset mainly involving the analysis of neutral analytes over a narrow range of pH 7–9.0. The most extensive review of real applications has been presented by Sandra [58]. Applications are shown for the separation of hydrophobic samples including carotenoids and tocopherols, triglycerides, vegetable oils, and margarines [59,60]. The use of CEC for these samples was based on the increased efficiencies that could be achieved with CEC over the conventional method of analysis using HPLC. However, no optimization was carried out to use the electrophoretic mobility of any charged species to enhance the selectivity.

The analysis of acids and bases has been investigated and reported by a number of workers [61,62]. For acids a mobile phase has to be used that will separate the acids in their ion-suppressed mode, i.e., two pH units below their pK_a. The disadvantage is the reduced EOF because of the suppression of the silanol ionization on the silica surface. In an attempt to maintain an EOF above 1.0 mm/s, the use of mixed-mode reversed-phase CEC has been developed incorporating sulphonic acids. For basic compounds, the same approach has been taken in analyzing these analytes in the ion-suppressed mode. For silica-based supports this has the problem of silica instability at high pH. However, success has been reported for basic drugs with pK_a of 8 at a pH of 9.3 on silica-based columns [61]. An extended discussion of the use of CEC and applications in the pharmaceutical industry has been given by Euerby and Gillott [63].

6. CONCLUSIONS AND POTENTIAL OF SILICA COLUMNS IN CEC

The driving force behind the application of CEC so far has been theoretical plate number, N. The classic separation [64] of tipredane diastereoisomers by high-resolution CEC was an early result of the availability of 3-μm-diameter bonded-silica-particle packed columns yielding up to 100,000 plates. With such high efficiencies, CEC should result in real advantages in terms of the column peak capacity, P, the number of peaks which can be separated in a chromatogram between realistic retention factor limits:

$$P = 1 + \frac{N^{1/2}}{4} \ln(1 + k) \tag{9}$$

Even though peak capacity may represent an overestimate of resolution when applied to real mixtures, it still allows meaningful comparison of CEC with HPLC. Very substantial improvements over HPLC are possible in CEC with 3-μm particles, and more especially if the full capability of, say, 1.5-μm-particle columns can be realized. Values of N of 2×10^5 are then expected, and for k = 10, P is now >250, taking CEC into a resolution domain which conventional HPLC cannot approach. In fact, most HPLC separations are generally achieved on the basis of selectivity, but the higher plate numbers available in CEC on columns with similar packings may offer substantial advantages for very complex mixtures of natural products and of biological compounds, as has been described in Section 5. There is an analogy here with the progress of gas chromatography, where the introduction of fused-silica capillary columns offered the resolution necessary to make routine the analysis of complex mixtures of fossil fuels and environmental origins.

The above discussion is based on silica column technology which is either available or almost available. In fact, in the early days of CEC, Knox and Grant deduced [31] that for typical mobile-phase ionic strengths between 1 and 10 mM, columns with particle diameters as small as 0.5 μm would not result in significant loss of EOF velocity; if such columns could be satisfactorily packed, plate numbers in excess of 5×10^5 and peak capacities beyond 400 should be possible. A more recent analysis by Knox [65], using considerations of double-layer overlap and a version of the Knox equation originally proposed for the plate height (H) in HPLC, also suggests that the diffusion-limited minimum value of H is obtained when particle diameters are ≤1 μm. As pointed out in Section 2, however, such materials must be as nearly monodisperse as possible if columns are to pack satisfactorily to yield plate numbers above 5×10^5.

Luo and Andrade have reexamined [66] the potential of CEC by comparing the effect of the conclusions of the Rice-Whitehead theory [35] of double-layer overlap on the determination of minimum d_p with those which result from more recent treatments of the velocity profile in electroosmotic flow. They concluded that, for ionic strength <10 mM, the particle size can again be less than 1 μm, and that plate numbers up to 1×10^6 should be theoretically possible. An obstacle to the realization of such efficiencies in CEC is, however, the consequence of the recognition [6] by Giddings that there is no satisfactory mathematical description of pore structure in particle packings. Accordingly, Luo and Andrade concluded [67] that the full potential of CEC would not be realized for randomly packed particle beds.

Clearly, at the time of writing, much research and development is necessary before the column technology will be available which will allow CEC to operate to its full potential. Unfortunately, the applications of CEC so far demonstrated have been sufficient to persuade only a few separation scientists to use it routinely, and extension of the technique is slow. However, as far as silica-based columns are concerned, enough is now known to prompt the manufacture of packings tailored specifically to take advantage of the unique combination in CEC of separation by partition and electromigration, and to adjust EOF generation and separation properties independently; problems of column durability and robustness need to be addressed, while column frits and couplings need more attention. Only then is CEC likely to have a significant impact in analysis.

ACKNOWLEDGMENTS

The assistance of Richard Carney, Maria Cikalo, Mark Robson, Stéphanie Roulin and Katherine Sealey in the preparation of this chapter is gratefully acknowledged.

REFERENCES

1. Dittmann MN, Rozing GP. In: Capillary Electrochromatography. Bartle KD, Myers P, eds. Cambridge, UK: Royal Society of Chemistry, 2001: Chap. 5. pp 64–86.
2. Bartle KD, Myers P, eds. Capillary Electrochromatography. Cambridge, UK: Royal Society of Chemistry, 2001.
3. Krull IS, Stevenson K, Mistry K, Swartz ME, eds. Capillary Electrochromatography and Pressurised Flow Capillary Electrochromatography. New York: HNB Publishing, 2000.
4. Deyl Z, Svec F, eds. Capillary Electrochromatography. Amsterdam: Elsevier, 2001.
5. Unger KK, Huber M, Walhagen K, Hennessy TP, Hearn MTW. Anal. Chem. 2002; 74:201A.
6. Giddings JC. Unified Separation Science. New York: Wiley, 1991.
7. Bartle KD, Myers P. J Chromatogr A 2001; 916:3.
8. Golay MJE. In: Desty DH, ed. Gas Chromatography. London: Butterworths, 1958.
9. Bruin GJM, Tock PPH, Kraak JC, Poppe H. J Chromatogr 1990; 517:557.
10. Horvath CS, Lin H-J. J Chromatogr 1976; 126:401.
11. Horvath CS, Lin H-J. J Chromatogr 1978; 149:43.
12. Dittmann MM, Wienand K, Bek F, Rozing GP. LC-GC 1995; 13:800.
13. Tallarek U, Rapp E, Scheenan T, Bayer E, Van As E. Anal Chem 2000; 72:2292.
14. Wen E, Asiaie R, Horvath CS. J Chromatogr A 1999; 855:349.
15. Knox JH. J Chromatogr A 1999; 831:3.

16. Colon LA, Maloney TD, Fermier AM. J Chromatogr A 2000; 887:43.
17. Tan C. US Patent S453163, 1995.
18. Fermier AM, Colon LA. J Microcolumn Sep 1998; 10:439.
19. Reynolds KJ, Maloney TD, Fermier AM, Coton LA. Analyst 1998; 123:1493.
20. Boughtflower RJ, Underwood T, Paterson CT. Chromatographia 1995; 40:313.
21. Angus PDA, Demarest CS, Catalano T, Stobaugh JE. J Chromatogr A 2000; 887: 347.
22. Robson MM, Roulin S, Raynor MW, Shariff SM, Bartle KD, Clifford AA, Myers P, Euerby MR, Johnson CM. Chromatographia 1996; 43:313.
23. Roulin S, Dmoch R, Carney R, Bartle KD, Myers P, Euerby MR, Johnson C. J Chromatogr A 2000; 887:307.
24. Boughtflower JR, Underwood T, Paterson CJ. Chromatographia 1995; 40:329.
25. Carney RA, Robson MM, Bartle KD, Myers P. J High Resolut Chromatogr 1999; 22:29.
26. Lord GA, Gordon DB, Myers P, King BW. J Chromatogr A 1997; 768:9.
27. Rapp E, Bayer E. J Chromatogr A 2000; 887:367.
28. Bartle KD, Carney RA, Cavazza A, Cikalo MG, Myers P, Robson MM, Roulin SCP, Sealey K. J Chromatogr A 2000; 892:279.
29. Walhagen K, Unger KK, Hearn MTW. J Chromatogr A 2000; 894:35.
30. Myers P. In: Bartle KD, Myers P, eds. Capillary Electrochromatography. Cambridge, UK: Royal Society of Chemistry, 2001: Chap. 3. pp 33–41.
31. Knox JH, Grant IH. Chromatographia 1987; 24:135.
32. Adam Th, Ludtke S, Unger KK. Chromatographia 1999; 49:S49.
33. Sealey K, Bartle KD, Myers P. Unpublished measurements, 2001.
34. Li D, Remcho VT. J Microcolumn Sep 1997; 9:389.
35. Rice CL, Whitehead R. J Phys Chem 1965; 69:4017.
36. Yang C, El Rassi Z. Electrophoresis 2000; 21:1977.
37. Smith N, Evans MB. J Chromatogra 1999; A 887:41.
38. Lee M. J Chromatogra Library 2001; 62: Chap. 6.
39. Zou H, Huang X, Ye M, Luo Q. J Chromatogra 2002; A 954:5–32.
40. Svec F, Peters EC, Sykora D, Frechet JMJ. J High Resolut Chromatogra 2000; 23:3.
41. Svec F. J Chromatogra Library 2001; 62: Chap. 6.
42. Jiang T, Jiskra J, Claessens HA, Cramers CA. J Chromatogra 2001; A 923:215–227.
43. Ishizuka N, Minakuchi H, Natanishi K, Soga N, Nagayma H, Hosoya K, Tanaka N. Anal Chem 2000; 72:1275.
44. Wistuba D, Schurig V. Electrophoresis 2000; 21:3152.
45. Chirica G, Remcho VT. Electrophoresis 1999; 20:50.
46. Tang Q, Lee ML. J Chromatogra 2000; A 887:265.
47. Ratnayake CK, Oh CS, Henry MP. J High Resolut Chromatogr 2000; 23:81.
48. Kato M, Dulay MT, Benett BD, Chen J, Zare RN. Electrophoresis 2000; 21:3145.
49. Cikalo MG, Bartle KD, Myers P. J Chromatogr A 1999; 836:35.
50. Shaw DJ. Electrophoresis. London: Academic Press, 1969.
51. Knox JH. Chromatographis 1988; 26:329.
52. Hunter RJ. Zeta Potential in Colloid Science. London: Academic Press, 1981.
53. Wright PB, Lister AS, Dorsey JG. Anal Chem 1997; 69.

54. Euerby MR, Gilligan D, Johnson CM, Roulin SCP, Myers P, Bartle KD. J Microcolumn Sep 1997; 9:373.
55. Jiskra J, Claessens HA, Cramers CA. J Sep Sci 2002; 25:569.
56. Djordjevic NM, Fitzpatrick F, Houdiere F, Lerch G, Rozing GP. J Chromatogr A 2000; 887:245.
57. Djordjevic NM, Fowler PWJ, Houdiere F, Lerch G. J Liq Chromatogr 1998;21: 2219.
58. Dermaux A, Sandra P. In: Bartle KD, Myers P, eds. Capillary Electrochromatography. Cambridge, UK: Royal Society of Chemistry, 2001: Chap. 8. pp 125–145.
59. Ferraz V. Doctoral dissertation, University of Ghent, 1995.
60. Ross G, Dittmann MM, Bek F, Rozing GP. Int Lab 1996; 5:10A.
61. Gillott NC, Euerby MR, Johnson CM, Barrett DA, Shaw PN. Anal Commun 1998; 35:217.
62. Cikalo MG, Bartle KD, Robson MM, Myers P, Euerby MR. Analyst 1998; 123: 87T.
63. Euerby MR, Gillott NC. In: r Bartle KD, Myers P, eds. Capillary Electrochromatography. Cambridge, UK: Royal Society of Chemistry, 2001: Chap. 7. pp 107–124.
64. Euerby MR, Johnson CJ, Bartle KD, Myers P, Roulin SCP. Anal Commun 1996; 33:403.
65. Knox JH. Abstr Analytica 2000, Munich, April 2000.
66. Luo A-L, Andrade JD. J Microcolumn Sep 1999; 11:682.

8

Effects of Pore Flow on Separation Efficiency in Capillary Electrochromatography with Porous Particles

Remco Stol
Organon, Oss, The Netherlands

Wim Th. Kok
University of Amsterdam, Amsterdam, The Netherlands

1. INTRODUCTION

High-pressure liquid chromatography (HPLC) has grown into a powerful separation technique routinely applied for the separation of a variety of compounds in a variety of matrices [1]. Despite its success, HPLC may still be regarded as a relatively inefficient technique, certainly when compared to, e.g., gas chromatography (GC) or capillary electrophoresis (CE).

The separation efficiency of a column for liquid chromatography and the relation with the mobile-phase velocity $\langle u \rangle$ can be described by the Van Deemter equation, which in lumped terms reads [2]

$$H = A + \frac{B}{\langle u \rangle} + C_m \langle u \rangle + C_s \langle u \rangle \tag{1}$$

The height equivalent to a theoretical plate H is composed of several independent contributions that can be added linearly. The A term represents the contribution of flow disturbances over the cross section of the column, the B term represents axial diffusion, and the C_m and C_s terms account for the mass transfer kinetics in the mobile and in the stationary phase, respectively. Detailed expressions for the respective C-term coefficients have been derived [1]. The theory has indicated the way to higher separation efficiencies: a downscaling of the diffusional dimensions within the column. Substantiation of this trend in practice can be found in the development of open-tubular LC [3,4], and of ultra-high-pressure HPLC with 1-μm-sized particles [5,6].

A relatively new format of liquid chromatography is capillary electrochromatography (CEC). In CEC the mobile phase is driven by electroosmosis instead of by pressure, i.e., the mechanical pump is replaced by an electrical power source [7–10]. The use of support materials with highly charged surfaces such as silica, and the application of high field strengths (>1000 V/m) ensures the generation of a high electroosmotic flow (EOF). As will be discussed later, the major advantage of CEC is the possibility to perform liquid chromatography (LC) much more efficiently than is possible with a pressure-driven flow.

Since the pioneering work of Knox et al. on CEC [9,10], porous silica particles have been used as the column packing material in the majority of research studies and applications. Porous silica has a number of characteristics that make it suitable for use in CEC. These are a large surface area, a high surface potential at moderate pH values, which allows the generation of a high EOF, and the commercial availability of materials with various surface chemistries. However, other support materials, such as polymeric phases [11] and alternative inorganic base materials [12], are also applicable in CEC.

Under well-chosen conditions a high EOF can develop through very narrow channels, which permits the use of small, submicrometer-sized particles [9,10]. The use of such small particles is expected to lead to a significant reduction of the A and C terms in Eq. (2). Moreover, EOF may be created through the pores of the stationary-phase particles [13–23]. It has been thought that full electrical double-layer overlap occurred within the pore volume of ordinary silica particles, and that intraparticle EOF could be developed only through the pores of macroporous particles [13,16]. However, it has been found that at conditions typical for CEC, that is, at ionic strengths of ~10 mM and with pore sizes of ~10 nm, a high intraparticle EOF may already develop [16]. This intraparticle EOF has important consequences for the separation efficiency that may be achieved in CEC, as will be discussed here in detail.

2. MAGNITUDE OF ELECTROOSMOTIC FLOW THROUGH POROUS MEDIA

2.1. Electrical Double-Layer Overlap and Pore Flow

When a surface is brought into contact with an ionic solution, charge will develop on that surface [24–26]. In the case of a silica surface, which is the base material for most chromatographic media and capillaries used in CEC, the surface gains a negative charge due to deprotonation of acidic silanol groups. This negative charge is compensated for by a small excess of positive counterions in the mobile phase next to the surface. A schematic overview of such an electrical double layer is shown in Figure 1.

When a high voltage is applied parallel to the surface, the (positive) counterions migrate toward the negative electrode. The migrating ions take along their hydration layer and the surrounding solution, and an electroosmotic flow (EOF) is created. The surface charge and with that the electric potential at the shear plane (the zeta potential), the ionic strength, and the dielectric properties of the mobile phase, and the applied electric field strength, determine the magnitude of this EOF. A certain length scale, the so-called Debye length (δ), can characterize the thickness of the double layer. For typical conditions in CEC the Debye length is of the order of 1–10 nm. When the flow channel size is not very large compared to the Debye length, the electrical double layer from the opposing walls overlaps in the center of the channel. Under these conditions the surface potential cannot fully develop and the EOF decreases correspondingly.

Several theories relating surface charge, zeta potential, and channel size with the magnitude and the flow velocity distribution of the EOF through the channel have been described [21,22,27]. The various models vary in complexity and detail, but most of the details are beyond the scope of the present discussion.

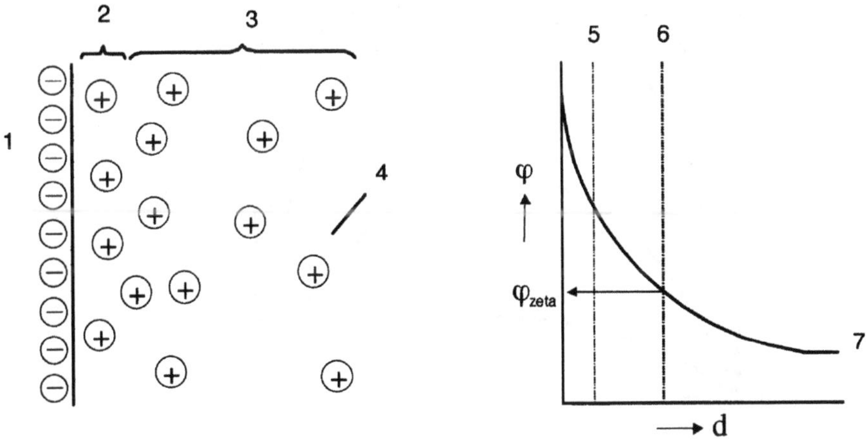

FIGURE 1 The electrical double layer and the potential distribution at the surface: 1, fixed surface charges; 2, Stern layer of "fixed" charges; 3, Gouy or diffuse charge layer; 4, counterions; 5, Helmholtz plane; 6, plane of shear; 7, potential distribution in the electrical double layer.

Using the relatively simple theory of Rice and Whitehead [27], the EOF flow profiles through narrow-sized channels can be calculated readily. Some results are shown in Figure 2.

For relatively wide channels with negligible electrical double-layer overlap ($r/\delta > 10$), a nearly flat flow profile is expected. It has often been stated that when the channel size and the Debye length are of similar dimensions ($r \approx \delta$), complete electrical double-layer overlap occurs and the EOF is negligible. However, when $r \approx \delta$, a significant EOF can still be created; the EOF velocity in the central part of the channel is approximately 20% of that in an infinitely wide channel. Only at conditions where $r/\delta << 1$ is the EOF fully inhibited by double-layer overlap [25]. It should be noted here that the approximations made by using the Rice and Whitehead theory at $r/\delta < 10$ may lead to significant errors in the calculation of the velocity distribution and magnitude of the EOF [17] compared to more sophisticated models.

Relevant for the discussion of the effects of pore flow in CEC is the total or average EOF through narrow channels. The effect of electrical double-layer overlap on EOF is usually expressed in an electroosmotic flow screening factor β, which is defined as the ratio of the EOF velocity to that obtained without double-layer overlap, as can be found from the Smoluchowski equation:

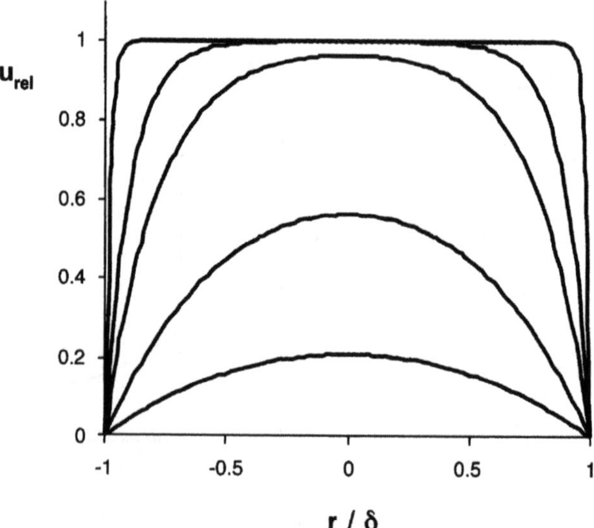

FIGURE 2 Electroosmotic velocity profiles in narrow channels with r/δ values of 1, 2, 5, 10, and 50, respectively (lower to upper curves).

$$u_{\mathrm{EOF}} = \frac{\varepsilon_0 \varepsilon_r \xi E}{\eta} \tag{2}$$

where u_{EOF} is the electroosmotic velocity, ε_0 is the permittivity of vacuum, ε_r is the relative permittivity of the medium, ζ is the zeta potential, and η is the viscosity of the medium. According to the model of Rice and Whitehead, the screening factor for narrow (cylindrical) channels can be found as a function of the ratio of the channel radius r and the double-layer thickness δ:

$$\beta = \frac{I_2(r/\delta)}{I_0(r/\delta)} \tag{3}$$

In Figure 3 the calculated electroosmotic flow screening factor is shown as a function of r/δ. The double-layer thickness, and with that the degree of double-layer overlap, is determined by the ionic strength of the solution. In Figure 4, the electroosmotic flow screening is shown for channels of varying diameter as a function of the ionic strength of an aqueous mobile phase. From the figure it can be seen that a substantial EOF may still be generated in very narrow channels, with diameters in the nanometer range. This allows the use of submicrometer-diameter particles in CEC, without the necessity to develop customized equipment as would be required to operate such columns with a pressure-driven flow [5,6]. Requirements are imposed on the ionic strength of the mobile phase, but these are not very strict; through capillaries of 1 μm in

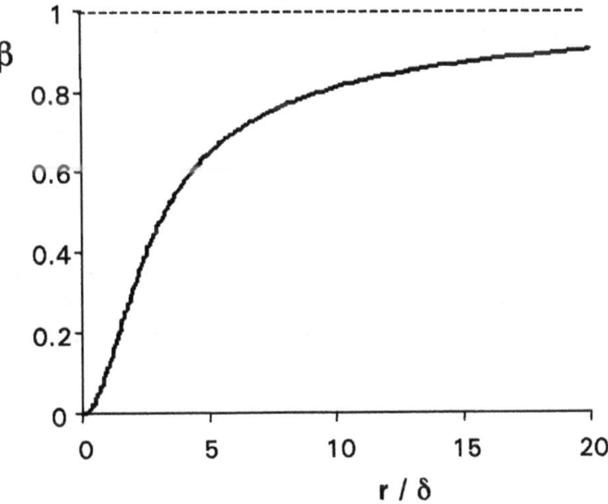

FIGURE 3 The EOF screening factor as a function of r/δ.

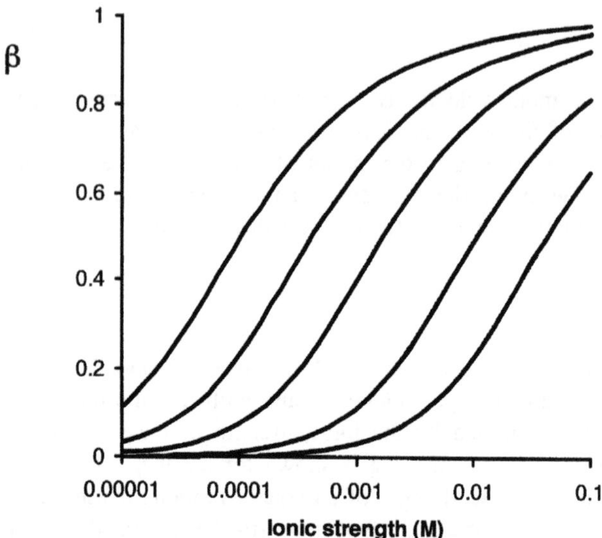

FIGURE 4 The EOF screening factor as a function of the ionic strength for a capillary with i.d. of 100, 50, 25, 10, 5, and 1 nm (from left to right).

diameter, the EOF can be >50% of its unscreened value at ionic strengths higher than ~0.1 mM.

When the electrical double layer in the channels between the particles making up the packed bed is small, a high and homogeneous EOF can be created through the column. This leads to a reduction of the *A*-term contribution to the total plate height to a level that is significantly lower than with a pressure-driven flow [9,10,16]. A significant improvement of the separation efficiency can thus be expected by changing the driving force from pressure to electroosmotic.

In electrochromatography with porous particles, a β factor for the pore volume (β_{in}) as well as for the interstitial volume (β_{out}) can be defined. (Note: For monolithic or continuous columns, only a single EOF screening factor can be defined for the pore volume.) An important parameter in the discussion of the effects of pore flow on the separation efficiency is the pore-to-interstitial flow ratio ω, which is defined as [18]

$$\omega = \frac{\beta_{in}}{\beta_{out}} \tag{4}$$

In all but a few of the CEC experiments that have been reported in the literature, the flow channels through the particles are much smaller than those between the particles. Under conditions at which a significant pore flow can be expected, the double-layer overlap in the interstitial channels is negligible, and β_{out} approaches 1. In this case the pore-to-interstitial flow ratio ω is equal to β_{in}.

2.2. Measurement of Pore Flow

A number of papers have appeared in which the importance of pore flow in CEC is discussed [13–22]. Often, these discussions are supported by rather indirect experimental evidence for the existence and the magnitude of the pore flow, such as by its assumed effect on the separation efficiency.

In previous work [15,17,28] we have evaluated the relation between the flow ratio ω, the pore size, and the ionic strength of the solution experimentally, by means of size-exclusion electrochromatography (SEEC). In SEEC the transport rates of the (neutral) macromolecules depend directly on ω. As in conventional, pressure-driven SEC, the separation in SEEC is based on the differential accessibility of the (stagnant) mobile phase in the pores of the particles for macromolecules of different sizes. However, with increasing pore flow ratio in SEEC, the velocity difference between the mobile-phase fractions inside and outside the particles decreases. The retention ratio τ_i (the retention time relative to a low-molecular-mass marker) for a probe molecule in SEEC is given by

$$\tau_i = \frac{\varepsilon_{out} + \varepsilon_{in} \cdot f_i \cdot \omega}{\varepsilon_{out} + \varepsilon_{in} \cdot f_i} \tag{5}$$

where ε_{out} and ε_{in} are the interstitial and the pore volume fractions of the column, respectively, and f_i is the pore volume fraction accessible for the probe molecule i. This accessible pore volume fraction f_i can be found for the different probe molecules from pressure-driven SEC experiments on the same column.

The Rice and Whitehead model was used in conjunction with a pore connectivity model to predict the pore flow and to compare the results with the experimental data. The pore connectivity model used describes the pores in the stationary-phase particles as channels of different diameter placed in series [17]. The varying degree of electrical double-layer overlap in such a nonuniformly sized channel results in pressure gradients being generated within the pores. This results in pressure-generated flows and the resulting average flow velocity is the weighted average of the velocity in the different parts of the pores, with a flow resistance ($1/d_{p,j}^2$) as the weight factor [17]. A satisfactory correlation between the experimental and the calculated retention data was found (see Figure 5).

In a more detailed study, the influence of the pore size and the ionic strength on the pore-to-interstitial flow ratio was investigated with this SEEC method. Measurements were performed with ionic strengths in the range from 10 μM to 100 mM and with particles with pore sizes ranging from 5 to 100 nm

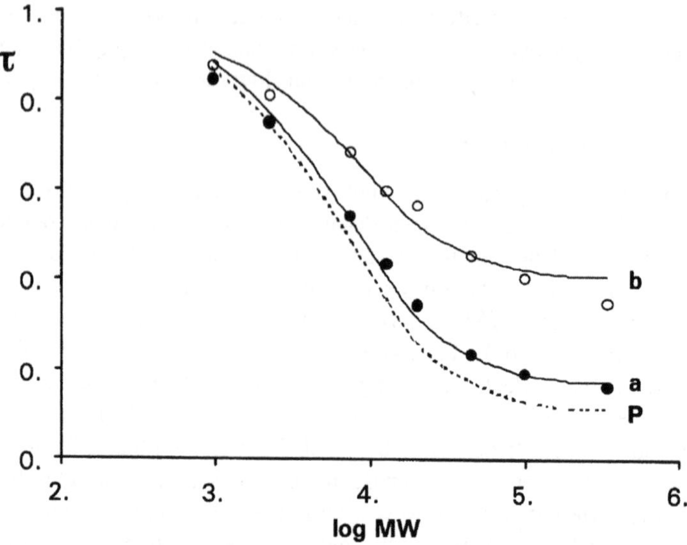

Figure 5 SEEC calibration curves used for the determination of the pore flow ratio ω. Polystyrene standards separated in DMF with 1 mM (a) or 10 mM (b) LiCl. Stationary phase: Lichrosorb 100–10. ● or ○, experimental values; ——, prediction using model; - - - - -, prediction for pressure-driven SEC (P).

[29]. Some of the results are shown in Figure 6. It can be seen that a high pore flow can be generated even with the 5- and the 10-nm-pore silica particles as typically used in LC. For narrow pores and high ionic strength values, good agreement between theory and experiments was found. Surprisingly, the EOF through larger-pore particles (30–100 nm) is found to be even higher than is predicted by the Rice and Whitehead model [27]. Also, when more accurate and complex (nonlinearized) electrokinetic models are used, unrealistic high surface charges and surface potentials have to be assumed to model the experimental data. A high EOF is thus more easily created through nanosized channels than can be expected on the basis of current theories.

Tallarek et al. determined the properties of electroosmotic flow in open and packed capillary columns using pulsed field gradient NMR [20]. They determined the dynamics of the pore and interstitial EOF in columns packed with gigaporous particles at a number of different ionic strengths. They were able to distinguish the two flow regimes by measuring and comparing the apparent diffusivities of the solvent molecules within the pores using pressure-driven LC and CEC. They demonstrated that the apparent diffusivity of the solvent molecules contained within the pore volume was much higher in CEC than that

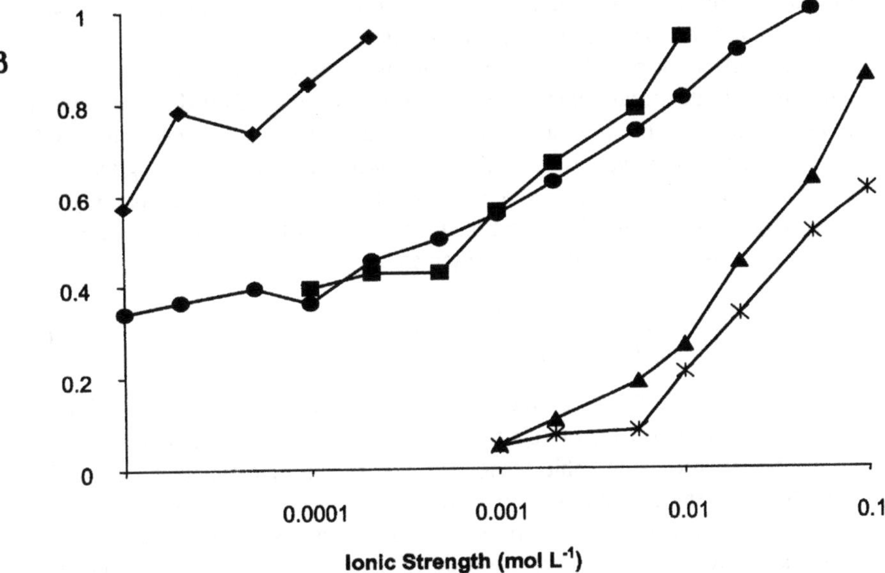

FIGURE 6 Experimentally determined intraparticle EOF screening factors for particles with different pore sizes as a function of ionic strength in DMF. Nominal pore diameters: ◆, 100; ■, 50; ●, 30; ▲, 10; ×, 5 nm.

determined for a pressure-drive flow. Moreover, they were able to show that the apparent pore diffusivity increased linearly with the applied field strength in CEC, thereby providing clear evidence for the existence of pore flow and its relation to the electrical field strength. Unfortunately, the method could be used only on a relatively large-volume scale, and the technique cannot easily be applied for the much smaller columns and particles commonly used in CEC.

3. EFFECTS OF PORE FLOW ON PEAK BROADENING IN CEC

3.1. The Enhanced Diffusion Effect

A high electroosmotic flow through the stationary-phase particles may be created when the appropriate conditions are provided. This pore flow has important consequences for the chromatographic efficiency that may be obtained in CEC. From plate height theories on (pressure-driven) techniques such as perfusion and membrane chromatography, it is known that perfusive transport may strongly enhance the stationary-phase mass transfer kinetics [30–34]. It is emphasised that the conditions for obtaining a high pore flow in CEC are much more favorable compared to the pressure-driven variants.

One effect of pore flow is that it enhances the mass transfer rate between the pore and interstitial volumes. Instead of by molecular diffusion only, which is by nature slow in solution, mass exchange occurs also by perfusive EOF. This effect can be treated as a form of stimulated diffusion. Following the original treatment for pressure-driven LC according to Rodrigues et al. [31], the plate height contribution from stationary-phase mass transfer resistance $H_{C,s}$ in the presence of pore flow can be written as

$$H_{C,s} = \frac{1}{30} \cdot \frac{\varepsilon_i}{\varepsilon_o + \varepsilon_i} \cdot \frac{(1 + k'')^2}{(1 + k')^2} \cdot \frac{d_p^2}{D_m^*} \cdot \langle u \rangle \tag{6}$$

where ε_i and ε_o are the pore and interstitial column volume fractions, k'' is the intraparticle retention factor ($\varepsilon_s K/\varepsilon_i$), k' is the effective retention factor in the conventional chromatographic sense, and d_p is the particle diameter. The difference between the conventional plate height equation for chromatography and Eq. (6) is that the diffusion coefficient of the solute inside the stationary phase is replaced by an effective diffusion coefficient D_m^*. This parameter describes the reduction of $H_{C,s}$ as a result of through-pore migration. For spherical particles the effective diffusion coefficient can be calculated as a function of the intraparticle velocity u_i as

$$D_m^* = D_m \cdot \frac{\nu_{\text{in}}}{18} \cdot \frac{1}{[\coth(\nu_{\text{in}}/6) - (6/\nu_{\text{in}})]} \tag{7}$$

where D_m is the diffusion coefficient of the solute in the mobile phase and ν_{in} is the reduced intraparticle velocity given by

$$\nu_{\text{in}} = \frac{u_{\text{in}} d_p}{D_m} \tag{8}$$

The increase in efficiency by this effect can be very significant, especially for slowly diffusing solutes and at high perfusion rates. In Figure 7 the efficiency enhancement factors are shown for molecules with different diffusion coefficients, calculated for particles with a diameter of 7 μm. For the typical low-mass solutes separated in reversed-phase CEC ($D_m \approx 5 \times 10^{-9}$ m²/s), an up to 30% reduction of the $H_{C,s}$ contribution to the total plate height can be expected. For high-mass, slowly diffusing solutes the mass transfer enhancement is even higher and the $H_{C,s}$ term can be reduced to approximately 5% of its original value in the absence of pore flow. It must be noted, however, that this is related to the relatively large particles used in these calculations. The effect becomes relatively unimportant for the total efficiency for small-sized particles.

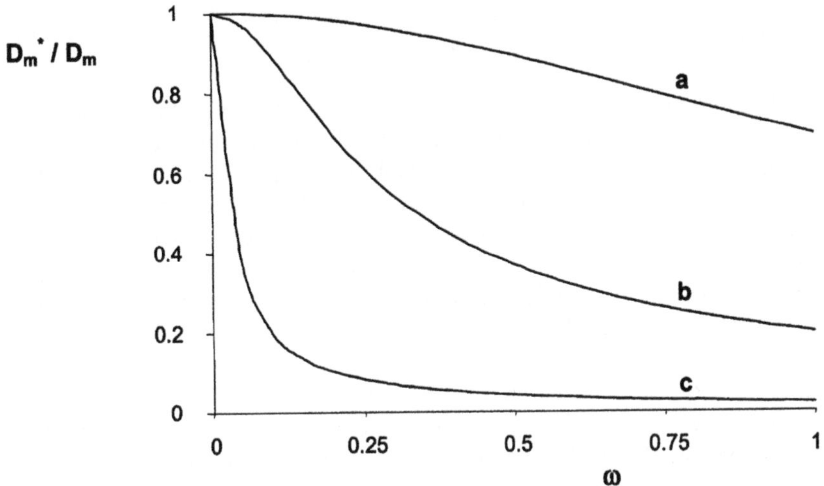

FIGURE 7 The stationary-phase mass transfer enhancement for nonsorbed tracers. Diffusion coefficients: (a) 5×10^{-9} m^2/s; (b) 10^{-9} m^2/s; and (c) 10^{-10} m^2/s. Other values used in calculations: $d_p = 7.0$ μm; $u_o = 2.0$ mm/s.

This is the result of both a lower effect on the diffusivity and a lower C_s contribution to the total plate height. Moreover, for retained solutes ($k' > 0$), the gain in mass transfer is less pronounced [18]. Nevertheless, the enhanced diffusivity effect of pore flow is highly interesting since it indicates that CEC may have a high potential for solutes with a low diffusion coefficient.

3.2. The Equilibrium Effect

In addition to the enhanced diffusivity effect, another issue needs to be taken into account when considering stationary-phase mass transfer in CEC with porous particles. The velocity difference between the pore and interstitial space may be small in CEC. Under such conditions the rate of mass transfer between the interstitial and pore space cannot be very important for the total separation efficiency, as the driving mechanism for peak broadening, i.e., the difference in mobile-phase velocity within and outside the particles, is absent. This effect on the plate height contribution $H_{C,s}$ has been termed the equilibrium effect [35].

How to account for this effect in the plate height equation is still open to debate. Using a modified mass balance equation and Laplace transformation, we first arrived at the following expression for $H_{C,s}$, which accounts for both the effective diffusivity and the equilibrium effect [18]:

$$H_{C,s} = \frac{1}{30} \cdot \frac{\varepsilon_{\text{in}}}{\varepsilon_{\text{out}} + \varepsilon_{\text{in}}} \cdot \frac{1}{1 + \omega \cdot \varepsilon_{\text{in}}/\varepsilon_{\text{out}}} \cdot \frac{(1 - \omega + k'')(1 + k'')}{(1 + k')^2} \cdot \frac{d_p^2}{D_m^*} \cdot \langle u \rangle \tag{9}$$

Without perfusive flow ($\omega = 0$) the equation reduces to the classical expression for $H_{C,ts}$ as used in pressure-driven LC. Equation (9) predicts some of the expected behavior of systems in which a high pore flow is present. For nonsorbed compounds (k'' and k' equal to 0) the C_s term contains a factor $(1 - \omega)$ in the numerator. This implies that peak broadening is not related to the flow velocity itself, but depends on the flow velocity difference between the pore and interstitial volume. For unretained compounds the C_s term vanishes with fully perfusive flow ($\omega = 1$). This result is as expected: when the transport rate of a tracer inside the particles matches that outside the particles, there cannot be a contribution to peak broadening related to mass exchange. For retained solutes ($k'' \neq 0$), however, the velocity within the particles is still different from that between the particles, and consequently mass transfer effects will occur.

Unfortunately, Eq. (8) is not fully satisfactory since it results in negative values for the C_s term at high pore flow ($\omega + k'' > 1$). Such high flow ratios may be created when a pressure-driven flow is directed against the electroosmotic flow. Also, when the interior of the particles has a much higher surface potential than the exterior, surface flow ratios > 1 may be expected. Of course, negative contributions to the total plate height are physically impossible and Eq. (9) cannot be valid.

The problem of negative values for $H_{C,s}$ arises from the uncoupling of the enhanced diffusion and the equilibrium effects. This argument led us to investigate alternative expressions that allow inclusion of the pore-to-interstitial flow ratio in the C-term equation for CEC. We tackled the problem by generating a solution for the pore flow problem for a slab-formed stationary zone. The form factor in the solution for a flat geometry was than replaced by that for a spherical geometry, as found by others. This rather intuitive modification of the C_s term in the presence of pore flow resulted in the following expression for $H_{C,s}$ [35]:

$$H_{C,s} = \frac{1}{30} \cdot \frac{\varepsilon_{\text{in}}}{\varepsilon_{\text{out}} + \varepsilon_{\text{in}}} \cdot \frac{(1 + k'' - \omega)^2}{(1 + k'')^2\,(1 + \omega\varepsilon_{\text{in}}/\varepsilon_{\text{out}})^2} \cdot \frac{d_p^2}{D_m^*} \tag{10}$$

This expression is more satisfying than Eq. (9) in that it predicts more or less expected behavior at all possible pore-to-interstitial flow ratios. Therefore, at the moment Eq. (10) appears to be the most appropriate solution for the C_s term for capillary electrochromatography in the presence of pore flow obtained so far.

3.3. Pore Flow and Flow Homogeneity

It has been shown that the velocity differences between the pore and interstitial spaces may be very small. Apart from the effect on the mass transfer, it also has implications for the flow velocity distribution over the column cross section, i.e., for the A-term contribution to the total plate height. The A-term contribution may be regarded as the sum of all factors that disturb the homogeneity of the flow in the column. Due to the flat electroosmotic flow profile between the particles, the A-term contribution in CEC is already significantly lower than with a pressure-driven flow.

The differences in flow profiles over the column cross section with various elution modes are shown schematically in Figure 8. With a pressure-driven mobile phase, the flow direction and the flow velocity depend on the interstitial channel size. Both the direction and velocity vary locally with position throughout the column (Figure 8A). When the same column is operated in nonperfusive CEC mode (e.g., when the column is packed with nonporous particles) the flow velocity is more homogeneous between the particles (Figure 8B). While this may already give a strong improvement of the separation efficiency, the local direction of the flow still differs with position in the column, giving rise to a certain level of A-term peak dispersion. When the column is packed with porous particles and operated at fully perfusive conditions, both the flow velocity and direction are homogeneous throughout the column (Figure 8C).

For CEC in the absence of pore flow, an A term equal to $1d_p$ has been proposed, while at fully perfusive conditions values of $0.2d_p$ have been reported [18,23]. In the limit of fully perfusive conditions ($\omega = 1$), the disturbances of

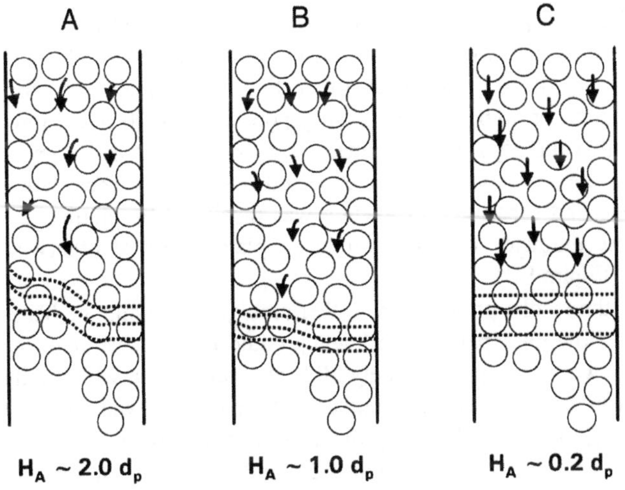

FIGURE 8 Illustration of the interstitial flow profiles in (A) pressure-driven chromatography; (B) electrochromatography with nonporous particles; and (C) electrochromatography at fully perfusive conditions. The length of the arrow represents the local velocity.

the flow cannot be scaled to the dimensions of the particles. The characteristic size is that of the structure making the skeleton of the particles. The observed relation between the relative perfusion rate and the A-term contribution can possibly be modeled as

$$H_A \approx (1 - \omega) \cdot d_p + \omega \cdot \varepsilon_{sk} \cdot d_p \tag{11}$$

where ε_{sk} is the column volume fraction of solid material (silica).

Another approach is to scale the A-term contribution to a so-called effective particle diameter as has been proposed by Vallano and Remcho [19]. They defined an effective particle diameter from the perfusive EOF velocity within each volume fraction of the pore size distribution of the particles. Their model allows inclusion of both the intraparticle EOF and a pore size distribution. However, both this approach as well as Eq. (10) are more or less empirical.

4. THE PORE FLOW EFFECT IN PRACTICE

4.1. CEC with (Macro-) Porous Particles

The various effects that have been predicted for electrochromatography with perfusive flow may allow for a strong increase of the separation efficiency. The most pronounced effects are expected at high values of ω, which can be created by the use of large-pore-sized particles and the use of high-ionic-strength mobile phases.

The first study on the effects of pore flow on the separation efficiency was reported by Li and Remcho [13]. They investigated the separation efficiency of columns packed with particles having pore diameters in the range of 6–400 nm and at ionic strengths up to 500 mM. They found that a significant perfusive EOF could be generated through particles having pore sizes >200 nm. Moreover, they observed that the high-velocity side of the H–u curves had a lower slope at conditions where perfusive EOF was expected (Figure 9A). The relation between the pore size and separation efficiency was displayed by plotting the logarithm of the pore diameter versus the reduced plate height at otherwise identical conditions (Figure 9B).

Unfortunately, the results proved to be rather ambiguous. While the beneficial effects of pore flow on separation efficiency could be demonstrated, the increase in efficiency was only approximately 30% at conditions at which fully perfusive behavior can be expected. The highest efficiency that was reported was a reduced plate height of 1.3. This improvement is much less as predicted.

Later work on macroporous particles in CEC indicated a much stronger effect of pore flow on the separation efficiency [16]. In this study it was shown that at moderate ionic strengths in the range of 0.1–10 mM, fully perfusive behavior could be created with particles having pore sizes of 50–400 nm. Plate

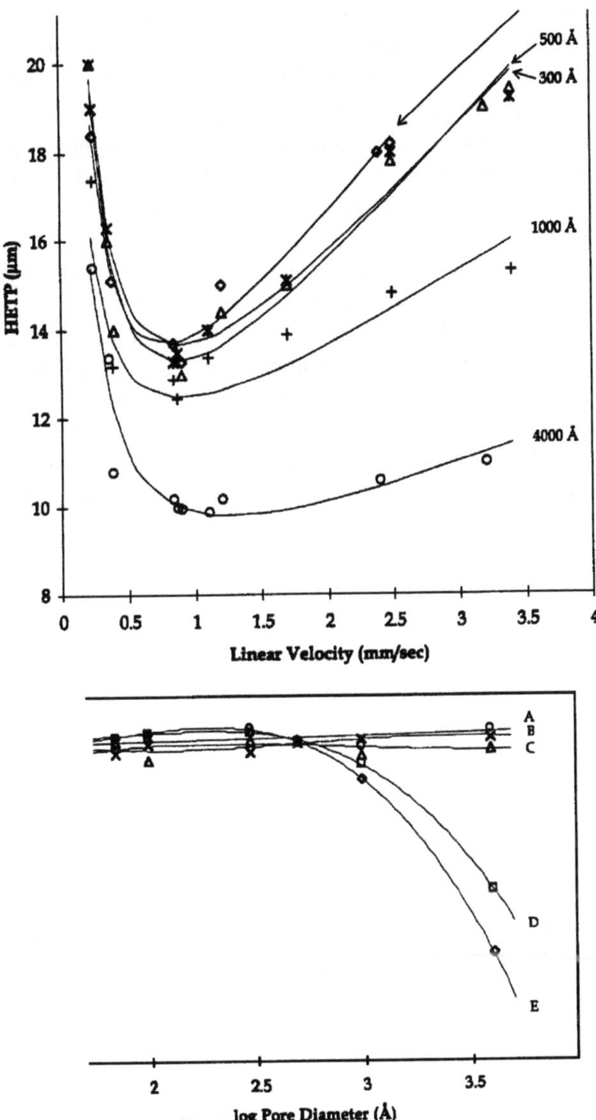

FIGURE 9 (Top) Plate height as a function of the linear velocity for columns packed with 7-μm-particle-size Nucleosil C_{18} with different pore size. (Bottom) Reduced plate height versus the logarithm of the pore diameter for various buffer concentrations: A, 10 mM; B, 50 mM; C, 100 mM; D, 300 mM; E, 500 mM. (Reproduced from Ref. 13, with permission.)

height curves were determined for a slightly retained compound for the various-pore-sized particles (Figure 10). The highest efficiencies were obtained with the 400-nm-pore-sized particles. Plate heights of only 2.3 μm, equivalent to 450.000 plates/m on 25-cm-long columns, were reported. This equals a reduced plate height of 0.3. In addition, the plate height curves were found to be approximately independent of the mobile-phase velocity at velocities higher than 1.0 mm/s, providing a technology to obtain very fast and efficient separations without the need to use small particles.

Other attractive features of fully perfusive stationary phases of relatively large particle diameter were also demonstrated. Due to the absence of stagnant mobile phase, the solutes migrate relatively fast through the columns. Mobile phase velocities >3.5 mm/s could be generated, which permitted the separation of five compounds in under 30 s, while generating 15,000 theoretical plates. The relatively large 7-μm particles are easy to pack. With a 72-cm-long column with 400-nm-pore particles, the separation of five components could be achieved in 12 min with a separation efficiency of 230,000 theoretical plates (see Figure 11).

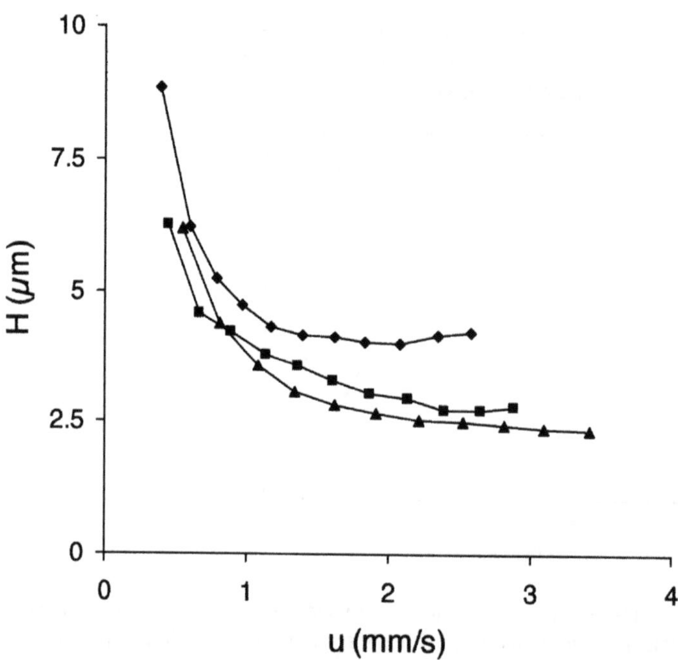

FIGURE 10 The effect of pore size on the *H–u* curves of fluorene in reversed-phase CEC. Columns packed with 10-μm reversed-phase particles with a nominal pore size of: ◆, 50 nm; ■, 100 nm; ▲, 400 nm.

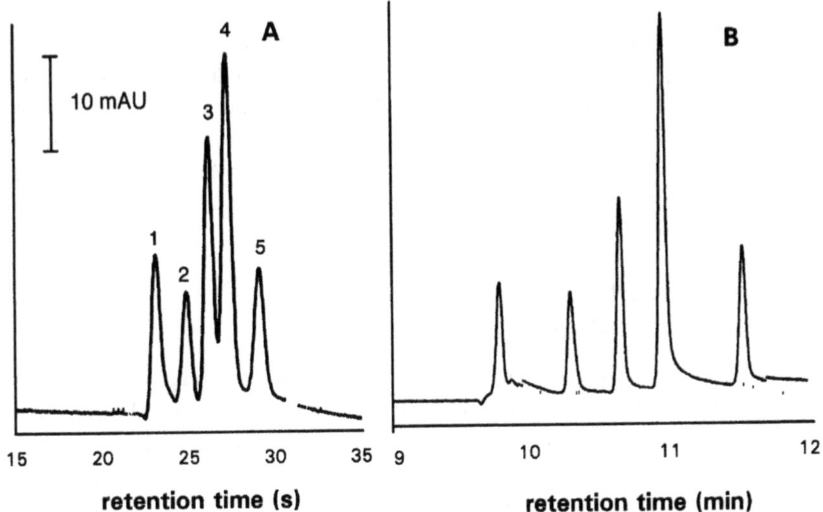

FIGURE 11 CEC with macroporous particles. (A) Column length 8.3 cm; N = 15,000. (B) Column length 72 cm; N = 230,000. Stationary phase, Nucleosil 4000-7 C_{18}; mobile phase, ACN/water (70/30); solutes, PAHs.

Vallano and Remcho reevaluated the use of macroporous particles in CEC and confirmed that reduced plate heights below unity could be generated when using particles with a pore size of 100 nm or larger [19]. They observed that the H–u curves suggested the presence of intraparticle EOF at ionic strengths as low as 1.0 mM with 100- and 400-nm pores. Additionally, they presented a model that can be used to estimate the efficiency improvement by and the magnitude of intraparticle EOF under a range of experimental conditions in terms of an effective particle diameter. The effective particle diameter can be used to obtain a good estimate of the A-term contribution in the presence of pore flow.

Al Rifaï et al. studied the chromatographic behavior of macroporous particles in reversed-phase electrochromatography and compared the results with pressure-driven LC with capillary columns as well as with standard-bore 4.6-mm-i.d. columns [23]. Using 400-nm-pore particles they obtained 650,000 plates/m in the CEC mode, corresponding to a reduced plate height of only 0.2. These high efficiencies were obtained for low retained solutes and provided a five-fold improvement of the optimal efficiency obtained in pressure-driven LC on the same column, and a 10-fold improvement compared to the 4.6-mm-i.d. column.

Macroporous particles have also been used in CEC to accomplish separation of enantiomers. Wikström et al. examined the use of vancomycine bonded to Lichrosphere diol silica particles with pore sizes in the range of 10–100 nm for chiral CEC [36]. They found that the 100-nm-pore-sized particles produced 4 times higher plate numbers than the 10-nm-pore particles. Remarkably, the 10-nm-pore particles had a lower particle diameter, so the efficiency enhancement factor due to pore flow must have been significant. In addition, the mobile-phase velocity dependency of the efficiency was found to be much lower with the 100-nm-pore particles. Both observations were explained by the occurrence of a high perfusive EOF.

4.2. Pore Flow Effects in Size-Exclusion Electrochromatography

The effects of pore flow in size-exclusion electrochromatography (SEEC) are even more apparent than in reversed-phase CEC. The solutes typically separated in SEEC are slowly diffusing macromolecules such as synthetic polymers. For these solutes the enhanced diffusion effect becomes relevant even at low pore flow velocities and at low pore-to-interstitial flow ratios.

Venema et al. studied SEEC with porous silica particles [14,15]. They separated narrow polystyrene standards on columns packed with particles of different pore size, and observed a significant improvement of the efficiency in SEEC over that in pressure-driven size-exclusion chromatography. Also, they observed that the efficiency improvement was more significant for large-pore particles and related this to a higher pore flow.

In a later report the effects of pore flow on the separation efficiency in SEEC were studied in more detail [28]. It was found that in both SEEC and pressure-driven SEC, the plate height curves were linear with the mobile-phase velocity, which is expected at mobile-phase velocities higher than the diffusion-controlled regime. It was found that both the *A*- and *C*-term contributions were significantly lower in the SEEC mode. Surprisingly, it was found that the most important gain in efficiency originated from a lower *C* term instead of increased flow homogeneity as has been commonly thought.

Already a low-pore EOF suffices to improve the efficiency by a factor 2–3. In SEEC a reduction in mass selectivity is found related to the occurrence of pore flow. Therefore, the pore flow ratio should be optimized to find a balance between improved efficiency and decreased selectivity.

5. OPTIMIZATION OF PARTICLE DIMENSIONS FOR HIGH-EFFICIENCY CEC

In electrochromatography with porous particles, both the mass transfer resistance and the flow inhomogeneity contributions to the total plate height are related to the (relative) perfusive EOF velocity. There are three effects that allow for the improved separation efficiency in the presence of pore flow, and

TABLE 1 Overview of the Three Efficiency-Enhancing Pore Flow Effects and Their Relations to Pore Flow and Particle Size

Pore flow effect	Relevant velocity parameter	Particle size dependency
Enhanced diffusivity[a]	$\sim 1/u_{in}$	$\sim d_p$
Equilibrium effect	$(1-\omega)^2$	d_p^2
Improved flow homogeneity	$1-\omega$	d_p

[a]For slowly diffusing compounds.

each effect requires its specific optimisation of experimental parameters, as shown in Table 1.

For slowly diffusing solutes, a significant improvement in the separation efficiency can be expected when a high intraparticle EOF is created. At low pore-to-interstitial flow ratio this may be accomplished by, e.g., the application of a high electrical field strength.

A more effective approach to enhance the separation efficiency is to create conditions at which a high pore-to-interstitial flow ratio is obtained. High ω values ensure both enhanced mass transfer kinetics and an improved flow homogeneity. This may be accomplished through the application of high-ionic-strength mobile phases and through the use of columns packed with large-pore-size particles. In order to obtain maximal efficiency, a pore-to-interstitial flow ratio close to unity is desired.

The total effect of the pore flow ratio is shown in Figure 12. the ratio was varied by using two different 7-μm particle types, with 50- and 400-nm pores, and mobile phases with different ionic strength values. Typically the reduced plate height decreased from ~1 at low ω to ~0.3 at ω values close to 1. The question that arises immediately is how large the pores and how high the ionic strength should be in order to achieve such high perfusive flow ratios. In Section 2 it has been described how the pore flow depends on the ionic strength (Figure 4). High perfusive flow ratios could be obtained with nearly all pore sizes (5–400 nm) typically used in liquid chromatography. Pore-to-interstitial flow ratios close to unity could be obtained with a pore size of 10 nm and larger. With 30-nm-pore-sized particles, fully perfusive conditions were obtained at intermediate ionic strengths of ~10 mM. The use of such particles and ionic strengths will not pose significant experimental problems in CEC.

6. CONCLUSIONS

It is clear that a significantly pore flow can be generated in electrochromatography when porous particles are employed as the stationary phase. This pore flow may strongly enhance the separation efficiency that may be obtained in CEC,

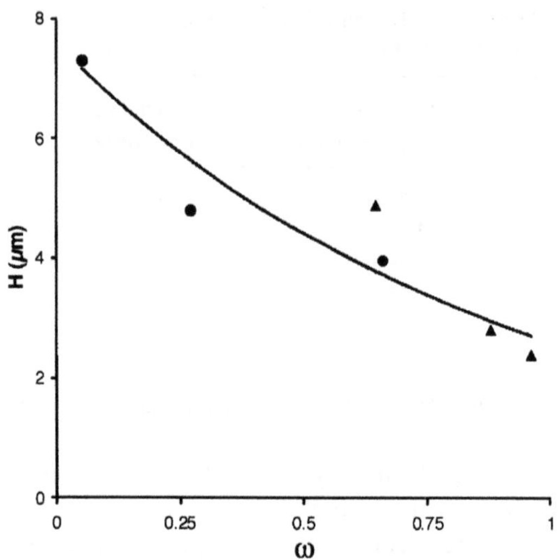

FIGURE 12 Plate height obtained with 7-μm reversed-phase particles as a function of pore flow ratio ω.

through a strong reduction of the stationary-phase mass transfer resistance and a further improvement of the flow homogeneity across the column cross section.

When the pore size, ionic strength, and electrical field strength are optimized for a high pore flow velocity and a high pore-to-interstitial flow ratio, reduced plate heights well below unity can be achieved in CEC of low-molecular-mass compounds. When such conditions can be created in combination with the use of small particles ($d_p < 1$ μm), plate heights below 1.0 μm will be possible.

REFERENCES

1. Katz E, Eksteen R, Schoenmakers P, Miller N, eds. Handbook of HPLC. Chromatographic Science Series Vol. 78. New York: Marcel Dekker, 1998.
2. Giddings JC. Dynamics of Chromatography, Part I: Principles and Theory. New York: Marcel Dekker, 1965.
3. Tsuda T, Novotny M. Anal Chem 1978; 50:632.
4. Swart R, Kraak JC, Poppe H. TRAC-Trend Anal Chem 1997; 16:332–342.
5. Giesche H, Unger KK, Esser U, Eray B, Truedinger U, Kinkel JN. J Chromatogr 1989; 465:39–57.
6. MacNair JE, Patel KD, Jorgenson JW. Anal Chem 1999; 71:700–708.
7. Pretorius V, Hopkins BJ, Schieke JD. J Chromatogr 1974; 9:23–30.
8. Jorgenson JW, Lukacs KD. J Chromatogr 1981; 218:209–216.

9. Knox JH, Grant IH. Chromatographia 1987; 24:135–143.
10. Knox JH, Grant IH. Chromatographia 1991; 32:317–328.
11. Peters EC, Petro M, Svec F, Frechet JMJ. Anal Chem 1998; 70:2296–2302.
12. Guo Y, Colon LA. Anal Chem 1995; 67:2511–2516.
13. Li D, Remcho VT. J Microcolumn Sep 1997; 9:389–397.
14. Venema E, Kraak JC, Tijssen R, Poppe H. Chromatographia 1998; 58:347–354.
15. Venema E, Kraak JC, Tijssen R, Poppe H. J Chromatogr A 1999; 837:3–15.
16. Stol R, Kok WTh, Poppe H. J Chromatogr A 1999; 853:45–54.
17. Stol R, Poppe H, Kok WTh. J Chromatogr A 2000; 887:199–208.
18. Stol R, Poppe H, Kok WTh. Anal Chem 2001; 73:3332–3339.
19. Vallano P.T, Remcho VT. Anal Chem 2000; 72:4255–4265.
20. Tallarek U, Rapp E, Van As H, Bayer E. Angew Chem Int Ed 2001; 40:1684–1687.
21. Grimes BA, Meyers JJ, Liapis AI. J Chromatogr A 2000; 890:61–72.
22. Liapis AI, Grimes BA. J Chromatogr A 2000; 877:181–215.
23. Al Rifai R, Demesmay C, Cretier G, Rocca JL. Chromatographia 2001; 53:691–696. Chap. 6, pp 316–394.
24. Hunter RJ. Foundations of Colloid Science. Vol. 1. Oxford, UK: Clarendon Press, 1986.
25. Hunter RJ. Zeta Potential in Colloid Science: Principles and Applications. London: Academic Press, 1988.
26. Dukhin SS, Derjauin BV. In: Matijevic E, ed. Surface and Colloid Science. New York: Wiley, 1974.
27. Rice CL, Whitehead R. J Phys Chem 1965; 69:4017.
28. Stol R, Kok WTh, Poppe H. J Chromatogr A 2000; 914:201–209.
29. Stol R, Poppe H, Kok WTh. submitted for publication, 2003.
30. Afeyan NB, Gordon NF, Mazsaroff I, Varady E, Fulton SP, Yang YB, Regnier FE. J Chromatogr 1990; 519:1.
31. Rodrigues AE, Lopes JC, Loureiro JM, Dias MM. J Chromatogr 1992;590:93.
32. Frey DD, Schweinheim E, Horvath C. Biotechnol Progr 1993; 9:273.
33. Hamaker KH, Ladisch MR. Sep Purif Meth 1996; 25:47.
34. Rodrigues AE. J Chromatogr B 1997; 699:47.
35. Poppe H, Stol R, Kok WTh. J Chromatogr A 2002; 965:75–82.
36. Wikstrom H, Svensson LA, Torstensson A, Owens PK. J Chromatogr A 2000; 869: 395–409.

9

Ultrashort-Column Capillary Electrochromatography

Takao Tsuda
Nagoya Institute of Technology, Nagoya, Japan

1. INTRODUCTION

If a very small separation system were available, we would like to set it up on the stage of a microscope. In such a case, all processes including injection, separation, and detection could be performed under microscopic observation. That would be very helpful for studying microorgans or other cultured substrates [1].

The miniaturization of analysis system, so-called micro total analysis systems (μTAS) or laboratories-on-a-chip, is now under development [2,3]. The microfabricated analysis system has significant potential in the areas of genomics, drug screening, and a variety of other clinical applications. Although the whole system can be miniaturized, the separation column is relatively long, generally 10–20 cm. This is due to the lack of separation ability of a short column. Short open and packed capillary columns can have many applications in the μTAS system. However, use of very short columns (1 cm long) has not been reported.

There are a few references about designing a small chromatographic system, including all operational units, especially separation column, injection, and detection. Ericson et al. have used a short packed column (effective length 4.5 cm) on a chip [4]. Dadoo and Zare [5] have demonstrated high-efficiency separation of five polycyclic aromatic hydrocarbons. Preparation of ultrashort packed and open-tubular capillary columns of lengths 1–2 cm have also been reported by Tsuda et al. [6,7].

In this chapter, system design and performance of ultrashort-column capillary electrochromatography (CEC) will be described. Preparation of ultrashort columns, and design of the pumping system, injector, and detector, will be discussed.

2. Electroosmotic Pumping

2.1 Use of Electroosmotic Flow for Pressurized Pumping

For developing miniaturized chromatographic systems, one possibility is to use a conventional pump with a split-flow device. Although such pumps are enough to run miniaturized chromatographic systems, their size is very large compared to the system itself. Based on the size of such miniaturized systems, the size of the mini pump should be 25 cm^3 or less. However, such small pumps are unavailable presently.

Alternatively, electroosmotic pumping could be used in this application. For generation of electroosmosis it is necessary to apply an electric voltage at both ends of the column. However, when we used electroosmotic flow instead of a pressurized pumping method, we encountered several problems. The hardest problem was evolution of gas at the electrode due to electrolysis. Elimination of this gas from the chromatographic system was one of the key challenges we faced in designing our system.

2.2 Advantages of Using Electroosmotic Flow

Flow profiles of electroosmosis in open-tubular capillary and packed capillary columns were observed by using a microscope–CCD–video system and a rectangular capillary (50×1000 μm) or rectangular capillary column or fused-silica capillary (75 μm in diameter) [8,9]. Rhodamine/methanol solution was used as a fluorescent sample solution [8–11].

Flow profiles of electroosmotic flow (EOF) in both open-tubular and packed capillary columns are shown in Figures 1 and 2, respectively [8,11]. It is generally known that the frontal profile of EOF is pluglike. We have examined experimentally, and found that the profile is reverse-parabolic, namely, the flow velocity in the center is less than at the edge by 0.4% [8]. An overlap of two successive flow profiles is shown in Figure 1B. It is evident that the electroosmotic flow profile deviates from plug to some extent. Despite this, there are still several advantages is using electro-driven flow, as the band broadening is still much smaller then with pressure-driven systems and this is especially beneficial for separation of compounds with small capacity factors ($k' < 2$).

2.3 Control of EOF by Applied Voltage

As is commonly known, the EOF velocity depends on the value of the gradient electric voltage and thus, EOF can be controlled by on/off control of applied

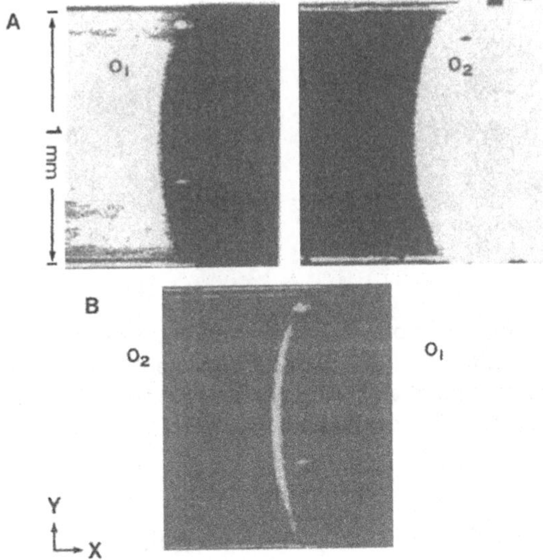

FIGURE 1 Frontal zone profile of electroosmotic flow in open-tubular capillary column. A rectangular capillary (1 mm x 50 μm and 16.4 cm long) was used. Colored sample: methanol solution of 0.1 mM Rohdamine 6G. Frontal zone profiles of O_1 (white) and O_2 (black) were successively taken. The period between two zones was 11.44 s. The distance between two frontal zones was 6.52 mm. Flow velocity of the center was 0.57 mm/s. The ratio of the flow velocities given by (flow velocity at half-radius)/(flow velocity at center) was 1.0027. The retarded speed of the flow velocity at the center compared to that at the corner was only 0.4%. Although the same scale was used for the X and Y axes, there are time intervals between O_1 and O_2 in (A). Figure (B) was obtained by the combination of O_1 and O_2 (overlapping two frontal zone profiles at the corner). Therefore the right and left profiles correspond to O_1 and O_2, respectively. Applied voltage, 1.59 kV; current, 120 nA.

voltage. Experiments were performed to examine this and the progress of the frontal zone under application of pulsed electric field was observed. It follows from Figure 3 that the response period is less than 1/15 s. The real response period may be faster, because this phenomenon has been observed using a CCD–camera–video recording system with a resolution of 1/30 s. The time lag would be reduced if we were to use a high-speed video camera system. In any case, it follows from the results presented here that application of voltage along the column can generate electroosmotic flow and that this flow can be easily controlled by applying voltage. The direction of flow is also reversable by exchanging the polarity of the applied voltage through the electrodes. It was found that the growth and decay are very rapid, less than 1/15 s (which is equal to the

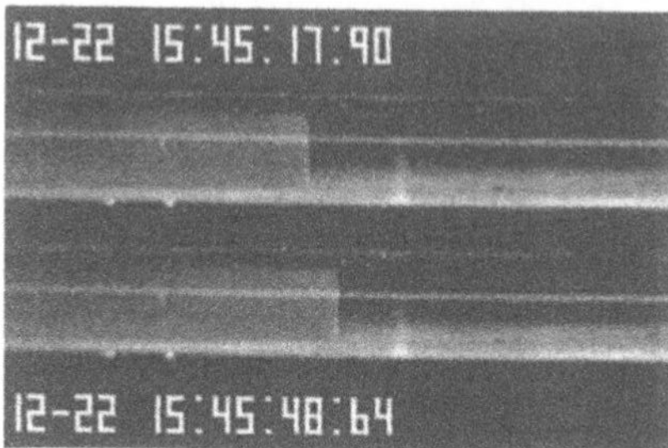

FIGURE 2 Frontal zone profile of electroosmotic flow in packed capillary column. The capillary column was packed with silica gel (particle diameter 5 μm). The medium was cyclohexanol. As cyclohexanol and silica gel have similar refractive indexes, the column looked transparent. The lower photo was taken 31 s after the upper one.

time resolution in this experiment). This is a unique characteristic of electroosmotic flow.

3. PREPARATION OF ULTRASHORT COLUMNS

3.1 Preparation of Packed Ultrashort Capillary Columns

Water-glass was used for making a frit [12] at one of the ends of a short fused-silica capillary tubing (its length is less than 4 cm), followed by slurry packing performed the same way as ordinary slurry packing. The frit should be stable to the pressure used during the packing process. Methanol (100%) was used as slurry solvent. After slurry packing, a high-density polyethylene frit was pressed and inserted into the end of the capillary column. An example of protocol for preparing a short capillary column is described in the following section. Several milligrams of packing and 0.4 mL of methanol were mixed well in a 1-mL vial. The slurry mixture was sucked into a gas-tight 250-μL microsyringe and pressed into a fused-silica capillary tube which had an end frit. Then the open end of the column was capped with a high-density polyethylene frit of approximately 0.5 mm thickness [13]. The column design is shown in Figure 4, and the setup for these ultrashort capillary columns are shown in Figures 5 and 6. One of the chromatograms obtained using an ultrashort packed capillary column is shown in Figure 7 [6].

3.2 Preparation of Open-Tubular Capillary Columns

An open-tubular capillary column has the simplest design and is easy to use due to the small resistance to flow. Although use of open-tubular capillary columns

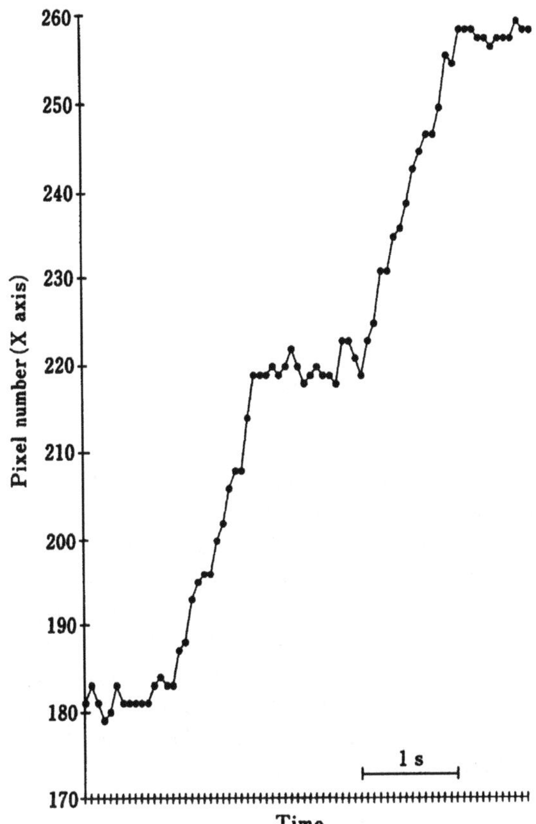

Figure 3 Progress of frontal zone under application of pulsed electric field. A pulsed electric field was applied in 2-s cycles (electric field was applied for half of the period, and was stopped for the other half). Frontal zone positions were measured from the digital picture on a CRT. The perpendicular line shows the direction of movement of the frontal zone, and 1 mm in length is equal to 167 pixels. The horizontal line shows the time, and the minimum time scale observed is equal to one-fifteenth. A round open-tubular capillary column 75 μm in diameter and 28 mm in length was used. Applied voltage, 150 V. Colored sample: methanol solution of 1 mM Rhodamine 6G.

has a long history, in most cases they lack efficiency and their sample-loading capacity is relatively low due to the relatively small surface area of the stationary phase. Several efforts to overcome these deficits have been reported in recent years. Reagents such as sodium hydroxide and ammonium hydrogen difluoride have been used to etch the inner wall of the capillary [14,15]. Colon et al.

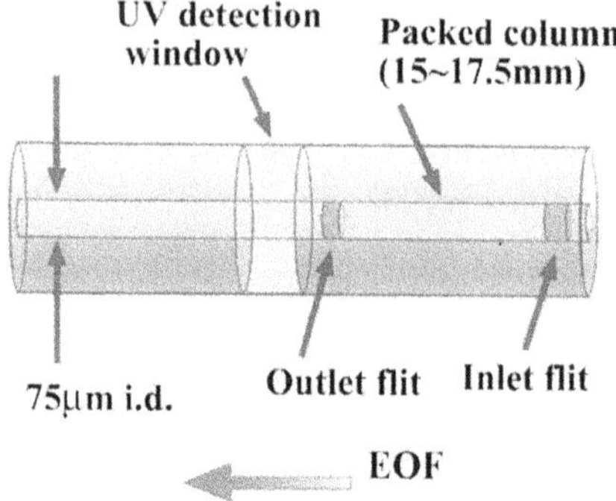

FIGURE 4 Schematic diagram for ultrashort packed column with UV window.

[16] and Freitag et al. [17] prepared the silica layer using a sol-gel method, and the column lengths used were generally from 40 to 100 cm.

In our study, a new silica-gel thin layer is formed through a polymerizing procedure utilizing silica-oligomers [18]. The electron scanning photographs of the inner wall are shown in Figure 8. It is seen that the newly formed silica-gel layer is very thin and is very porous. Further, it can be easily modified with alkyl silanes and the modified silica-gel stationary phase exhibits improved separation efficiency. A 2-cm-long ultrashort column modified with docosyl methyl dichloro silane is usable for separations. Figure 9 shows a chromatogram obtained using an ultrashort open-tubular capillary column [18].

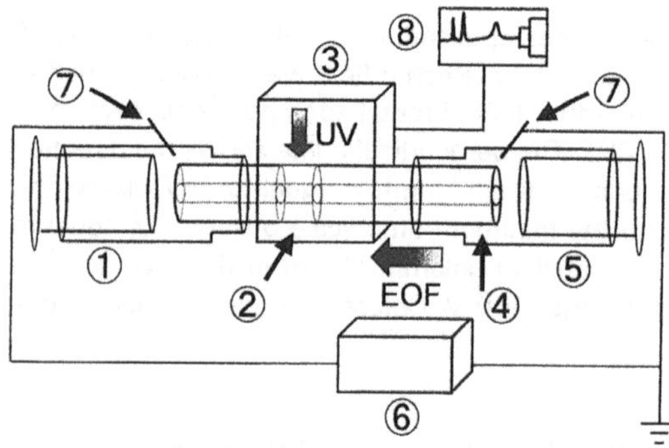

FIGURE 5 CEC system for ultrashort capillary column with UD detection: (1) outlet reservoir; (2) UV detection window; (3) UV detector; (4) ultrashort capillary column; (5) inlet reservoir; (6) high-voltage power supply; (7) Pt electrode; (8) recorder.

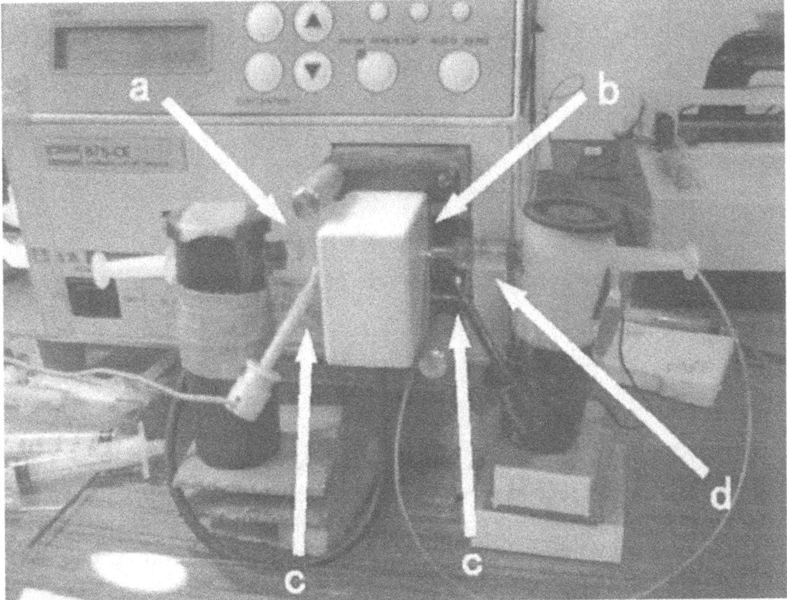

FIGURE 6 Photo of instrumentation for ultrashort capillary column with UV detection: (a) outlet reservoir; (b) UV detection part; (c) Pt electrode; (d) inlet reservoir.

4. SETUP OF CHROMATOGRAPHIC SYSTEM FOR SHORT COLUMNS

The schematic diagram for a microcapillary electrochromatograph is shown in Figure 8. The capillary electrochromatograph comprised a high-voltage power supply, UV detector, home-made short capillary column of 75-μm inner diameter (packed length 15.0 mm and total length 36.0 mm), and two reservoirs. These short columns were packed with cation-exchange supports (particle diameter 5 μm, IC-CATION-SW, Tosoh, Yokohama, Japan). Frits were made at both ends of the packed segment. Reservoirs at both ends of the capillary were made from a 2.5-mL polyethylene syringe and this reservoir was called a "syringe-type reservoir." By using the syringe-type reservoir, the ultrashort capillary column could be set horizontally in the apparatus. Both ends of a capillary column were set into reservoirs through the rubber septum. The polyethylene syringe-type reservoir was pierced with a platinum wire (outer diameter 0.5 mm) as an electrode. The reservoir at the inlet side was always grounded for safety. The UV detector was set at 210 nm. To set the ultrashort capillary column, the holder around the UV sensor in the detector head was cut and shortened. The

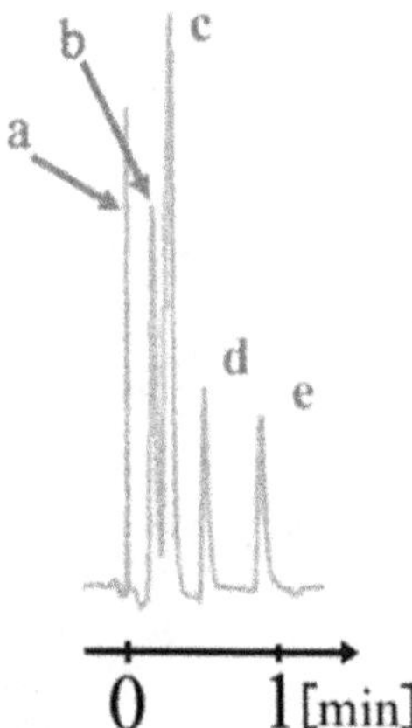

FIGURE 7 CEC chromatogram with ultrashort packed capillary column. Column: 75 μm in inner diameter and 15 mm in length pcked with IC-CATION-SW. Eluent: 30% methanol aqueous solution containing 30 mM KH_2PO_4 and 25 mM EGTA. Aplied voltage, −3.0 kV. Injection, −3.0 kV for 0.5 s. Samples: (a) uracil; (b) adenine; (c) cytosine; (d) dopamine; (e) serotonin.

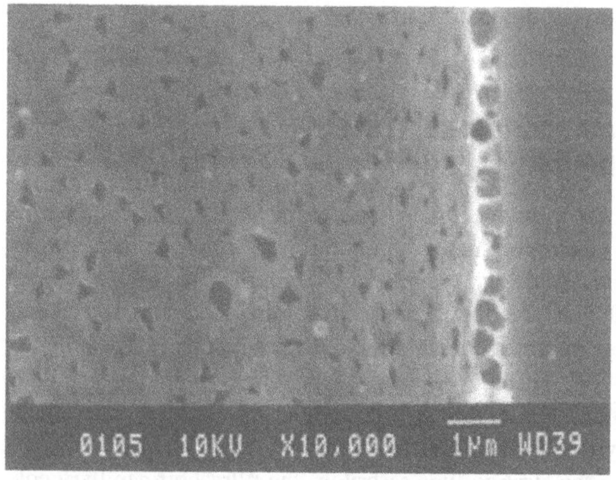

FIGURE 8 Scanning electron micrograph of silica-gel thin layer formed on the inner wall of capillary tubing. There are three parts in the photo. The right part is a section of fused-silica capillary glass body, and the center is a cross-sectional view of a section of silica-gel thin layer foamed on the inner wall, thickness ca. 0.3 μm. On the surface of the silica-gel layer, two-thirds of the photo, there are a lot of holes. From the photo, the silica-gel layer has a nano-cell structure.

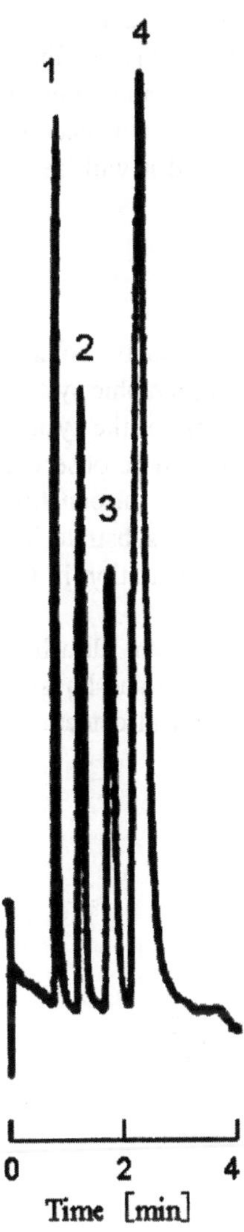

FIGURE 9 Separation with ultrashort open tubular capillary column. Column: effective length 1.7 cm, whole length 4.7 cm, inner diameter 30 μm, stationary phase C22. Eluent: mixture of 20 mM phosphate buffer (pH 7.0) and methanol (1:1). Applied voltage: −2.1 kV. Current: 2.4 μA. Detection: UV 210 nm. Sample: thiourea (1), naphthalene (2), diphenyl (3), and fluorene (4).

width of the modified detector head was 27 mm (the original width was 50 mm). The detection point was right after the packed segment of the capillary. A photograph of the arrangement for the ultrashort column chromatographic system using a UV detector is shown in Figure 9. This system will become simpler when we use a UV fiber to pass light for detection, and it will be set up on the stage of a microscope.

5. SETUP ON THE STAGE OF A MICROSCOPE

The concept of a miniaturized chromatographic system of an ultrashort column on the stage of microscope is shown in Figure 10 [1]. A chromatographic system is constructed on a slide glass for a microscope such that the size of the system is 2.5 cm wide and 7.5 cm long. The chromatographic system can be observed from both downward and upward directions. Therefore, the sample introduction process can be controlled under microscopic observation. If the substrate has color or fluorescence, it is possible to observe the process of separation in the ultrashort column.

The two reservoirs are set at the two ends of the column. An ultrashort column is kept through a small hole at the central plate. The plate and housing are made by acrylic plastic plate. For detection, an electrochemical detector is

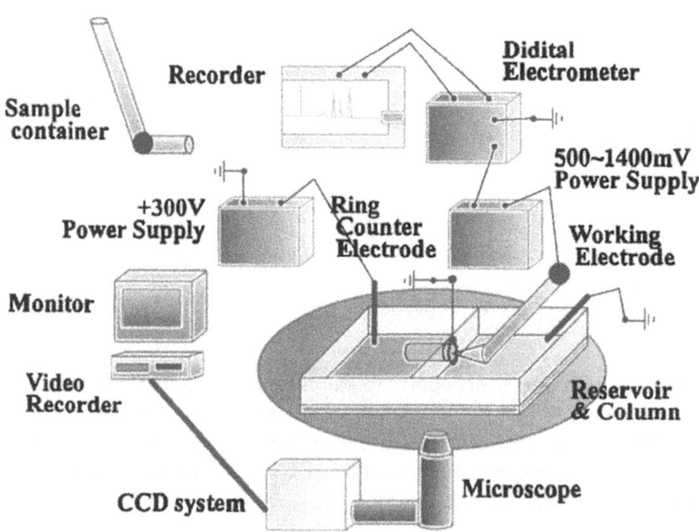

FIGURE 10 Concept of miniaturized chromatographic system on the stage of a microscope.

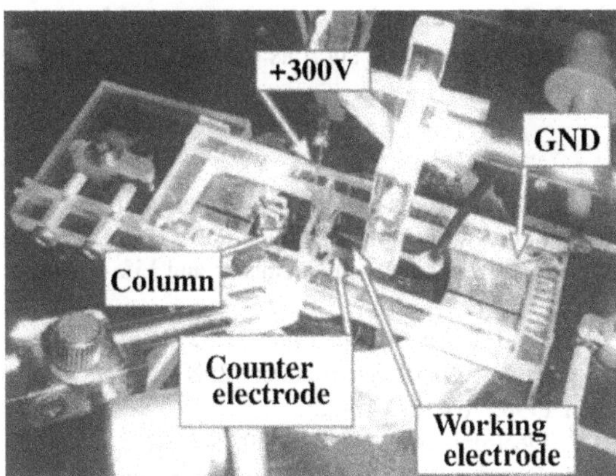

FIGURE 11 Photograph of chromatographic system constructed on a slide glass for a microscope.

used. When we use an optical fiber for leading UV light, we can use it at the end of the ultrashort column. These settings are shown in Figure 11 [13].

The sample chamber is made of a silicone rubber tube and has two parts: a holder and the sample chamber (volume 16 μL). Injection is performed without shutting off the applied voltage. When the mouth of the sample chamber is transferred to cover the end of the capillary column during applied voltage along the column, the sample solution is immediately injected electrokinetically into the column. In our design, the length of the extruded part of the capillary column in the inlet-side reservoir is about 10 mm [6]. Therefore, any instability in the baseline of the chromatogram due to on/off operation of the voltage was not observed. As the reservoir on the inlet side of the capillary column is always grounded, the operator is safe, without risk of electric shock. The injection period is defined as the period for the end of the capillary column to be covered with E_1.

6. CONCLUSION

Preparation of our ultrashort capillary column and other tools such as injection and detection are presented in this chapter. These devices can be used for developing a miniaturized chromatographic system, which can be used as an alternative separation method for either a conventional liquid chromatographic system or a micro total analysis system.

The results discussed here show that geometric miniaturization opens up further possibilities for solving a difficult problem that might not be solved by traditional tools. The miniaturized chromatographic system presented here can be useful for analysis of trace biological samples, such as single-cell analysis.

REFERENCES

1. Sakaki T, Kitagawa S, Tsuda T. In: Proceeding of the 20th Symposium on Capillary Electrophoresis Nov. 29–Dec. 1, Awaji, Japan. 2000:94–95.
2. Manz A, Becker H, eds. Microsystem Technology in Chemistry and Life Sciences. Berlin, Germany: Springer-Verlag, 1998.
3. Jacobson SC, Ermakov SV, Ramsey JM. Anal Chem 1999; 71:3273–3276.
4. Ericson C, Holm J, Ericson T, Hjertén S. Anal Chem 2000; 72:81–87.
5. Dadoo R, Zare RN. Anal Chem 1998; 70:4787–4792.
6. Sakaki T, Kitagawa S, Tsuda T. Electrophoresis 2000; 21:3088–3092.
7. Tsuda T, Matsuki S, Munesue T, Kitagawa S, Ochi H. J Liq Chromatogr 2003; 26: 697–707.
8. Tsuda T, Kitagawa S, Dadoo R, Zare RN. Bunski Kagaku 1997; 46:409–414.
9. Tsuda T, Ikedo M, Jones G, Dadoo R, Zare RN. J Chromatogr 1993; 99:201–207.
10. Tsuda T, ed. Electric Field Applications in Chromatography, Industrial and Chemical Processes. Weinheim, Germany: VCH, 1995.
11. Tsuda T. Bunseki 1999; 4:335–339.
12. Kitagawa S, Inagaki M, Tsuda T. Chromatography 1993; 14:39R–43R.
13. Sakaki T. Micro-electrochromatography constructed on a stage of microscope. Master's thesis, Nagoya Institute of Technology, Nagoya, Japan, March 2001.
14. Pesek JJ, Matyska MT. J Chromatogr A 1996; 736:225–264.
15. Pesek JJ, Matyska MT. J Chromatogr A 2000; 887:31–41.
16. Guo Y, Colon LA. Anal Chem 1995; 67:2511–2516.
17. Constantin S, Freitag R. J Chromatogr A 2000; 887:253–263.
18. Matsuki S. Electrochromatography using an open-tubular capillary column having a stationary phase with large surface area. Master's thesis, Nagoya Institute of Technology, Nagoya, Japan, March 2002.

10

Microstructure and In-Silico Developments for High-Sensitivity Proteomics Research

Thomas Laurell, Johan Nilsson, and György Marko-Varga
Lund University, Lund, Sweden

1. INTRODUCTION

Both gene and protein analyses require a large range of sophisticated analytical technology developments that address the specific part of, for instance, sequence information or protein structure details. This biological information is vital in the overall life science research that currently has reached a number of milestones such as the human genome map as well as a number of other microorganism and species sequence maps. Today it is also possible to clone, express, and make crystal structures from proteins of interest to be used for further modeling work that is fundamental in the understanding of protein structure and the possible interactions and affinities that these proteins can undergo. These steps can be performed in large scale with the ability to produce milligram amounts of protein, sufficient to make additional structure characterization with modern analytical technology such as capillary chromatography and gel-based separations and mass spectrometry for protein structure and sequence elucidation. In this respect, DNA molecules are considered easier to handle bioanalytically when compared to proteins. The long spiraling DNA ladder molecule with the famous double helix is composed of only four basic constituents. Proteins, on the other hand, are much more complicated biomacromolecules that can fold up into intricate and often unpredictable forms and shapes, with 20 amino acids as building blocks. This aspect by itself makes sequencing of proteins a much harder task. In addition, the set of proteins a cell uses is constantly changing with time, which adds additional challenges. Posttranslational modification, which is used by the cell to balance between active and inactive forms of the protein, occurs continuously, which makes the proteome a moving target, where one finds that the actual expression, regulation, and composition in the cell are dynamic events.

Posttranslational modification mechanisms are also a way for the cell to keep intracellular biological systems in balance. As an example, a single modification such as a phosphorylation can alter a biologically active target compared to its inactive latent form. The stoichiometry of these events are key requisites, for instance, in the mechanistic understanding of cell signaling, as exempified below. The kinetics of these events are crucial; some proteins are modified in less than a minute, while others can persist in the cell for hours or even days without undergoing structural changes. Finally, the major difference in approaching proteomics research which makes a big difference to the field of genomics, is that there is no polymerase chain reaction (PCR) technology for proteins that makes sequence amplifications possible.

1.1. Directed Government and Private Research Efforts

Both academic research and pharmaceutical drug development are looking for the "holy grail" of proteomics. Pharma companies are continually seeking ways to reduce the expense and duration of drug discovery and development, while academic groups focus on trying to understand the mechanisms that govern the way proteins are expressed and regulated. In this approach, information about a target protein's 3-D structure to design for biopharmaceutical modulation, as well as the development of small-molecule compounds that bind to the protein and alter its biological activity, are of key importance.

Certainly, the impact of the human genome initiative has sparked renewed interest in the application of protein structure determination, resulting in structural proteomics reaching all the way to basic biological research as well as drug discovery and development.

In the last years, several countries, including the member states in the European Union, the United States, Canada, and Japan, have been active in organizing programs that have been government-funded to improve the techniques of structural protein determination. Additionally, the discovery of novel structure settings whereby the understanding of protein and protein complex formations and alterations will be better understood are also objectives for these programs. These initiatives are fueling the growth of a high-throughput structural proteomics industry, in which biotech and pharma companies have demonstrated new, higher-throughput methods of protein sequence and structure determinations. A very important component of this research is the advances that are being made in information technology (IT). A strong software capability is required for full development and exploitation of structural proteomics. One should not underestimate that structural proteomics will be a useful tool in the emerging discipline of systems biology. An example is the newly established nonprofit organization, Institute for Systems Biology (Seattle, WA, USA), which has a strong position in protein science and technology, with an entrepreneurship ability that has an impact on the proteomics society globally.

One urgent issue at the moment is the handling of the wealth of data generated by proteomics studies, resulting in large numbers of both protein annotations and peptide sequences. The task to fish out the important information and try to link this biological information to an understandable biological evolution is a great challenge, especially as we start to reach the medium- and low-abundant protein expression levels.

2. PROTEOMICS RESEARCH FIELDS

Proteomics research also has applications within areas such as target validation, drug screening, and the discovery of diagnostic markers. Proteomics may be divided into three categories: expression proteomics, functional proteomics, and structural proteomics. Functional proteomics focuses on the high-throughput determination of protein function. Expression proteomics handles the information gathered in studies in which data generated from gel or chromatography separations is followed by mass spectrometry identification. Finally, structural proteomics verifies the validity of key target proteins that have been selected as interesting candidates in a study that has proven to have a functional link to the biology being addressed. Structural proteomics using 3-D structure analysis will make specific binding and folding hypotheses possible.

Current functional proteomics studies emphasize the determination of protein–protein interactions (see below), but also include the study of small-molecule and ligand–protein interactions, changes in protein phosphorylation states, activation during signal transduction (phosphoproteomics), protein localization, and enzyme activity. Proteomics is a collection of scientific approaches and technologies aimed at characterizing the protein content of cells, tissues, and whole organisms. An important goal of proteomic studies is to understand the biological roles of specific proteins and to use this information to identify key regulatory proteins, as presented by the schematic illustration in Figure 1 [1–3]. In this respect, both liquid-phase and gel-based separations are currently used by proteomics research groups. These two high-resolution technologies are highly complementary separation techniques. It seems also at this point that within mass spectrometry, laser desorption, and electrospray ionization are complementary ways of generating protein sequence and structure identities. Although Fourier transform ion cyclotron resonance mass spectrometry has a fantastic potential, it is still not yet in a format which allows efficient high-speed protein identifications to be made. Smith et al. [4] have put forward a strategy whereby

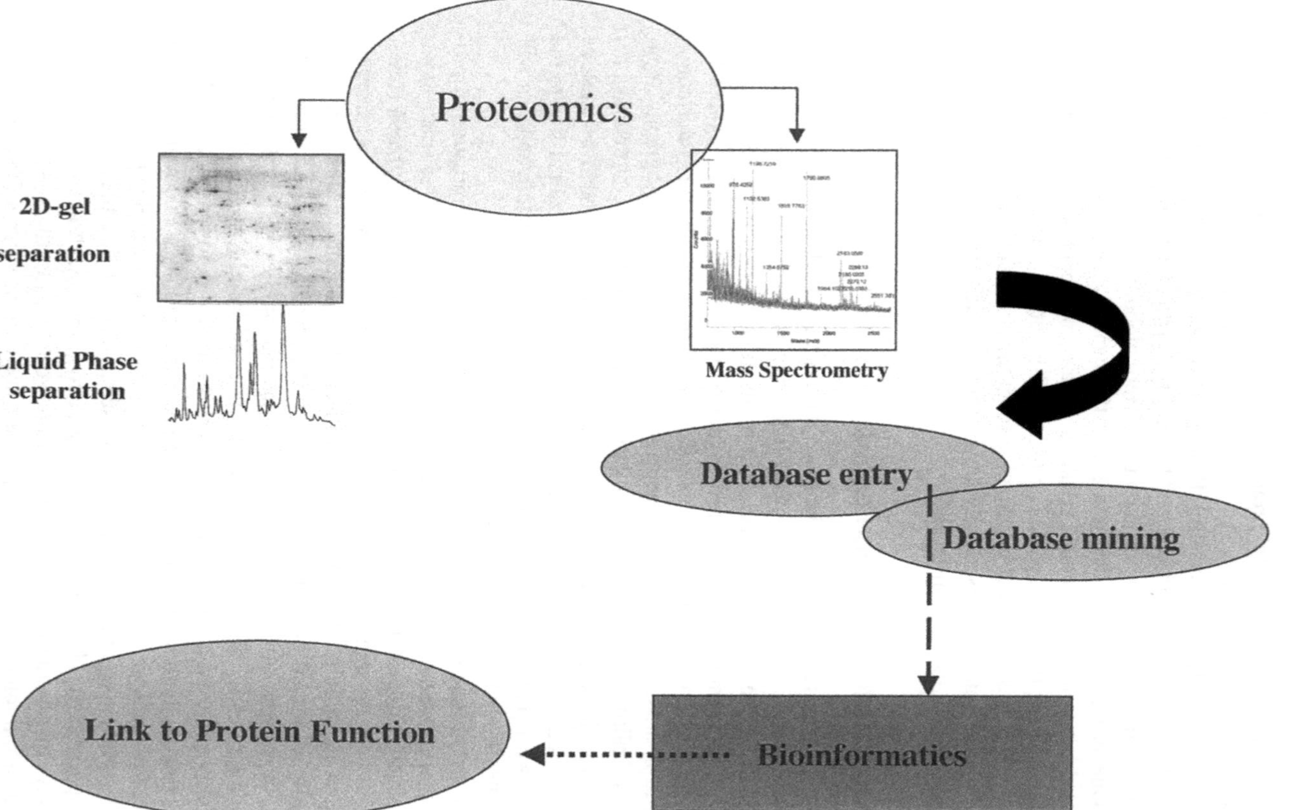

FIGURE 1 Schematic illustration of the proteomics process comprising the utilization of both gel-base and liquid-phase separations interphased to mass spectrometry analysis followed by database search mining, annotation, and a final link to the functional role of proteins.

it would be possible to characterize significantly more than 105 components with mass accuracies better than 1 ppm. This will build upon development of the accurate mass tags, produced by global protein enzymatic digestions.

Proteome research has become one of the most rapidly expanding research areas. By comparison to genomics it offers significantly more complexity, which also results in more technical challenges to the research field. This is contrary to the genomics field, in which the technology is more or less established and instrumental sequencing platforms are readily available. The sample preparation part of the technology whereby DNA and RNA is isolated is also a rather straightforward, standard procedure that automated robotic instrumentation can perform. This is in most situations not the case in protein expression research. Fully automated instrument platforms are available that can handle 2-D gel separations, image analysis, and excision of gel spots and subsequent mass spectrometry analysis. However, these robotic platforms can only handle the high-abundant expression area of the proteome. On the other hand, multidimensional chromatography is used by interfacing to on-line, or off-line to MS, reaching low-abundant protein expression. These levels can be reached only with the prerequisite that sufficient biological material is at hand. Even with sufficient starting material, the limitation is still that the weaker spots or the proteins at low levels are still difficult to identify for both gel-based as well as liquid-phase separations.

2.1. Protein–Protein Interactions

Proteins seldom operate on their own as single biomacromolecules: they form complexes and complex variatiants. Human cells are also more demanding to make in-vitro developments from, since slow proliferation and cell cycle turnover are rate-limiting. This has led to a move toward utilization of microorganisms to study protein complexes from a human cell perspective. Most popular are adaptations of the yeast two-hybrid assay, which constitute the most frequently used technologies for high-throughput, systematic determination of protein–protein interactions. These investigations in yeast can be made by high-throughput genome-wide two-hybrid studies. The yeast two-hybrid assay is a method for determining whether two proteins can interact physically with each other. However, there are limitations, such as the fact that interacting proteins must be able to enter the yeast nucleus in sufficient quantities to interact and assemble the transcriptional activator, a requirement that is not met by all proteins. Additionally, interacting proteins may not be transcriptional activators themselves, limiting the type of proteins that can be analyzed. Trans-membrane regions of proteins are also difficult to study. With the exception for all the limitations inherently present in microorganisms such as the yeast two-hybrid system, databases are available that are very helpful to drug discovery, enabling hypothesis building and making decisions on the best of choices for picking out potential drug targets.

2.2. Biological Regulations

The link between biological changes within a cell or organ and the actual "snapshot" expression status of genes and proteins at that very moment is a massive challenges for life science scientists. In this respect, human diseases are in many cases closely linked to the signal transduction pathways in cells. Cell signaling is fundamntal in all living organisms. Typically, signal transduction can operate through various molecules, receptors, and signaling pathways. Each cell and cellular component must detect outside impulses through receptors and ion channels, which are positioned throughout the cell membrane. In most cases the mechanism involves a collaboration between contiguous cells and internal molecules. It compiles a complex system of communication, known as signal transduction. The very function of cell signaling and the detailed understanding of the various members of this chain in the cascade offers the promise and possibility of better treatments for disease. There is an intense activity within research groups trying to develop tools to map key regulating targets known to be associated with various diseases. Another area of intense development is the establishment of ion channel assays permitting the passage of, e.g., calcium, potassium, sodium, and other ions. In these cases, G-protein-coupled receptors respond to external signals by interacting with G proteins whereby cell-surface enzymes penetrate the surface of the cell membrane in response to outside stimuli.

3. HUMAN PROTEOME STUDY

Our group has been actively involved in studying human primary target cells that are involved in pulmonary diseases such as fibrosis [5–8]. The elucidation of protein expression in human primary cells upon growth factors such as TGF-β has been of major interest to us. TGF-β stimulation will induce a large number of proteins within the cell, and one aim is to map proteins which are posttranslationally modified or otherwise regulated by the TGF-β signaling pathway. Figure 2 shows the signal transduction pathway that we and others have investigated in some detail [7–10]. It presents the SMAD pathway, whereby TGF-β as the ligand for the TGF-β2 receptor binds, followed by a dimerization with TGF-β2 receptor triggering the intracellular signaling. The details of these experiments are presented below.

3.1. Disease Pathways

It is known that SMAD proteins, identified as signaling mediators of the TGF-β superfamily, are involved in cell growth, morphogenesis, development, and

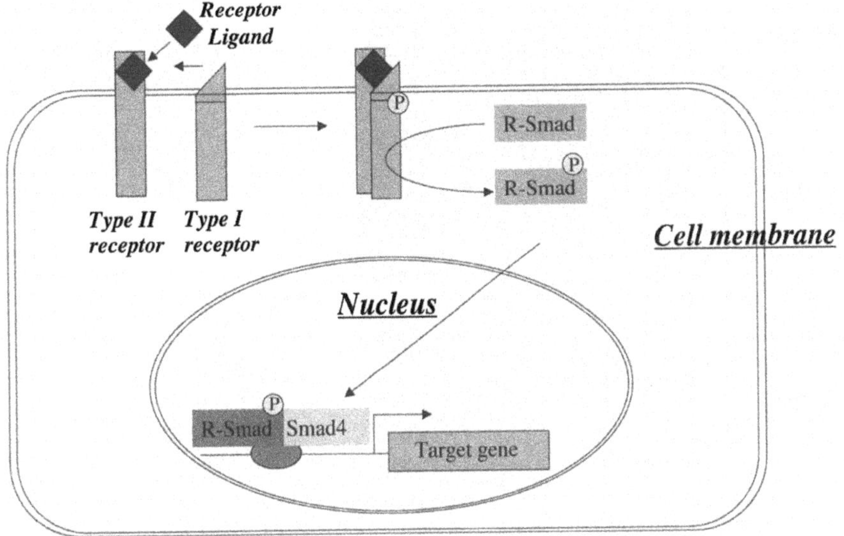

Figure 2 SMAD signaling pathway, in which ligand binding activates the dimer formation of type I and type II receptors, which initiate intracellular signaling.

immune responses [11,12]. It has been reported that eight mammalian SMAD proteins have been identified [12].

The SMAD family is classified into three major groups: the pathway-restricted SMADs (SMAD 1, 2, 3, and 9); the common mediator SMAD (SMAD 4); and inhibitory SMADs (SMAD 6 and 7). The pathway-restricted SMADs become phospohorylated by serine/threonine kinase receptors and then associate with the common SMAD. The heteromeric SMAD complexes then translocate to the nucleus where they bind to DNA and induce transcription of downstream targets. In mammalian cells, SMAD 4 forms complexes with SMAD 2 and SMAD 3 after activation of TGF-β or activin type I receptors [13,14], whereas it forms complexes with SMAD 1 [14,15] and SMAD 5 and possibly with SMAD 9 after activation of BMP type I receptors.

In order to study the changes in protein expression in consequence of TGFβ1 activation by the TGF-β1 receptors, we used human fetal fibroblasts. In this initial study, protein expression patterns were compared between cells in a resting state in comparison to activated cells. The proteins were separated by 2-D PAGE and the protein patterns were analyzed by image analysis using the PDQUEST software (Bio-Rad). A representative expression profile of the human fibroblast proteome is shown in Figure 3, with corresponding protein spots at an expression density of approximately 1700 spots.

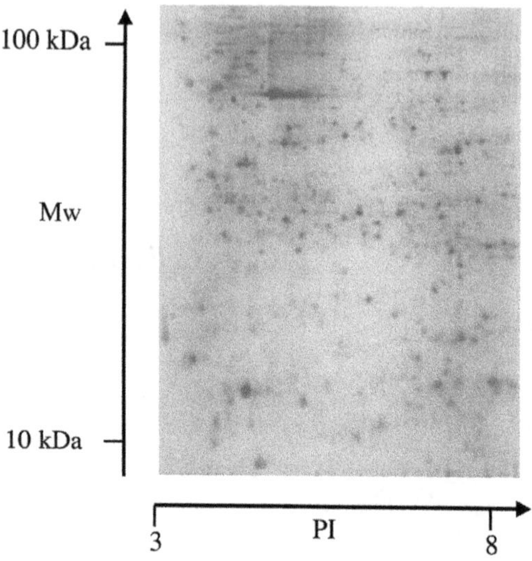

FIGURE 3 Two-dimensional gel electrophoresis expression map generated from primary human cells using immobiline strips with a PI range of 3–10 and a molecular size window of 8–120 kDa.

Comparison of member gels revealed a number of proteins that were expressed at different levels between the samples [4–6]. The hierarchic expression process that was made is shown in Figure 4, starting with the member gels comprising the resting state of cell samples to be compared with the activated cell expressions. Finally, after image analysis, a set of reference gels is produced, from which the final master gel is made. The master itself is a synthetic gel that holds all the experimental diversity and information from all the member gels.

The human primary cells were washed and the amount of cells were estimated to 1.3×10^6 cells. Solubilisation was performed using 7 M urea, 2 M thiourea, and 4% CHAPS. The resulting gels are depicted in Figure 3 and 4.

3.2. Protein Separations

Two-dimensional gel separations were performed on the cell lysates generated. We used Immobiline Dry strips (180 mm, pH 3–10 nonlinear), rehydrated in solubilization solution also containing 10 mM dithiothreitol (DTT), and 0.5% IPG 3-10 buffer. The isoelectrophofocusing (IEF) step was performed at 20°C in a IPGphor. We found it to be important to equilibrate the strips for 10 min using 65 mM DTT, 6 M urea, 30% (w/v) glycerol, 2% (w/v) sodium dodecyl

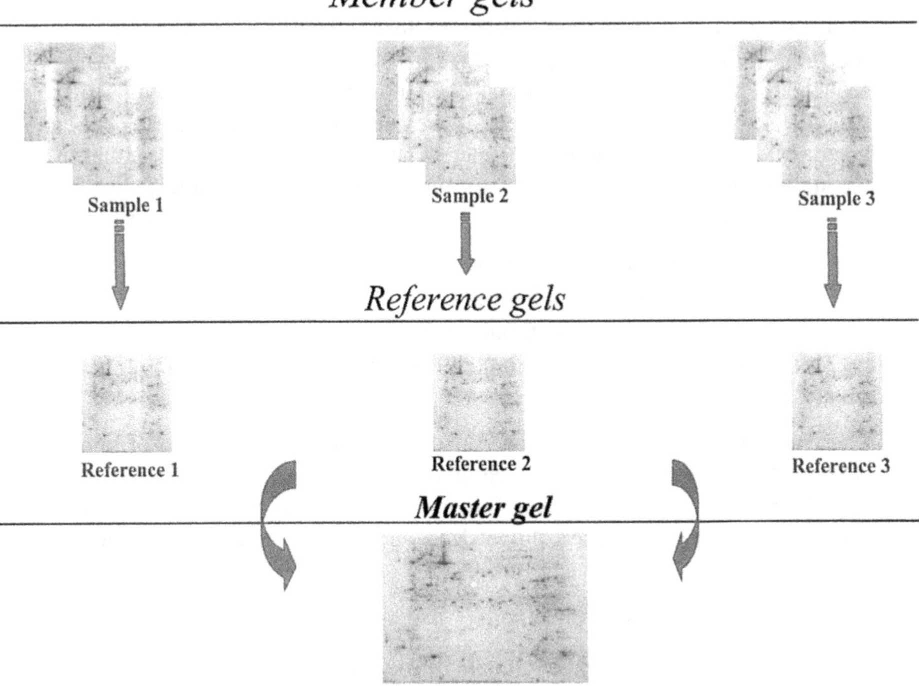

FIGURE 4 Hierarchtic image analysis processing of a subset of primary human cells whereby the groups, clustered from the first set of a reference image, are compared. These reference images are next compared to one another by reference set analysis, which finally forms the master image that holds all the information within all of the participating member gels.

sulfate (SDS), and 50 mM Tris-HCl pH 8.8. A second equilibration step was also carried out for an additional 10 min in the same solution except for DTT, which was replaced by 300 mM iodoacetamide. The strips were next applied onto a 14% homogeneous Duracryl SDS gel. The strips were overlaid with a solution of 1% agarose in electrophoresis buffer (kept at 60°C). We then ran the electrophoresis in a Hoefer DALT gel apparatus at 20°C and constant potential, 100 V for 18 h. In these resulting proteome maps we used postgel fluorescent staining, a fast and simple staining technique that is ready within a few hours and compatible with mass spectrometry. It also allows sensitivities comparable to silver staining, but with an increased linear dynamic quantification range.

3.3. Sample-Handling Procedures in Proteomics Studies

In the search for proteins with properties that influences important biological functions we have come to a point where both separation efficiency as well as sequencing ability and sensitivity can be automated to a great extent. Experimental in-vivo and in-vitro models as well as small-animal experiments and human clinical material can easily generate peptide mass sequence data that fills the autosamplers and 2-D gel tanks for higher-capacity expression studies.

In principle, the most common ways of separating protein samples include:

Two-dimensional gel electrophoresis
One-dimensional electrophoresis
Multidimensional liquid chromatography
Various types of chromatographic fractionation
Affinity probe isolations

These techniques are also often used in combination. A high number of samples is generated in these studies, most commonly by 2-D gel spot excising or capillary chromatography fractionation. Either way, there is a great demand to handle fractions that vary in sample volume sizes, and that can be enriched in terms of

Volume reduction
High-density protein preparation

These requirements are highly valuable for both laser desorption ionization and electrospray ionization mass spectrometry analysis.

These sample preparation processes chosen will make a big difference in the final sequence readout, since often several-orders-of-magnitude improvements can be achieved [16–20] by the use of microtechnology and microfluidic systems. In our experience it is fair to say that often one is left with no protein ID at the end of the day if no sample preparation is applied, or, if the sample preparation methodology and technology are not performed in an optimal way. However, if made with careful experimental design using cutting-edge technology, the problematic protein spot/chromatographic sample fraction will result in identities that have significant ID hits in the database query protocol.

Sample enrichment by solid-phase extraction has several purposes. It not only makes the proteome sample amendable for analysis to a greater extent, it also makes structural information greater.

Typically, e.g., in excised 2-D gel samples, one often has problems in MS analysis with (a) interfering acryl amide contamination or other types of interferents present in the gel that are dissolved upon extraction of the proteins, and (b) interfering salt that hampers sequencing and peptide mass fingerprint generation by mass spectrometry analysis. Finally, but not the least important aspect, is microextraction enrichment whereby the actual detection amplification is obtained. However, success in the last step of the sample preparation procedure also requires that the two previously mentioned analysis limitations are solved (a and b above) to an optimal extent.

4. MICROSTRUCTURES AS ANALYTICAL TOOLS

Small does not just mean that one uses less consumables and expensive chemicals and biochemical reagents in analytical measurements. Small will also lead to increased sensitivity and reduced reaction time. However, in order to utilize the full power of miniaturization, the proper requirements for adapting analysis in miniaturized units needs to be addressed and solved.

4.1. Lab-on-a-Chip

Lab-on-a-chip has its origin in the μ-TAS idea, first postulated by Michael Widmer, at that time with Ciba-Geigy, Basel, Switzerland [21]. Today there are a great deal of follow-ups and developments on this very idea, including instrumentation as well as dedicated assay designs that is being developed and used by companies, of whom probably the most well recognized are Caliper and Agilent. Microchip and microconduit units have been developed and are commercially available whereby fast and ready-to-use instrumental kits can be used [22–24]. There is even a scientific journal, *Lab-on-a-Chip*, since a few years ago, as well as a μTAS-section within *Sensors & Actuators B*.

4.2. Microfluidic Chips for Peptide and Protein Analysis

The microfluidics area is expanding rapidly due to the possibility of handling large sample sets in which only minute amounts of the sample is required [25].

Microstructures for integrated peptide and protein analysis address in most cases the analysis of endogenous markers, amino acids, enzymes, and antibodies, but also lately a broad window of proteins identified in proteomic studies [26–28]. To a large extent, developments made utilizing microfluidic chips have been performed by the separation of peptide- and protein-containing samples with electromigration techniques. Both zone electrophoresis and the use of sieving matrices have been applied [29]. One makes use of the electrostatic charge the protein has at a given pH and separates them as either positively or negatively charged biomacromolecules. Isoelectric focusing is another powerful technique whereby microchannel systems are suitable for protein separations. It has been presented by several groups [30–32], and the interface to mass spectrometry for sequence identification is a part which requires special attention. Typically, but not necessarily, fast assay readouts can be obtained by CE on a chip [33]. Various coating procedures have been investigated, being a main variable in order to obtain high resolution with reasonable separation cycles [34,35].

Methods similar to two-dimensional gel electrophoresis, which currently has the highest resolving power of intact protein separations, have been developed for capillary electrophoresis chip separations [36]. Yao et al. presented an SDS microseparation system for the analysis of intact proteins [37]. An alternative to this approach is the microchip separation devices for digested protein samples, the resulting peptide mixtures separated by electrochromatography, as reported by Slentz [38]. The supports used in these units were not traditional silica- or polymer-based beads, but rather a small micromachined stationary phase within the microchannel. These monolithic microphones compile an array of thousands of small cylindrically shaped channels. A version of the monolithic support structure was also demonstrated with polystyrene sulfonic acid [38]. Clearly the use of polymeric materials has an advantage in terms of manufacturing costs and a more simplified processing procedure as compared to etched silicon. Coyler showed an application whereby serum proteins were separated by microchip capillary electrophoresis with postcolumn derivatization and detection, all integrated within the microchip [28].

Micellar electrochromatography was applied by Ramsey's group on a two-dimensional separation system with a microfabricated device, using two different separation mechanisms, electrochromatography as the first dimension and capillary electrophoresis as the second dimension [39]. For detection of proteins, fluorescently labeled products from tryptic digests of β-casein were analyzed in 13 min with this system [39].

4.3. Microstructure Developments for Protein Amplification Detection

The ability to handle low sample volumes has become essential in order to gain high yield in sensitivity amplification by different types of protein enrichment technologies. This thereby provides several benefits:

1. It makes it possible to use small amounts of biological sample material.
2. It offers the ability to perform complementary protein identification as well as functional and structural assays on a single sample.
3. It increases analysis throughput and array analysis.
4. It provides improved sensitivity.

These abilities that new technology offer is of high importance in, for instance, clinical studies, where in many situations there are only limited amounts of, e.g., biofluids such as synovial fluid, or tissue biopsies. There is also an ethical aspect of what sampling steps actually can be performed in human clinical studies in order not to risk the health and well-being of the patients. It is also a fact that when clinical studies are being done, some of the patient groups, for instance, the healthy, do not have the same pathological changes within a certain organ. As a result, the sampling of a biofluid that might be central to the study is difficult to perform. Taking the above into account, the need for minute sample analysis technology is essential, and is why microstructure developments in our microtechnology research group have a central role.

The microstructure material most commonly used by our group is monocrystalline ⟨100⟩ or ⟨110⟩ oriented silicon. Generally, wet etching is performed by oxidizing wafers at high temperature, whereby a protecting layer of silicon dioxide is formed. Next follows a process step in which the formed silicon dioxide layer is coated with photoresist by spin coating. A mask with the desired pattern is then placed in proximity to the wafer and is illuminated with UV light. After additional processing steps including high-temperature exposure and wet etching by KOH, the highly porous silicon morphology, the desired properties we utilize for bioanalytical purposes are obtained by anodizing the silicon wafer in an electrochemical etching process. The large number of variable porous structures that can be developed is obtained by controlling the anodization voltage, current, and illumination of the sample. The final step of the process involves removal of the remaining silicon dioxide using HF. Figure 5 shows examples of porous silicon structures at varying magnifications, where the space and three-dimensional orientation is best illustrated in the lower image capture.

The gain in absolute surface that is generated by the various etching techniques is shown in the side views of the silica structures in Figure 6. The three pictures in Figure 6 (a–c) show an absolute increase in surface area, accessible as an active biosurface for protein analysis.

The three-dimensional porous silicon depends on the process that generates pores, which vary from nanometers to micrometers in size. If one makes calculations on the surface area gain that is obtained by varying microstructure geometries as shown in Figure 6, it is clear that a highly diffrentiated/fractalized surface is desired. This can to a certain extent be accomplished by means of photolitographic techniques, however, as the fractal degree reaches structural dimension below micrometer level, microfabrication processing costs become a major issue from a commercial point of view. In order to address this issue, our group has introduced the use of nanoporous silicon in-situ fabricated in microfluidic chip components, generating high-surface-area microchips for biochemical analysis. As an alternative to this technique, the use of microbeads loaded into microfluidic chips has also proven to be a successful way of obtaining high-surface-area microfluidic components. A number of groups have developed and introduced different chip designs to perform on-line digestion [40–43] prior to both laser desorption and electrospray mass spectrometry. The rapid enzyme catalysis, which occurs within minutes rather than hours when performed in batch analysis at macro scale, was proven for a number of different proteins [40–43].

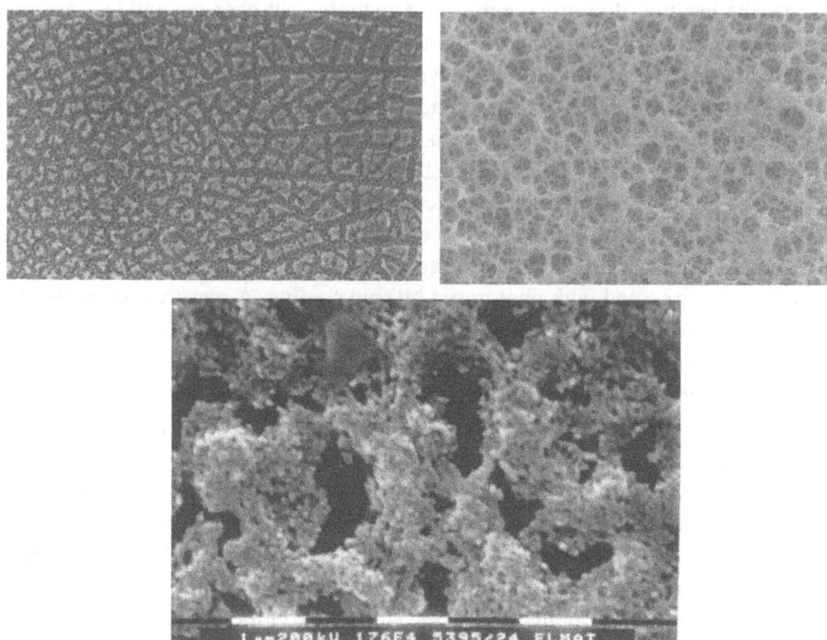

Figure 5 Silicon structure morphologies utilized in proteomics, pictures seen from above. The resolution of every morphological structure is increased from left to right.

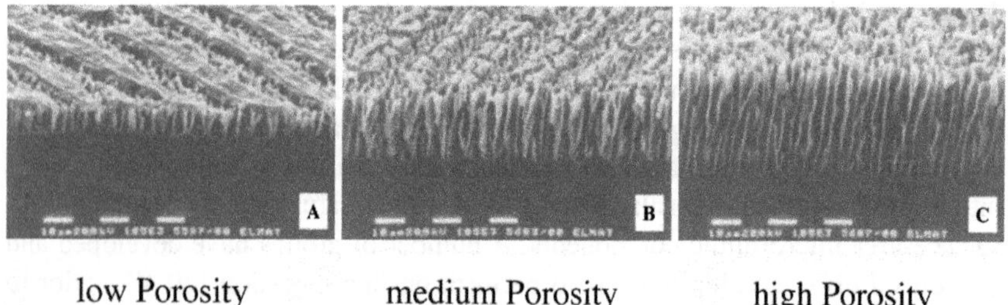

Figure 6 Varying monocrystalline porosities generated by the etching manufactiring process whereby characteristic low-, medium-, and high-porosity structures can be processed.

4.4. Dedicated Silicon Microextraction Chip

An evident application in protein-related work at the chip level is the use of solid-phase microextraction for sample clean-up and enrichment of dilute samples. The original silicon microextraction chip (SMEC), investigated in [16], had a weir as a bead-trapping structure and was fabricated in silicon. This chip is referred to as the weir-SMEC henceforth. A principle drawing of this chip is shown in Figure 7a. A new and optional design for a microextraction chip was realized having a standing wall grid in the flow channel as the bead-trapping structure (Figure 7b), henceforth called grid-SMEC. The use of ⟨110⟩ oriented silicon enabled the microfabrication of the vertical grid structure utilizing conventional KOH wet etching [20].

It should be noted that the bead-trapping chip was fabricated using only one photo lithography and anisotropic wet etching step. A scanning electron micrograph of the grid of standing walls composing the bead trap is shown in Figure 8.

The final grid-SMEC with the packed bed and the exiting capillary is shown in Figure 9. The seven standing walls were spaced 16 μm apart; each wall was 13 μm wide and 200 μm high. To collect the effluent from the chip, a capillary was glued into the flow channel immediately after the bead trap, minimizing the dead volume of the structure. The individual beads of the pack-

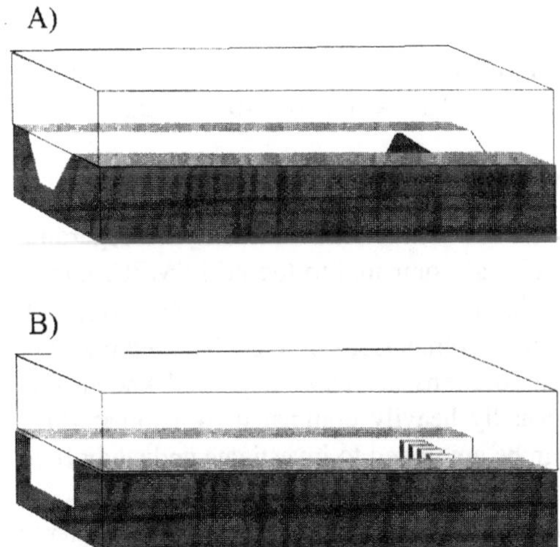

FIGURE 7 Picture of silicon microextraction chip (SMEC) structures with weir configuration (top figure), and the grid-SMEC unit (bottom figure).

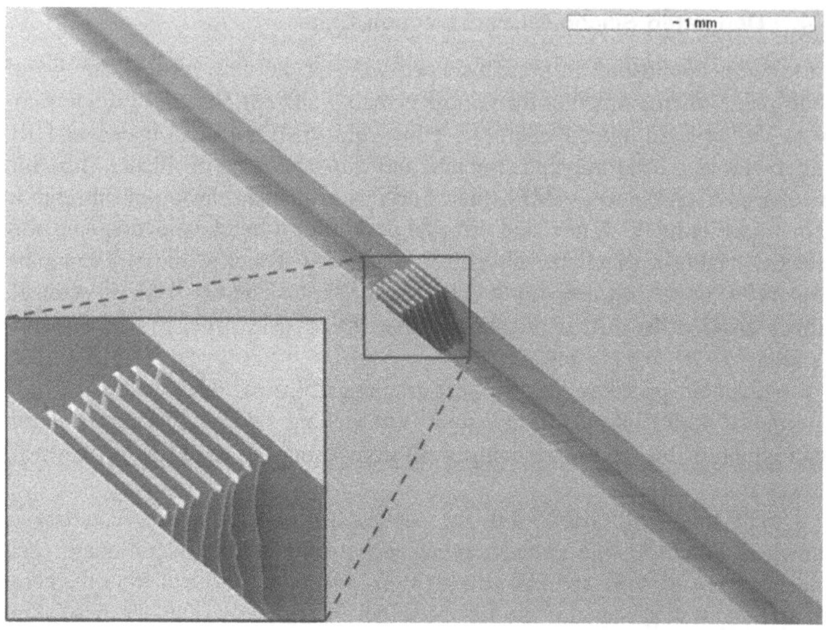

FIGURE 8 Close-up of the grid-SMEC with the seven standing walls within the width of the 200-μm flow channel into which the 20–50-μm-particle diameter beads are packed.

ing material are also resolved in the photo (Figure 9), where the 20-μm-o.d. beads form a packed formation in the capillary with about 10–20 beads.

4.5. Microfluidic Modeling and Dispersion Characteristics

In order to further investigate and understand the unfavorable results regarding sample elution time of the weir-SMEC as compared to the grid-SMEC, microfluidic modeling of the flow profile immediately after the weir was performed. When the size of fluidic channels is in the dimensions of a few hundred micrometers or smaller and aqueous-based solutions are used, it is well known that pressure-driven fluid transport is usually heavily dominated by laminar flow. Therefore microfluidic modeling can be employed to investigate and study fluid flow characteristics in microchip designs prior to fabrication.

An immediate analysis of the two chip structures indicated that the large changes in channel geometry of the weir-SMEC would cause a difference in dispersion effects. This hypothesis was investigated by means of microfluidic modeling, comparing the flow profiles of the two SMECs. The software package

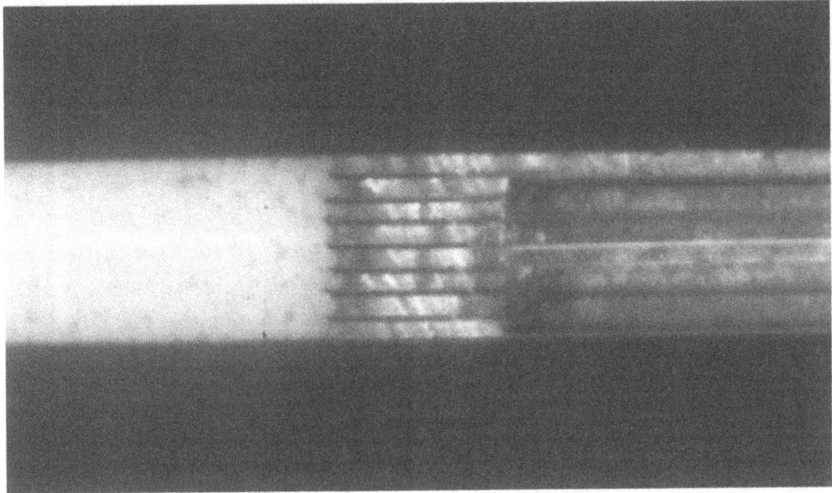

Figure 9 Close-up of packed-bed weir-SMEC using 20-μm-particle-sized polymeric support material. The homogenious packing structure of the packed bed can be seen on the left-hand side lining up close to the eight-channel microfluidic grid structure of the chip.

used was the ANSYS 5.7.1—FLOTRAN module. The modeling results clearly showed differences in flow profiles when passing through the bead trap and into the exciting capillary. As seen in Figure 10, the dark blue zones indicate low flow velocity zones, whereas yellow and red indicate high linear flow rates [20]. The weir-SMEC clearly demonstrates large stagnant zones immediately after the bead trap.

The box inserts show the spatial distribution of a sample plug at different time intervals after displacement from the packed bed, indicating a large dispersion. In the transition region from the thin bead trapping passage(10 μm high) to the exit zone (the steep slope of the bead trapping wall, 200 μm deep), the rapid change in physical dimensions of the flow channel causes an immediate dispersion of the sample plug. The stagnant zones are perfused very slowly, being a major cause for the long elution time of the sample plug.

In contrast, the grid-SMEC offers channel geometries with less dramatic dimensional changes and thus displays a more homogeneous flow profile with reduced dispersion, as shown in Figures 11 and 12. Modeling of the flow profiles in the grid-SMEC also verified this hypothesis. Figure 11 shows a 2-D lengthwise side view and Figure 12 shows a lengthwise top view of the flow profile in the grid-SMEC. Note that the dynamic range in the color maps of these figures are individually normalized between the grid-SMEC and the weir-

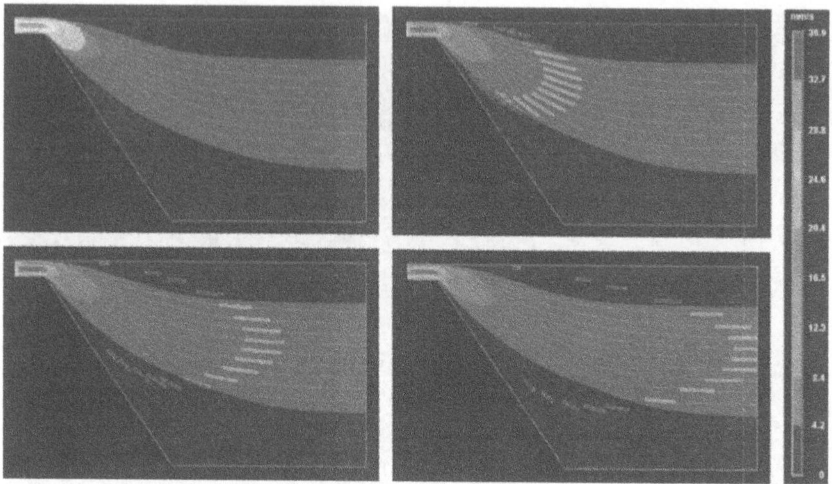

FIGURE 10 Time-lapse sequence (side-view cross section) of a sample plug, transported through the bead trap of the weir-SMEC and into the exiting flow channel. Four time points during the passage of the sample are monitored in these images. The length of each legend indicates the flow velocity.

SMEC. In the grid-SMEC the maximum linear velocity was found to be 8.4 mm/s, whereas the highest linear flow rate in the weir-SMEC was almost four times higher (30.5 mm/s), which also illustrates the problems arising when mixing microfluidic structures with large dimensional changes of the flow channels.

The above example demonstrates the power of utilizing modeling tools when designing microstructures for high-performance bioanalytical systems. Trivial design errors are easily avoided and preliminary optimization of a microfluidic structure may thus be accomplished "in silico," prior to extensive and expensive processing rounds in the microfabrication laboratories.

4.6. Experimental SMEC Evaluation

In order to evaluate experimentally the dispersion characteristics for the two SMEC designs, one weir-SMEC and one grid-SMEC were assembled with the same length, (10 cm), same capillary connectors, and packed with equal amounts of beads using an 80-nL bead volume. Each SMEC was then equilibrated and loaded with the respective peptide and peptide samples.

The resulting data were generated from loading of the two microchip configurations using a peptide mixture of Ang I, ACTH clip 1-17, ACTH clip 18-39, and ACTH clip 7-38 with a level of 2 fmol/μL (2 nM), respectively. This

FIGURE 11 Time-lapse sequence (side-view cross section from the center of a slit) of a sample plug as it is transported through the bead trap of the grid-SMEC and into the exiting flow channel. The red color indicates the higher flow rates of the grid region. Same four flow image of sample passage as in Figure 10.

was followed by a washing step, after which the chip was emptied by pressing air through the structure. Subsequently, the peptides were desorbed from the SMECs with a flow rate of 1.0 μL/min, in which the desorbing agent comprised a mixture of both the organic modifier and acidified matrix. Next, the displaced peptides were deposited from the end of the capillary as discrete spots at 30-s intervals onto the stainless steel MALDI target plate. The crystallized spots were subsequently analyzed by MALDI-TOF MS. These MALDI spectra are shown in Figure 13. Mass spectra A result from the first 30 s, which holds the fraction of 0.5 μL of eluate collected. Mass spectra B result from the eluate collected between 30 and 60 s, spectra C represent the eluate collected between 60 and 90 s, and mass spectra D the eluate collected between 90 and 120 s.

As is easily concluded from Figure 13, the peptides eluted from the weir-SMEC were present in mass spectra A, B, C, and D, while the same elution from the grid-SMEC only gave rise to peaks in mass spectra A, B, and C. Thus, the weir-SMEC gave a larger dispersion, displaying the eluted peptides with an elution time of 2 min. However, the grid-SMEC provided a more concentrated sample plug, with the entire sample plug eluting through the chip within 1.5

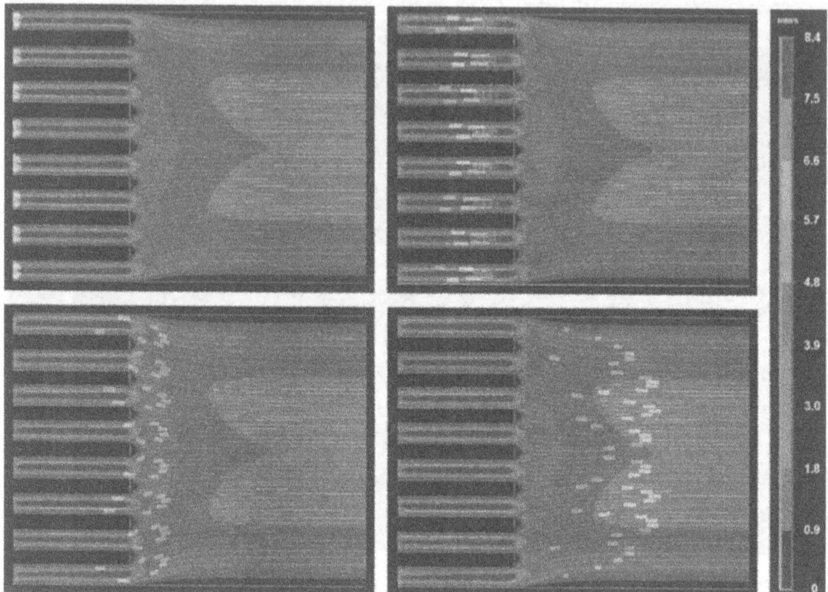

FIGURE 12 Time-lapse sequence (top-view cross section) of the liquid front as it is transported through the grid bead trap and into the exiting flow channel. The length of each legend indicates the flow velocity. Same four flow image of sample passage as in Figure 10.

min. Note that the peaks for two of the peptides disappeared in spectrum C for the grid-SMEC, whereas all four peptides are detected in mass spectrum C for the weir-SMEC.

5. PROTEIN IDENTIFICATION BY MASS SPECTROMETRY

Over the past 10 years, there has been a rapid development of various mass spectrometry techniques in order to sequence, identify, and characterize proteins and peptides. Matrix-assisted laser desorption/ionization time-of-flight (MALDI-TOF) has become the most preferably used technique for a fast processing of biological samples.

The interface of protein separation to protein detection is in many cases a time-consuming and laborious exercise. We developed in this human study a simple and highly efficient process protocol whereby we could advance the protein samples that had been excised from the 2-D gel and digested. The digests were run through the SMEC preparation and analyzed by MALDI-TOF

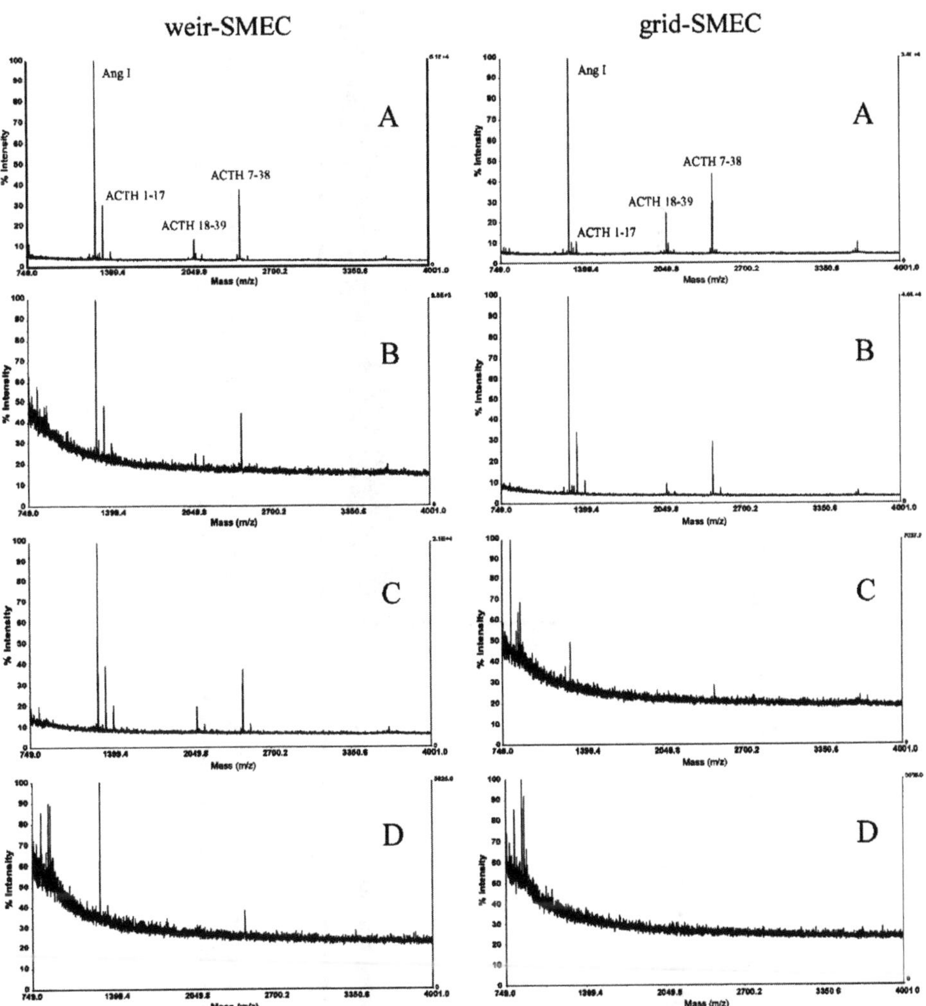

FIGURE 13 Mass spectra of peptide mixture [Ang I, ACTH clip 1-17, ACTH clip 18-39, and ACTH clip 7-38 with a level of 2 fmol/μL (2 nM), respectively] enriched on and eluted from the weir- (left) and grid-SMEC (right), respectively. Each spectrum, A–D, corresponds to 30-s collection of the eluent from each SMEC at intervals 0–30, 30–60, 60–90, and 90–120 s, respectively, using a peptide mixture.

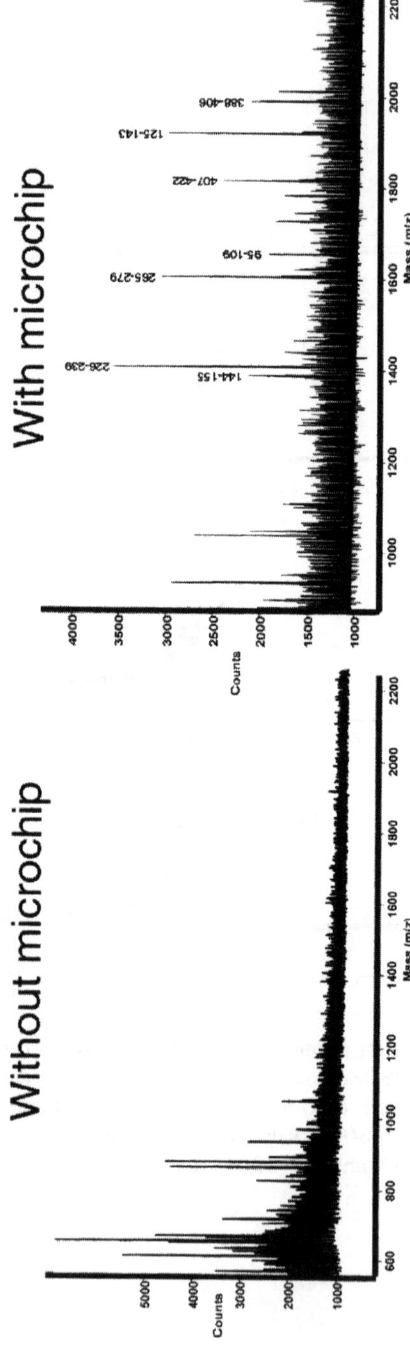

FIGURE 14 MALDI-TOF MS spectra generated from a 2-D gel spot excised from a human proteomic study, in which the corresponding spectrum is shown without SMEC sample preparation (to the left) and with SMEC sample preparation (to the right), illustrating the power of microchip enrichment.

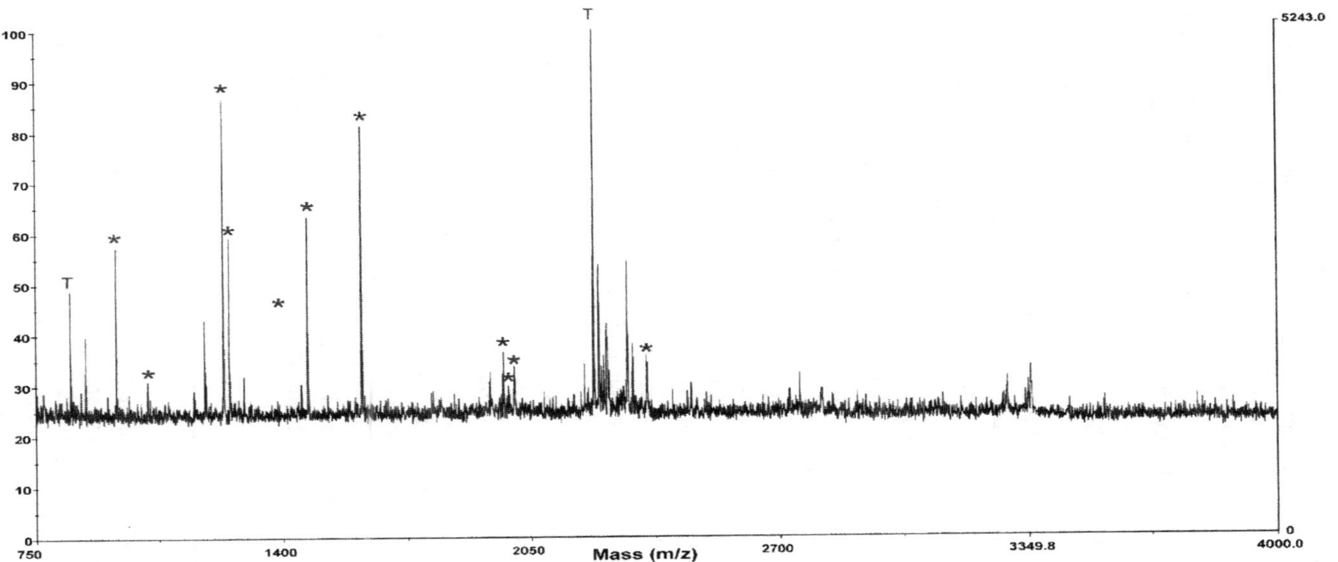

FIGURE 15 MALDI spectrum from a 2-D gel spot excised from a human proteomic study in which the corresponding spectrum of the cathepsin D precursor could be identified after using SMEC micropreparation sample preparation followed by elution and spotting onto the MALDI target plate and MALDI analysis. The peptide mass fingerprinting revealed the identity of the protein using the Mascot bioinformatic software and the Swissprot protein database. The (*) indicates the peptide masses corresponding to the cathepsin D precursor, and (T) the trypsin peptide fragments that were used for internal mass calibration.

MS. The amplification effect is exemplified by the preparation of ATP synthase, in which the spectrum without microchip preparation results in a few unidentifiable peaks around the molecular weight of 850 (see Figure 14, left spectrum). The rest of the spectrum does not show any peaks above the noise level. With microchip preparation we can identify seven peptide masses that corresponds to the successful identity of ATP synthase (see Figure 14, right spectrum). The low level of expression of this protease is also seen from the weak but fully isotope resolved peptides in the spectrum.

The resulting mass spectra from the human cell lysate samples are shown in Figures 15 and 16, in which the peptide mass fingerprints generated with the use of microchip extraction are presented. In the first example we identified cathepsin D, a protease known to be involved in a number of diseases (see Figure 15). In the latter example, 11 peptides were found to correspond to the identity of annexin II as shown in the MS spectrum (see Figure 16). The indication is also made for the internal standards in this spectrum, indicated with T, having molecular masses of 742 and 2021, respectively. These two internal standards are enzymatic products of the tryptic enzyme trypsin itself. With an optimal trypsin addition these two peptides show up consistently in the MS spectrum. The trypsin level needs to be kept low enough not to generate too high peptide signals, which otherwise result in suppression effects that discriminate the ionization of peptides from the analyte protein of interest. The use of too low levels of trypsin, on the other hand, results in low catalysis yield with too few analyte peptides and low levels thereof that make protein identification impossible.

We found in this study that the protein sequence coverages for these human gene products varied between 16% and 41%. The accuracy of the sequence masses were all below 50 ppm. The number of peptides found from the protein identities varied from 4 to 14. The sequence coverage of each protein depends on the size of the corresponding protein, and the number of peptides mass analyzed. It is not the actual sequence coverage that gives the basis for accepting or rejecting an accurate annotation; rather, the variance in accuracies of the respective peptide mass sequences is a good indicator of the solidity of the protein ID. The Mascot software from Matrix Sciences was used in these data-

---

FACING PAGE

Figure 16 Resulting MALDI spectrum from a 2-D gel spot excised from a human proteomic study in which the corresponding spectrum of annexin V was identified with the same approach as in Figure 15. The magnification (at the right top) shows eight of the identified peptide sequence masses at low abundances but still with a statistically significant peptide annotation.

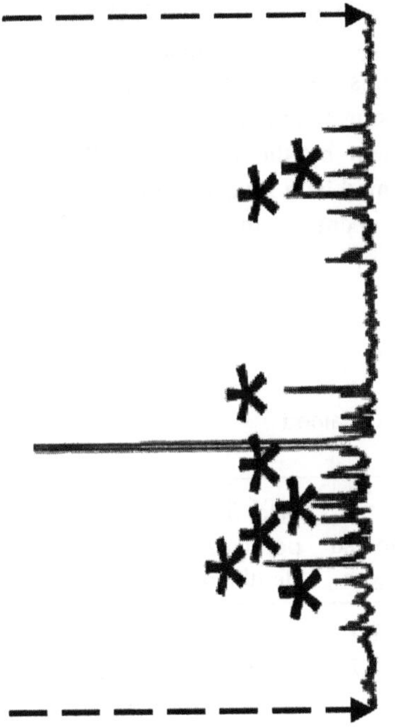

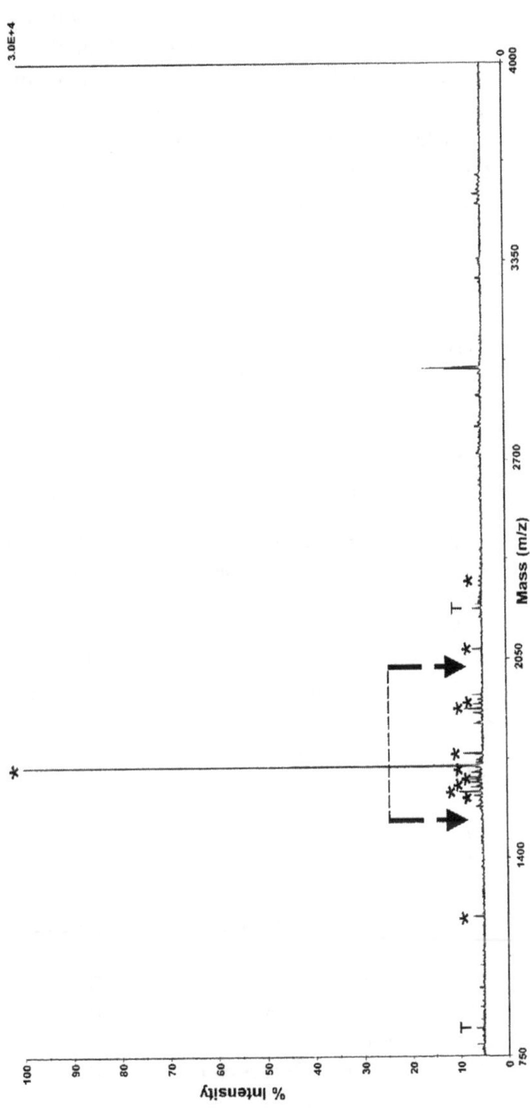
3.0E+4
% Intensity
Mass (m/z)
750
1400
2050
2700
3350
4000
0
10
20
30
40
50
60
70
80
90
100
T
T

base searches. The scoring for each and every protein is given in Table 1. Additional identities were made with the chip technology from annotated 2-D gel samples. It should be pointed out that these proteins were in most cases not readily detectable by MALDI mass fingerprinting using sample preparation with only the dried-droplet technique and no enrichment by the microchip. The knowledge of disease-associated pathways as presented here involves important biological information at the molecular level that opens up the possibility of identifying the best targets for therapeutic intervention.

Table 1 Identification Data on 20 2-D Gel Spots as Obtained by Sample Purification Using the SMEC

Spot number	Identified proteins	Accession number	Mowse score	No. of peptides identified	Sequence coverage (%)
A12	Sorting nexin 4	0952119	52	6	18.2
B12	Calunenin precursor	043852	97	10	21.3
C12	NA	—	—	—	—
D12	Ubiqutin carboxy-terminal hydrolase L1	AAD09172 (P09939)	83	8	14
E12	Cathepsin D precursor	P07339	97	10	29.4
F12	Phosphoglycerate mutase	P18669	127	12	20
G12	Triosephosphate isomerase	P00938	138	11	34.4
H12	NA	—	—	—	—
A11	Vimentin	P08670	159	14	48.3
B11	NA	—	—	—	—
C11	Glutathione transferase omega 1	P78417	49	4	30.8
D11	Nuclear CHL	000299	67	6	20
E11	Serum albumin	P02769	142	15	39.5
F11	NA	—	—	—	—
G11	RHO GDP-dissociation inhibitor (RHO GD1)	P52565	62	6	17.1
H11	Annexin V (lipocortin V)	P08758	62	8	20
A10	Annexin V	P08758	162	13	41.9
B10	Annexin V	P08758	83	7	35
C10	40S Ribosomal protein (40P)	P08865	82	12	16.8
E10	NA	—	—	—	—

6. CONCLUDING REMARKS

Microstructure developments will have a major impact in the new era of proteomics. Micro fluidic modeling is a highly useful "in silico" technology to make the first attempts of hypothesis testing. The improved performance of the grid-SMEC as found by modeling was also verified by recorded mass spectra of model peptides. Future work will also include improved modeling to provide absolute dispersion factors of the microstructures. Further modeling of the SMEC design and subsequent microchip implementation is expected to give us guidance for novel microstructure developments.

Miniaturization developments are also in line with the ambition that the Human Proteome Organization (HUPO) program has. One of the goals in the frame program being set is to generate human antibodies to all of the 35,000 human gene products that will serve as the reagents in human protein microchip devices [44]. The mission of HUPO is to strive for consolidation of national and regional proteomics organizations into a worldwide structure as well as to engage in scientific and educational proteomics activities and to assist in the coordination of public proteome initiatives that stretch geographically over Europe, Asia, Oceania, and the Americas [45]. The first world congress of the Human Proteome Organization was held in Paris, France, in November 2002 where the details of large-scale proteome analysis, bioinformatics, and human immuno reagent were outlined.

Additional challenges will be to map most of the human proteins that will hold around 1 million different types, including the structural variations that occur posttranslationally. This number will exclude all the T-cell receptors as well as all antibodies.

Additional value will be provided by understanding of proteins not only considered as expressed sequences of molecules but rather as three-dimensional structures of biomacromolecules. The recent introduction of thousands of 3-D models of proteins into the EXPASY database will be of invaluable help.

In the near future we will witness the breakthrough of high-throughput proteomics expression studies in which additional information on structure, interactions, and functional aspects of large number of proteins will be made. This, since the true understanding of the cell function is the key information to be gleaned from the high number of experiments performed, in order to get the precious information we need to make a quantitative description. This description needs to include the stoichiometry, kinetics, and activation energy of each protein present in a protein complex in a cellular pathway [46].

REFERENCES

1. Wilkins M, Williams KL, Appel RD, Hochstrasser DF, eds. Proteome Research: New Frontiers in Functional Genomics. Berlin, Germany: Springer-Verlag, 1997.

2. Westermayr R, Vaven T, eds. Proteomics in Practice. Weinheim, Germany: Wiley-VCH, 2002.
3. Aebersold R,Goodlett DR. Chem Rev 2001; 101:269–295.
4. Smith RD, Anderson GA, Lipton MS, Pasa-Tolic L, Shen Y, Conrads TP, Veenstra TD, Udseth HR. Proteomics 2002; 2:513–523.
5. Malmström J, Larsen K, Hansson L, Köfdahl C-G, Norregaard Jensen O, Marko-Varga G, Westergren-Thorsson G. Proteomics 2002; 2:394–404.
6. Malmström J, Westergren-Thorsson G, Marko-Varga G. Electrophoresis 2001; 22: 1776–1784.
7. Bratt C, Lindberg C, Lindberg C, Marko-Varga G. J Chromatogr A 2001; 909: 279–288.
8. Westergren-Thorson G, Malmström J, Marko-Varga G. J Pharm Biomed Anal 2001; 24:815–824.
9. Wrana IJ, Attisano L, Wieser R, Ventura F, Massague J. Nature 1994; 370:341–347.
10. Henis YI, Moustakas A, Lin HY, Lodish HF. J Cell Biol 1994; 126:139–154.
11. Roberts AB, Sporn MB. In: Sporn MB, Roberts AB, eds. Peptide Growth Factors and Their Receptors. Part 1. Heidelberg, Germany: Springer-Verlag, 1990:419–472.
12. Heldin C-H, Miyazono K, ten Dijke P. Nature 1997; 390:465–471.
13. Lagna G, Hata A, Hammati-Brivanlou A, Massagué J. Nature 1996; 383:832–836.
14. Yingling JM, Wang X-F, Bassing CH. Biochim Biophys Acta 1995; 1242:115–136.
15. Kretzschmar M, Liu F, Hata A, Doody J, Massagué J. Gene Dev 1997; 11:984–995.
16. Laurell T, Marko-Varga G, Ekström S, Bengtsson M, Nilsson J. Rev Mol Biotechnol 2001; 82:161–175.
17. Ekström S, Malmström J, Wallman L, Löfgren M, Nilsson J, Laurell T, Marko-Varga G. Proteomics 2002; 2:413–421.
18. Ericsson D, Ekström S, Bergqvist J, Nilsson J, Marko-Varga G, Laurell T. Proteomics 2001; 1:1072–1081.
19. Ekström S, Nilsson J, Helldin G, Laurell T, Marko-Varga G. Electrophoresis 2001; 22:3984–3992.
20. Bergkvist J, Ekström S, Wallman L, Löfgren M, Marko-Varga G, Nilsson J, Laurell T. Proteomics 2002; 2:422–429.
21. Widmer M. In: Widmer HM, Verpoorte E, Barnard S, eds. Anal Meth Instr. Proceedings 2nd International Symposium Miniaturisation in Total Analytical Systems. 1996:3–8.
22. Bousse L, Mouradian S, Minalla A, Yee H, Williams K, Dubrow R. Anal Chem 2001; 73:1207–1212.
23. Mouradian S. Curr Opin Chem Biol. 2001; 6:51–56.
24. Wang J. Nucleic Acid Res 2000; 28:3011–3016.
25. Verporte S. Electrophoresis 2002; 23:677–712.
26. Chiem N, Harrison JD. Anal Chem 1997; 69:373–378.
27. Wang J, Ibanez Chatrati MP, Escarpa A. Anal Chem 2001; 73:5323–5327.

28. Cheng SB, Skinner CD, Taylor S, Attiya S, Lee WE, Picelli G, Harrison DJ. Anal Chem 2001; 73:1472–1479.
29. Coyler CL, Tang T, Chiem N, Harrison DJ. Electrophoresis 1997; 18:1733–1741.
30. Huang T, Wu X-Z, Pawliszyn J. Anal Chem 2000; 72:4758–4761.
31. Dolnik V, Liu S, Jovanovich S. Electrophoresis 2000; 21:41–47.
32. Bruin GJM. Electrophoresis 2000; 21:3931–3951.
33. Mao Q, Pawlyiszin J. Analyst 1999; 124:637–641.
34. Oleschuk RD, Loranelle L, Schulz-Lockyear LL, Ning Y, Harrison DJ. Anal Chem 2000; 72:585–590.
35. Wen J, Lin Y, Xiang F, Matson DW, Udseth HR, Smith RD. Electrophoresis 2000; 21:191–197.
36. Rocklin RD, Ramsey RS, Ramsey JM. Anal Chem 2000; 72:5244–5249.
37. Yao S, Anex DS, Caldwell WB, Arnold DW, Smith KB, Schultz PG. Proc Natl Acad Sci USA 1999; 96:5372–5377.
38. Slenz BE, Penner NA, Lugowska E, Regnier F. Electrophoresis 2001; 22:3736–3743.
39. Gottschilk N, Jacobson SC, Culbertson CT, Ramsy JM. Anal Chem 2001; 73: 2669–2674.
40. Li J, LeRiche T, Thremblay T-L, Wang C, Bonneil E, Harrison DJ, Thibault P. Cell Mol Proteomics 2002; 2:157–169.
41. Wu CC, MacCoss MJ. Curr Opin Mol Therapeut 2002; 4:242–250.
42. Ekström S, Ericsson D, Önnerfjord P, Bengtsson M, Nilsson J, Marko-Varga G, Laurell T. Anal Chem 2001; 73:214–219.
43. Bengtsson M, Ekström S, Wallman L, Nilsson J, Laurell T, Marko-Varga G. Talanta 2002; 56:341–353.
44. Paik T. Oral presentation at Proteomics Symposium, From Genome to Proteome—Functional Proteomics, Siena, Italy, September 2–5, 2002.
45. Hanash S, Celis JE. Mol Cell Proteomics 2002; 1:413–414.
46. Nees KE, Lee JC. Mol Cell Proteomics 2002; 1:415–420.

11

Micro Chemical Processing on Microchips

Yoshikuni Kikutani
Kanagawa Academy of Science and Technology, Kanagawa, Japan

Takehiko Kitamori
The University of Tokyo, Tokyo, and Kanagawa Academy of Science and Technology, Kanagawa, Japan

1. INTRODUCTION

Micro analytical systems using electrophoresis separations are major topics of this book. Indeed, it is true that a combination of microchip electrophoresis and laser-induced fluorescence (LIF) gives an extremely powerful tool for separation and analysis of biomaterials [1,2]. However, outstanding reduction in amount of reagents and incomparable enhancement of throughput are needed not only in bioanalyses such as genome technology alone. They are also highly demanded in wide branches of chemical and biological technologies, for example, in various analytical techniques including environmental analysis, diagnostics, bioassays, etc. [3–8], and in various scale chemical production [9–11]. To address all these demands, not only aqueous but organic solutions should be controlled, both ionic and neutral species should be separated, and not only fluorescent but also nonfluorescent molecules should be detected. Combination of electrophoretic specimen handling, electroosmosis-driven fluid control, and fluorescent detection techniques are not sufficient to fulfill all these requirements.

In this chapter, we report our recently developed methodology comprising micro unit operations (MUOs) [12], continuous-flow chemical processing (CFCP) [12], and thermal lens microscopy (TLM) [13]. With this methodologies, a variety of chemical and analytical process could be integrated onto a small microchip. We will explain these key technologies—MUO, CFCP and TLM—and introduce some applications such as rapid and highly sensitive wet analysis of heavy metal ions, diagnostic immunoassays, and rapid and high-yield chemical synthesis.

2. MICRO UNIT OPERATIONS AND CONTINUOUS-FLOW CHEMICAL PROCESSING

Just as for chemical plants [14,15], design and construction of a micro chemical system is done by proper combination of unit operations, such as mixers, reactors, etc. [12]. However, simple miniaturization of conventional unit operations is not effective and sometimes does not work at all, because many physical parameters, e.g., heat and mass transfer efficiencies, or size of the specific interfacial area, are significantly different in micro dimensions [9,12,16]. To realize equivalent functions in microspaces, it is necessary to create new operation units, namely, micro unit operations (MUOs), to take into account the characteristics of microspaces [12]. In capillary electrophoresis, efficient heat transfer through capillary walls allows higher voltages without unwanted temperature increase, resulting in rapid sample separation [17]. In similar ways, microspace characteristics make chemical operations more efficient than conventional-scale ones. As illustrated in Figure 1, we have successfully developed a number of fundamental MUOs, such as mixing and reaction [18–20], two- and three-phase formations [16,21–23], solvent extraction [16,21–27], solid-phase extraction [28,29], heating [30,31], and cell culture [32].

These MUOs were based on pressure-driven flow control, which allows wider choice of fluids than electroosmotic flow (EOF) does. Many organic solvents and even gaseous samples can be treated with this pressure-driven approach [16]. Low Reynolds number is one of the characteristics of microchannels, resulting in a tendency to form a laminar flow instead of a turbulent one [16,33–37]. We have developed a stabilization technique for laminar flow by using microchannels with special cross sections [12]. In this technique, ridgelike structures at the bottom of the channel are fabricated along the streamlines (Figure 2), and the ridges act as stream guides. With the aid of stabilized multiphase laminar flow (MPLF), we can realize elemental unit operations that could hardly be carried out in EOF-driven systems, such as solvent extraction from aqueous to organic phase. Figure 3 shows a typical example of MPLF [12].

MPLF is also useful for system construction [12]. MUOs can be connected by a continuously flowing MPLF network, and molecules transported by spontaneous motion, namely, diffusion and distribution among different liquids. It is not necessary to use electric field to control motion of molecules, thus we can handle a variety of chemical species regardless of their charge. Once the channel circuits are properly designed, chemical species are conveyed from one MUO to another, and sequential chemical processes can be carried out automati-

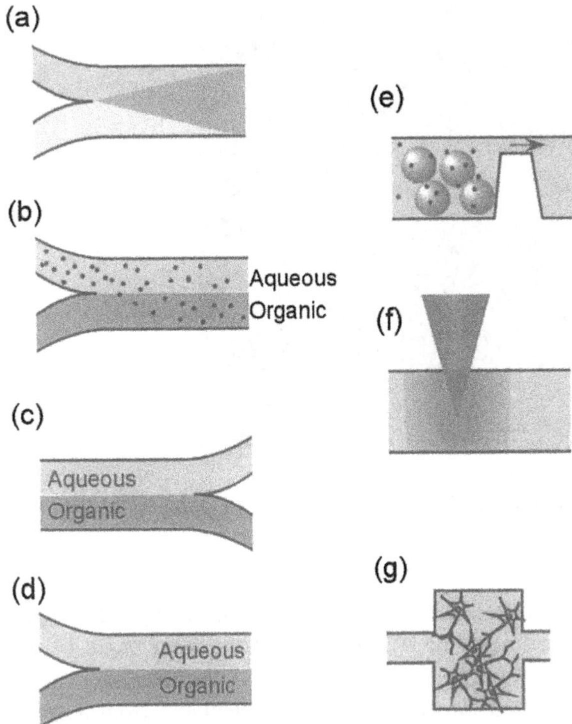

FIGURE 1 Schematic illustrations of micro unit operations: (a) mixing and reaction; (b) solvent extraction; (c) phase separation; (d) two-phase formation; (e) solid-phase extraction; (f) heating; (g) cell culture.

cally. We call this system construction methodology continuous-flow chemical processing, and we have already succeeded with the integration of relatively complex chemical systems onto a single microchip using this approach. Some examples of CFCP will be presented later in this chapter.

3. THERMAL LENS MICROSCOPE

Detection is one of the most difficult problems in micro chemical processes. Because sample volume becomes extremely small in such systems, an ultrasensitive detection method is indispensable. For example, limited sample channel thickness causes very small signal-to-background ratio in absorption spectroscopy, thus only very concentrated samples can be analyzed.

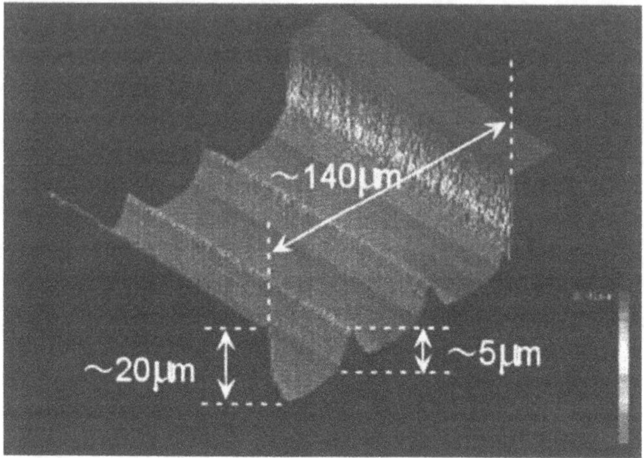

FIGURE 2 Three-dimensional image of the guide structure to stabilize three-phase laminar flow.

In laser-induced fluorescence (LIF), the problem is solved by using high-intensity laser light and fluorimetry [2,38,39]. Because the wavelength of the fluorescent emission light is slightly longer than that of the excitation light, signal intensity can be measured directly rather than calculated from a small difference between incident and transmitted beams. LIF method is so sensitive that it can be used even for single-molecule detection [39]. However, fluorescence spectroscopy has some drawbacks, such as that almost all molecules release the energy upon excitation as heat, rather than fluorescent light, so LIF is applicable only to fluorescent molecules.

We have developed a novel ultrasensitive detection method, thermal lens microscopy (TLM), for nonfluorescent species [13]. TLM is photothermal spectroscopy under an optical microscope. Our thermal lens microscope (TLM) has a dual-beam configuration: excitation and probe beams [13]. The wavelength of the excitation beam is selected to coincide with an absorption band of the target molecule and that of the probe beam is chosen to be where the sample solution (both solvent and solute) has no absorption. For example, in determination of methyl red dye in water, cyclohexane, and *n*-octanol, a 514-nm emission line of an argon-ion laser and a 633-nm emission line of a helium–neon laser were used as excitation and probe beams, respectively [21]. Figure 4 shows the configuration and principle of TLM [13]. The excitation beam was modulated at 1 kHz by an optical chopper. After the beam diameters were expanded, the excitation and probe beams were made coaxial by a dichroic mirror just before they were introduced into an objective lens whose magnification and numerical aper-

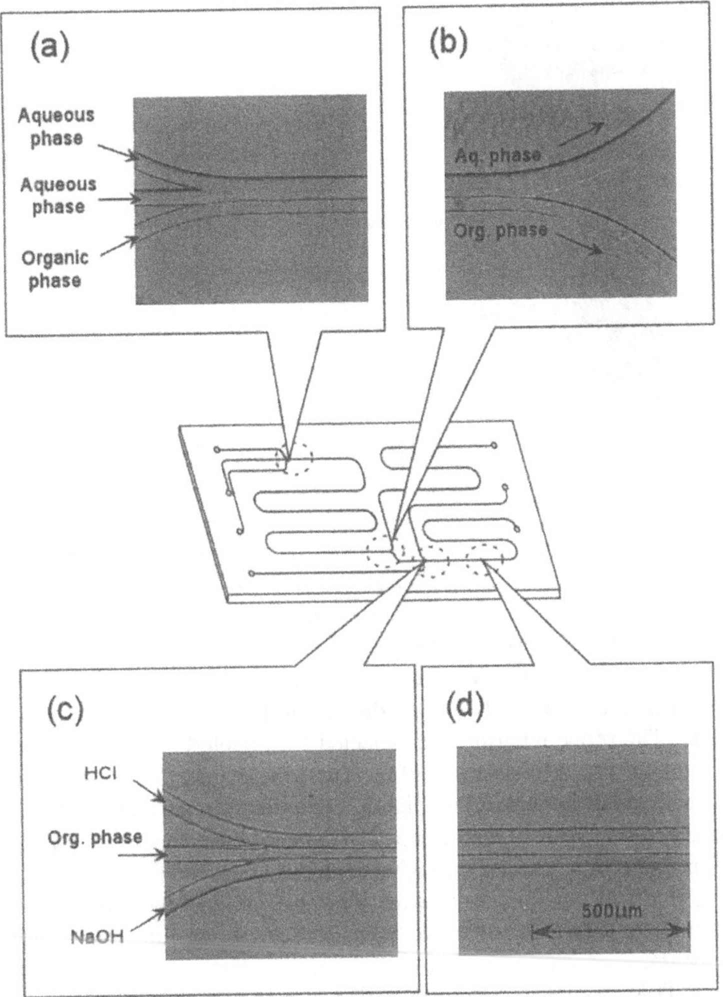

FIGURE 3 Photographs of the multiphase laminar flow in the microchannels: (a) confluence of water and *m*-xylene; (b) phase separation part; (c) confluence of HCl, organic phase, and NaOH; and (d) three-phase flow.

ture were 20-fold and 0.46, respectively. Beam waist and confocal length of the excitation beam were 1.3 μm and 10.8 μm, respectively. Samples were set in the center of the confocal volume of the excitation beam. Due to the spatial profile of laser intensity and thermal diffusion after light absorption, intermittent irradiation of the excitation beam produced a repeated spatial temperature pro-

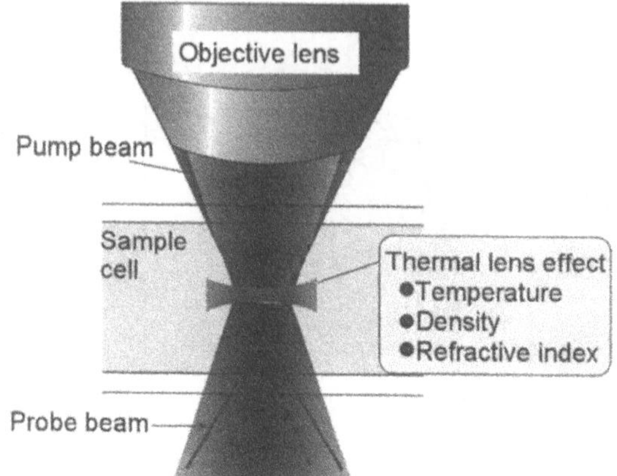

FIGURE 4 Operating principles of thermal lens microscope (TLM).

file in/around the waist. Since the refractive index of solvents depends on temperature, the temperature profile acted as a transient optical lens. When the focal point of the probe beam deviated from the center of the temporal lens along the axis, the focal point of the probe beam was shifted by the temporal lens and, therefore, probe beam intensity after passing through a counter objective lens, an interference filter, and a pinhole was modulated at the same frequency as the optical chopper. The probe intensity was detected by a photodiode and the output current was fed into a lock-in amplifier. The synchronous signal with the chopper was recorded as a thermal lens signal. Generally, it is difficult to detect thermal lens effect for optical microscopes of which chromatic aberration is compensated because light passing through a center of a lens would not be bent. In order to optimize the thermal lens setup, the focal point of the probe beam was set 10 μm lower than that of the excitation beam by controlling divergence of the probe beam.

By using TLM, we have demonstrated high sensitivity, high applicability, and high spatial resolution in various applications. As to sensitivity, TLM can detect single solute molecules in a detection volume, where limit of detection (LOD) is estimated by time-averaged value [40,41]. With regard to the high applicability and high spatial resolution, TLM can be applied to determination of target solute in free solution in an optical cell [42], DNA detection in polymer matrix [43], dye molecule determination in a single cell [44], and detection of nanometer-sized single particles in solution [45]. Limit of detection values of TLM for various samples are summarized in Table 1. A desktop-sized TLM is shown in Figure 5 (Institute of Microchemical Technology, Japan).

TABLE 1 Limit of Detection (LOD) of TLM for Various Samples

LOD	Solute	Solvent (media)	Ref.
0.4 molecules	Pb-porphyrine	Benzene	[40,41]
48 molecules	Dye	Water	[42]
7 molecules	DNA	Poly-acrylamide	[43]
78,000 molecules	Dye	Single cell	[44]
1 particle	Ag colloid	Water	[45]

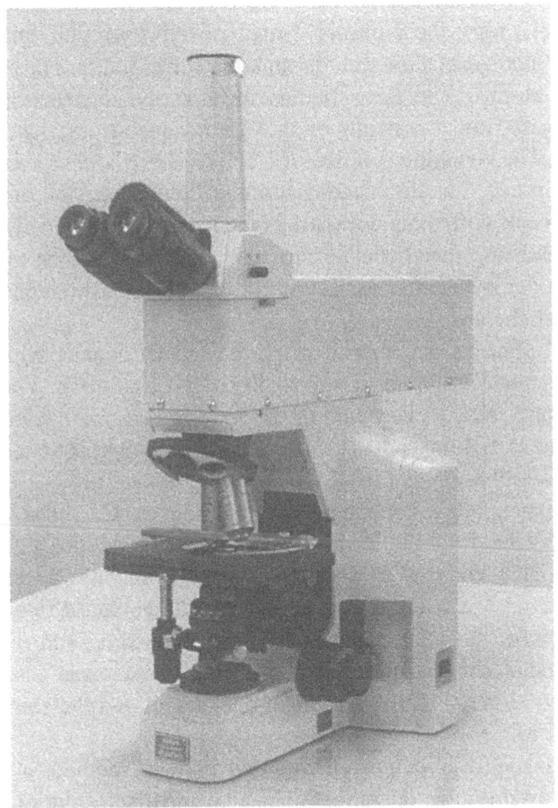

FIGURE 5 A desktop thermal lens microscope ITLM-10 (Institute of Microchemical Technology, Inc., Japan).

4. MICRO CHEMICAL REACTION AND EXTRACTION SYSTEM: HEAVY METAL ION ANALYSIS

As described above, the design, fabrication, control, and detection methods were established for integrated microchemical systems. We have applied the techniques to a variety of chemical systems.

The first example chosen here is environmental trace heavy metal analysis in water, made possible with the microchip. Operational principles are described in Figure 6 [12] for cobalt analysis. The design and photo of the chip is shown in Figure 7. A total chemical process for wet analysis were decoupled into seven types and 13 steps of unit operations, such as chelating, extraction, washing, etc., and translated into MUOs. The CFCP design in Figure 6 was reconstructed from those MUOs. The microchip consists of two main areas, a reaction and extraction area and a washing (decomposition and removal) area. First, the sample solution containing Co(II) ions, the 2-nitroso-1-naphtol (NN) solution, and *m*-xylene are introduced at a constant flow rate through the corresponding three inlets using microsyringe pumps. The three liquids meet at the intersection point, and a parallel two-phase flow, consisting of an organic/aqueous interface, forms in the microchannel. The chelating reaction of Co(II) and NN and extraction of the Co(II) chelates proceed as the reacting mixture flows along the microchannel. Since the NN reacts with other coexisting metal ions, such as Cu(II), Ni(II) and Fe(II), these coexisting metal chelates are also extracted into the *m*-xylene. Therefore, washing is needed after extraction for the decomposition and removal of coexisting metal chelates.

The coexisting metal chelates decompose when they make contact with hydrochloric acid, and the metal ions are dissolved in the HCl solution. The decomposed chelating reagent, NN, is dissolved in the sodium hydroxide solution [46]; in contrast to the coexisting metal chelates, the Co chelate is stable in HCl and NaOH solutions and remains [46].

In the latter (washing) area, the *m*-xylene phase containing the Co chelates and the coexisting metal chelates from the former (reaction and extraction) area is interposed between the HCl and NaOH solutions, which were introduced through the other two inlets at a constant flow rate. Then a three-phase flow, HCl/*m*-xylene/NaOH, forms in the microchannel. The decomposition and removal of the coexisting metal chelates proceed along the microchannel in a similar manner as described above. Finally, the target chelates in *m*-xylene are detected downstream by TLM.

The advantages of this approach compared with conventional methods are simplicity and omission of troublesome operations. The acid and alkali solutions cannot be used simultaneously in the conventional washing method, but this becomes possible by using three-phase flow in the microchannel. This chemical processing corresponds to the integration of eight MUOs on a microchip, two-phase formation, mixing and reaction, extraction, phase separation, three-phase

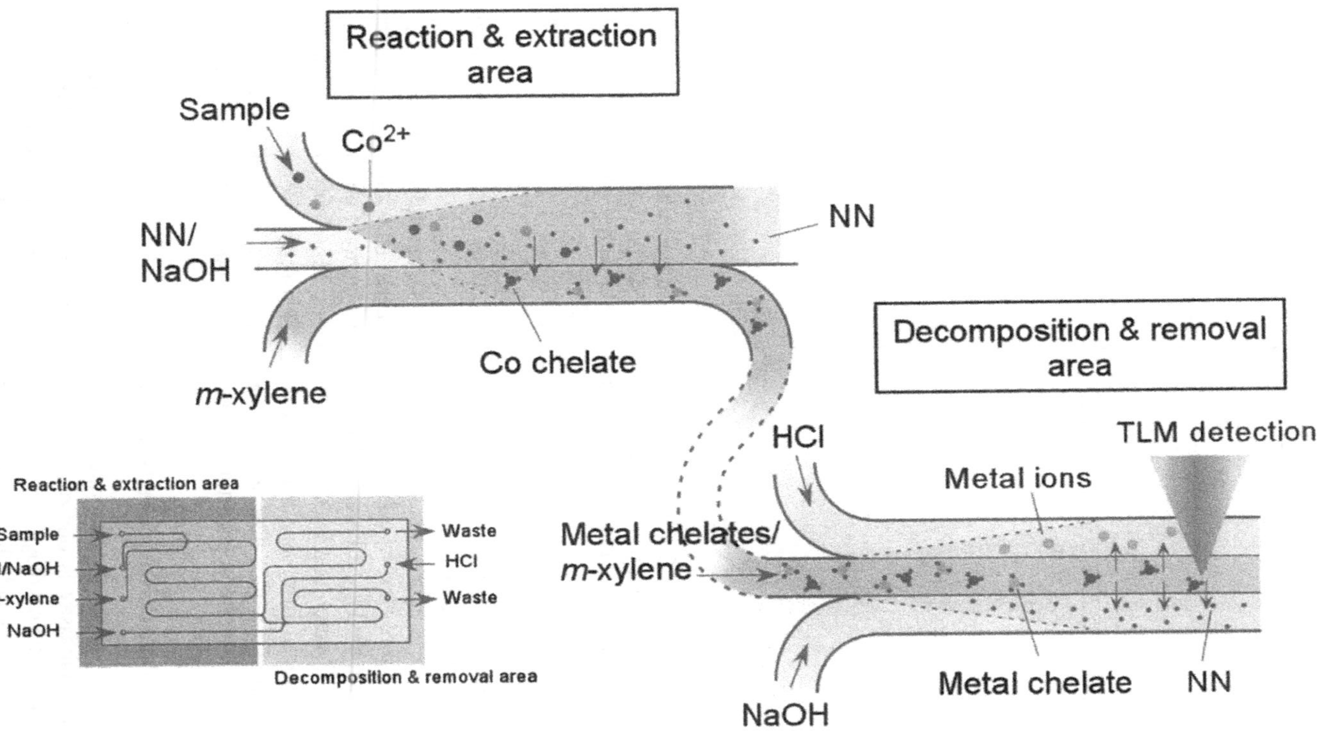

FIGURE 6 Schematic illustration of Co(II) determination by combining MUOs.

(a)

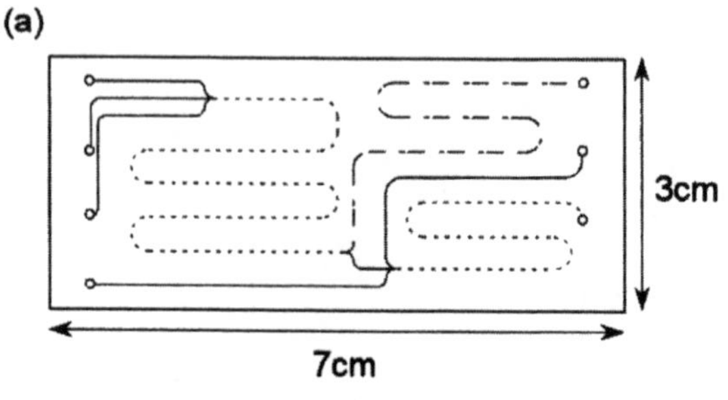

(b)

FIGURE 7 Microchip for Co(II) analysis. (a) Schematic illustration. The solid, dotted, and chain lines are 50 μm wide and 20 μm deep, 140 μm wide and 20 μm deep, and 90 μm wide and 20 μm deep, respectively. (b) Photograph.

formation, decomposition of coexisting metal chelates, removal of metal ions, and removal of reagents.

Cobalt in an aqueous solution mixture was successfully determined. Even at zmol (10^{-21} mol) levels, cobalt could be extracted and determined. The calibration line showed good linearity and the determination limit obtained from 2σ reached to 0.13 zmol, that is, 78 chelate molecules. More important, the analysis time was reduced from 2–3 h to only 50 s. This kind of drastic reduction in analysis time and device size, even for a complicated chemical procedure like this, anticipates future application to mobile advanced analytical equipment.

5. MICRO IMMUNOASSAY SYSTEM

The second example of integrated microsystems is for immunoassay [28,29, 47,48]. Immunoassay is one of the most important analytical methods and is widely used in clinical diagnoses, environmental analyses, and biochemical studies because of its extremely high selectivity and sensitivity. Enzyme-linked immunosorbent assay (ELISA) or other immunosorbent assay systems, in which antigen and antibodies are fixed on a solid surface, are applicable to many analytes with high sensitivity and are used in many fields including clinical diagnosis.

The conventional heterogeneous immunosorbent, however, requires relatively long assay times, and also involves tedious liquid-handling procedures and large quantities of expensive reagents. Moreover, realization of point-of-care (POC) testing is difficult with conventional immunoassay systems, since rather large devices are necessary for automated diagnosis systems. To overcome these drawbacks, a microchip-based system was developed. Integration of analytical systems into a microchip should bring about enhanced reaction efficiency, simplified procedures, reduced assay time, and reduced consumption of samples, reagents, and energy.

Recently, several papers about integration of heterogeneous immunoassay systems into microscale devices have been published [28,29,47–49]. In Ref. 28, the possibility of immunosorbent assay on a microchip was reported. In this system, antigen–antibody reactions were performed on surfaces of micro beads packed in a microchannel with a dam structure (Figure 8). The reaction time necessary for an antigen–antibody reaction was reduced to 1/90 in the integrated microsystem, due to size effects in the liquid microspace. This effect is brought about by an increase in the specific interface area and reduction of the diffusion distance (Figure 9). An increase in the specific interface area corresponds to increased reaction field. The specific interface area of 50 μL of solution in the

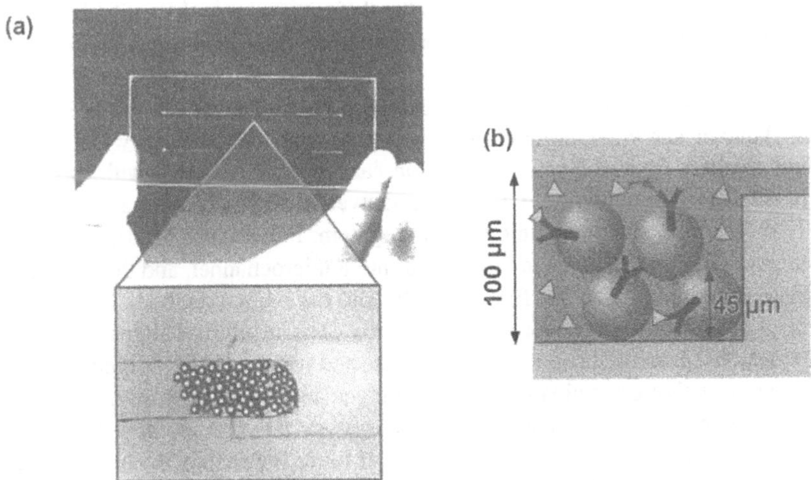

FIGURE 8 (a) Immunoassay microchip with microbeads. (b) Cross section image of the reaction area.

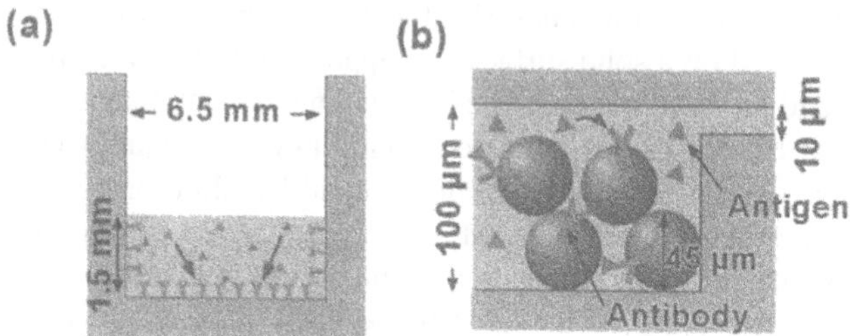

FIGURE 9 Schematic illustrations of antigen–antibody reaction. (a) Microtiter plate. (b) Microchip.

microtiter plate well (0.65 mm in diameter) was estimated to be 13 cm^{-1}, whereas that of the microchannel [11 beads (45 μm in diameter) in 100 μm × 100 μm × 200 μm channel space] was 480 cm^{-1}. Therefore that of the microchannel was 37 times larger than that of the microtiter plate, and the reaction rate may be increased by this larger reaction field.

In a conventional microtiter plate assay, a 1.5-mm movement would be necessary for the most distantly located antibody molecule to react with the antigen fixed on the surface of the well, since the liquid depth was 1.5 mm. On the other hand, the liquid phase of the microchannel filled with polystyrene beads was much smaller. The longest distance from an antibody molecule to the reaction-solid surface may be less than 20 μm. Diffusion time is proportional to the squares of the diffusion distance, so the diffusion time of the antibody molecule to the antigen in the microchip would be more than 5600 times shorter than the conventional method.

This system was applied to a microchip-based clinical diagnosis system [29]. Human carcinoembryonic antigen (CEA), one of the most widely used tumor markers for serodiagnosis of colon cancer, was evaluated with this system. An ultratrace amount of CEA was dissolved in serum samples and successfully determined in a short time with this system. Polystyrene beads, precoated with anti-CEA antibody, were introduced into a microchannel, and then a serum sample containing CEA, the first antibody, and the second antibody conjugated with colloidal gold were reacted successively. The resulting antigen–antibody complex, fixed on the bead surface, was detected using a TLM. A highly selective and sensitive determination of an ultratrace amount of CEA in human sera was made possible by a sandwich immunoassay system that requires three antibodies for an assay. Detection limit dozens of times lower than in conven-tional ELISA was achieved (Figure 10). Moreover, when serum samples from 13 patients were assayed with this system, there was a high correlation ($r = 0.917$) with conventional ELISA assay. System integration reduced the time necessary

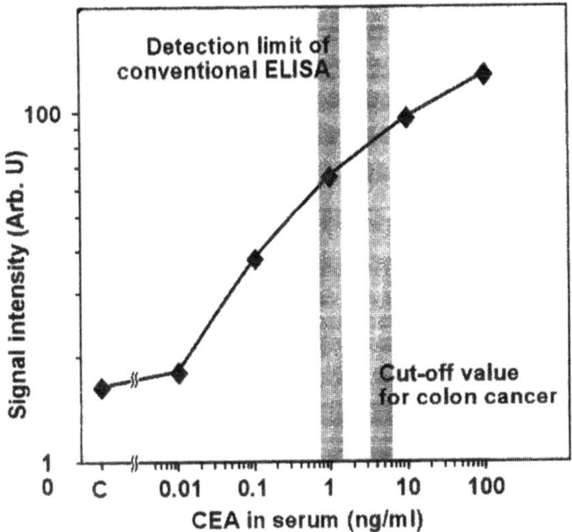

FIGURE 10 Calibration curve for CEA assay from human sera.

for the antigen–antibody reaction to ~1%, thus shortening the overall analysis time from 45 h to 35 min. Moreover, troublesome operations required for conventional heterogeneous immunoassays could be substantially simplified. This microchip-based diagnostic system was the first μ-TAS shown practical usefulness for clinical diagnoses with short analysis times, high sensitivity, and easy handling.

In microchip systems, higher integration is thought to be easily realized by multichannels. To realize higher-throughput analyses, a microchip system which can process several samples simultaneously was reported [49]. In this integrated system, the chip had branched multichannels and four reaction and detection regions; thus the system could process four samples at a time using only one pump (Figure 11). Interferon-γ was assayed by three-step sandwich immunoassay, with the system coupled to a TLM as a detector. The biases of the signal intensities obtained from each channel were within 10%, and CVs (coefficients of variance) were almost the same level as for the single straight channel assay. The assay time for four samples was 50 min instead of 35 min for one sample in the single-channel assay; hence higher throughput was realized with the branched structure chip.

Simultaneous assay of many samples may also be achieved by simply arraying many channels in parallel on a chip. This approach, however, needs many pumps and capillary connections and a high degree of integration seems to be difficult. On the other hand, a microchip with branched microchannels

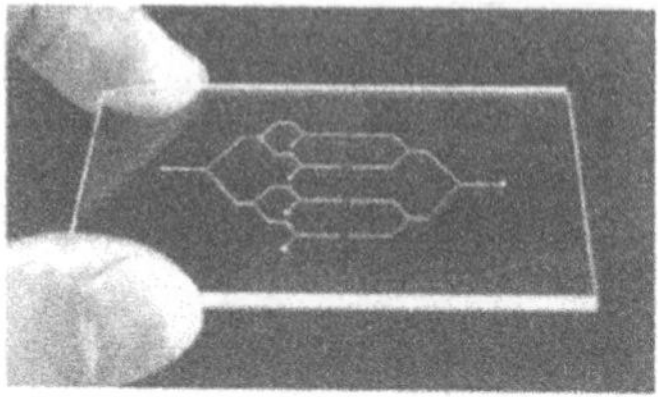

Sample injection

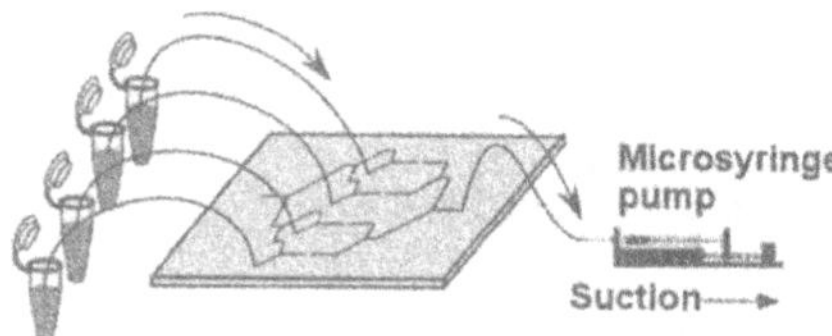

Reagent injection

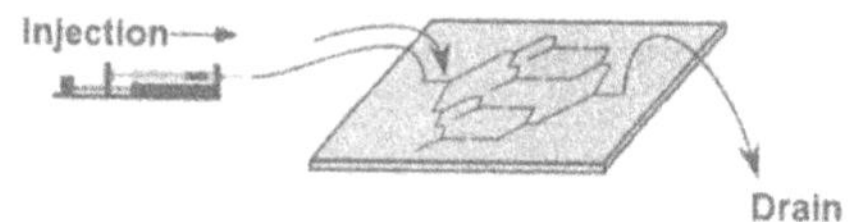

FIGURE 11 Simultaneous determination setup using branched multichannel immunoassay chip.

seems to be suitable for simultaneous assay. By branching multichannels, the numbers of pumps and capillary connections required for the system should be reduced. An automated multichannel immunoassay system with much higher throughput can be realized by development of multiple fluidic control devices in the near future.

6. PILE-UP MICROCHIP CHEMICAL PLANT

The third example of CFCP systems is a micro chemical plant [50–52]. Development of an organic synthesis system using microchannel chips is one of the most promising applications in microchip technologies, due to the realization of efficient chemical reactions by exploiting advantages provided by liquid microspace, and the versatile optimization of microchannel design by micro fabrication technologies [8–11]. In order to realize practical and efficient organic synthesis systems, the following two factors are considered to be important:

1. Choice of appropriate chemical reaction, which effectively exploits advantageous features of liquid microspace such as short diffusion distance, large specific interface, and small heat capacity
2. Design of microchannel structures for respective synthesis purposes

Here we describe a phase-transfer synthesis as an example of a suitable micro-scale synthesis reaction, in which the large specific interfacial area and short molecular diffusion distance play important roles not only for effective main reaction, but also for avoiding an undesirable side reaction [50].

As described earlier, we have been focusing on the multiphase flow formed by introducing different types of solvents such as aqueous and organic solutions. When the organic and aqueous phases were introduced through the two inlets of the microchannel by syringe pumps, a stable liquid/liquid interface formed as seen in the photographs of Figure 3. In this case, a large specific interfacial area could be obtained without any stirring. By exploiting the large specific interfacial area provided by organic and aqueous phases, demonstration of a fast and high conversion synthesis involving phase transfer should be possible. In addition, stable two-phase flow also allows us easy separation of two phases inside the microchannel under continuous flow condition by splitting the reaction channel into two channels at its end. Thus we chose the phase transfer reaction as a model synthetic reaction to demonstrate two-phase laminar flow.

In order to realize our concept in an actual experiment, we carried out a preliminary investigation of a phase-transfer diazocoupling reaction as one simple example (Figure 12). Previously, a diazocoupling reaction in a microchip has been successfully demonstrated for synthesizing a small amount of azobenzene for combinatorial synthesis purposes [53]. In contrast, a continuous multiphase flow-based system is expected to synthesize large amounts of chemicals

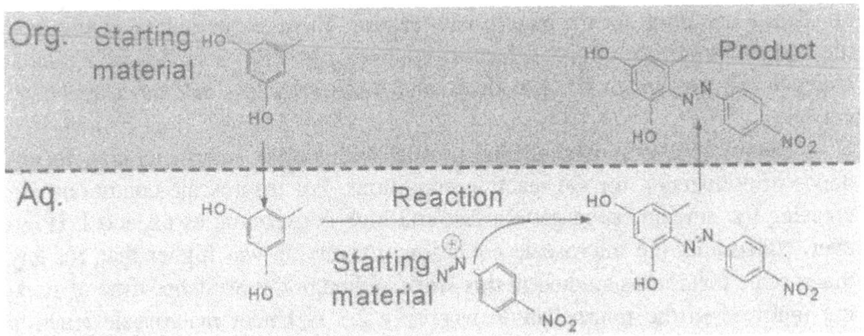

FIGURE 12 Phase-transfer diazocoupling reaction

with high conversion rate. Rapid phase transfer of starting material and the produced chemical species across the liquid/liquid interface play important roles to realize both the fast chemical reaction and isolation of the chemical species produced.

We also compared the reaction efficiency of microscale and macroscale reaction conditions. The microscale reaction was performed by introducing ethyl acetate containing 5-methyl resorcinol (10^{-3} M) and an aqueous phase containing 4-nitrobenzene diazonium tetrafluoroborate (10^{-4} M) through the two inlets of the microchip under continuous flow condition. Volume flow rates of the organic and aqueous phases were fixed at 10 μL/min each. In this case, linear flow rate was estimated to be 1.3 cm/s. When the organic phase made contact with the aqueous phase, distribution of the resorcinol derivative started in the aqueous phase, and reacted with the diazonium salt. The resulting electrically neutral main product was redistributed into the organic phase. In the case of macroscale reactions, 10-mL volumes of each of the reagent solutions (organic and aqueous solutions) were poured into a glass vessel (3.5 cm in diameter) and stirred. In this case, stirring conditions were varied to evaluate the effect of mixing and specific interfacial area on reaction time and conversion. The experimental conditions are shown in Figure 13. Letting the reaction mixture stand without stirring gave a long molecular diffusion distance and small specific interfacial area. Weak stirring provided effective molecular diffusion condition, but with almost the same specific interfacial area as without stirring. Strong stirring provided effective reaction condition with respect to both. In all the experimental conditions, conversion of diazonium salt was evaluated by standard reversed-phase high-performance liquid chromatography (HPLC) using octadecylsilyl (ODS) column. Methanol eluent was used with a flow rate of 1 mL/min. For the microscale reaction, simple introduction of organic and aqueous phases into the microchannel provided a stable liquid/liquid interface, and the specific interfacial area had a larger value than those of the strong stirring and no stirring condition for the macroscale reaction. In our experimental conditions, the specific interfacial area (surface-to-volume ratio, S/V) provided by a microchannel 250 μm wide, 100 μm deep, and 3 cm long was calculated to be 80 cm^{-1}.

Figure 13 also shows reaction profiles and specific interfacial area dependence of conversion for the reaction conditions. For macroscale conditions, increasing the stirring speed gave a fast and high conversion, as expected. However, concerning the microscale condition, conversion was higher than for any macroscale conditions studied in this work, although the residence time of starting matrices in the microchannel was only 2.3 s. Under macroscale reaction conditions, insoluble precipitate species of a side product were observed visually, the amount of which depended on the mixing conditions (without stirring >

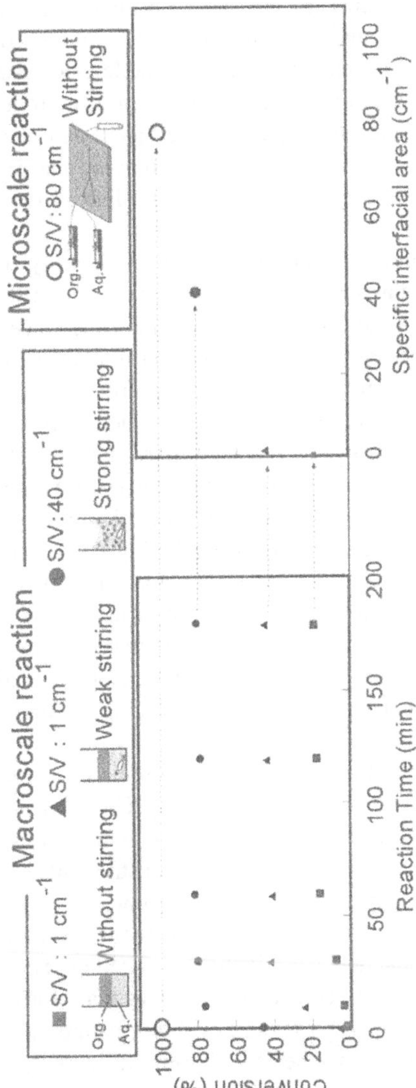

FIGURE 13 Reaction conditions and results obtained with phase-transfer diazocoupling reaction under microscale and macroscale condition.

weak stirring > strong stirring). In contrast, the microscale reaction condition gave no precipitate species and the conversion was close to 100%.

For the diazocoupling reaction, an undesirable side reaction of the main product and a second diazonium salt to forming a bisazo product is well known [54]. Thus in our case, the large specific interfacial area and short molecular diffusion distance played important roles in removing the main product from the aqueous phase to the organic phase, which allowed the undesirable side reaction to be avoided.

Although the amount of starting material of 5-methyl resorcinol was in excess compared to the diazonium salt and the organic phase still included residual starting material, close to 100% conversion rate was successfully achieved.

Choosing an appropriate synthetic reaction is crucial to realize practical and truly meaningful organic synthesis using microchips. In the next section, combinatorial synthesis and massive synthesis of chemicals using similar phase-transfer synthesis are described. In order to apply such an effective microreaction to practical use, development of multireactor systems becomes inevitable. Operating microreactors in parallel, we multiplied the total productivity in proportion to the number of reactors. This numbering-up technology is expected to result in a very effective means for production adjustment [9–11,51,55]. In conventional chemical engineering, scaling up is a time- and cost-consuming process due to heat and mass transfer problems that have to be considered repeatedly for all testing plants. Flexible adjustment of production can be done with the numbering-up concept, by simply increasing or decreasing the number of reactors to meet demands.

To demonstrate the applicability of the numbering-up technology, a 10-layered pile-up microreactor was fabricated [51]. The fabrication of the reactor was carried out using conventional photolithography only, wet etching and thermal bonding techniques. No special facilities or instruments were required (Figure 14). After the microchannels were etched on one side of a glass plate, holes with a diameter much larger than that of the microchannels were drilled and 10 glass plates with microchannels were laminated together. The glass surface opposite the channel-etched surface becomes a cover plate for the microchannels of another glass plate. A plain glass plate with no features was placed to cover the top plate. They were sandwiched by alumina weights, and heated with the same conditions that had been used for fabrication of single-layered glass microchips [16]. Figure 15 shows a photograph of the 10-layered pile-up reactor. An amide formation reaction was carried out between amine in aqueous solution and acid chloride in organic solution using the reactor. We observed that the maximum throughput for the 10-layered pile-up reactor was 10 times larger than that of a single-layered one. Productivity of the pile-up reactor for the reaction was as high as 1 g/h, suggesting that many conventional plants producing fine

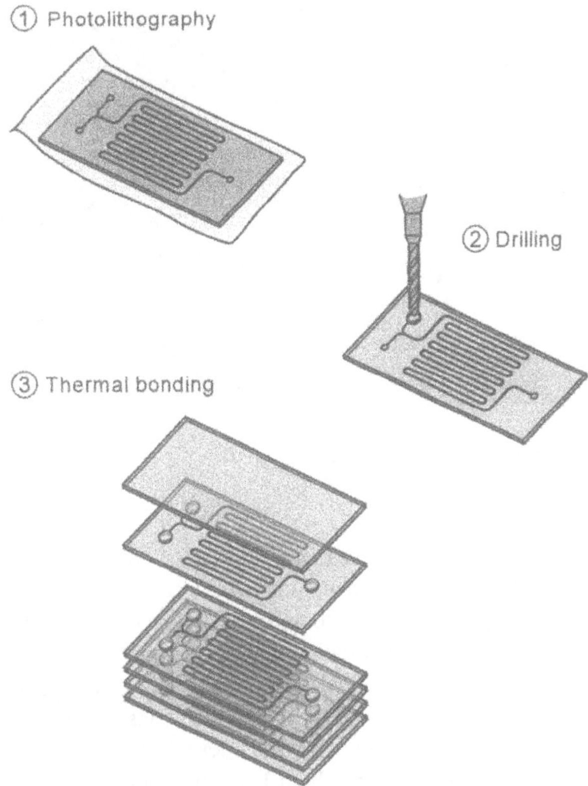

Figure 14 Fabrication procedure for the pile-up microreactor. (1) Photolithography: Conventional photolithography/wet etching methods were applied. The back side of the glass plate was covered with polyolefin tape during the HF treatment. (2) Drilling: Penetrating holes were drilled at the inlet and outlet ports of the microchannel circuit. (3) Thermal bonding: The required number of glass plates with microchannels and one cover plate were laminated and bonded thermally at 650°C.

chemicals can be replaced by microreactors through the numbering-up technology.

In the numbering-up technology, the same reaction is carried out in every microchannel. If different reagents are supplied in different combinations for all the microchannels, many different products can be obtained at once [52]. This is another eagerly expected application of microreactors and it would be very effective, especially for combinatorial synthesis in drug discovery [8]. For the

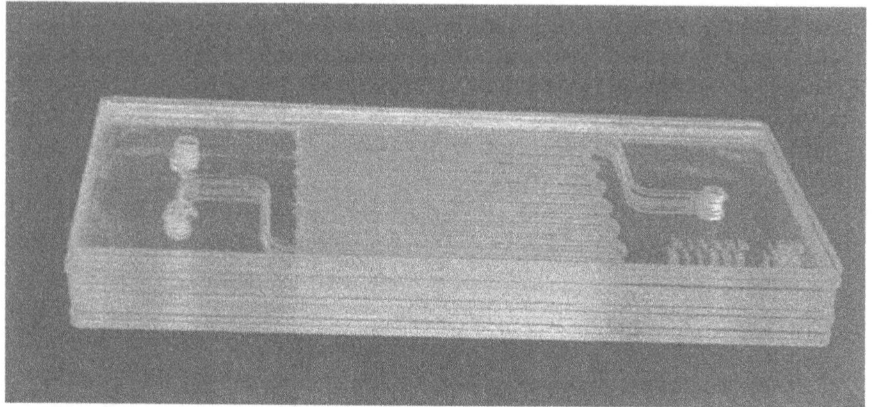

FIGURE 15 Photograph of a 10-layer pile-up microreactor.

integration of combinatorial parallel multimicroreactor systems onto a single chip, it is necessary to fabricate complicated channel networks containing many branches and joints [52]. Topologically, such a complicated network contains two-level crossing structures: a three-dimensional structure that cannot be fabricated by conventional photolithography followed by bonding of a cover plate [52,56]. Therefore, we have used three layers of glass plates to construct three-dimensional channel structures. Two glass plates with microchannels on one side were prepared and another glass plate, in which holes were drilled at the necessary points, was sandwiched between the two channel-etched surfaces. The three glass plate layers were laminated and thermally bonded. With this simple a technique, a three-dimensional channel network as illustrated in Figure 16 could be constructed within a single microchip. Integrity of the three-dimensional channel network was confirmed by actually operating 2×2 parallel phase-transfer amide syntheses.

7. CONCLUSION

A unique micro integration methodology has been shown for a wide variety of chemical systems. The methodology comprises MUOs, CFCP, and TLM. Some of its applications, which were hardly realized with conventional microsystems based on electrophoresis and LIF detection, were explained with examples. All these examples proved that micro chemical systems drastically reduce chemical processing time and improves efficiency. The micro chemical systems are concluded to be promising as flexible, smart, and mobile advanced chemical systems in near-future chemical technology.

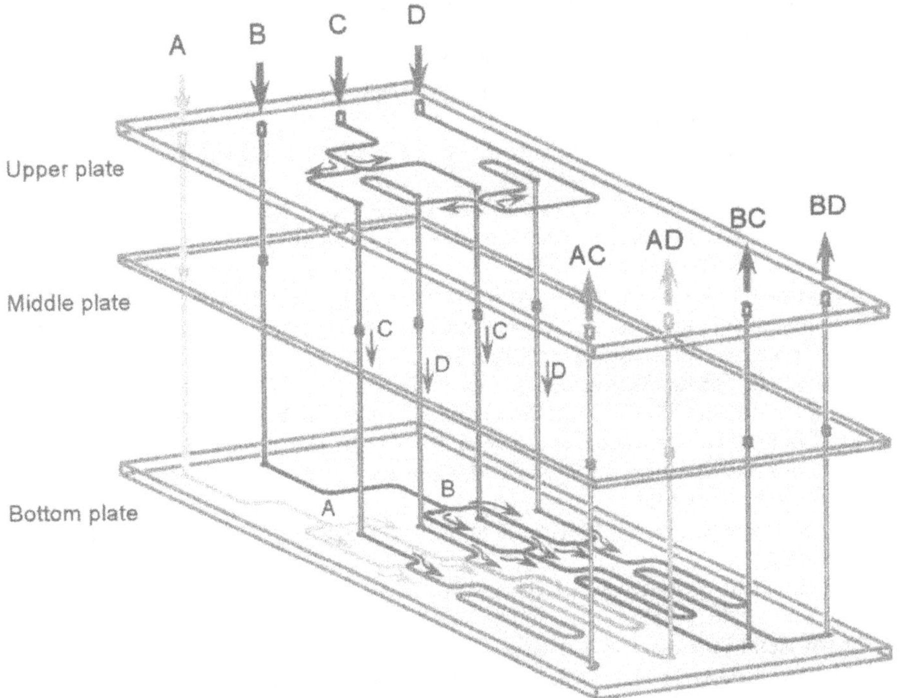

FIGURE 16 Schematic representation of a three-dimensional microchannel circuit for combinatorial chemistry.

REFERENCES

1. Bruin GJM. Electrophoresis 2000; 21:3931.
2. Ambrose WP, Goodwin PM, Jett JH, Orden AV, Werner JH, Keller RA. Chem Rev 1999; 99:2929.
3. Manz A, Becker H, eds. Microsystem Technology in Chemistry and Life Sciences. Berlin, Germany: Springer-Verlag, 1999.
4. van den Berg A, Olthuis W, Bergveld P, eds. Micro Total Analysis Systems 2000. Dordrecht, The Netherlands: Kluwer Academic, 2000.
5. Ramsey JM, van den Berg A, eds. Micro Total Analysis Systems 2001. Dordrecht, The Netherlands: Kluwer Academic, 2001.
6. Reyes DR, Iossifidis D, Auroux P-A, Manz A. Anal Chem 2002; 74:2623.
7. Auroux PA, Iossifidis D, Reyes D, Manz A. Anal Chem 2002; 74:2637.
8. DeWitt SH. Curr Opin Chem Biol 1999; 3:350.
9. Ehrfeld W, Hessel V, Löwe H. Microreactors New Technology for Modern Chemistry. Weinheim, Germany: Wiley-VCH, 2000.

10. Haswell SJ, Middleton RJ, O'Sullivan B, Skelton V, Watts P, Styring P. Chem Commun 2001, 391.
11. Jensen KF. Chem Eng Sci 2001; 56:293.
12. Tokeshi M, Minagawa T, Uchiyama K, Hibara A, Sato K, Hisamoto H, Kitamori T. Anal Chem 2002; 74:1565.
13. Uchiyama K, Hibara A, Kimura H, Sawada T, Kitamori T. Jpn J Appl Phys 2000; 39:5316.
14. Levenspiel O. Chemical Reaction Engineering. 3rd ed. New York: Wiley, 1999.
15. Fogler HS. Elements of Chemical Reaction Engineering. 3rd ed. Upper Saddle River, NJ: Prentice Hall, 1999.
16. Hibara A, Tokeshi M, Uchiyama K, Hisamoto H, Kitamori T. Anal Sci 2001; 17:89.
17. Rouessac F, Rouessac A. Chemical Analysis Modern Instrumentation Methods and Techniques. Chichester, UK: Wiley, 2000.
18. Sato K, Tokeshi M, Kitamori T, Sawada T. Anal Sci 1999; 15:641.
19. Sorouraddin HM, Hibara A, Proskurnin MA, Kitamori T. Anal Sci 2000; 16:1033.
20. Sorouraddin HM, Hibara A, Kitamori T. Fresenius J Anal Chem 2001; 371:91.
21. Surmeian M, Hibara A, Slyadnev M, Uchiyama K, Hisamoto H, Kitamori T. Anal Lett 2001; 34:1421.
22. Hisamoto H, Horiuchi T, Tokeshi M, Hibara A, Kitamori T. Anal Chem 2001; 73: 1382.
23. Hisamoto H, Horiuchi T, Uchiyama K, Tokeshi M, Hibara A, Kitamori T. Anal Chem 2001; 73:5551.
24. Tokeshi M, Minagawa T, Kitamori T. Anal Chem 2000; 72:1711.
25. Sato K, Tokeshi M, Sawada T, Kitamori T. Anal Sci 2000; 16:455.
26. Tokeshi M, Minagawa T, Kitamori T. J Chromatogr A 2000; 894:19.
27. Minagawa T, Tokeshi M, Kitamori T. Lab Chip 2001; 1:72.
28. Sato K, Tokeshi M, Odake T, Kimura H, Ooi T, Nakao M, Kitamori T. Anal Chem 2000; 72:1144.
29. Sato K, Tokeshi M, Kimura H, Kitamori T. Anal Chem 2001; 73:1213.
30. Tanaka Y, Slyadnev MN, Hibara A, Tokeshi M, Kitamori T. J Chromatogr A 2000; 894:45.
31. Slyadnev MN, Tanaka Y, Tokeshi M, Kitamori T. Anal Chem 2001; 73:4037.
32. Tamaki E, Sato K, Tokeshi M, Sato K, Aihara M, Kitamori T. Anal Chem 2002; 74:1560.
33. Kim H-B, Ueno K, Chiba M, Kogi O, Kitamura N. Anal Sci 2000; 16:871.
34. Kamholz AE, Weigl BH, Finleyson BA, Yager P. Anal Chem 1999; 71:5340.
35. Brody JP, Yager P. Sens Actuators A 1997; 58:13.
36. Weigl BH, Yager P. Science 1999; 283:346.
37. Kenis PJA, Ismagilov RF, Whitesides GM Science 1999; 285:83.
38. Ocvirk G, Tang T, Harrison DJ. Analyst 1998; 123:1429.
39. Hill EK, de Mello A. Analyst 2000; 125:1033.
40. Tokeshi M, Uchida M, Uchiyama K, Sawada T, Kitamori T. J Luminescence 1999; 83–84:261.
41. Tokeshi M, Uchida M, Hibara A, Sawada T, Kitamori T. Anal Chem 2001; 73: 2112.

42. Sato K, Kawanishi H, Tokeshi M, Kitamori T, Sawada T. Anal Sci 1999; 15:525.
43. Zheng J, Odake T, Kitamori T, Sawada T. Anal Chem 1999; 71:5003.
44. Harada M, Shibata M, Kitamori T, Sawada T. Anal Sci 1999; 15:647.
45. Mawatari K, Kitamori T, Sawada T. Anal Chem 1998; 70:5037.
46. Kitamori T, Suzuki K, Sawada T, Gohshi Y, Motojima K. Anal Chem 1986; 58: 2275.
47. Tokeshi M, Sato K, Kitamori T. Riken Rev 2001; 36:24.
48. Sato K, Yamanaka M, Takahashi H, Tokeshi M, Kimura H, Kitamori T. Electrophoresis 2002; 23:724.
49. Bernard A, Michel B, Delamarche E. Anal Chem 2001; 73:8.
50. Hisamoto H, Saito T, Tokeshi M, Hibara A, Kitamori T. Chem Commun 2001; 2662.
51. Kikutani Y, Hibara A, Uchiyama K, Hisamoto H, Tokeshi M, Kitamori T. Lab Chip 2002; 2:193.
52. Kikutani Y, Horiuchi T, Uchiyama K, Hisamoto H, Tokeshi M, Kitamori T. Lab Chip 2002; 2:188.
53. Salimi-Moosavi H, Tang T, Harrison DJ. J Am Chem Soc 1997; 119:8716.
54. For the diazocoupling reaction with resorcinol and 4-nitrobenzene diazonium, see Coffey S, ed. Rodd's Chemistry of Carbon Compounds. Vol. 3, Part C. New York: Elsevier Scientific, New York, 1973:136 and references cited therein.
55. Ehrfeld W, Golbig K, Hessel V, Löwe H, Richter T. Ind Eng Chem Res 1999; 38: 1075.
56. Anderson JR, Chiu DT, Jackman RJ, Cherniavskaya O, McDonald JC, Wu H, Whitesides SH, Whitesides GM. Anal Chem 2000; 72:3158.

12

Micropreparative Applications and On-Line Sample Treatment

Julia Khandurina
Torrey Mesa Research Institute, San Diego, California, U.S.A.

1. INTRODUCTION

Electric field-mediated separations in capillary and microchannel dimensions represent excellent analytical tools, but they can also be used in a number of micropreparative applications. Miniaturization of chemical and biochemical sample processing steps, such as fractionation and isolation, purification, derivatization, and chemical reactions, provides benefits in terms of speed, reagents consumption, and cost. Moreover, building integrated systems comprising multiple sequential or parallel processing steps, both analytical and preparative, enables high-throughput implementations at the large scale to meet the fast-growing demands in industrial pharmaceutical and research environments. In this chapter, the most recent advances in capillary electrophoresis-based preparative methods are surveyed. Special attention has been given to microscale fraction collection applications and online sample preparation, as well as emerging novel microfabricated devices to further increase the levels of integration and multifunctionality in chemical and biochemical processing technology. Some novel interesting technological and scientific developments, which do not have immediate practical implementations but are potentially promising, are also described.

2. MICROPREPARATIVE FRACTION COLLECTION IN CAPILLARY ELECTROPHORESIS

Traditional preparative separations and purification of proteins and DNA by slab-gel electrophoresis employs time- and labor-consuming methods based on visualization and excision of the bands of interest from gels or blots for further processing. In recent years, computer-driven cutting robots have been introduced, mostly for 2-D protein gels. However, both manual and robotic band/spot excision from the gels or blots lack accuracy and remain time-limiting and low-throughput procedures. On the other hand, electric field-mediated separations in capillary and microchannel dimensions can be used successfully in micropreparative applications, despite being considered primarily as analytical tools. The feasibility of capillary-based micropreparation and fraction collection following electrophoretic separation has been demonstrated for DNA fragments, peptides, proteins, and oligosaccharides. Automated single- and multicapillary systems have been developed, capable of collecting hundreds of samples with good precision [1–8], as described in the following sections. Very recently, microfabricated fluidic devices have also been employed for rapid and high-precision fraction collection, benefiting from their multifunctionality and possibility of electrokinetic manipulation of analyte zones during separation.

2.1. DNA Applications

The excellent resolving power of gel or polymer solution-filled narrow-bore channels readily provides sequencing-grade separation of very similar size DNA molecules, usually detected by laser-induced fluorescence [9]. Collection of any individual or all separated fragments by capillary electrophoresis (CE) proved to be feasible and provided enough material for such downstream processes as polymerase chain reaction (PCR), sequencing, or cloning [10–12].

Early proof-of-concept experiments on capillary electrophoresis-based microscale preparative separations employed open-tubular setups with a sweep liquid at the outlet end of the capillary [13]. The separated sample components were transferred into a fraction collection device through a conventional liquid chromatography detection cell. To enable fraction collection without interrupting the electrophoresis process, porous glass junction and on-column frits were introduced [14,15]. Successful collection of small amounts of oligonucleotides was first reported in the late 1980s [16]. The identity of the collected fractions was verified by dot-blot assay. Later, capillary electrophoresis-based fraction collection of larger dsDNA fragments [4,6,7] was also demonstrated. Magnusdottir et al. [7] collected CE-separated DNA fragments by direct transfer from the capillary outlet to a positively charged membrane, driven by a synchronous motor. This approach to CE with direct blotting was faster and required fewer manual steps than conventional blotting from gels. In addition, it resulted in higher resolution, and each DNA fragment was collected into a very concentrated spot on the membrane, due to the small surface of the capillary outlet and the collection device design, which induced refocusing of the electric field lines across the membrane. The proposed method made further analysis of very small amount of DNA fractions (picogram range) possible, e.g., by hybridization.

Fraction collection can be accomplished by using either constant electric field or a voltage programming method. For efficient isolation of fractions without contamination from closely migrating bands, precise time control is essential during the fraction collection process. Because of its high separation performance, capillary gel electrophoresis (CGE) often yields very narrow peaks, representing a real challenge for fraction collection procedures. This problem was successfully addressed by decreasing the electric field strength during the collection step, thus slowing down migration rates by electric field programming during fractionation of small ssDNA fragments (40–60-mers) [1]. Separation and collection of multiple PCR products was also reported using capillary gel electrophoresis, and the collected fractions were reamplified by PCR and analyzed again by CGE [8]. The sheath flow-supported approach allowed researchers to avoid frequent interruption of the electric field when the separation capillary was moved from one collection vial to the next. Another interesting development was the introduction of two detection points near the end of the separation capillary, which significantly improved collection precision over the regularly used single-detection-window setup [2].

To accommodate large-scale applications, multicapillary fraction collection systems were developed [6,17]. A 16-capillary instrument reported by Irie et al. [6] was capable of independently collecting and automatically sorting fractions through the corresponding sampling capillaries into 16 sample trays, movement of which was controlled individually according to the detection signals (Figure 1). DNA fragments with 1 base difference in length were successfully separated and collected. In addition, differentially expressed DNA fragments were automatically sorted by comparative analysis, where two similar cDNA fragment groups, labeled by two different fluorophores, were analyzed in the same gel-filled capillary. Thus, automatic DNA fraction collectors are useful for gene hunting research, drug discovery, and DNA diagnostics. Another approach introduced a replaceable sieving matrix in conjunction with agarose microwell sampling plates, in which the 12-capillary array was dragged through the agarose gel plate, leaving the separated fragments in the microwells without any interruption of the electric current [17] (Figure 2).

Precise timing is very important for an efficient fraction collection, to avoid cross-contamination of closely migrating peaks. Microfabricated fluidic devices allow electrokinetic manipulation of analyte zones during separation, as selected zones can be redirected from a separation channel to a side channel for collection or other assays [18,19]. The applicability of microchip electrophoresis for rapid and high-resolution fraction collection, using a monolithic simple cross glass device, has been demonstrated [20]. Various size DNA fragments were separated and collected by simply redirecting the desired portions of the detected sample zones to the corresponding collection wells using appropriate voltage manipulations. The efficiency of sampling and collection of the fraction of inter-

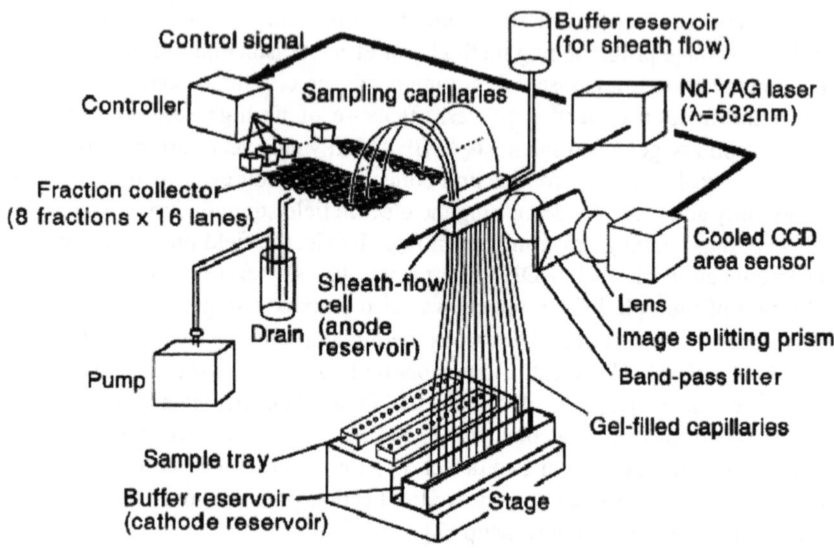

FIGURE 1 Schematic drawing of an automatic DNA fragment collector, comprising 16 gel-filled capillaries for DNA sizing, a sheath flow cell for detecting and collecting DNA fragments, and 16 sampling capillaries for transporting DNA fragments to a fraction collector consisting of 16 identical lanes. DNA fragments migrate in the gel-filled capillary from the buffer reservoir (cathode) to the sheath flow cell (anode) under negative high voltage applied. They are further transported through the sampling capillaries from the sheath flow cell to the vials set on the fraction collector. This transfer is performed by the flow of the buffer solution supplied from the buffer reservoir by gravity. Multiple fluorophores attached to the different DNA fragment groups are detected using an image-splitting prism combined with bandpass filters and a cooled CCD area sensor. (Reprinted from Ref. 6 with permission.)

est was enhanced by placing a cross-channel configuration at or immediately after the detection point. Upon the detection of the band of interest, the potentials were reconfigured to the collection mode in such a way that the selected sample zone migrated to the collection well, and the rest of the analyte components in the separation channel were stopped or slightly reversed, in this way increasing the spacing between the zone being collected and the one immediately following. By this means, precise isolation of spatially close consecutive zones can be facilitated. The separation/collection cycle was repeated until all required fractions were physically collected. The amounts of collected DNA fractions were enough for further downstream processing, such as conventional PCR-based analysis [20]. An integrated fraction collection system, based on

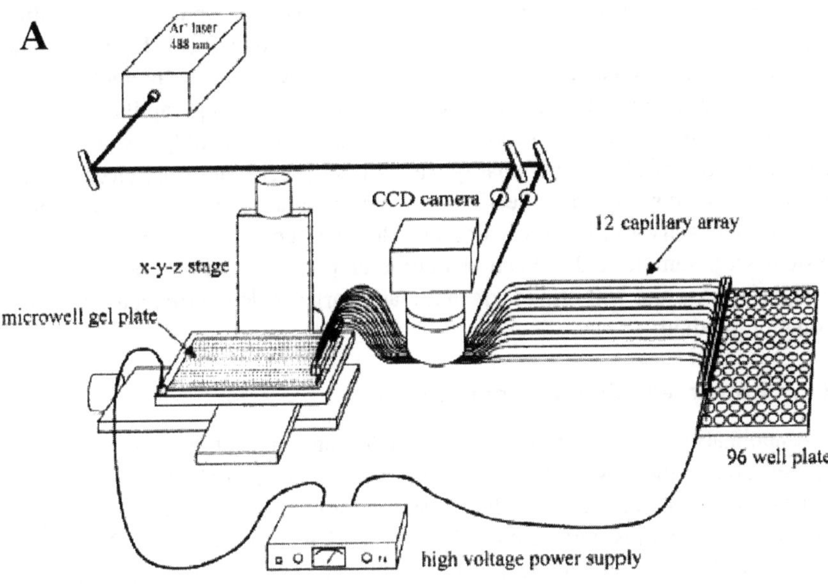

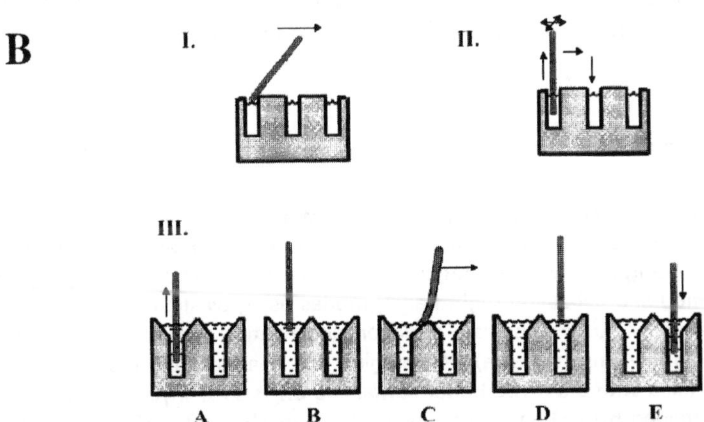

FIGURE 2 (A) Design of a capillary array instrument with two-point detection and side-entry illumination for high-sensitivity and high-resolution multicapillary fraction collection applications. (B) Scheme of the collection cycle using an agarose gel plate with wedge collection wells. (From Ref. 17 with permission.)

a micromachined microfluidic cross-connector module coupled to fused-silica capillaries, has also been demonstrated by the same authors [21] (Figure 3). The microfluidic cross-connector was machined in acrylic substrate by microdrilling a precisely centered flat end and through channels, enabling capillary connections with essentially zero-dead-volume joints. The assembly featured integration of microfluidic functionality, facilitating precise and fast elecrokinetic manipulations for fraction collection, and long separation pathways, while minimizing necessary fluidic reservoirs and electrical connections. Interfacing the system with sample and collection microtiter plates by robotic positioning is straightforward and will provide an automated large-scale micropreparative fractionation method.

2.2. Protein and Peptide Applications

The application of CE for preparative separations of peptides and proteins is limited due to the low preparative capacity of capillary columns. In addition, adaptation of an analytical capillary system to a preparative one is not straightforward and requires certain modifications of CE instruments [2,22]. Several procedures for fraction collection from a capillary have been developed recently, as reviewed in Ref. 22. For continuous fraction collection in CE it is necessary to modify the capillary outlet to complete the electrical circuit. Karger and coworkers achieved this by using a coaxial sheath liquid interface to transport the sample components leaving the capillary exit into the collection microcapillary

---

FACING PAGE

FIGURE 3 High-precision micropreparative separation system. (A) Micromachined acrylic cross-connector coupled with four fused-silica capillaries. (B) Schematics and corresponding images of electrokinetic sample manipulations during separation (left panel) and collection (right panel) modes using plastic cross-connector/capillary assembly. Arrows indicate direction and relative migration velocities of buffer/sample flow (not to scale). For imaging, the channels and reservoirs 2, 3, and 4 were filled with 0.5 × TBE buffer containing 0.2% poly(vinylpirrolidone). Reservoir 1 contained 50 μM sodium fluorescein dye in TBE buffer. (C) Collection of multiple fractions with cross-connector/capillary assembly. Separation and collection distances were 5 cm. Electrophoretic traces of PCR product mixture (1). Detection of the first (2), second (3), third (4), and fourth (5) peaks, each followed by voltage reconfiguration to collection mode, during which larger DNA fractions retained in the separation channel. For better visualization, traces 3, 4, and 5 were offset along the *x* axis by the values corresponding to the times of potential reconfiguration for the collection of the preceding bands. (Reprinted from Ref. 21 with permission.)

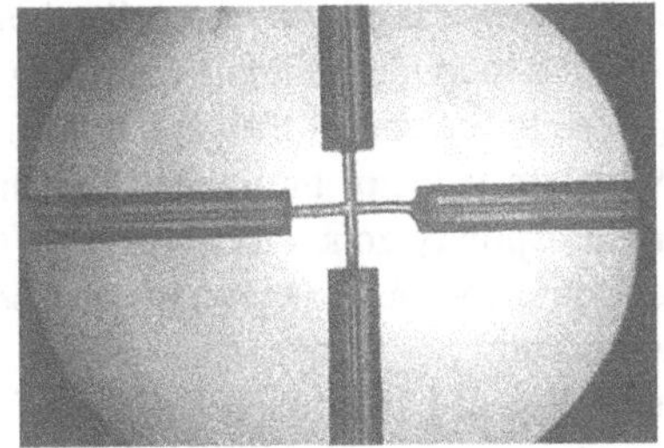

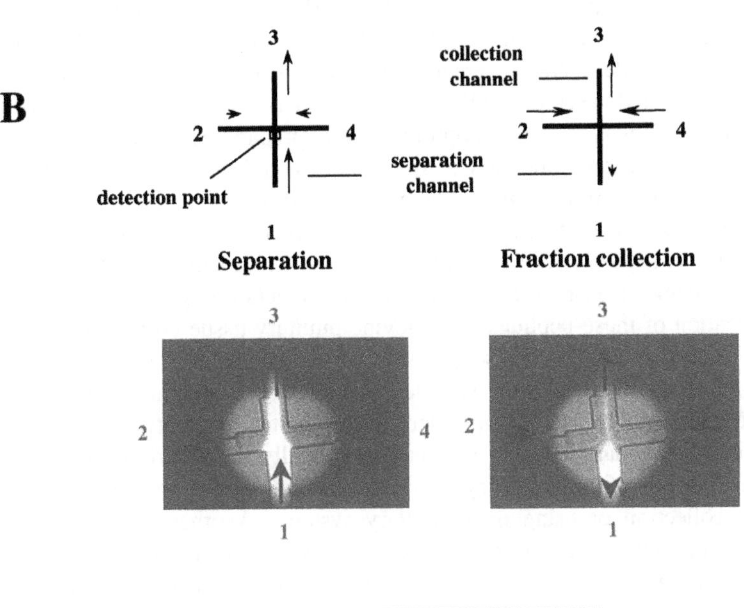
B
3
collection
channel
2
4
detection point
separation
channel
1
Separation
Fraction collection
3
2
4
1

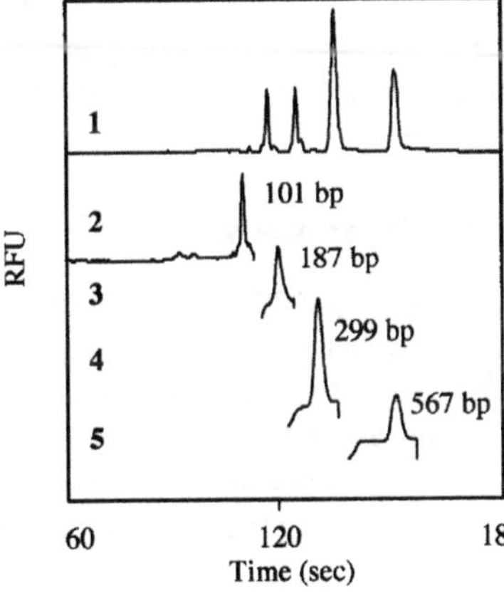
1
2
101 bp
RFU
3
187 bp
4
299 bp
5
567 bp
60
120
180
Time (sec)

vials [2]. In this device the precise timing of the collection steps was based on measuring the velocity of each individual sample zone between two UV-absorption detection windows close to the end of the capillary. This design eliminated the need for a constant analyte velocity throughout the entire column. Consequently, sample stacking in capillary zone electrophoresis (CZE) with large injection volumes and isoelectric focusing of proteins and peptides could be utilized with high collection accuracy.

Multiple sequential CE-based fraction collection of peptides and glycopeptides from complex tryptic maps has been demonstrated by Boss et al. [3,23]. In this case, the resolution was compromised due to heavy sample loading and reduction of voltage (to control Joule heating). The collection buffer was optimized by addition of 12% methanol, thus improving collection yields. Sequentially collected peptide fractions were analyzed by Edman sequencing and MALDI MS (matrix-assisted laser desorption ionization mass spectrometry) to verify peptide identity. It was shown that optimization of preparative separation parameters enabled consistent collection of single pure nonglyco- and glycopeptides by CE followed by further analysis using other methods.

Micropreparative CE has also been applied to isolate opioid peptides, endorphins, and enkephalins from complex biological fluids and tissues [24]. Capillary inner diameter, sample load, and fraction collection have been optimized for purification of these peptides from bovine pituitary tissue homogenate. The purity of the collected peptides was confirmed by analytical CZE of reinjected fractions. Sufficient quantity of leucine enkephalin was obtained for further characterization by CE-MS. The problem of the limited preparative capacity of CE (typically less than 1 μg per run) can be partially facilitated by increasing the inner diameter of the capillary at the expense of separation efficiency, by repetitive collection or using multicapillary systems. Another solution to the preparative capacity problem could be converting analytical capillary separations to a continuous free-flow arrangement, as described in Refs. 25 and 26. An obvious advantage of free-flow electrophoresis compared to capillary zone electrophoresis from the preparative separation point of view is that the former allows continuous sample feeding and fraction collection. This can significantly enhance system performance.

2.3. Other Preparative Separations

Karger's group demonstrated preparative isolation of individual peaks of a fluorescently derivatized oligosaccharide ladder by capillary zone electrophoresis [2] (Figure 4). The authors used two detection windows for more accurate timing of the fraction collection and CZE sample stacking of large injection volumes in 200-μm-i.d. capillaries. A sheath flow collector was employed to maintain permanent electric current during the collection, as well as a semipermeable

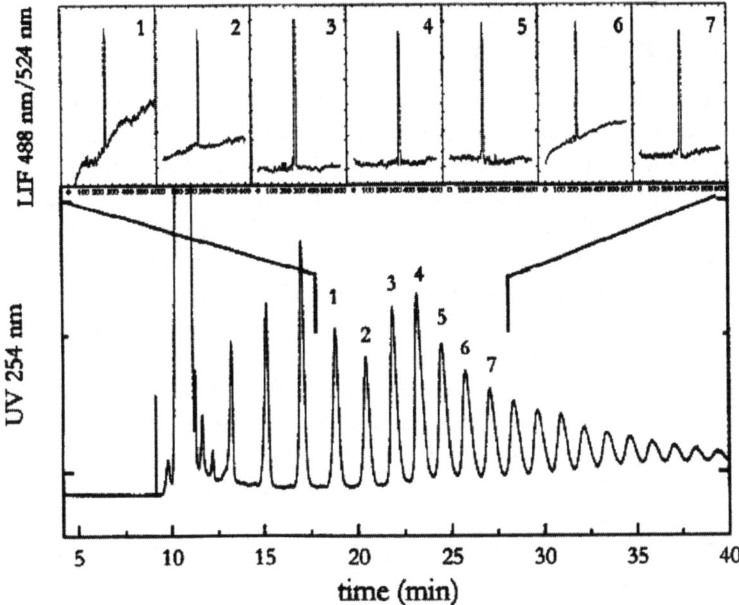

FIGURE 4 Isolation of specific peaks from an APTS-labeled glucose ladder with collection timing based on velocity determination from the two-point detection scheme. Conditions: injection 1 μL; capillary 200 μm i.d. × 30 cm; detection UV at 254 nm; field 330 V/cm. Reinjected fractions: electrokinetic injection, 10 s at 400 V/cm; capillary, 50 μm i.d. × 25 cm, 20 cm to the detection; LIF excitation, 488 nm emission, 524 nm detection; field 330 V/cm. (Reprinted from Ref. 2 with permission.)

membrane to close the system inlet to suppress the backflow associated with the sheath flow droplet pressure [2].

Chankvetadze et al. have demonstrated the potential of flow-counterbalanced capillary electrophoresis (FCCE) in chiral and achiral micropreparative separations [27]. Unlimited increase of separation selectivity can be achieved for binary mixtures, such as (±)-chlorpheniramine with carboxymethyl-β-cyclodextrin chiral selector, or α- and β-isomers of a asparatame dipeptide. The carrier of the chiral selector or pseudo-stationary phase, electroosmotic flow (EOF), pressure-driven flow, or hydrodynamic flow can be used as a counterbalancing flow to the electrophoretic mobility of the analyte or vice versa, resulting in dramatic changes of the effective mobilities of the sample mixture components [28]. This approach can be used for micropreparative CE, stepwise separations, and fraction collection of multicomponent mixtures [27].

3. MICROSCALE ON-LINE SAMPLE PREPARATION

3.1. Capillary Electrophoresis

Of all alternative technologies developed for DNA analysis, electrophoresis in multiplexed capillary columns or microchannels appears to be the most suitable for real applications in high-speed, high-throughput production settings. The cost of the analysis can be significantly reduced as a direct result of automation and parallelization. Typically, only 5–20 nL of sample is injected into the capillary, and subnanoliter amounts in microfluidic chips, at concentrations identical to or less than those loaded into wells of slab gels. However, in order to be able to actually inject the samples, much larger volumes (over 1 μL) are required for handling, not to mention additional solution amounts needed for robotic liquid-handling workstations. Therefore, one must miniaturize the sample preparation steps prior to CE/microchip analysis and interface it directly to the capillary/microchannel array in order to take advantage of the much lower sample requirements. It is anticipated that developing new technologies will result in reducing sample preparation by 100-fold in volume, increase in speed by 10-fold, and multiplexing up to 1000 different samples at a time [29], versus traditional sample preparation technologies.

The small sample requirement of CE, as low as ~5 nL, leads to significant reduction in the amount of DNA template and reagents, resulting in a substantial lowering of the cost per base in DNA sequencing and PCR-based analysis. Miniaturization of cycle sequencing as well as PCR (total volume ~1.5 μL) in a capillary and microwell formats has been demonstrated [30–34,35]. A capillary reactor has additional advantages of higher reaction speed due to its small heat capacity, and compatibility with highly multiplexed parallel capillary array electrophoresis. Numerous attempts have been made to develop robotic stations to perform purification and sample preparation of sequencing reactions and loading in slab-gel electrophoresis [36,37]. Complicated combinations of multiple movements of robotic arms and precise liquid dispensing is bulky, costly, and limited in terms of miniaturization and adequate coupling to microseparation methods. An alternative concept is to purify reaction products on-line by chromatographic separation, such as size-exclusion chromatography [38], which has unique features of high recovery, desalting ability, suitability for pressure application, and integration with CE.

Yeung's group has developed an integrated system for multiplexed on-line sample preparation prior to electrophoretic analysis in a capillary array format [29]. In one implementation, on-line Sanger reaction followed by size-exclusion chromatographic purification (desalting), injection and sizing of DNA sequencing fragments in uncoated fused-silica capillaries filled with replaceable polyethylene oxide sieving matrix has been accomplished. The key to the successful operation of such a complex instrument is the thorough optimization of all the important parameters (temperature, reaction volumes, buffer composition, special additives to suppress large surface-to-volume ratio effects, etc.) and synchronization of the processing and analysis steps (valving and switching, chromatographic purification, injection, rinsing, and regeneration of the column) [30,33] (Figure 5). A detailed description of all the processing steps and instrument design can be found in Ref. 29.

The same group has also demonstrated an integrated on-line PCR capillary gel electrophoresis (CGE) system for DNA typing and disease diagnostics [33]. To integrate and automate all the steps, from real biological samples to final readable results, two applications were investigated: the four short tandem repeat (STR) loci CTTv for DNA fingerprinting, and a DNA probe for human immunodeficiency virus (HIV-1) diagnostics. The system was constructed to

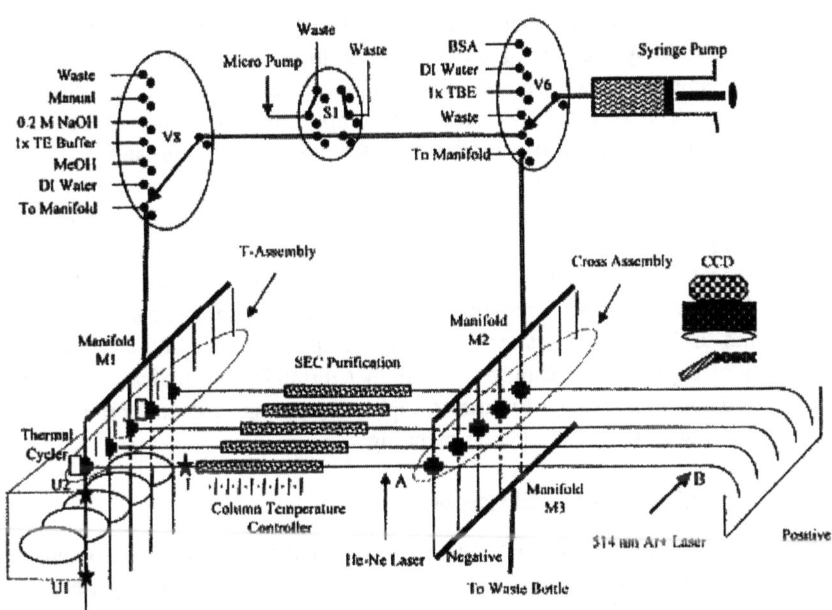

FIGURE 5 Schematic of a complete multiplexed and integrated instrumental design with eight capillaries. Stars at I, U1, and U2 represent the multiplexed freeze/thaw valves. The T-assembly is made up of eight pieces of commercial junctions stacked together. These connect to the manifold M1, the SEC (size-exclusion chromatography) purification columns, and the reaction loops. The cross-assembly is made of eight pieces of standard crosses packed together with built-in heaters. V8 is an eight-position motorized titanium valve with a center port. S1 is a two-position motorized PEEK valve. V6 is a six-position motorized PEEK valve. (Reprinted from Ref. 33 with permission.)

allow one-step ss- and ds-DNA fragment analysis after PCR [31]. Simultaneous amplification of more than one DNA region of interest in one reaction mixture reduces labor, time, cost, and cross-contamination, since sample handling is minimal. On the other hand, it requires fine tuning of the reactions parameters, including temperature and time of denaturation, annealing, and extension steps, to ensure efficient amplification. For accurate fragment size determination, a calibration using co-injection with appropriate size standards is also required. In addition, incorporation of DNA extraction and purification from real samples (biological fluids) in an on-line instrument would save time, labor, and minimize the exposure of workers to infectious samples. An automated and integrated system for DNA typing directly from blood samples has been developed [34], based on eight capillary array electrophoresis and microfluidics (Figure 6).

Three STR loci, vWA, THO1, and TPOX, were co-amplified simultaneously in fused-silica capillaries, using a hot-air thermocycler and certain modifications of standard protocols necessary for direct typing from blood (Figure 7). A programmable syringe pump and a set of multiplexed liquid nitrogen freeze/thaw switching valves were employed for liquid handling in the liquid distribution network. The system fully integrated sample loading, PCR, addition of a standard, on-line injection of samples and standards, separation, and detection. Regeneration and cleaning of the entire system prior to subsequent runs were also integrated in the instrument. The total time for one run was 2.5 h, however, overlapping the processing steps, e.g., by performing cleaning and regeneration of the separation capillaries during the reaction step, and subsequent PCR setup during the electrophoretic separation of the first set of samples, can shorten the cycle time to 1.3 h. The system is readily compatible with future upgrade to hundreds of capillaries to achieve even higher throughput. Applications can be extended using this concept to other sample preparation protocols prior to CE, such as drug screening and peptide mapping.

3.2. Ultrathin-Layer Gel Electrophoresis

Guttman and co-workers introduced membrane-mediated sample micropreparation procedures in conjunction with ultrathin-layer (≤190-μm) slab-gel electrophoresis [39]. As little as 0.2 μL of samples and reagents can be spotted manually or robotically onto the tabs of a membrane loader, incubated if necessary, and then placed into the cassette in intimate contact with the straight edge of the ultrathin-layer agarose or composite agarose/linear polyacrylamide gel. Under applied electric field strength, the sample molecules, e.g., DNA or proteins, migrate into the gel, forming finely defined, sharp bands. This procedure facilitates the introduction of a large number of minute amounts of samples (96 parallel analysis lanes) into microgels, eliminating the need to form individual wells and enabling sample pretreatment and loading to be accomplished easily

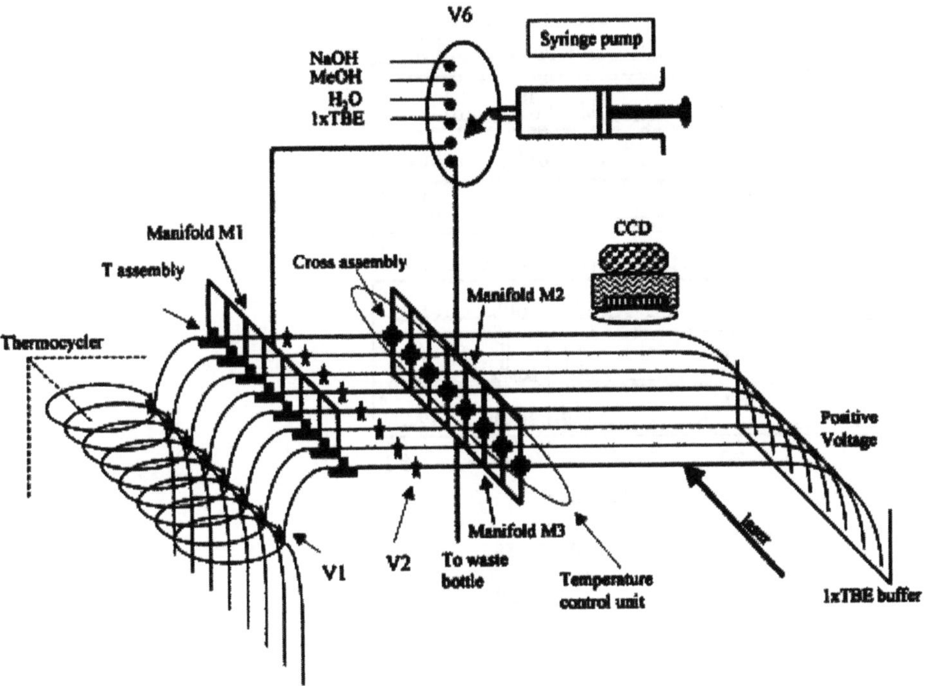

FIGURE 6 Schematic of a complete multiplexed and integrated instrumental design for PCR analysis with eight channels. Different functional capillaries were connected by a T-assembly and a cross-assembly. Stars represent the freeze/thaw valves V1 and V2. A syringe pump and a six-way selection valve were used to load and distribute liquids for various purposes. A CCD camera was used to collect fluorescence from the eight capillaries simultaneously. The temperature control unit at the cross-assembly was used to denature DNA prior to injection. (Reprinted from Ref. 34 with permission.)

on the benchtop, outside the separation setup. These small-pore-size microfibrous membranes serve as a micropreparation platform, e.g., for performing restriction digestion of DNA molecules by endonucleases followed by direct injection and rapid ultrathin gel electrophoretic analysis of the resulting fragments [40]. Complete digestion of several nanograms of target DNA was accomplished on the membrane at room temperature in a few minutes with a single or a combination of different restriction enzymes using only submicroliter volumes of the reagents. The entire processing time, including sample micropreparation (digestion), loading into multilane format, and electrophoretic analy-

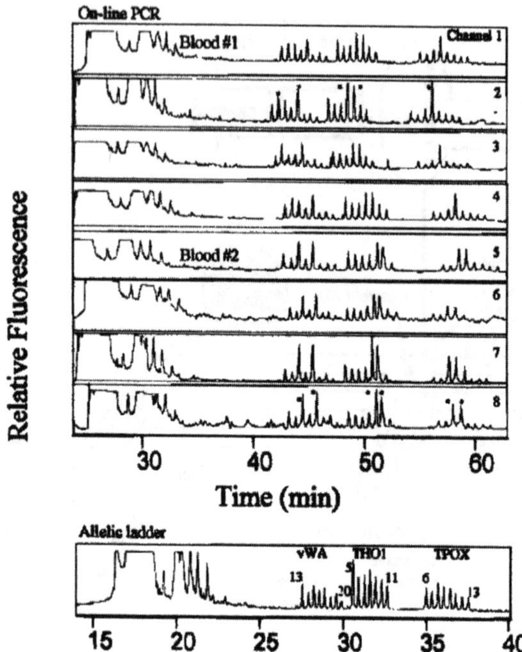

Figure 7 Electropherograms showing simultaneous genotyping in eight channels from blood. Injection: $T = 90°C$, voltage 150 V/cm, effective length 60 cm, matrix 1.5% M_n 8,000,000 and 1.45% M_n 600,000 PEO solution. Laser: 15-mW 488-nm argon-ion. Detection: CCD through a 520-nm long-pass filter, exposure time 300 ms. Capillaries in the array are labeled from 1 to 8 according to the proximity to the excitation laser. Two blood samples were analyzed. Channels 1–4 show sample 1 and channels 5–8 show sample 2. The lower electropherogram is the vTT allelic ladder by itself. The genotype (asterisks) was defined by an increase in the relative intensities of the peaks within each locus when the sample is present. (Reprinted from Ref. 34 with permission.)

sis, required less than 20 min. In another application of the same system, fast, sensitive, high-performance and high-throughput analysis of proteins (14–116-kDa range) was demonstrated, using noncovalent labeling and sodium dodecyl sulfate (SDS) composite agarose/linear polyacrylamide gels [41]. The method eliminated time-consuming sample preparation by pre- or postseparation staining/labeling. Membrane-mediated sample handling and injection is readily automated (robotic spotting) and can be applied to most high-throughput applications, such as automated DNA sequencing [42], DNA restriction digestion [40], protein analysis, etc.

3.3. Microfabricated Devices for On-Line Sample Pretreatment

Over the past few years, microfabricated devices have attracted a great deal of attention as a possible means of increasing the throughput and mass sensitivity of analytical procedures via miniaturization. Sample pretreatment (e.g., desalting, preconcentration, digestion) and CE-based separation in microdevices prior to MS analysis are of special interest due to the wide applicability of mass spectrometry for the detection and characterization of biological samples. Large-scale analysis of chemical libraries, expressed proteins, etc., requires development of high-throughput procedures for sample pretreatment and delivery to MS instruments. A number of successful efforts have been reported to couple microfluidic devices to mass spectrometers [43–51]. Such combination enables automated sample delivery and enhanced MS analysis efficiency by, for instance, sample processing/enrichment/cleanup and fractionation prior to detection. These devices can transport the analyte fluid electrokinetically or by pressure, and generate electrospray via an attached capillary or more complex emitter couplings. Enzymatic digestion was monitored in real time with high detection sensitivity (0.1–2 pmol/μL of loaded sample) on a hybrid microchip nanolectrospray device by peptide mass fingerprinting [51]. The same research group used a microfabricated, electrically permeable thin glass septum to generate electrospray by electroosmotically pumping the solutions through field-free channels, past the point where the CE electric potential was applied [50]. Automated MS analysis was demonstrated by Ekstrom et al. [4], utilizing a porous microfabricated digestion chip, integrated with a sample pretreatment robot and a microdispenser for transferring reaction products to a MALDI target plate.

A new design for high-throughput microfabricated electrospray ionization time-of-flight MS (CE/ESI-TOF-MS) with automated sampling from a microwell plate has been reported by Karger's group [52,53]. The assembly combined a sample loading port, separation channel, and a liquid junction for coupling the device to the MS with a miniaturized subatmospheric electrospray interface (Figure 8). The microdevice was attached to a polycarbonate manifold with external reservoirs equipped for electrokinetic and fluid pressure control, which simplified the microdevice design and enabled extended automatic operation of the system. A computer-activated electro-pneumatic distributor was used for sample loading and washing the channels. Automated CE/ESI-MS analysis of peptides and protein digests was successfully demonstrated.

A very recent series of publications by Locascio and others [54–56] demonstrated successful applications of various plastic materials, such as polydimethylsiloxane (PDMS), polymethylmethacrylate (PMMA), copolyester, and their combinations, for assembling integrated fluidic structures to perform on-line sample pretreatment by affinity dialysis and concentration for fast and sensi-

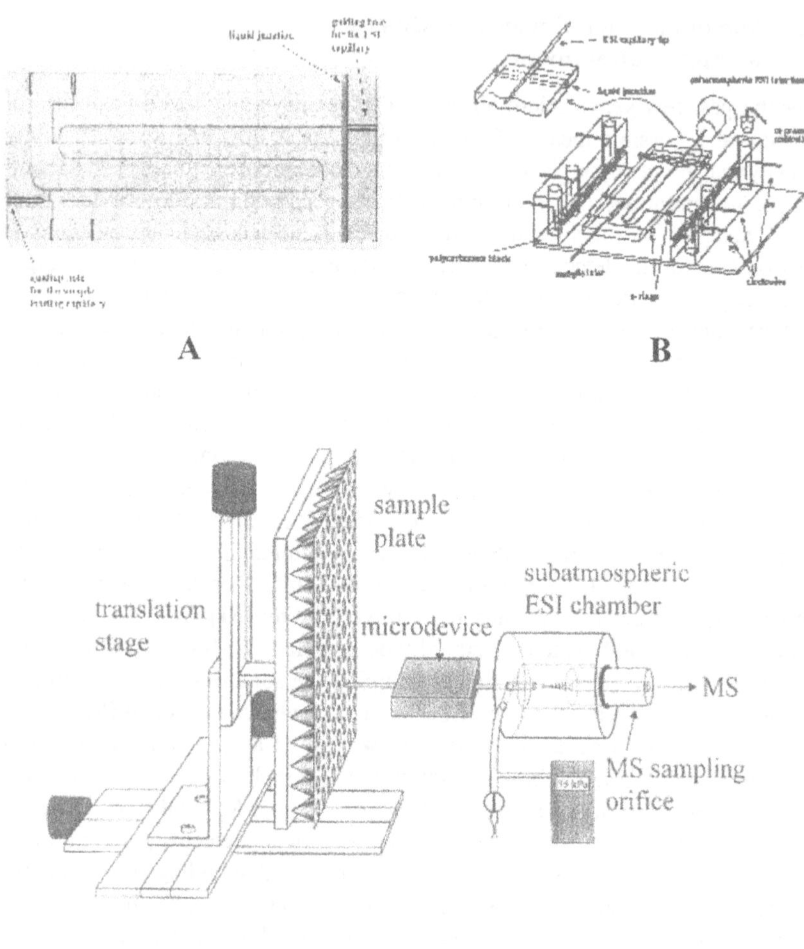

FIGURE 8 Automated high-throughput microchip CE/ESI-MS system. (A) Photograph of the glass microchip (5 × 2 cm). (B) Diagram of the microdevice design, including the microchip and polycarbonate manifold. (C) Overall design of the coupling of the microwell plate sample delivery system equipped with a microdevice for high-throughput separation-MS analysis. (Reprinted from Ref. 53 with permission.)

tive ESI-MS analysis of different compounds. Multilayer devices were assembled from separate plastic pieces imprinted with microchannels by silicon template imprinting and capillary molding techniques. Three-dimensional fluidic devices were constructed by sandwiching a polyvinylidene fluoride membrane between the substrate layers with appropriate channels and simply clamping the assembly together using the good adhesive properties of the polymeric materials. Access holes were drilled in the plastic to interconnect fluidic layers and interface with external inlet/outlet lines. These devices performed affinity capture, concentration, and direct identification of the targeted compounds by ESI MS (coupled to the chip through the capillary and microdialysis junction), as well as miniaturized ultrafiltration of affinity complexes of antibodies [54]. The analyte solutions were pumped through the channel countercurrently to the buffer flow in the adjacent channel (via the semipermeable membrane). By this means it was possible to perform dialysis and separation of aflatoxin affinity complexes with their antibodies from unbound compounds, followed by passing the solution of affinity complexes against an air counterflow in another fluidic layer for water evaporation and analyte concentration. Another, similar device performed ultrafiltration of affinity complexes of phenobarbital antibody and barbiturants, including sequential loading, washing, and dissociation steps. These microfluidic devices significantly reduced dead volume and sample consumption, and increased MS detection sensitivity by one to two orders of magnitude. An integrated platform, based on similar PDMS/membrane assemblies, was also reported [55] for rapid and sensitive protein identification by on-line digestion and consequent analysis by ESI-MS, or transient capillary isotachophoresis/capillary zone electrophoresis with MS detection. A miniaturized membrane reactor was fabricated, using a porous membrane with adsorbed trypsin separating two adjacent channels. Due to the large surface-to-volume ratio of the porous membrane structure, extremely high catalytic turnover was achieved.

4. INTEGRATED MICROREACTORS

Miniaturization of chemical and biochemical processes has significant advantages with respect to cost, throughput, kinetics, safety, and scale-up. Microreactors are especially in great demand in biochemical processing in the pharmaceutical industry for acceleration and efficiency increase of preclinical drug discovery, chemical development, manufacturing, and mass screening. Typically, chemical synthesis devices handling submicroliter volumes are considered as microreactors, and they require integration of specially designed components, such as micromixers, micropumps and valves, microreaction chambers, miniaturized heat exchangers, microextractors, and microseparators, depending on the applications. Microreactors enable on-site production at the point of demand. Informative overview of the most recent progress in this area can be found in a number of publications [57–59]. In this chapter we restrict ourselves to a brief outline and some interesting developments of micropreparative devices employing electrokinetic phenomena. One of the most important features of a microreactor is its remarkably large surface-to-volume ratio. Thus, extremely rapid and highly exothermic reactions can be carried out under isothermal conditions, and higher selectivity and more precise kinetic information, compared to conventional-scale methods, are available, due to very low mass transfer distances [57]. Microreactors can be gas or liquid phase. The fundamental design and operation problem of the latter is adequate mixing and fluidic control. To overcome laminar flow limitations and increase diffusive mixing, various configurations have been constructed [57,58]. Pumping of the fluids within the chips has been achieved by a variety of means including pressure, electroosmotic, electrophoretic and electrohydrodynamic flows, and their combinations.

Application of electroosmotic flow has certain advantages over pressure-driven flow, namely, flat flow profile, experimental simplicity, absence of moving parts and valves, minimal backpressure effects, separation efficiency increase, and the possibility of controlling multiple channels with just a few electrodes. Other reported methods of fluid control have been electrohydrodynamic pumping, creating gradients in surface pressure with small voltages and surfactant molecules with redox groups, use of hydrophobic/hydrophilic barriers on the surface [59], and thermocapillary pumping [60–62].

Chemical synthesis in microreactors can be conducted in both continuous-flow and batch modes, with the first being typically utilized for synthesis of one product and the second for parallel processing. Some illustrative examples are given in Refs. 57–59.

Electrophoretic microfluidic chips feature a number of microreactor characteristics and have been used for conducting chemical and biochemical reactions in channels and microfabricated chambers, mixing reagents, microextraction and microdialysis, post- and preseparation derivatizations, etc. The most recent achievements are reviewed in Ref. 63 and other similar publications. These integrated microdevices perform PCR amplification, cell sorting, enzymatic assays, protein digestion, affinity-based assays, etc. In this section we describe such integrated microsystems and the most recent advances in this field.

4.1. Multifunctional Devices

Construction of integrated devices, combining sample preparation, processing, and analysis steps, has been a central driving force in the development of micro total analysis systems (μ-TAS). Among the latter are PCR-based DNA amplification, restriction digestion, other enzymatic assays, analyte preconcentration, filtering, dialysis, and various pre-/postderivatization and detection methods. The ultimate integration of all of these steps into a single microchip has not been realized so far, although higher-level system integration and complex microdevices are rapidly emerging.

Currently, an interesting commercial product is being developed to provide 96-channel microfluidic plastic LabCards [64] for high-throughput miniaturized biochemical assays, DNA analysis, etc., addressing the whole range of common laboratory procedures such as mixing, incubation, metering, dilution, purification, capture, concentration, injection, separation, and detection. Other integrated microchips for clinical applications have been designed and prototyped [60,61,65]. A dual-function microchip [65] has integrated two key steps in the analytical procedure—cell isolation and PCR. White blood cells were isolated from whole blood on 3.5-μm-feature-sized weir-type filters, etched in silicon-glass chips (4.5-μL volume), followed by direct PCR amplification of isolated genomic DNA. Modification of filter size and shape and/or using specific capture agents can be employed to increase selectivity of the devices.

A filter chamber array, microfabricated in silicon and sealed with glass, enabled real-time parallel analysis of three different samples on beads in a volume of 3 nL on a 1-cm^2 chip [66]. Each filter chamber contained microfabricated pillars to trap and localize reacting particles. Single-nucleotide polymorphism (SNP) analysis by solid-phase pyrosequencing has been performed in this chamber, using biotinylated primers attached to streptavidin-coated beads. Allele-specific pyrosequencing is based on the difference in DNA polymerase extension reaction of match and mismatch primers hybridized to the target DNA. The nucleotide incorporation results in the release of pyrophosphate, which is enzymatically converted to ATP, and detectable by luciferase-generated light. Passive fluidic valves, consisting of hydrophobic patches of plasma-deposited octafluorocyclobutane, were incorporated in the device and served as physical barriers between liquids, to allow controllable sequential loading of different samples without their interference. The device is reusable, enables parallel sample handling, and is amenable to the implementation of complex biochemical assays on beads.

A simple, elegant method of thermocapillary pumping, based on surface tension changes under localized heating, was incorporated into a microchip structure for guiding nanoliter solution droplets in a controllable manner through the channels, mixing and incubating them to perform enzymatic reactions [60, 61]. This integrated system also incorporated thermocycling chambers, microfabricated gel electrophoresis channels, and a set of doped-diffusion diode elements, fabricated in Si, for detection of β-radiation (^{32}P-labeled DNA). The thermocapillary pumping idea was recently also developed by Troian et al. [62], who proposed configurable thermofluidic arrays to pattern flow at the microscale level. The development of this approach will enable "pixelating" the chip surface to precisely control the temperature pixel by pixel and, therefore, surface tension and viscosity of the liquid droplets. By this means, it will be possible to move, mix, and/or otherwise manipulate small volumes.

Another integrated device, developed recently in Burns's group, employed photodefinable polyacrylamide gels as a sieving medium for DNA electrophoresis with locally controlled gel interface and electrode-defined sample compaction and injection technique [67] (Figure 9). The latter approach was based on integrated microfabricated electrodes and helped achieve sample compaction without migration into the gel, enabling control over the size and application of the sample plug.

A complex PMMA-based fluidic device was built by Soper's research group [68], which coupled capillary nanoreactors to microseparation platforms (electrophoresis chips) for the generation of sequencing ladders and PCR products. The nanoreactors consisted of fused silica capillary tubes with a few tens of nanoliters of reaction volume, which can be interfaced wiht the chips via connectors micromachined in PMMA, using deep X-ray etching. A DNA tem-

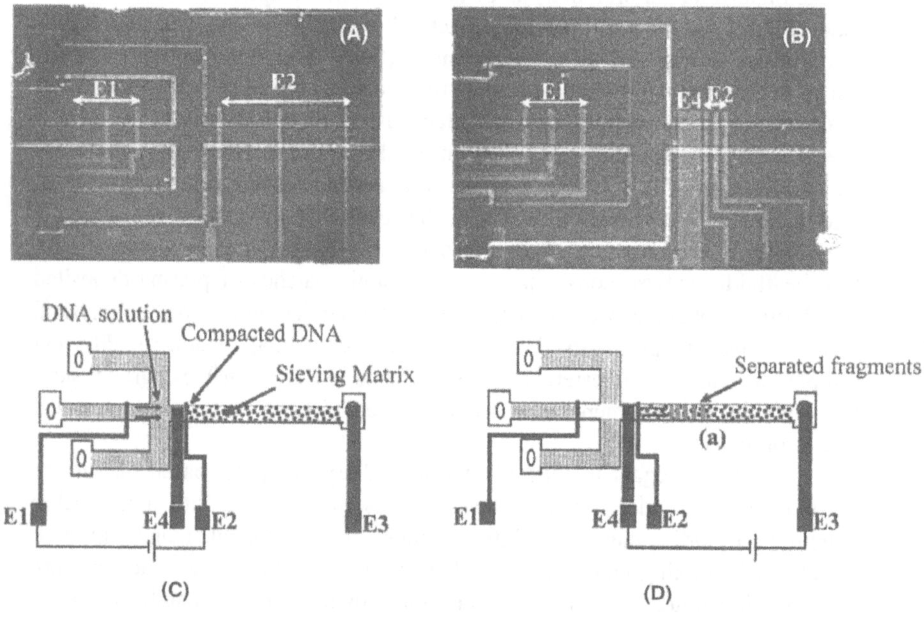

FIGURE 9 Electrode arrangements in microfabricated sample injection and separation system. (A) 50-μm thin electrodes (E1, E2) are used for sample compaction and separation. (B) A thick electrode (E4) is introduced to allow the use of higher voltages during the sample release and separation phase. (C), (D) Schematic of operation of electrode-defined sample compaction, release, and subsequent separation for the electrode system shown in (B). (Reprinted from Ref. 67 with permission.)

plate was immobilized to the nanoreactor walls via a biotin–streptavidin linkage. Following fast thermocycling in a special chamber oven, the DNA fragments were directly injected (electrokinetically) into the microchip device for fractionation. A dual fiber-optic component was micromachined into the plastic chip platform for integrated laser-induced fluorescence detection, and consisted of two channels, accommodating laser light delivery and emission collection fibers (Figure 10). These results look promising in the perspective of the fabrication of automated devices, which consume minute amounts of costly reagents.

4.2. Integration of PCR and Electrophoresis

DNA processing integrating microscale PCR amplification followed by electrophoretic analysis is considered as another example of micropreparative applications involving electrokinetic phenomena. Since PCR has become a key tool in modern molecular biology, a number of groups have investigated microchip-based PCR systems in both disposable microchamber and microfluidics formats. The obvious benefits of such miniaturization come from the improved thermal energy transfer compared to conventional macrovials, resulting in greatly increased speed of thermal cycling, reduced amount of expensive reagents used, and the possibility of creating versatile, multifunctional, integrated systems. However, potential problems, associated with PCR downscaling, such as surface-related effects, become greatly enhanced with shrinking reaction volumes and consequently increasing surface-to-volume ratio. Various biocompatible PCR-friendly surfaces have been investigated by passivation of silicon, fused silica, glass, or plastic walls via coatings by chemical modification and/or physical adsorption. The other issue is related to handling of small volumes of liquids and the incorporation of a miniaturized DNA amplification unit into the chain of sample processing and analysis, to develop truly automated and high-throughput technology. The major milestones of the development in this field have been discussed elsewhere [9,12]. In this chapter we mention only a few recent interesting implementations. A number of studies have been done to optimize microscale DNA amplification. Two possible embodiments can be envisioned, such as closed microchamber and open format, both of which can be multiplexed to an array. Fast, real-time PCR analysis was demonstrated in silicon microchambers by several groups [65,69–71]. DNA amplification can be coupled with special microfluidic cartridges designed to carry out several sequential steps of DNA extraction from different sample types (tissues, cells, whole blood, soil, food, etc.) prior to the PCR [72]. DNA amplification in open format has not yet been reported, but other chemical and enzymatic reactions, conducted in nano- and pico-vials, have shown promising results [73–76].

There have been several reports on integration of DNA amplification and electrophoretic separation on a single microfabricated chip [77–82]. These de-

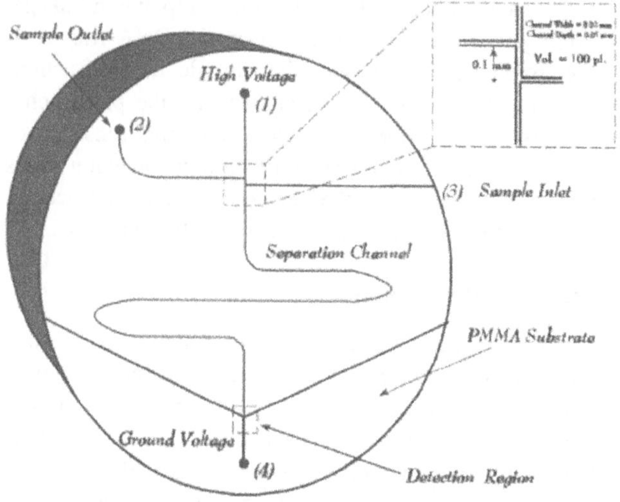

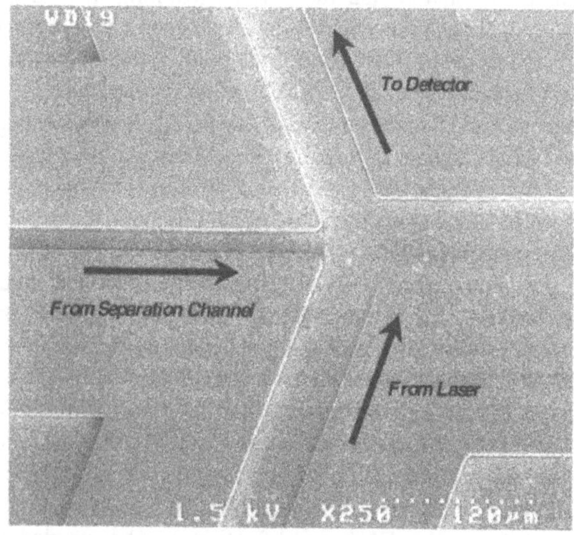

FIGURE 10 (Top) Topographical layout of the microelectrophoresis chip fabricated in PMMA. The channels were 50 μm in depth and varied in width (20 or 50 μm). The injector contained a volume of 100 pL. (Bottom) SEM image of dual fiber-optic component micromachined in PMMA. (Reprinted from Ref. 68 with permisison.)

vices had small reaction wells, which were thermocycled to generate the amplicons, followed by the injection/separation/detection steps in the interconnected microchannel network. An integrated system combining fast on-chip DNA amplification by local thermocycling and preconcentration technique, followed by microchip electrophoretic sizing, was reported by Khandurina et al. [80]. Electric field-mediated DNA fragment preconcentration was accomplished prior to injection by means of a microfabricated porous membrane structure [83]. This method enabled reduction of the PCR cycle number and total analysis time (less than 20 min) [80]. A microchip device, developed by Mathies's group, was coupled to a special microchamber with fast heating/cooling capabilities for effective DNA amplification [77]. Recently the same group developed an integrated monolithic system incorporating several 280-nL PCR chambers etched into a glass structure, connected to microfluidic valves and hydrophobic vents for sample introduction and immobilization during the cycling (Figure 11) [81,82]. The low thermal mass and use of thin-film heaters enabled cycle times as fast as 30 s. The amplified products were injected directly into the corresponding gel-filled microchannels and detected by laser-induced fluorescence. The excellent detection sensitivity attained (20 copies/μL) suggested that it might be possible to perform stochastic single-molecule PCR amplification.

An interesting alternative to a microchamber reaction format has been proposed by Manz's group, who demonstrated a continuous-flow PCR system [84]. A PCR cocktail was pumped continuously through a serpentine glass channel, periodically passing the three different temperature zones to perform denaturation, annealing, and amplification steps. Although total reaction time was relatively long (50 min for 20 cycles), multiple simultaneous reactions can be carried out by sequential introduction of separate reactions in each loop.

Ultrashort thermal cycling times of 17 s per cycle have been achieved by employing IR radiation-mediated heating and compressed-air cooling to amplify DNA in a capillary or on-chip microchambers by Landers's group [85–87]. A tungsten lamp, thermocouple feedback, and computer interface were used to accurately control the temperature by the combination of light intensity and air flow.

A novel thermocycling system, based on a capillary equipped with bidirectional pressure-driven flow and in-situ optical position sensors, has been presented by Ehrlich's group [88]. A 1-μL PCR mix droplet was controllably moved between three different heated zones in a 1-mm-i.d. oil-filled capillary. A light-scattering detector was used to control the droplet position. Thirty PCR cycles were completed in 23 min with good amplification efficiency. The maximum possible speed with this arrangement was estimated to be theoretically as high as 2.5 min to complete the reaction. A hybrid PDMS–glass microchip for functional integration of DNA amplification and gel electrophoresis was recently reported [89]. Thermoelectric heating/cooling control was used in this

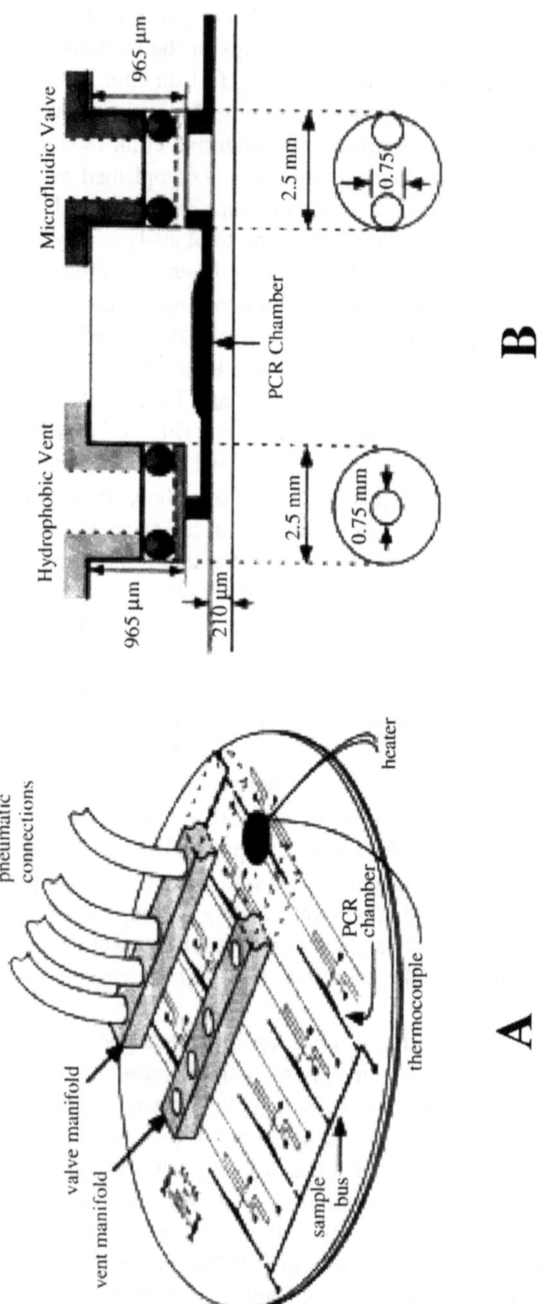
pneumatic connections
valve manifold
vent manifold
PCR chamber
thermocouple
heater
sample bus
A
Hydrophobic Vent
Microfluidic Valve
PCR Chamber
965 µm
965 µm
210 µm
2.5 mm
2.5 mm
0.75 mm
0.75
B

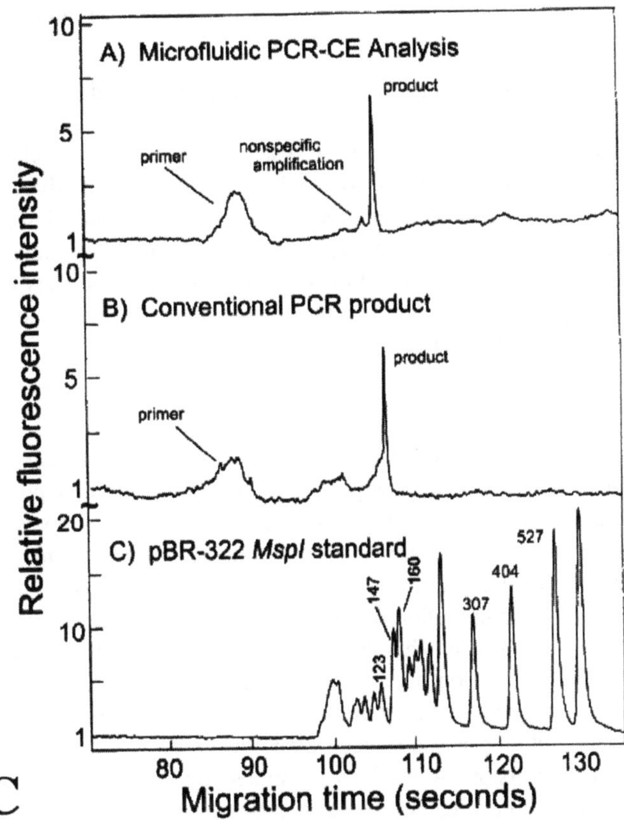

Figure 11 Monolithic integrated microfluidic PCR amplification and CE analysis system. (A) Schematic of PCR-CE device. The PCR chambers are connected to a common sample bus through a set of valves. Hydrophobic vents are used to locate the sample and to eliminate gas. The chambers are connected directly to the cross-channel of the CE system for product injection and analysis. Two aluminum manifolds, for the vents and valves, are placed into the respective ports and clamped in place using vacuum. The manifolds are connected to the external solenoid valve for pressure and vacuum actuation. Thermocycling is accomplished using a resistive heater and a miniature thermocouple below the 280-nL chamber. (B) Detailed schematic of microfluidic valves and hydrophobic vents. Sample is loaded from the right by opening the valve by vacuum and forcing the sample under the membrane by pressure; vacuum is simultaneously applied at the vent to evacuate the air from the chamber. The sample stops at the vent, and the valve is pressure-sealed to enclose the sample. Dead volume for the valves and vents is approximately in the nanoliter range. (C) Microfluidic PCR-CE analysis of an M13 amplicon (20 starting copies/µL). 20 cycles of 95°C for 5 s, 53°C for 15 s, and 72°C for 10 s. Electropherograms from the top: chip PCR-CE analysis, positive control with identical PCR mix using a Peltier thermal cycler; pBR322 *MspI* DNA ladder for size comparison. (Reprinted from Refs. 81 and 82 with permission.)

case. Such devices can be disposable due to their inexpensive and relatively simple fabrication.

It is interesting to note similar trends of system integration in capillary electrophoresis. An automated capillary array machine for DNA typing directly from blood samples [34], as well as a nanoreactor-based cycle sequencing instrument [90], have been developed by Yeung's group. Microfluidic elements and a microthermocycler were incorporated to co-amplify STR loci or carry out sequencing reactions, as well as on-line CZE purification of the products, loading/injection, and regeneration/cleaning system (see Section 3.1).

4.3. Single-Cell-Based Microreactors

Microfluidic manipulations at the level of a single cell can be considered a further development of the microreactor approach and offer exciting new possibilities in micropreparative biotechnology and bioanalysis. Single cells represent extremely complex natural microreactors, and studies at this level help investigate important in-vivo processes. A number of studies have been focused lately on using optical forces for manipulation of small objects (cells, bacteria, beads, microdroplets, etc.) in liquid phase. This technique, known as optical tweezers, or laser trap, allows selecting, capturing, and guiding single cells or microorganisms. A new microsystem, presented by Reichle et al. [91], is based on the combination of optical and chip-based dielectrophoretical trapping of cells and beads with micrometer precision. A bead was captured by the laser tweezers and brought into contact with a single cell held in a dielectrophoretic field cage. The latter was constructed with eight specially arranged microfabricated electrodes, which generated a high-frequency AC electric field. These manipulations enabled direct cell handling to study intercell ligand–receptor interactions and to conduct measurements on a single-cell level (Figure 12). Another microfluidic system [92] has been designed for high-speed separation/isolation of a single living cell or microorganism in the presence a large number of other cells in the solution. This was achieved by integrating laser trapping and dielectrophoretic forces. Flow velocity within the microchannel was adjusted to balance the optical force on the cell. A target entrapped in the laser focal point can be transported through the microchannel system and isolated. This is a promising technique for screening of microbes, pure cultivation of cells, etc. A microreactor system, based on precise optical manipulation of water droplets in oil emulsion, has also been described [93]. Extremely small quantities of samples (down to the level of a single DNA molecule) can be brought into contact by the fusion of two microdroplets induced by the optical pressure of the laser beam.

A research group at Göteborg University has demonstrated a microfluidic device for combinatorial electrofusion of liposomes and cells [94]. Optical trapping was used to transport individual liposomes and cells through microchannels

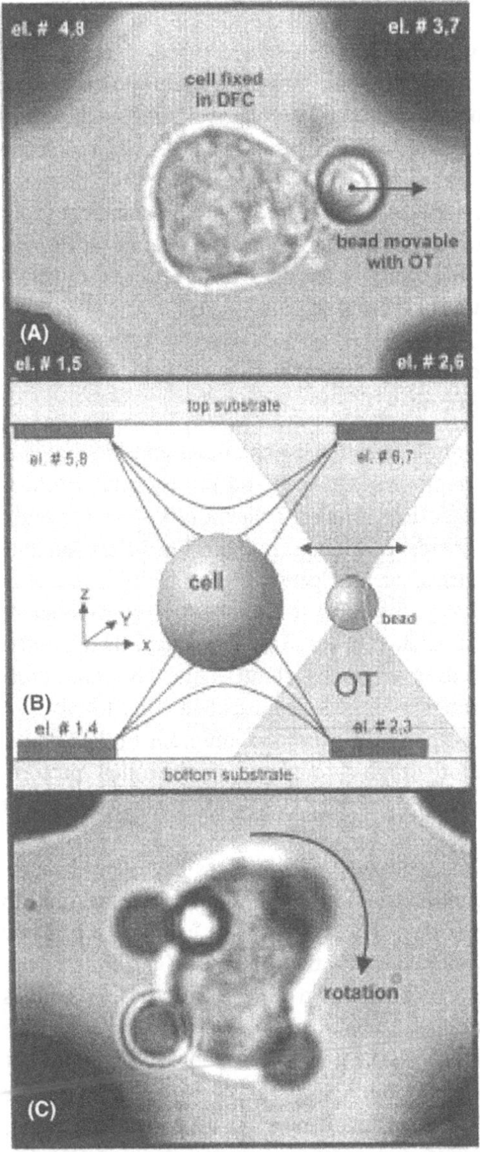

FIGURE 12 Combination of dielectrophoretic field cage (DFC) and optical tweezers (OT) for the measurement of bead-cell adhesion: (A) 4.1-μm polystyrene particle trapped with laser tweezers (right) in contact with T-lymphoma cell (~15 μm in diameter). Cell and bead were brought into contact. The time for stable adhesion was measured. (B) Schematic representation of the experimental system used to measure the adhesion forces between bead and cell with the cell trapped in a DFC and the bead trapped in the laser focus of the OT. (C) Probing different surface regions of the cell for bead–cell adhesion (five beads are attached to a single cell). (Reprinted from Ref. 91 with permission.)

into the fusion container, where selected liposomes were fused together using microelectrodes. A large number of combinatorially synthesized liposomes with complex compositions and reaction systems can be obtained from small sets of precursor liposomes. This device can also facilitate single cell–cell electrofusions (hybridoma production).

The exploitation of these new microscale chemical and biochemical tools, based on microreactor chips, will eventually reap great rewards and bring new opportunities in micropreparation and analysis for drug discovery and development, biotechnology, as well as fundamental studies.

4.4. Liquid-Handling Issues in Micropreparative High-Throughput Applications

In spite of all the advantages of miniaturization in chemical and biochemical processing, an interface between microscale devices and the external world is paramount to further automate sample introduction/collection process for high-throughput applications, and to provide continuous monitoring of the analyte molecules. In research-based industries, such as genomics, proteomics, metabolomics, and especially drug discovery, the need to carry out large number of complex experiments poses a real challenge in terms of efficiency, data quality, and cost. For example, in primary drug screening alone, major pharmaceutical companies test-score their new biological targets against compound libraries of hundreds of thousands of compounds per year, generating millions of data points in search for new targets. Therefore, high-throughput systems that perform large-scale sample preparation, reactions, and analysis, using minor amounts of reagents, are in great demand.

The so-called Sipper Chip Technology [95] addresses this issue by performing experiments in a serial, continuous-flow fashion at the rate of 5000 to 10,000 experiments per channel per day. The system uses capillaries to draw nanoliters of reagents from microwell plates into the channels of the chip, where they are mixed with the target biomolecules, and a series of processing steps is carried out to determine if the compound of interest binds the target. A range of assays, including fluorogenic kinase assays with electrophoretic separation of the reaction products from substrates, mobility shift assays, and cellular assays using Ca^{2+}-sensitive fluorescent indicators with 50–100 cells per data point, have been demonstrated, using various chip designs, as summarized in Ref. 96.

Another possible solution to this problem has been implemented by Attiya et al. [97]. The device contained a large sample introduction channel with a volume flow resistance $>10^5$-fold lower than that in the analysis microchannels. This approach enabled interfacing the large sample introduction channel with an external pump (up to 1-mL/min flow rate) for pressure-driven sample delivery without perturbing the solutions and electrokinetic manipulations within the rest of the microchannel network. On-chip mixing, reaction, and separation of ovalbumin and anti-ovalbumin were demonstrated with good performance and reproducibility. Such a strategy for decoupling the electrokinetic flow in the microchannels from the external environment extends the applicability of microchip analysis and provide a useful alternative to mechanical-valve flow control.

Shi et al. [98] introduced a pressurized capillary array system to simultaneously load 96 samples into 96 sample wells of a radial microchannel array electrophoresis microplate for high-throughput DNA sizing (Figure 13). As a result, 96 samples were analyzed in less than 90 s per microplate, demonstrating the power of microfabricated devices for large-scale and high-performance nucleic acid characterization.

Researchers in the drug discovery field have also been searching for microfluidic-based solutions to address the drastically growing scale of combinatorial chemistry libraries and screening processes [96]. The goal is the fabrication of a chemical microprocessor that combines the reactions and screening on a single microdevice. One of the latest prominent industrial developments of the kind is a device which comprises multiple reaction units on one chip [99]. These units are enclosed but connected to the outside environment by fluidic channels [100]. Two different reagents are fed into each reaction well, where the reaction takes place, followed by spectroscopic detection/identification. This biochip uses nanoliter-range reaction volumes and can be fabricated using appropriate materials accommodating various chemical conditions. Three-dimensional microfluidic structures for complex integrated processing are currently being developed to address main applications in both chemical synthesis and genomic screening.

5. CONCLUSIONS AND OUTLOOK

The driving force behind the development of micropreparative technologies is the raising demands of geno-, proteo-, and other newly emerging -omics in modern biotechnology for automation and miniaturization of sophisticated chemical and biochemical processes, to enable the necessary scale of the production and research and the fast pace of progress in these fields, including drug discovery and medical diagnostics. The challenge is now to keep up with the explosion of the biological information and the advances in information systems. Upstream preparation, on-line treatment, and delivery of high-quality samples must be better, faster, and cheaper to serve the downstream applications spinning genomic and proteomic information out of nucleic acid- and protein-based technologies at ever-accelerating rates. The recent success in shrinking of bioanalytical instrumentation to the micro- and nanoscale calls for the corresponding miniaturization in preparative and sample handling techniques. Capillary electrophoresis, microfabricated fluidic devices, and other microscale separation

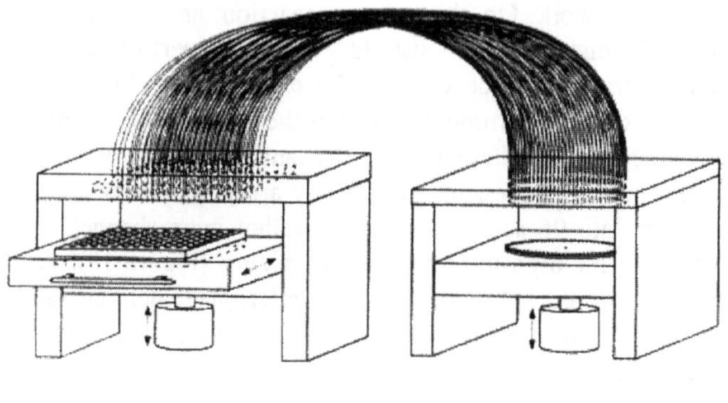

A

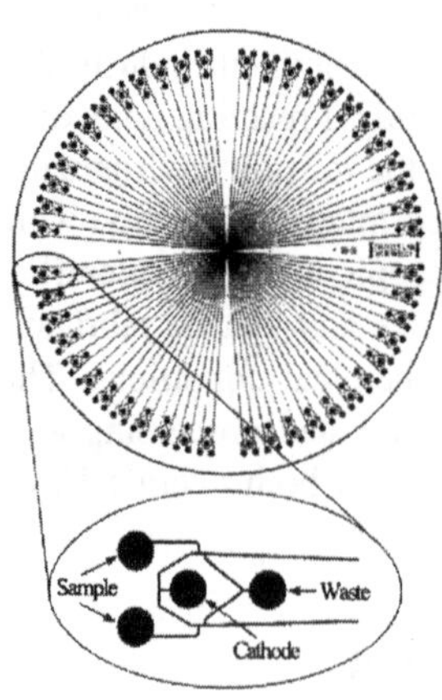

B

FIGURE 13 (A) 96-sample capillary array loader. Pressurization of the microtiter dish chamber to 21 kPa (~3 psi) is used to transfer 96 samples (transfer volume ~1 μL) to the sample reservoirs of the radial microplate. (B) Mask pattern for the 96-channel radial capillary array electrophoresis microplate. The substrate is 10 cm in diameter. Channels are 110 μm wide and 50 μm deep. The distance from the injector (double-T) to the detector is 3.3 cm. (Reprinted from Ref. 98 with permission.)

and microreactor approaches, described above, hold a tremendous potential to meet these goals.

ACKNOWLEDGMENT

The author acknowledges the kind support of Syngenta Research and Technology.

REFERENCES

1. Guttman A, Cohen AS, Heiger DN, Karger BL. Anal Chem 1990; 62:137–141.
2. Minarik M, Foret F, Karger BL. Electrophoresis 2000; 21:247–254.
3. Boss HJ, Rohde MF, Rush RS. Anal Biochem 1995; 231:123–129.
4. Ekstrom S, Opperfjord P, Nilsson J, Bengsson M, Laurell T, Marko-Varga G. Anal Chem 2000; 72:266–293.
5. Guttman A, Cohen AS, Paulus A, Karger BL, Rodriguez H, Hancock WS. In: Shafer-Nielsen C, ed. Electrophoresis '88. New York: VCH, 1988:51.
6. Irie T, Oshida T, Hasegawa H, Matsuoka Y, Li T, Oya Y, Tanaka T, Tsujimoto G, Kambara H. Electrophoresis 2000; 21:367–374.
7. Magnusdottir S, Heller C, Sergot P, Viovy JL. Electrophoresis 1995; 18:1990–1993.
8. Kuypers AW, Willems PM, van der Schans MJ, Linssen PC, Wessels HM, de Bruijn CH, Everaerts FM, Mensink EJ. J Chromatogr 1993; 621:149–156.
9. Bruin GJM. Electrophoresis 2000; 21:3931–3951.
10. Dolnik V, Liu S, Jovanovich S. Electrophoresis 2000; 21:41–54.
11. Kutter JP. Trends Anal Chem 2000; 19:352–363.
12. Sanders GHW, Manz A. Trends Anal Chem 2000; 19:364–378.
13. Hjerten S, Zhu MD. J Chromatogr 1985; 327:157–162.
14. Huang X, Zare RN. J Chromatogr 1990; 516:185–189.
15. Wallingford RA, Ewing AG. Anal Chem 1987; 59:1762–1766.
16. Cohen AS, Najarian DR, Paulus A, Guttman A, Smith JA, Karger BL. Proc Natl Acad Sci USA 1988; 85:9660–9663.
17. Minarik M, Kleparnik K, Gilar M, Foret F, Miller AW, Sosic Z, Karger BL. Electrophoresis 2002; 23:35–42.
18. Hong JW, Hagiwara H, Fuji T, Machida H, Inoue M, Seki M, Endo I. In: Ramsey JM, van den Berg A, Eds. Micro Total Analysis Systems 2001. Dordrecht, The Netherlands: Kluwer, 2001:113–114.
19. Effenhauser CS, Manz A, Widmer HM. Anal Chem 1995; 67:2284–2287.
20. Khandurina J, Chovan T, Guttman A. Anal Chem 2002; 74:1737–1740.
21. Khandurina J, Guttman A. J Chromatogr 2002; 979:105–113.
22. Kasicka V. Electrophoresis 2001; 22:4139–4162.
23. Boss HJ, Rohde MF, Rush RS. Pept Res 1996; 9:203–209.
24. Lee HG, Desiderio DM. Anal Chim Acta 1999; 383:79–99.
25. Kasicka V, Prusik Z, Sazelova P, Jiracek J, Barth T. J Chromatogr 1998; 796: 211–220.

26. Kasicka V. Electrophoresis 1999; 20:3084–3105.
27. Chankvetadze B, Burjanadze N, Bergenthal D, Blaschke G. Electrophoresis 1999; 20:2680–2685.
28. Chankvetadze B. J Chromatogr A 1997; 792:269–295.
29. Yeung ES. In: Heller MJ, Guttman A, eds. Integrated Microfabricated Biodevices; New York: Marcel Dekker, 2002:87–127.
30. Tan H, Yeung ES. Anal Chem 1997; 69:664–674.
31. Zhang N, Yeung ES. J Chromatogr B 1998; 714:3–11.
32. Swerdlow H, Jones BJ, Wittwer CT. Anal Chem 1997; 69:848–855.
33. Tan H, Yeung ES. Anal Chem 1998; 70:4044–4053.
34. Zhang N, Tan H, Yeung ES. Anal Chem 1999; 71:1138–1145.
35. Shandrick S, Ronai Z, Guttman A. Electrophoresis 2002; 23:591–595.
36. Williams PE, Marino MA, Del Rio SA, Turni LA, Devaney JM. J Chromatogr A 1994; 680:525–540.
37. Hawkins TL, McKernan KJ, Jacotot LB, MacKenzie JB, Richardson PM, Lander ES. Science 1997; 276:1887–1889.
38. Hagel L, Janson J-C. J Chromatogr Libr 1992; 51A:A267-A307.
39. Guttman A, Ronai Z. Electrophoresis 2000; 21:3952–3964.
40. Guttman A, Ronai Z, Barta C, Hou Y-M, Sasvary-Szekely M, Wang X, Briggs S. Electrophoresis 2002; 23:1524–1530.
41. Csapo Z, Gerstner A, Sasvary-Szekely M, Guttman A. Anal Chem 2000; 72: 2519–2525.
42. Gerstner A, Sasvari-Szekely M, Kalas H, Guttman A. BioTechniques 2000; 28: 628–630.
43. Wen J, Lin Y, Xiang F, Matson DW, Udseth HR, Smith RD. Electrophoresis 2000; 21:191–197.
44. Licklider L, Wang X, Desai A, Tai Y-C, Lee T. Anal Chem 2000; 72:367–375.
45. Figeys D, Pinto D. Electrophoresis 2001; 22:208–216.
46. Xiang F, Lin Y, Wen J, Matson DW, Smith RD. Anal Chem 1999; 71:1485–1490.
47. Li J, Thibault P, Bings NH, Skinner CD, Wang C, Colyer C, Harrison D J. Anal Chem 1999; 71:3036–3045.
48. Bings NH, Wang C, Skinner CD, Colyer CL, Thibault P, Harrison D. Anal Chem 1999; 71:3292–3296.
49. Zhang B, Liu H, Karger BL, Foret F. Anal Chem 1999; 71:3258–3264.
50. Lazar IM, Ramsey RS, Jacobson SC, Foote RS, Ramsey JM. J Chromatogr A 2000; 892:195–201.
51. Lazar IM, Ramsey RS, Ramsey JM Anal Chem 2001; 73: 1733–1739.
52. Zhang B, Foret F, Karger BL. Anal Chem 2000; 72:1015–1022.
53. Zhang B, Foret F, Karger BL. Anal Chem 2001; 73:2675–2681.
54. Jiang Y, Wang P-C, Locascio LE, Lee CS. Anal Chem 2001, 73, 2048–2053.
55. Gao J, Xu J, Locascio LE, Lee CS. Anal Chem 2001; 73:2648–2655.
56. Ross D, Johnson TJ, Locascio LE. Anal Chem 2001; 73:2509–2515.
57. Chovan T, Guttman A. Trends Biotechnol 2002; 20:116–122.
58. Ehrfeld W, Hessel V, Lehr H. In: Manz A, Becker H, eds. Microsystem Technology in Chemistry and Life Sciences. Berlin, Germany: Springer-Verlag, 1999: 233–252.

59. DeWitt SH. Curr Opin Chem Biol 1999; 3:350–356.
60. Burns MA, Johnson BN, Brahmasandra SN, Handique K, Webster JR, Krishnan M, Sammarco TS, Man FP, Jones D, Heldsinger D, Mastrangelo CH, Burke DT. Science 1998; 282:484–487.
61. Burns MA, Mastrangelo CH, Sammarco TS, Man FP, Webster JR, Johnson BN, Foerster B, Jones D, Fields Y, Kaiser AR, Burke DT. Proc Natl Acad Sci USA 1996; 93:5556–5561.
62. Troian SM, Darhuber AA, Davis JM, Reisner WW, Wagner S. ACS National Meeting, San Diego, April 1–5, 2001; Paper 213.
63. Khandurina J, Guttman A. J Chromatogr 2002; 943:159–183.
64. http://www.aclara.com; ACLARA Biosciences. October 2002.
65. Wilding P, Kricka LJ, Cheng J, Hvichia G, Shoffner M, Fortina P. Anal Biochem 1998; 257:95–100.
66. Anderson H, van der Wijngaart W, Stemme S. Electrophoresis 2001; 22:249–257.
67. Brahmasandra SN, Ugaz VM, Burke DT, Mastrangelo CH, Burns MA. Electrophoresis 2001; 22:300–311.
68. Soper SA, Ford SM, Xu Y, Qi S, McWhorter S, Lassiter S, Patterson D, Bruch RC. J Chromatogr A 1999; 853:107–120.
69. Belgrader P, Benett W, Hadley D, Richards J, Stratton P, Mariella R, Milanovich F. Science 1999; 284:449–450.
70. Northrup MA, Benett B, Hadley D, Landre P, Lehew S, Richards J, Stratton P. Anal Chem 1998; 70:918–922.
71. Taylor TB, Wimm-Deen ES, Picozza E, Woudenberg TM, Albin M. Nucleic Acids Res 1997; 25:3164–3168.
72. http://www.cepheid.com; Cepheid. October 2002.
73. Litborn E, Emmer A, Roeraade J. Anal Chim Acta 1999; 401:11–19.
74. Litborn E, Emmer A, Roeraade J. Electrophoresis 2000; 21:91–99.
75. Clark RA, Ewing AG. Anal Chem 1998; 70;1119–1125.
76. Bernhard DD, Mall S, Pantano P. Anal Chem 2001; 73:2484–2490.
77. Woolley AT, Hadley D, Landre P, deMello AJ, Mathies RA, Northrup MA. Anal Chem 1996; 68:4081–4086.
78. Waters LC, Jacobson SC, Kroutchinina N, Khandurina J, Foote RS, Ramsey JM. Anal Chem 1998; 70:5172–5176.
79. Waters LC, Jacobson SC, Kroutchinina N, Khandurina J, Foote RS, Ramsey JM. Anal Chem 1998; 70:158–162.
80. Khandurina J, McKnight TE, Jacobson SC, Waters LC, Foote RS, Ramsey JM. Anal Chem 2000; 72:2995–3000.
81. Lagally ET, Simpson PC, Mathies RA. Sensors Actuators B 2000; 63:138–146.
82. Lagally ET, Medintz I, Mathies RA. Anal Chem 2001; 73:565–570.
83. Khandurina J, Jacobson SC, Waters LC, Foote RS, Ramsey JM. Anal Chem 1999; 71:1815–1819.
84. Kopp MU, deMello AJ, Manz A. Science 1998; 280:1046–1048.
85. Giordano BC, Copeland ER, Landers JP. Electrophoresis 2001; 22:334–340.
86. Huhmer AFR, Landers JP. Anal Chem 2000; 72:5507–5512.
87. Oda RP, Strausbauch MA, Huhmer AFR, Borson N, Jurrens SR, Craighead J, Wettstein PJ, Eckloff B, Kline B, Landers JP. Anal Chem 1998; 70:4361–4368.

88. Chiou J, Matsudaira P, Sonin A, Ehrlich D. Anal Chem 2001; 73:2018–2021.
89. Hong JW, Fujii T, Seki M, Yamamoto T, Endo I. Electrophoresis 2001; 22:328–333.
90. Xue G, Pang HM, Yeung ES. J Chromatogr A 2001; 914:245–256.
91. Reichle C, Sparbier K, Muller T, Schnelle T, Walden P, Fuhr G. Electrophoresis 2001; 22:272–282.
92. Arai F, Ichikawa A, Ogawa M, Fukuda T, Horio K, Itoigawa K. Electrophoresis 2001; 22:283–288.
93. Katsura S, Yamaguchi A, Inami H, Matsuura S-I, Hirano K, Mizuno A. Electrophoresis 2001; 22:289–293.
94. Stromberg A, Karlsson A, Ryttsen F, Davidson M, Chiu DT, Orwar O. Anal Chem 2001; 73:126–130.
95. http://www.calipertech.com/products/throughput_sipper.html; Caliper Technologies, Inc. October 2002.
96. Khandurina J, Guttman A. Curr Opin Chem Biol 2002; 6:359–366.
97. Attiya S, Jemere AB, Tang T, Fitzpatrick G, Seiler K, Chiem N, Harrison DJ. Electrophoresis 2001; 22:318–327.
98. Shi Y, Simpson PC, Schere JR, Wexler D, Skibola C, Smith MT, Mathies RA. Anal Chem 1999; 71:5354–5361.
99. http://www.orchidbio.com; Orchid Biocomputer, Inc. October 2002.
100. Leach M. Drug Discovery Today 1997; 2:253–254.

13

NMR Detection in Capillary Electrophoresis and Capillary Electrochromatography

Dimuthu A. Jayawickrama, Andrew M. Wolters, and Jonathan V. Sweedler
University of Illinois, Urbana, Illinois, U.S.A.

1. INTRODUCTION

Quantitative and qualitative analyses of mixtures are critical and often prerequisites in the chemical and biological sciences. The increased growth of biochemical and medical sciences in recent years has demanded faster and more versatile separation techniques. High-performance liquid chromatography (HPLC), gas chromatography (GC), and slab-gel electrophoresis (SGE) are well-established techniques to analyze complex mixtures. Researchers often look for unprecedented levels of chemical information regarding mixtures in the shortest possible time. For this purpose a number of hyphenated techniques, such as liquid chromatography-mass spectrometry (LC-MS), liquid chromatography-nuclear magnetic resonance (NMR), spectroscopy (LC-NMR), and gas chromatography-mass spectrometry (GC-MS) have been developed and commercialized. The on-line detection with MS and NMR allow total structure elucidation and increased throughput by eliminating physical isolation and characterization of components [1]. Because of the performance offered by such systems, hyphenated techniques are increasingly employed in routine analytical work [2,3].

As described in previous chapters, capillary electrophoresis (CE) and capillary electrochromatography (CEC) provide rapid and high-efficiency separation in a small-volume capillary format. Because of these features, capillary separation techniques are increasingly used in research and industry [4–8]. In recent years CEC has emerged as an analytical technique which often combines the separation efficiency of CE and the selectivity of LC [9–11]. The introduction of small-volume (~1-μL) NMR probes to record high-resolution NMR spectra has enabled coupling between capillary separation and information-rich detection of NMR spectroscopy [12,13]. The first CE/NMR experiment was reported in 1994 [13], with a number of improvements on CE/ and CEC/NMR hyphenation appearing since in the literature.

1.1. NMR Spectroscopy

NMR is a ubiquitous and indispensable tool for elucidating molecular structures, determining impurities, and studying molecular dynamics. NMR is also used to analyze simple mixtures without physical separation, and to measure molecular properties and bulk properties of the medium. The nondestructive nature of NMR permits the sample to be used for further investigation. As a noninvasive technique, NMR is often used to study molecular binding and to screen potential drug candidates. Therefore, despite its low sensitivity, NMR has become an essential analytical tool in academic and industrial environments. However, the inherent insensitivity causes detection limits of NMR to be a few orders below that of other standard analytical techniques [14]. At present, the limit of detection achieved by NMR in concentration terms is in the millimolar range.

A number of strategies have been used to enhance NMR sensitivity. For example, higher-field magnets are used because the sensitivity increases with 7/4th power of the magnetic field strength. Magnets based on superconducting technology with gigahertz field strengths are becoming available; unfortunately, these instruments are high in cost. The search for higher field strengths with novel superconducting materials is a continuous effort [15]. Another approach to increasing NMR sensitivity is to use the spin-polarization-induced nuclear Overhauser enhancement (SPINOE) technique [16]. In this method a sensitivity improvement of ~50-fold or more can be achieved. However, the applications are limited mostly to nonaqueous samples. A similar technique, dynamic nuclear polarization (DNP), uses saturating electron spins coupled to nuclear spins to increase NMR signal intensity [17]. NMR detection with DNP has been implemented in chromatographic separations [18]. However, additional instrumentation to generate microwave power and a source of electron radicals is required. Coils made with superconducting materials [19] and cryogenically cooled coils [14] can also improve the sensitivity by minimizing receiver coil noise. One inexpensive method to improve NMR sensitivity is to use a miniaturized radiofrequency (RF) coil [12]. The RF coil in the probe is an important component of NMR instrumentation. It must be capable of delivering radiofrequency energy necessary to excite NMR active nuclei and collect signals from the sample.

Different probe designs have been developed, with several of them being specially designed for microscale separations [14].

1.2. NMR as an On-Line Detector

Watanabe and Niki introduced the coupling of NMR to LC as an on-line detector [20]. After these initial stopped-flow experiments, Bayer et al. [21] reported the first continuous-flow LC-NMR experiment. However, a number of impediments associated with LC-NMR hindered routine analytical application for a number of years. Since then, new instrumentation and analytical methodologies for LC-NMR have been developed and commercialized. The development of high-field-strength magnets, better solvent-suppression techniques, more sensitive small-diameter transmitter/receiver coils, on-column sample preconcentration, and expanded flow cells have improved the sensitivity of LC-NMR.

LC-NMR has become a vital analytical tool in combinatorial chemistry for separation and identification libraries of small organic molecules [22] and peptides [23]. Stereochemical analysis of drugs in biological fluids [24], separation and identification of terpenoid mixtures [25], analysis of nitroaromatic compounds containing environmental samples [26], and analysis of polymer mixtures have been performed with LC-NMR. The conventional LC-NMR with column dimensions 4.6 (i.d.) $\times$ 150 mm requires millimeter volumes of eluants. NMR solvent-suppression techniques remove large solvent signals and improve the dynamic range of the signal. However, the spectral features of NMR signals in the proximity of solvent resonances are also affected. As a result, vital spectral information required to deduce structures, especially of minor analytes, is lost. One remedy is to use deuterated solvents. The high cost associated with deuterated solvents prohibits routine applications of large-volume deuterated mobile phases. Miniaturized separation techniques such as CE and CEC provide the advantage of utilizing microliter volumes.

Although not at capillary level nor a physical separation, Johnson and He first introduced the combination of electrophoresis and NMR (E-NMR) to study electrophoretic mobilities and diffusion coefficients [27]. Sweedler and co-workers described the first application of microcoil NMR probe as an on-line detector for CE [13,28]. In 1998 Bayer, Albert, and co-workers reported the first CEC/NMR experiments which also included CE/NMR and capillary LC/NMR results [29].

2. THEORY AND INSTRUMENTATION OF CE/CEC-NMR

Instrumental schematics for typical CE-NMR and CEC-NMR systems are shown in Figures 1A and 1B. CE is performed by applying a voltage across the capillary and separating analyte ions based on their electrophoretic mobilities,

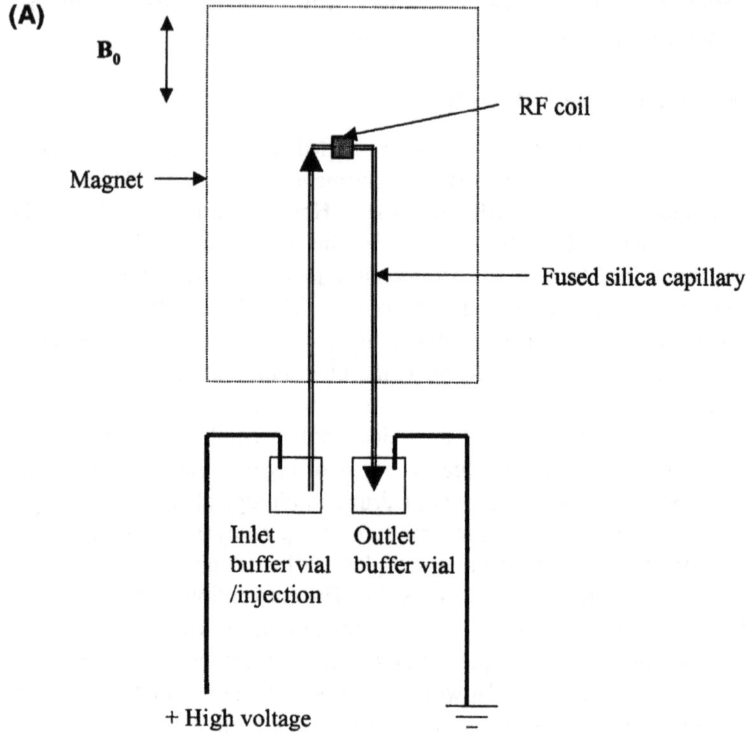

FIGURE 1 Instrumental schematic for (A) CE-NMR and (B) CEC-NMR.

where they eventually migrate past an on-column detector. In CEC the separation is performed using an LC-type column under CE conditions, so that the flow is based on electroosmotic flow instead of pressure flow. Often in CEC, low pressures (>30 bar) are used to prevent air bubble formation at frits and/or achieve faster separation [30]. For CE/ and CEC/NMR, magnetic objects must be kept several meters from the magnet, depending on the magnetic field strength. Thus, nonmagnetic plastic vials and Pt electrodes can be positioned close to the magnet or within the magnet bore. The columns used in CEC may be positioned closer to or inside the magnet bore if they are nonmagnetic; obviously, steel-jacketed columns are problematic. Thus, long transfer lines to the NMR probe from the column are required and may degrade the effective separation efficiency. The newly introduced shielded magnets allow equipment to be moved within a meter of the magnet. Several of the recent developments leading to improved CE/ and CEC/NMR are discussed below.

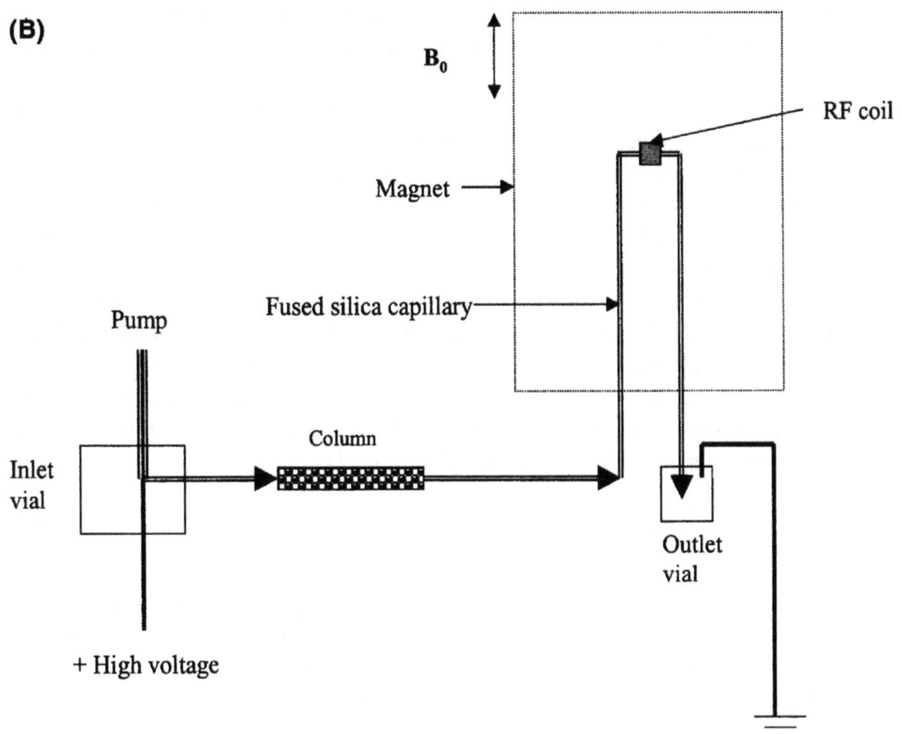

FIGURE 1 (*continued*)

2.1. NMR Microcoils

2.1.1. Rationale for Development of Microcoils

The term "microcoil" generally describes NMR active volumes of 1 μL or less. The main impetus for the development of minuscule NMR probes over the last decade has been to increase sensitivity and improve detection limits. The small sample size and straightforward hyphenation to capillary-scale separation techniques have enabled scientists to explore the application of NMR microcoils to solve complex analytical problems. The signal-to-noise (S/N) ratio, which is related to the sensitivity and the detection limit, is defined as the ratio of NMR resonance height to the root mean square (RMS) noise. Several parameters influence the NMR signal strength. Among them are parameters influenced by the nuclei being measured: precession frequency, natural abundance, gyromagnetic ratio, concentration, and the active volume. Acquisition parameters such as number of scans, acquisition time, and relaxation delays are also important in achieving good S/N. Mathematical manipulation of the free induction decay (FID) is routinely performed to improve the S/N. Also importantly, the design of the NMR probe can be optimized to increase the S/N. Maximum coupling between the probe and the sample is achieved by miniaturizing the NMR coil just to accommodate the sample. Consequently, the mass sensitivity or the minimum amount of analyte that can be detected is improved. The ratio of the volume available for detection (V_{obs}) to the total amount of volume (V_{tot}) needed for analysis is known as the observe factor (f_0) [14]. Ideally for maximum sensitivity, the f_0 should approach unity. However, to prevent magnetic susceptibility boundaries near the sample being measured, the sample often extends beyond the NMR observation region, V_{obs}. For example, a standard 5-mm NMR experiment requires ~750 µL of sample volume, although the NMR active volume is ~250 µL. The sensitivity of microcoil probes is best described with a volume independent parameter known as mass sensitivity (S_m) [14]:

$$S_m = \frac{\text{S/N}}{\text{mol } t^{1/2}} \tag{1}$$

where mol and t are number of moles within V_{obs} and the acquisition time, respectively. The use of S_m instead of concentration sensitivity (S_c) is more appropriate for mass-limited samples as measured with NMR microcoils and to compare probes. More details on high-resolution NMR spectroscopy of small volumes can be found elsewhere [14,31].

To date, two arrangements of NMR coils, the solenoidal radio frequency (RF) coil (Figure 2A) and the saddle-type (Helmholtz) RF coil (Figure 2B) have been employed as on-line NMR detectors with CE and CEC. Theoretical studies have shown that reduction of the diameter of the RF coils increases the coil sensitivity [32]. The miniaturized versions of saddle types are commonly used in commercial probes. As a major development, a saddle coil which houses 1.7-mm-diameter sample tubes has been introduced by Varian. Another significant contribution is the designing of an inverse coil to accommodate 3-mm-diameter sample tubes with a detection volume of 60 µL and a total volume of 140 µL [33]. Fabrication procedures hinder further reduction of diameter of saddle-type coils that are optimized for sample volumes smaller than ~1 µL.

The coil sensitivity of solenoidal probes is inversely related to the coil diameter [32]. The S/N per unit volume for solenoidal coils with diameter greater than 100 µm is approximated by Eq. (2) [34]:

$$\text{S/N} \propto \frac{\omega_0^{7/4}}{d_c} \tag{2}$$

where ω_0 is the Lamor frequency. Because S_m is proportional to S/N, smaller-diameter probes increase the mass sensitivity. For a given length and diameter,

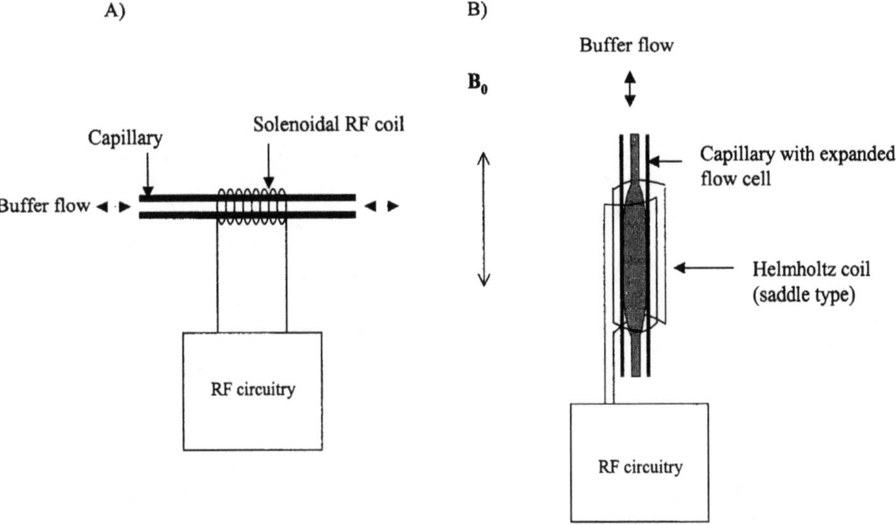

FIGURE 2 (A) Solenoidal RF coil wrapped directly around a fused silica capillary placed orthogonal to B_0. (B) Helmholtz coil with an expanded flow cell is placed in parallel to B_0.

solenoidal RF coils are about two or three times as sensitive as saddle-type coils [14,31].

2.1.2. Microfabrication to Construct Miniature RF Coils

Microfabrication allows the construction of miniature coil structures with accurate geometries and good mechanical stability, which are especially important for coils below 100 μm in diameter. A method based on microcontact printing (μCP) has been reported to microfabricate solenoidal microcoils [35]. A coil with a spiral structure is printed directly onto the capillary by electrodeposition of copper (Figure 3A). High-resolution NMR spectra with linewidth smaller than 1 Hz can be obtained with this fabricated microcoil. However, the sensitivity of the coil is lower than what has been reported with wound solenoidal microcoils. This difference was attributed to high dc resistance of electrodeposited copper and a small cross-sectional area. Planar microcoils, as shown in Figure 3B, with an outer diameter of ~200 μm, have been designed with gallium arsenide material which is more suitable for coupling with preamplifiers [14,36]. Although a planar coil is ideal for chip-based CE/CEC-NMR, magnetic susceptibility issues and detection at picoliter volumes are significant challenges.

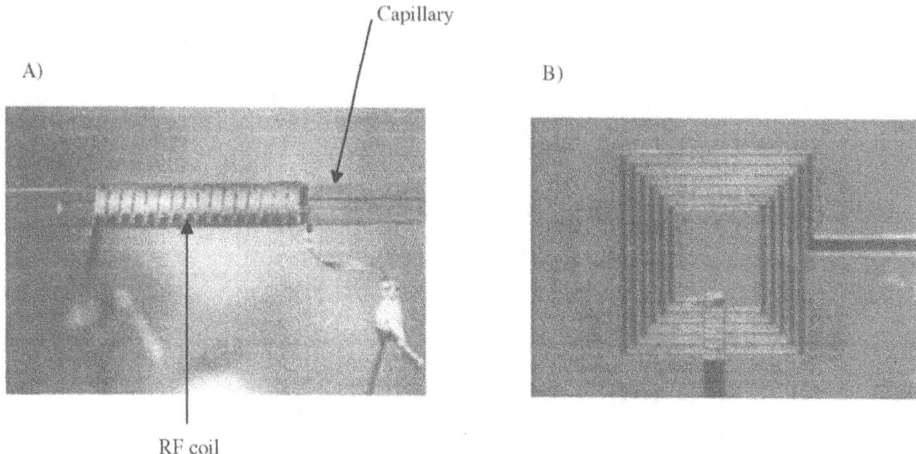

Figure 3 (A) Optical micrograph of a 16-turn microfabricated RF coil (70-μm wire separated by 30 μm) printed on a fused silica capillary (75-μm i.d./324-μm o.d.). the length of the coil 1.6 mm with an enclosed volume of 8 nL. (B) Scanning electron micrograph of a planar microcoil (200-μm o.d./97-μm i.d.). (Reproduced with permission from Ref. 14. Copyright 1999 American Chemical Society.)

2.1.3. Multiple Microcoil Probes

Because of their miniature size (~1 mm length), more than one solenoidal coil (and, hence, more than one sample) can easily be placed within the homogenous region of the static magnetic field [37–40]. Acquiring data from multiple samples simultaneously greatly increases NMR throughput, and has a number of advantages for separations. Electrical concerns, chiefly NMR signal cross-talk (bleed-through) among coils, must be minimized by appropriate probe design. Signal bleed-through can be minimized by detuning and optimizing shims for the active coil [41], orienting the coil and matching circuit right angle to the neighboring coil, and using shielded copper plates [37]. Special pulse sequences with selective excitation pulses can also reduce cross-talk [42]. Multiple sample probes are a new and rapidly changing area, and further advances are expected; the systems that have been developed to date that affect CE/NMR and CEC/NMR are described in greater detail in the following sections.

2.2. Interfacing CE/CEC and NMR

NMR coupling to CE or CEC does not require major modification to the existing CE or CEC instrumentation because the transceiver coil is either wrapped around the capillary or the capillary is housed inside the saddle coil. For example, the solenoidal coil can be coupled to the CE or CEC capillary by wrapping the Cu wire around the separation capillary. The capillary acts as both the sample holder and coil form. Generally, the number of turns of the RF coil controls the length of the NMR active volume. For example, a 1-mm-length coil wrapped around a 100-μm-i.d. capillary results in an ~8-nL volume. Larger observe volumes can be realized with expanded flow cells [29]. The close proximity of the coil to the sample greatly improves the fill factor, f_0. However, magnetic susceptibility effects degrade line shapes and S/N if f_0 is increased too much as the coil comes into closer contact with the sample. In theory, a sample enclosed in an infinite-length and perfectly uniform hollow cylinder experiences a uniform magnetic field [14]. In reality, however, broad NMR signals with S/N below the expected values are observed with samples in a narrow-walled capillary. The variation of magnetic susceptibilities of Cu, air, polyimide coating, and adhesives in the vicinity of sample contribute to the distortion of the static magnetic field. In addition, magnetic field inhomogeneity is also enhanced due to close proximity of sample and coil windings. The magnetic homogeneity can be improved using matching fluid with a volume magnetic susceptibility close to copper [12] and/or using zero-magnetic-susceptibility materials for coil construction [31]. Another approach is to use thicker-wall fused silica capillaries. As shown in Figure 4, the 200-Hz linewidth is reduced to 11 Hz by changing the capillary from 75-μm i.d., 145-μm o.d. to 75-μm i.d., 350-μm o.d. Microcoils placed at the magic angle (54.7°) to the static field also improve the line shape and linewidth. A combination of more than one of these approaches will further improve the NMR spectral linewidths and hence the S/N [14].

2.3. Electrophoretic Current-Induced and Flow Effects

Solenoidal coils with orthogonal orientation to the static magnetic field (B_0) form the most mass sensitive RF coils (Figure 2A). In this configuration, however, electrophoretic current generates an additional local magnetic field gradient, which perturbs the magnetic field homogeneity (see inset in Figure 5). The strength of this induced magnetic field (B_i) can be described by

$$B_i = \frac{\mu_0 i r}{2\pi R^2} \tag{3}$$

where μ_0 is the permeability constant, i is the electrophoretic current, r is the radial distance from the center of the capillary, and R is the capillary i.d. [14]. The distorted magnetic field is difficult to restore by shimming. The effect of induced magnetic field gradient leads to loss of scalar coupling, broader NMR lines, and poor S/N. As shown in Figure 5, the scalar coupling of the methyl triplet of triethylamine disappears with increasing voltage [43]. A significant

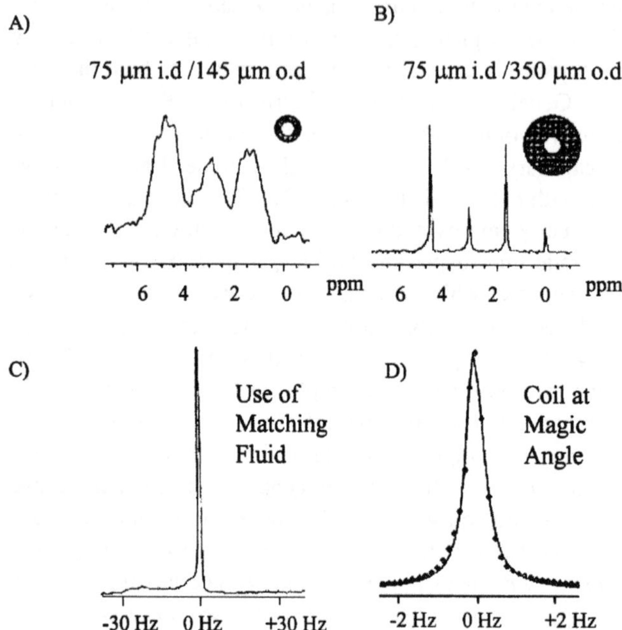

FIGURE 4 ^{1}H NMR spectra obtained at 300 MHz. (A) Without magnetic susceptibility matching fluid, linewidths for arginine are 200 Hz for thin-walled (75-μm i.d./145-μm o.d.) fused silica capillary. (B) For a thicker-walled (75-μm i.d./350-μm o.d.) fused silica capillary the linewidth is reduced to ~11 Hz. (C) Linewidth of HOD (10% H_2O/90% D_2O) is less than 1 Hz with magnetic susceptibility matching fluid. (D) Microcoil configure at magic angle. Linewidth of HOD (10% H_2O/90% D_2O) is 0.6 Hz. (Reprinted with permission from Ref. 14. Copyright 1999 American Chemical Society.)

loss of S/N and increase in NMR linewidth are also observed at higher voltages [13,44]. A voltage increase from 0 to 8 kV causes linewidth to increase from 1.5 to 15 Hz with corresponding S/N decreases from 147 to 19 [44].

In addition to the induced magnetic field, the migration rate can greatly influence the NMR signal intensity and linewidth. Unlike in LC, the rate of migration of each band in CE is characteristic of electrophoretic mobility and so each move past the detector at different rates. The effective longitudinal ($T_{1\text{eff}}$) and transverse relaxation ($T_{2\text{eff}}$) times of NMR nuclei in a flow system can be related to the residence time of the analyte (τ) in the detector and the relaxation times (T_1, T_2) under static conditions [45]:

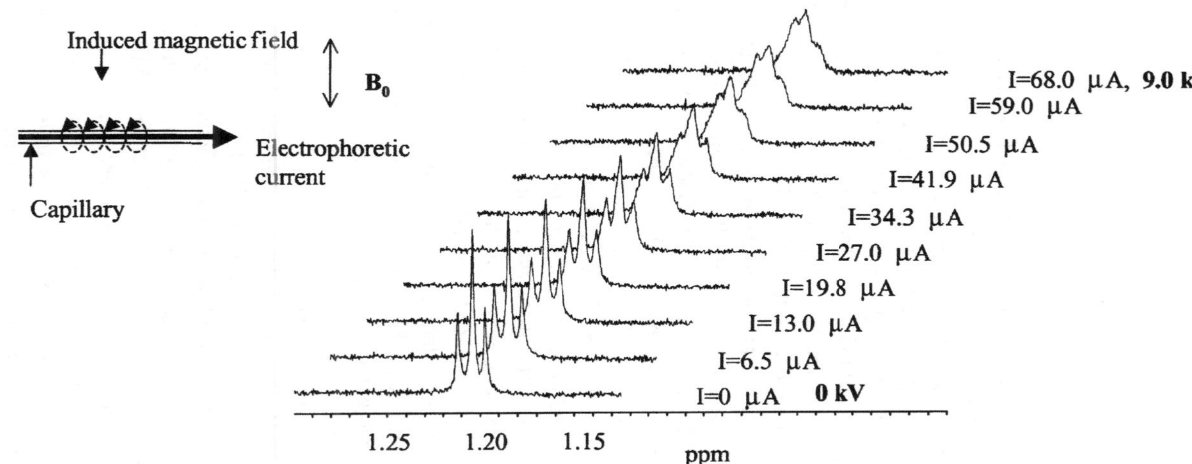

FIGURE 5 Microcoil CE-NMR spectra of triethylamine methyl peak in 1 M borate buffer with increasing applied voltage 0.0 to 9.0 kV by increments of 1.0 kV. (Reprinted with permission from Ref. 47. Copyright 2002 American Chemical Society.)

$$\frac{1}{T_{1\text{eff}}} = \frac{1}{T_1} + \frac{1}{\tau} \quad (4)$$

$$\frac{1}{T_{2\text{eff}}} = \frac{1}{T_2} + \frac{1}{\tau} \quad (5)$$

At fast flow rates, the saturation of signals is avoided because spins leave the NMR active region, thus leading to an increase in S/N. The NMR linewidth is related to transverse relaxation time and, in a flow system, the linewidth is further influenced by the residence time. As a result, short residence times at fast flow rates yield broader NMR signals with low S/N. Ideally, detection of each analyte should be optimized using suitable acquisition parameters to obtain NMR spectra with comparable S/N. In addition, the overall temperature change in buffer by Joule heating may also contribute to linewidth increase [44].

Instrumental and experimental conditions have been developed to obtain high-resolution NMR spectra with CE-NMR. To eliminate electrophoretic current-induced magnetic field gradient effects on NMR spectra, CE experiments can be performed at lower voltages than standard voltages. Often, the CE capillary has been designed to fit inside the magnet bore [14,44]. This method also prevents possible arcing to grounded RF coil circuitry. Two experimental approaches have been described, with the outlet vial inside the magnet in both methods. In one configuration, the inlet vial is outside to permit easy sample introduction. Because of the long length of the capillary, relatively low field strengths are used, reducing the induced magnetic field from the electrophoretic current. The microcoil can be oriented at the magic angle (54.7°) to B_0 to reduce the local field effects and improve the sensitivity [46]. The other approach has the inlet vial inside the magnet bore, which requires preloading the samples prior to the probe being inserted into the NMR magnet.

A simple and straightforward answer to the current-induced thermal and magnetic field gradient effects is stopped-flow experiments. This technique also allows the acquisition of time-consuming 2-D NMR experiments. However, termination of voltage may lead to loss of separation efficiency and reduced peak resolution due to diffusion over a longer period. Precautions must be taken to avoid unnecessary gravimetric flows which can affect the NMR intensities. A unique technique which acquires NMR spectra under quiescent conditions has been described [44]; In this method, once the analyte reaches the NMR microcoil observe region, the voltage is applied for 15 s followed by 1 min of no applied voltage during which NMR data acquisition occurs. Because the spectra are acquired under quiescent conditions, high-resolution NMR spectra with good S/N are obtained. Separation efficiencies of the order of 50,000 have been reported for arginine and triethylamine in this study.

A novel CE-NMR method with a dual-microcoil probe to acquire on-flow data has been described [43]. Figure 6 illustrates the instrumentation schematic for this dual-coil CE-NMR detection. The probe is fabricated by individually wrapping two coils around two separate outlet capillaries. At a given time the CE separation is performed only with one outlet capillary, keeping the second capillary floating or at open circuit. Figure 7 shows a series of NMR spectra obtained by alternating electrophoretic flow with two coils. The current-induced effects on spectra are eliminated as shown in Figure 7C, because only the spectra of sample in the electrically floating capillary are recorded.

The best sensitivity with a saddle coil is achieved with its axis parallel to B_0. Although the sensitivity is two to three times less than that of a solenoidal coil of the same dimensions, the electrophoretic current-induced magnetic field does not contribute to magnetic field inhomogeneity along the z axis. The effect of applied voltage on ^{1}H NMR spectra of lysine in phosphate buffer has been reported using saddle-type coils [29,47]. As shown in Figure 8, high-resolution NMR spectra with similar S/N are recorded at different voltages. Slight chemical shift change of water resonance is observed due to Joule heating.

2.4. NMR Observe Volume

Because of the difficulty of fabrication, saddle coils tend to be larger than solenoids, although a 2.5-μL saddle coil probe for static samples has recently been

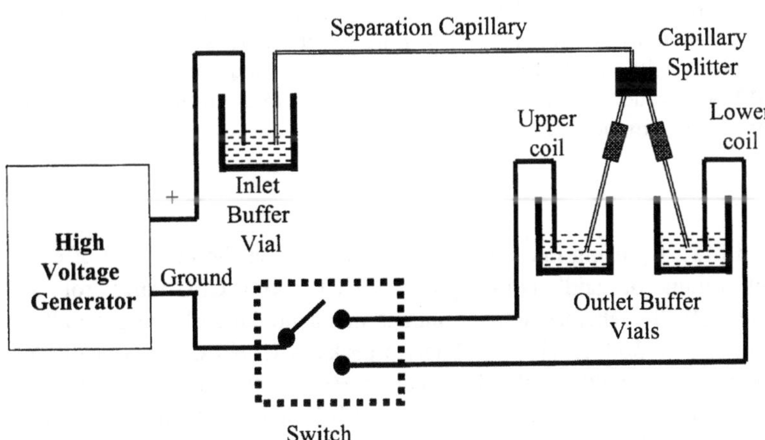

FIGURE 6 Instrumental schematic of CE with two-microcoil NMR detection showing the arrangement of the separation capillary, the two outlet capillaries, and the two NMR detection coils. (Reprinted with permission from Ref. 43. Copyright 2002 American Chemical Society.)

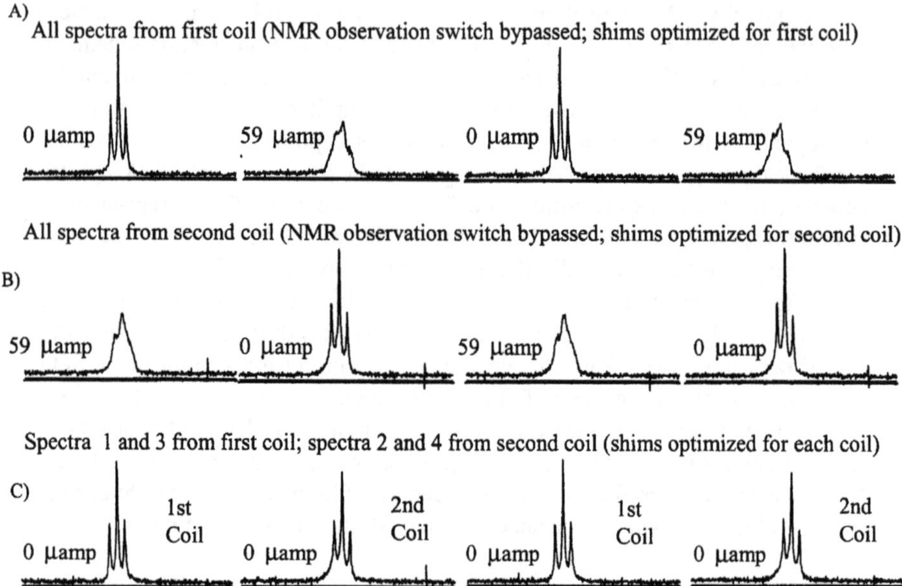

FIGURE 7 Arrays of two-microcoil CE-NMR spectra (LB = 0) of the methyl peak of triethylamine in 1 M borate buffer. Spectra acquired during alternation of electrophoresis flow between two outlet capillaries. (A) All spectra acquired from upper coil (shim settings optimized for upper coil; NMR observation switch bypassed). (B) All spectra acquired with lower coil (shim settings optimized for lower coil; NMR observation switch bypassed). (C) NMR spectra acquired from microcoil on outlet capillary without electrophoretic flow (shim settings optimized for active coil; NMR observation switch in-line). (Reproduced with permission from Ref. 43. Copyright 2001 American Chemical Society.)

described [48]. By using insets in the larger-volume saddle coils, the effective flow cell volume of saddle coils reported for CE and/or CEC range from 250 to 400 nL [29,49], whereas active volumes in solenoidal coils are generally in the range of 5 to 500 nL [13,44]. Analyte peaks in CE/CEC are typically 500 nL or less. For solenoidal coils, the small observe volumes and short residence times often preclude obtaining data with good S/N. The larger observe volumes of saddle-type coils greatly enhance the concentration sensitivity at the possible expense of separation efficiency. Separation efficiencies as high as 50,000 have been reported with CE-NMR using a 5-nL solenoidal microcoil [44]. Nevertheless, larger observe volumes and ability to record NMR data with applied volt-

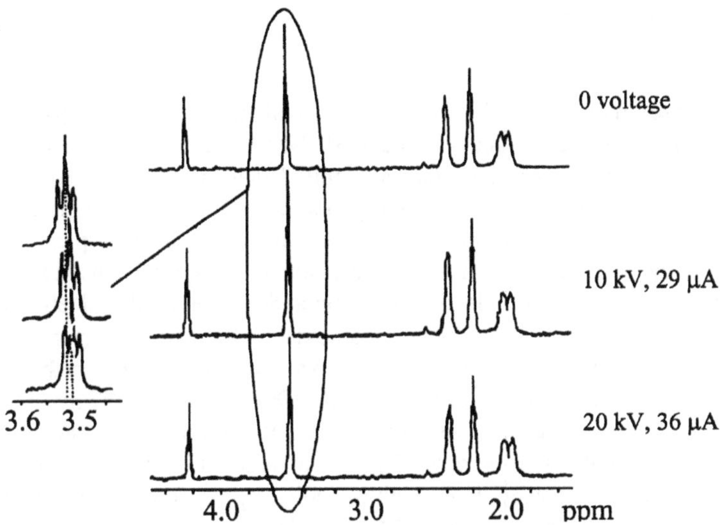

FIGURE 8 Static 600-MHz ^{1}H NMR spectra of lysine under CZE conditions: (1) without voltage, (2) with 10 kV (29 mA), (3) with 20 kV (29 mA). Inset shows slight change in chemical shift due to temperature. (Adopted with the permission from Ref. 29. Copyright 1998 American Chemical Society.)

age are attractive features of saddle-type coils. Just like saddle coils, a modification to solenoidal microcoil fabrication allows one to change the observe volume by changing the fused silica capillary [50]. Instead of wrapping the coil directly around the silica capillary, the coil is wrapped around polyimide tubing with suitable internal diameter. The fused silica capillary is then inserted into the polyimide tube and the NMR data are recorded in the usual manner. High-resolution spectra with linewidths of 1–2 Hz can be obtained with this new coil fabrication approach.

3. APPLICATIONS OF CE/CEC-NMR

Separation of simple mixtures [13,44] to more complicated biofluidic mixtures with CE-NMR and CEC-NMR have been reported [51,52]. Several instrumental modifications and methodologies have been described to use CE/CEC-NMR as a diagnostic tool [44,53–55]. Though still in its infancy, chip-based CE-NMR with microfabricated microcoils may be able to analyze picoliter volume samples [56,57].

3.1. Mixture Analysis

3.1.1. CE-NMR

The first reported analysis with CE-NMR was of a simple amino acid mixture containing arginine, cysteine, and glycine [13]. NMR spectra were recorded using a 5-nL-observe-volume cell under an applied field of 12.5 kV (260 V/cm). To obtain data with good S/N in a short period of time, a 20-nL sample was injected. The efficiency reported for cysteine was ~5000, well below typical for CE. However, the analysis of high-nanogram amounts is especially useful for mass-limited samples. To exclude the electrophoretic current-induced effects associated with the horizontally positioned solenoidal coil, the periodic stopped-flow CE-NMR method was used to analyze a mixture of arginine and triethylamine (TEA) [44]. Improved limits of detection (LODs) for arginine [57 ng (330 pmol)] the TEA [9 ng (88 pmol)] were reported using field-amplified stacking injection. The sensitivity was increased by two- to four-fold without compromising the separation efficiency. Even with periodic stopped flow, a separation efficiency as high as 50,000 was reported for arginine.

Novel instrumentation integrates capillary HPLC (cHPLC)-NMR, CE-NMR, and CEC-NMR and provides extreme versatility [29,49]. A number of CE-NMR and CEC-NMR applications performed with this integrated instrumentation have been described. In this system, saddle-type coils with 240–400-nL observe volumes increase the sensitivity and avoid current-induced magnetic field gradients. A mixture of lysine and histidine has been analyzed in a continuous-flow CE experiment with a detection limit of 336 ng (2.3 nmol) [29].

The ability to identify metabolites and biotransformed products in biological fluids during separation is a considerable challenge. The potential application of CE-NMR and CEC-NMR for analysis of metabolites in biofluids has been demonstrated [51,52]. For example, CE-NMR has successfully analyzed the major metabolites of paracetamol in human urine [51]. Two major metabolites, paracetamol glucuronide and paracetamol sulfate conjugates, as well as endogenous material (hippurate) have been characterized. Comparison of chemical shifts has confirmed the presence of these compounds. The estimated amount that can be detected in this study with a S/N of 3 is ~10 ng.

In CE the separation efficiency can often be improved by optimizing the buffer pHs. Bayer and co-workers have demonstrated on-line CE-NMR to analyze a mixture of two commonly found soft-drink ingredients, caffeine and aspartame, at two pH values [58]. Separation at pH 10 (using glycine buffer) has been achieved in under 70 min. However spectral overlap of the analyte and the glycine is unavoidable in this separation. Changing the buffer to formate (pH = 5) reverses the order of migration and increases migration time. Although spectral overlap between glycine and analytes are avoided, both caffeine and aspartame migrate closer to each other; also at pH 5, more intense NMR spectra have been recorded. The experimental conditions often are a compromise between separation efficiency and NMR spectral quality. Injection of larger analyte volumes and concentrations greater than buffer concentration to improve NMR sensitivity may degrade the separation efficiency CE [44].

Despite the high mass sensitivity compared to typical NMR spectroscopy, CE/CEC NMR still suffers from poor concentration sensitivity. As discussed earlier, on-line concentration methods have enhanced the sensitivity [44]. A recent study with capillary isotachophoresis (cITP)-NMR demonstrated dramatically improved concentration sensitivity [50]. Capillary isotachophoresis is a well-established technique to concentrate either positively or negatively charged analytes on capillary during electrophoresis. Using cITP, concentration sensitivity enhancements of two orders of magnitude have been reported for nanoliter-volume NMR. Figure 9A shows a proton NMR spectrum of 5 mM tetraethylammonium bromide (TEAB) acquired with a solenoidal coil with an observe volume 30 nL. The stopped-flow cITP-NMR spectrum of 8 μL of TEAB with initial concentration of 200 μM is shown in Figure 9B. The S/N recorded for TEAB under cITP condition is twice as high as for TEAB without preconcentration. Overall concentration enhancements of ~100-fold have been reported. Furthermore, the sample observation efficiency is also greatly increased, from 0.5% to 50%. As a result, high-S/N NMR spectra can be recorded in a short period of time. For example, the inset (Figure 9B) represents a stopped-flow COSY spectrum acquired in 22 min, which is ~10,000 times faster than would be possible without isotachophoresis. In addition, a mixture containing TEAB, Ala-Lys, methyl green (MeG), and trace impurity of MeG has also been analyzed with cITP-NMR. A considerable overlap of adjacent zones is evident from these migration times; however, with careful examination each migrating species can be identified. The length of the NMR observe volume appears greater than the length of analyte zones and results in overlapping bands. Designing smaller volume coils with fewer turns may allow the analysis of cITP zones with improved performance.

Electrophoresis is especially useful to separate charged species in the presence of neutral or oppositely charged species. Such applications with cITP-NMR are attractive to measure trace drug metabolites and synthetic organic products as demanded in the pharmaceutical industry. Wolters et al. recently extended the application of cITP-NMR for trace impurity analysis in the presence of 1000-fold excess of a neutral compound [41]. As shown in Figure 10, 200 μM (1.9 nmol) atenolol, a beta-blocker used for treatment of cardiovascular diseases, has been successfully isolated in the presence of 200 mM sucrose. At the pH of analysis, atenolol is positively charged and therefore is effectively separated from sucrose. The intensity of the signal is well below the intensity of ^{13}C satellite signals of anomeric protons of sucrose. This clearly demonstrates the difficulty of obtaining good S/N data for trace materials in a standard NMR

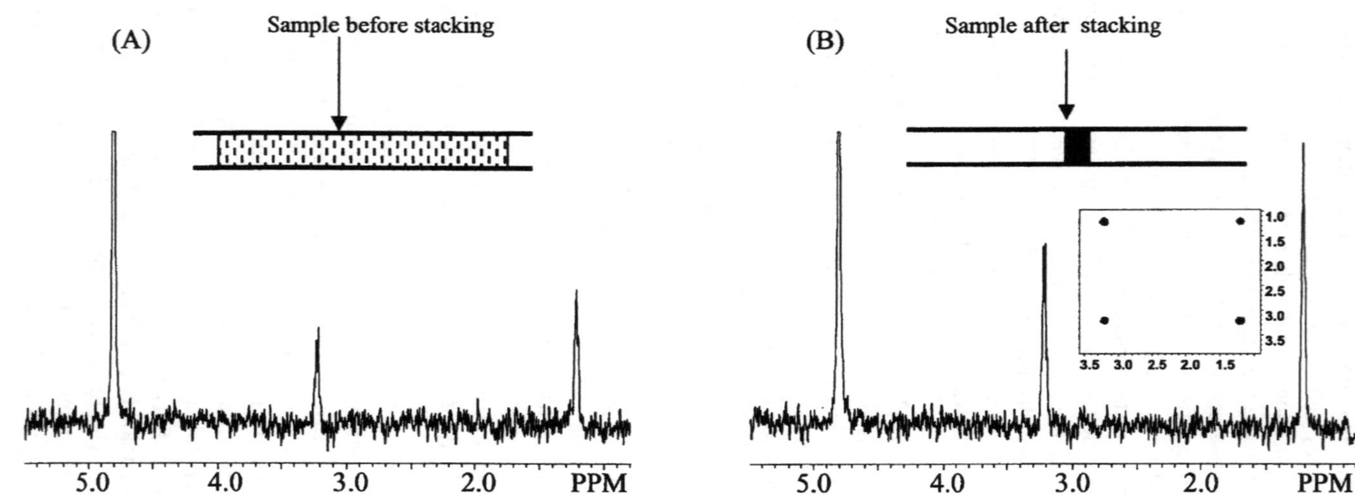

FIGURE 9 ^{1}H NMR spectra of TEAB. (A) Capillary filled with 5 mM TEAB without sample stacking (S/N of peak at 1.2 ppm = 13). (B) 8 μL of 200 μM TEAB injected; spectrum after sample stacking by cITP (S/N of peak at 1.2 ppm = 30). The inset shows a COSY spectrum. (Reproduced with permission from Ref. 50. Copyright 2001 American Chemical Society.)

experiment. Figure 10B illustrates the stopped-flow microcoil ^{1}H NMR spectrum of a 25 mM atenolol sample. An on-flow cITP-NMR spectrum of atenolol (200 μM injected concentration) which was obtained and processed under the same parameters is shown in Figure 10C. The spectra in Figures 10B and 10C are plotted at the same scale. Comparison of S/N of these two spectra indicates that the concentration of the focused atenolol band is ~40 mM. Therefore, 67% of the injected 1.9 nmol of atenolol occupies the NMR observe volume of 30 nL, almost reaching the theoretical limit of stacking efficiency.

Trapping peaks in cITP/NMR can be problematic because of the narrow lengths of the focused bands. New instrumentation with two solenoidal microcoils in a single probe enhances cITP/NMR for trace analysis. The first (scout) coil is used to detect the analyte, whereas the second coil is used for data acquisition (Figure 11). At any given time one coil is made active using an RF switch to select the coil, detuning the other coil's matching circuitry, and optimizing the shims for the coil of interest. Thus, the two coils are tuned independently of each other to minimize cross-talk. As illustrated in Figure 12, capillary-scale ITP significantly improves the performance of the most mass-sensitive NMR microcoils by allowing microliter samples to be concentrated and measured using the most sensitive 10–100-nL-volume microcoils. cITP-NMR has successfully separated and identified trace materials at ~0.1% in the presence of excess uncharged species. The low current, < 10 μA, during the cITP process permits NMR data acquisition on-flow without spectral degradation.

3.1.2. CEC-NMR

The instrumentation required to interface CEC and NMR is the same as required for CE-NMR, with the added benefit that the sample-loading capability of CEC is higher, making detection easier. A continuous-flow CEC-NMR experiment has been performed in parallel with CE-NMR to analyze a human urine sample [51]. The two-dimensional electropherogram is shown in Figure 13A. The high loading capacity of CEC (using a 500-nL sample injection) increases the S/N, decreasing the time required to record acceptable-quality NMR spectra. One-dimensional proton NMR spectra extracted from the two-dimensional electropherogram at the migration times of paracetamol glucuronide, paracetamol sulfate, and hippurate are shown in Figure 13B. The presence of paracetamol glucuronide has been confirmed by acquiring high-resolution NMR spectra in a stopped-flow 2-D TOCSY experiment (Figure 14) [52].

Gradient elution is an important and widely used technique to achieve faster and/or better separation of complex mixtures. The power of gradient elution has been realized with CEC-NMR while successfully analyzing an analgesic mixture containing caffeine, acetaminophen, and acetylsalicylic acid [58]. Under isocratic conditions (2 mM borate, 80% D_2O, and 20% CD_3CN), poorly separated peaks with a solvent peak-eluting close to acetaminophen are ob-

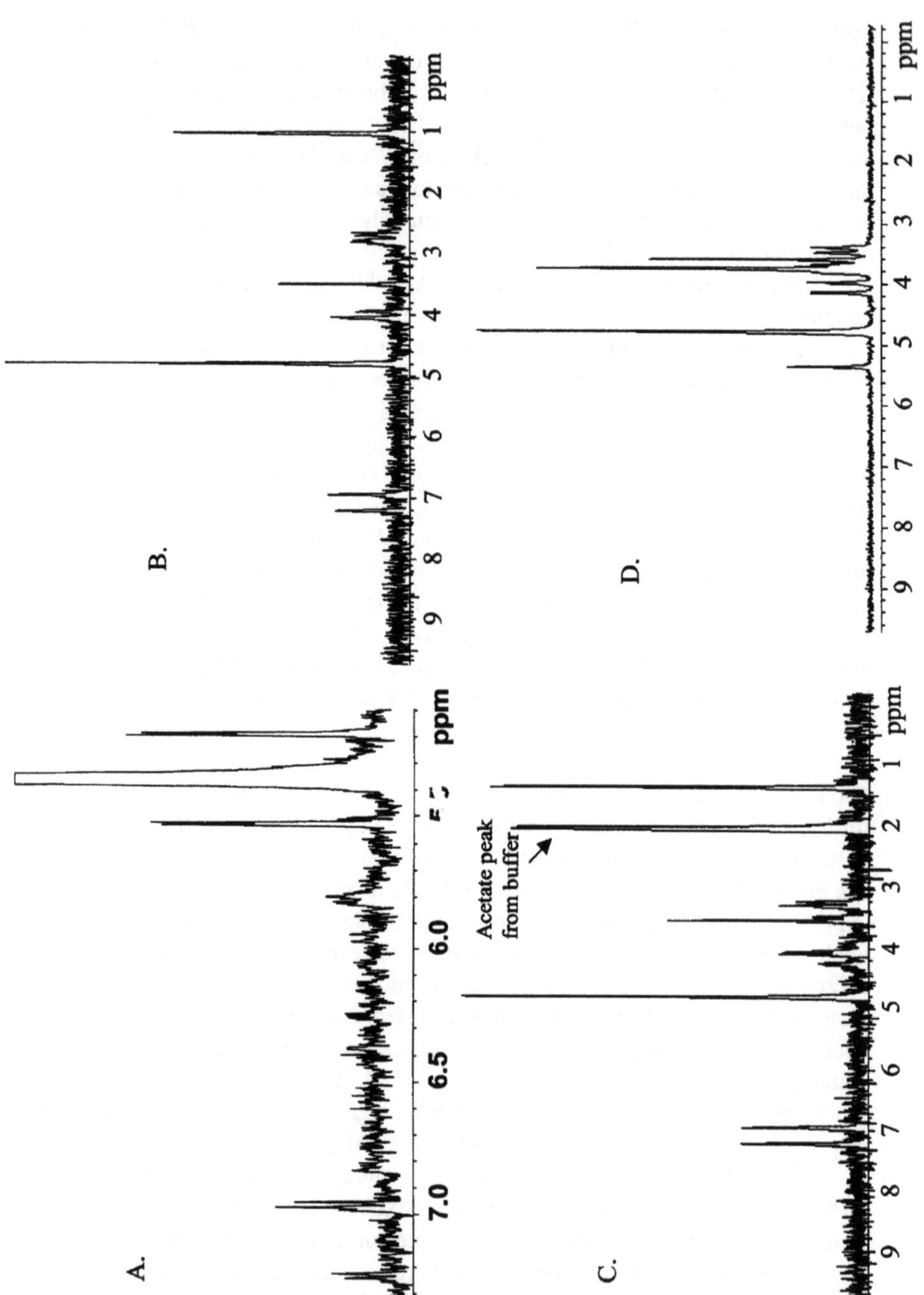
A.
B.
C.
D.
Acetate peak
from buffer
ppm

FACING PAGE

FIGURE 10 (A) ^{1}H NMR spectrum of the trace impurity sample (200 μM atenolol and 200 mM sucrose in 50% TE/D_2O) from 5-mm probe. The expanded and vertically increased area is shown. Microcoil ^{1}H NMR spectra shown in (B)–(D) recorded and processed with identical parameters. (B) Static NMR spectrum obtained with direct injection of 25 mM atenolol to the NMR microcoil. S/N of atenolol methyl peak is 21. (C) On-flow cITP-NMR spectrum of atenolol sample band at peak maximum during analysis of the trace impurity sample (200 μM atenolol and 200 mM sucrose in 50% TE/D_2O). No sucrose peaks can be observed. S/N atenolol methyl peak is 34. (D) Stopped-flow cITP-NMR spectrum of sucrose at peak maximum from the same experiment as in (C). (Adopted with the permission from Ref. 41. Copyright 1998 American Chemical Society.)

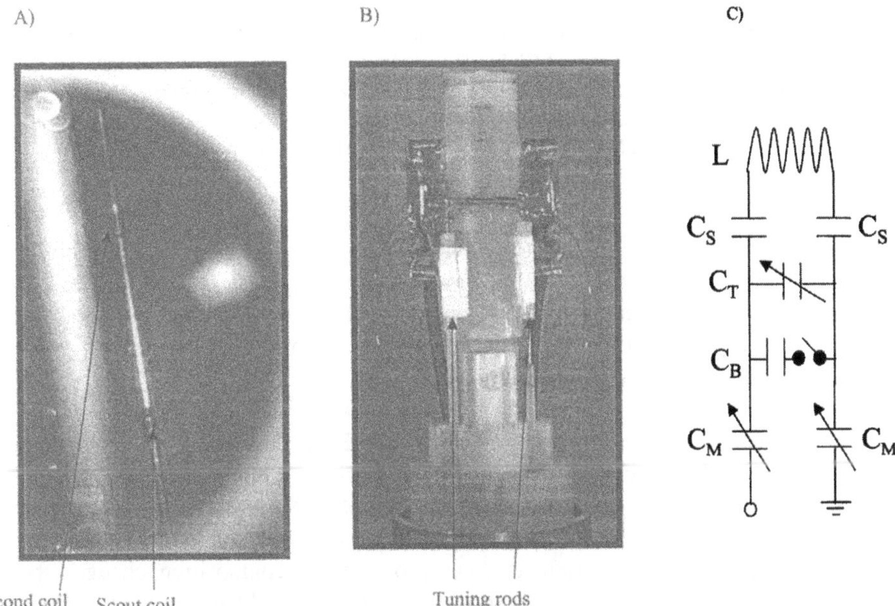

FIGURE 11 Description of dual microcoil probe. (A) Two coils wrapped around a polyimide sleeve (B) Dual-coil probe mounted on top of the probe head. (C) Schematic of balanced tank circuit used for each microcoil. Coil (L), series capacitors (C_S) 3.3 pF, tuning capacitors (C_T) 0.6–4.5 pF, matching capacitor (C_M) 0.6–4.5 pF, bridge capacitors (C_B) 24 pF. (Reproduced with permission from Ref. 41. Copyright 2002 American Chemical Society.)

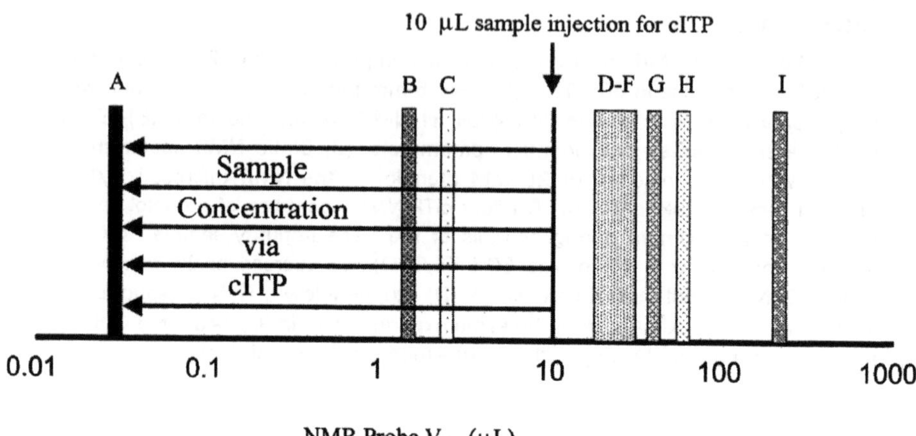

FIGURE 12 Comparison of detection volumes of various NMR probes. Volume is indicated within brackets: (A) CE-NMR microcoil (0.030 μL); (B) MRM flow probe (1.5 μL); (C) 1-mm Bruker probe (2.5 μL); (D) 1.7-mm Nalorac probe (20 μL); (E) Nalorac flow probe (24 μL); (F) 2.5-mm Bruker flow probe (30 μL); (G) Varian nanoprobe (40 μL); (H) 3-mm Varian probe (60 μL); (I) 5-mm probe (220 μL).

served. The large sample injection volume required to improve NMR sensitivity reduces the separation efficiency. However, by applying a solvent gradient of 0% to 30% CD_3CN in 25 min, the separation has been improved significantly. The total separation time has been reduced by one-third with the solvent gradient in comparison to isocratic elution.

One interesting feature in the chromatogram is the chemical shift change (~0.3 ppm) of residual protons of CD_3CN and HOD throughout the separation. Two main adverse effects from a solvent gradient elution on NMR spectral quality are line broadening and chemical shift changes [43,59]. The change in magnetic susceptibilities across the NMR observe volume contribute to the linewidth increase. The continuous change of solvent composition changes the chemical shifts of certain resonances. For example, solvent-sensitive protons such as the proton in HOD are influenced to a greater extent than the residual protons of CD_3CN. Although these effects do not interfere with spectral interpretation in this work, both line broadening and chemical shift changes can be obstacles in performing steep solvent gradients with on-line NMR detection.

Applied pressure during CEC can decrease the separation time [30]. In this study the pressurized capillary electrochromatography (pCEC) coupled to NMR separated and identified a mixture of unsaturated fatty acid esters. The analysis time has been reduced by factor of 10 compared to nonpressurized CEC.

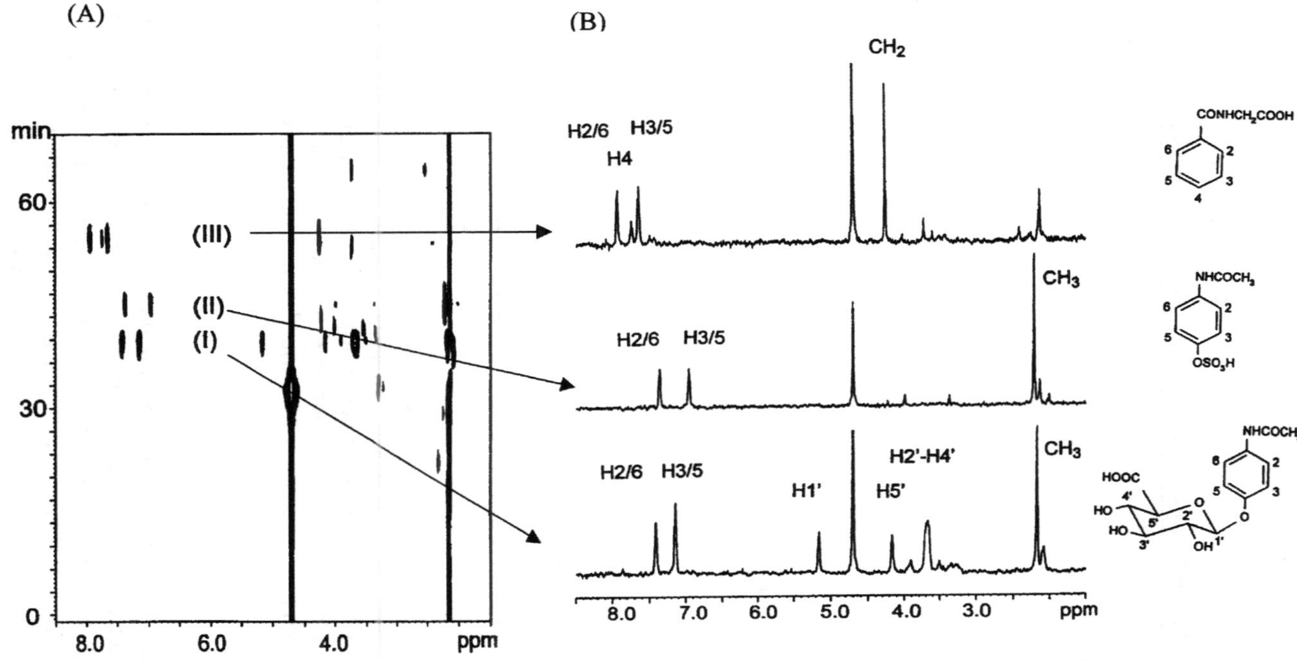

FIGURE 13 (A) Contour plot of a two-dimensional electropherogram recorded on-flow with CEC-NMR for separation of humane urine extract. (B) Single rows extracted from (I, II, and II) the continuous flow CEC-NMR from Figure 13A. (From Ref. 51; reproduced by permission from The Royal Society of Chemistry.)

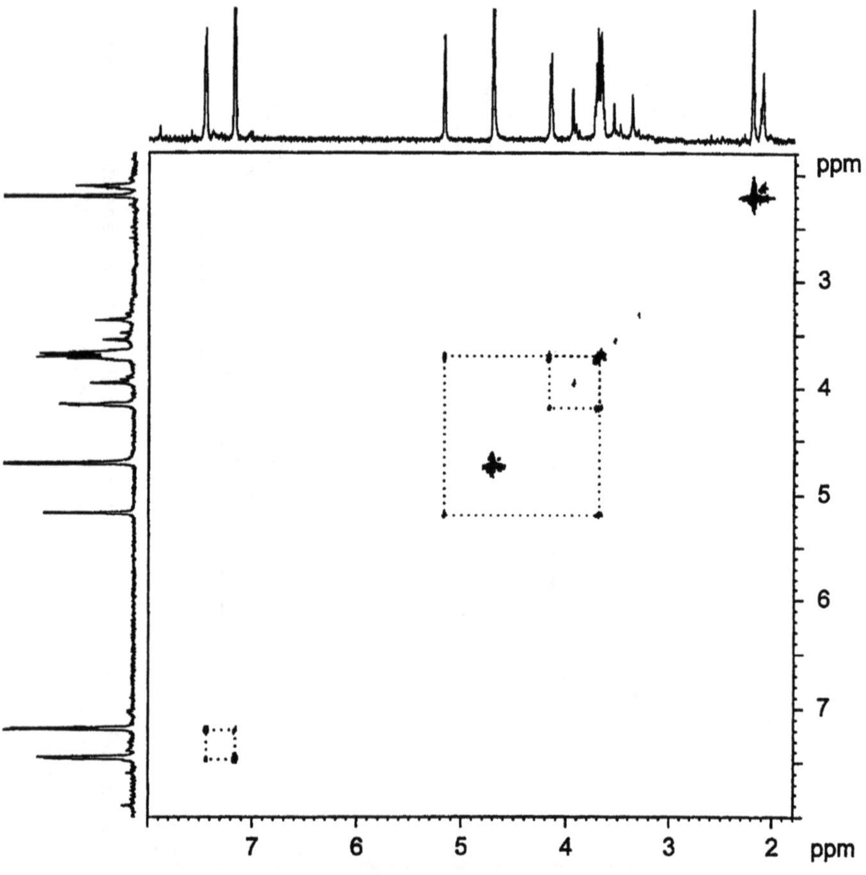

Figure 14 Stopped-flow TOCSY CEC-NMR spectra of paracetamol glucuronide. Acquisition parameters: number of 144 scans for each of 256 increments. Spectral width of 4716 Hz, number of points 4096. Processing parameters: zero filled and multiplied by apodization function of 3 Hz in both dimensions. (From Ref. 52; reproduced with permission from The Royal Society of Chemistry.)

The plate heights increase with increasing pressure in pCEC; however, the plate height recorded even at 100 bar is still below what has been recorded with cHPLC. Both cHPLC and pCEC have been used to separate a mixture of methyl esters of palmitoleic, oleic, eicosenoic, and erucic acids. Capillary HPLC operating at 16 bar has separated the mixture in 110 min, whereas pCEC (16 bar and 20 kV) analyzed the mixture in 13 min. However, a better resolution has been achieved with cHPLC. Further pressure increases during pCEC do not shorten the migration time significantly but do reduce the separation efficiency. The reduced separation time as in pCEC provides extra time to perform more information-rich multidimensional NMR experiments during the same total analysis time.

3.2. NMR as a Diagnostic Tool in CE and CEC

In addition to elucidation of molecular structures, NMR can also extract valuable information about physicochemical parameters. Because of the omnipresence of protonated solvents in CE/CEC, mobile-phase events can be monitored with NMR. Early studies using E-NMR involved the calculation of diffusion coefficients, electrophoretic mobilities, and viscosity [27]. Stagnant mobile-phase mass transfer kinetics and diffusion effects [60] and fluid mass transfer resistance in porous media-related chromatographic stationary phases [61] have been studied with NMR spectroscopy. NMR imaging of the chromatographic process [62] and NMR microscopy of chromatographic columns [63] have also been reported. Several applications of NMR to on-line studies of CE/ and CEC/ NMR are highlighted.

3.2.1. Injection Performance/Plug Profile

The injection performance and profile of a water plug in CE have been studied in a CE-NMR experiment [44]. Figure 15 shows a series of NMR spectra of HOD (10% H_2O in D_2O), with the inset illustrating the normalized S/N of HOD as a function of migration time. The leading parabolic-type concentration profile originates from the gravimetric injection, whereas the trailing flat profile is due to application of potential after the injection. A parabolic model has been developed, with the inset in Figure 15 illustrating the least-squares fit to the intensity of the first seven resonances. The calculated volume of injection of 10% H_2O in D_2O according to the Hagen-Poiseuille equation is 48 nL, a volume corresponding to a plug length of 12.3 mm. Diffusion and mixing also contribute to plug profiles. The contribution of diffusion is related to the root-mean-square displacement (σ) of a peak by $\sqrt{2Dt}$, where D is the diffusion coefficient and t is time. The calculated σ for HOD with a diffusion coefficient of 1.94×10^{-5} cm^2/s and migration time of 22 min is 2.9 mm. As shown in Figure 15, the length of the HOD plug is ~16 mm, consistent with contributions from the injection length combined with diffusional broadening. In addition to injection performance, other aspects such as flow effects related to CE performance can also be monitored.

3.2.2. Following the cITP Process

Knowledge of the pH of migrating bands can reveal valuable mechanistic information about the cITP process [54]. Measurement of the pH of migrating bands

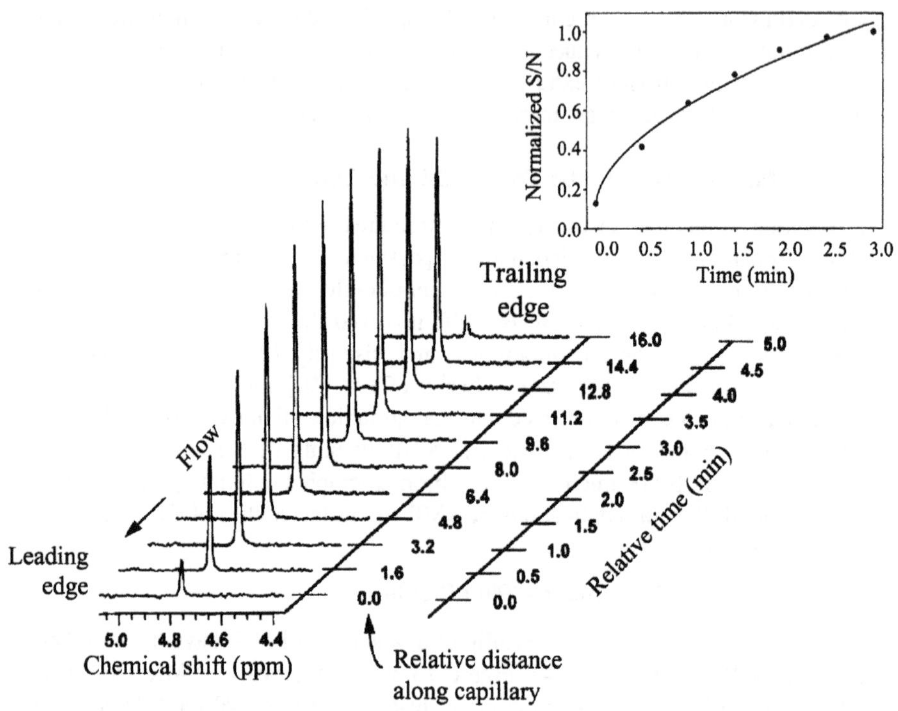

FIGURE 15 Observation of HOD plug through a solenoidal NMR microcoil. The plug is gravimetrically injected to a 50-cm-long capillary (100 μm i.d./235 μm o.d.) followed by application of a 7.0-kV voltage. Once the first HOD peak appears at the NMR coil, the voltage is reduced to 0.7 kV. The single-scan spectra show that the HOD concentration increases parabolically, followed by an abrupt trailing end. The relative time started at an offset of 22 min. The relative position along the capillary is computed from the migration time and velocity to provide an indication of the total plug length as it reaches the detector. (Reproduced with permission from Ref. 44. Copyright 1998 American Chemical Society.)

during cITP can be performed using NMR; this may be the first on-line cITP pH (pD) measurement, although pH measurements have been reported using off-line approaches [64,65]. The chemical shift of the nonexchangeable methyl peak of acetate in both the leading electrolyte (LE) and the trailing electrolyte (TE) are used to measure the pD. Because the solvents are deuterated, pD rather than pH more accurately describes the acidic or basic nature of the medium. The chemical shift of protons of the acetate group changes from ~2.07 to ~1.93 ppm in the range of pD 4–6, and is therefore ideal to probe pH in this range. The on-line and off-line pD measurements for the LE in this work are 5.18 ±

0.04 and 5.14 ± 0.02, respectively, and therefore indistinguishable. In general, chemical shift uncertainty is ±0.001 ppm. Figure 16A illustrates a sequence of NMR spectra of acetate peaks as the LE and sample (atenolol) progress through the NMR coil. In addition to the peak at 2.05 ppm, a second peak also appears around 2.08 ppm as the band passes through the coil. The acetate signal at 2.05 ppm from the LE completely disappears with the sample, while the acetate peak at 2.08 ppm gradually grows as a single acetate peak. The peak at 2.08 ppm represents the more acidic acetate from the TE. As shown in this figure, the acetate peak of the LE experiences an asymmetric broadening toward the downfield spectral region. The increase in linewidth of acetate (Figure 16B) during the appearance of atenolol limits accurate chemical shift measurements. As a result, pD measurements at the interface of LE/sample or sample/TE cannot be measured as accurately as in static systems. The line broadening of the analyte NMR bands appears due to magnetic susceptibility differences across the RF coil when multiple compounds in differing environments are present in the NMR probe observe volume.

Many theories and models are available to explain the cITP process [66]. Such information is vital to develop better separations with isotachophoresis.

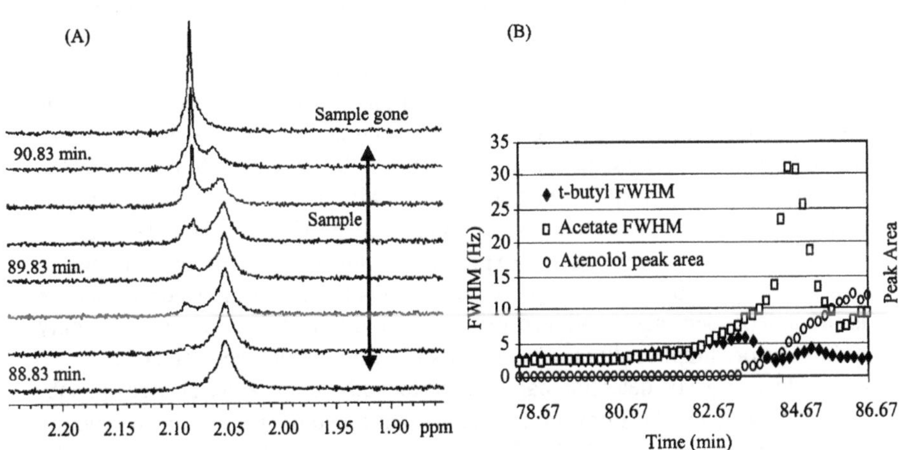

FIGURE 16 (A) Progression of cITP-NMR spectra displaying acetate chemical shift during passage of interface between focused sample band and TE through RF coil. Each spectrum consists of 8 scans acquired in 10 s. For display purposes, alternating spectra plotted (20-s time resolution). Focused sample band present from 88.83 min to 90.83 min and no longer present at 91.17 min. (B) Acetate and *tert*-butyl alcohol, full width at half-maximum (FWHM), in Hz and atenolol peak area as a function of run time (min) in cITP-NMR. (Reproduced with permission from Ref. 54. Copyright 2002 American Chemical Society.)

However, there is a lack of experimental approaches to study real-time on-line events in cITP. Using the ability of NMR to track individual components in a complex mixture, sample stacking, the behavior of the leading and trailing electrolytes, and buffer components have been followed by using NMR-observable electrolytes [54].

3.2.3. NMR Thermometry in CE and CEC

Intracapillary temperature often influences the pH of the buffer, pK_a, peak shapes, migration times, and overall efficiency in electrophoretic measurements [67]. Moreover, events such as protein conformation, aggregation, denaturing, and DNA/RNA separation are also affected by temperature. Knowledge about intracapillary temperature and thermal gradients is an important aspect of CE/CEC. The use of thermocouples located outside the capillary is a common method to measure intracapillary temperature [68,69], although the thermal response time is slowed by the thermal mass of the capillary, resulting in decreased accuracy in the temperature measurements. Internal temperature in CE has been calculated from micellar capacity factors [69], electroosmotic mobilities [70], buffer conductivities [70], absorption spectrum of Co(II) chloride [71], and water O-H stretching frequency [72].

Certain NMR chemical shifts are temperature-sensitive and are commonly used to probe temperature of samples [73]. The linear dependence of HOD proton frequency on temperature allows a straightforward determination of temperature [74]. Because CE/CEC uses predominately aqueous solvents, HOD is ideal to probe temperature in capillary electrophoretic studies. Lacey et al. introduced CE/NMR thermometry to measure intracapillary temperature [53]. Proton spectra of H_2O in phosphate buffer recorded with a saddle coil show the shift of resonance in the first 20 s after applying voltage (Figure 17A). The upfield shift of water resonance can be explained by shielding of the nucleus due to loss of hydrogen bonding. As the frequency dependence of the water resonance is pH-independent and combined pH, conductivity, and deuteration cause only a ≤2°C error, simple calibrations are possible using NMR thermometry.

On-line NMR detection can be used to study temperature changes in subsecond time intervals. Intracapillary temperature evolution in CE as a function of time with different applied voltages has been studied [53]. As expected, the temperature remains constant prior to application of voltage. Low voltages, such as 2 kV, cause only a small increase in temperature (~1°C), even after 15 min of applied voltage. However, for higher voltages the rate of increase of temperature is greater and the temperature reaches a higher steady-state value. The time to reach ambient temperature upon discontinuation of 15 kV is longer than 20 min. Also exciting, the temperature of a migrating plug of low-conductivity electrolyte has been followed. For example, the injection of a plug of 1 mM NaCl causes a band of lower conductivity that has a higher temperature (due to

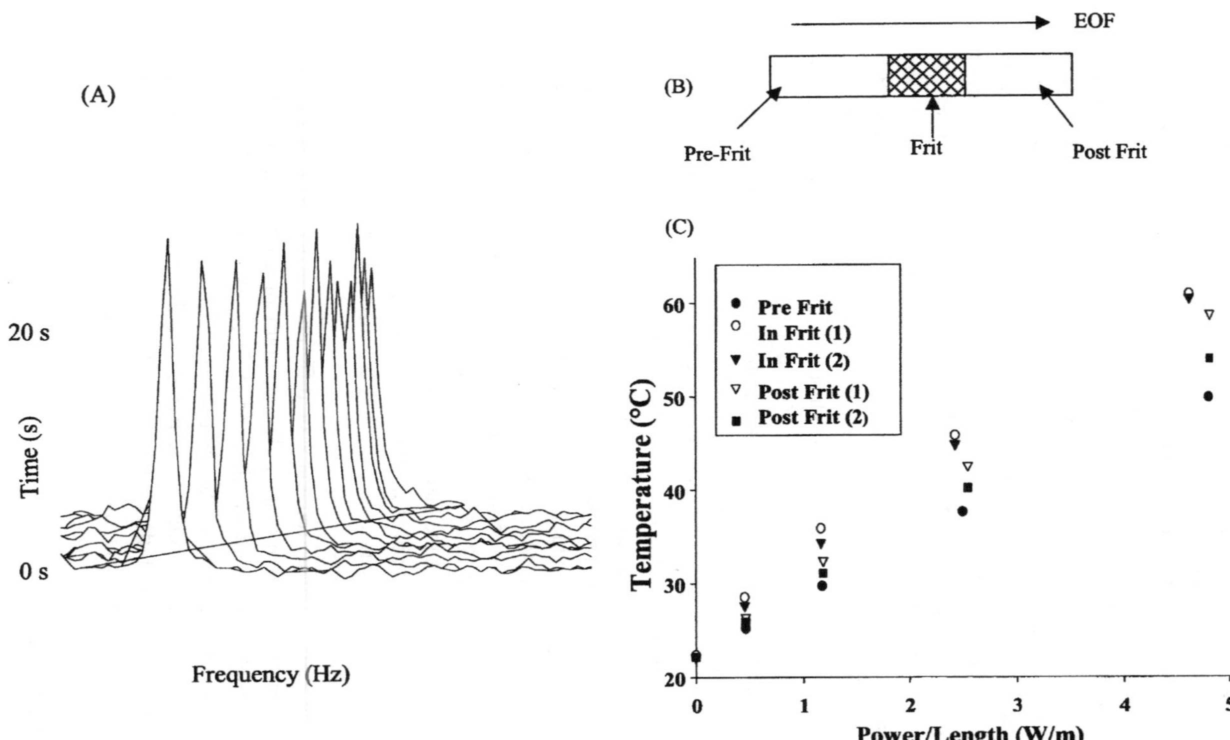

FIGURE 17 (A) Stacked plot of 40 NMR spectra of H_2O resonance during initial 20 s after power was applied across the capillary. The first spectrum coincides with the application of 12 kV across a 38-cm, 100-μm-i.d./360-μm-o.d. fused silica capillary filled with 50 mM borate buffer. (B) Schematic of a chromatographic frit placed within a fused silica capillary. (C) Plot of temperature in and around a chromatographic frit as a function of electric power dissipated in 100-μm-i.d./360-μm-o.d. capillary. (Reproduced with permission from Ref. 55. Copyright 2002 American Chemical Society.)

its higher resistance); even after migrating to the NMR detection point at the end of the capillary, the temperature in the plug remains 10°C higher than the background electrolyte temperature. Given the fact that many buffers have pHs that are temperature-sensitive, such migrating "hot" zones also likely result in local pH disturbances. Such effects are certainly worth studying using CE/NMR.

Knowledge about temperatures and/or temperature change in the region of a CEC frit or the start of the packed bed can generate valuable information about the regions of sharp discontinuities within the column. The on-line temperatures in and around chromatographic frits have been measured using similar instrumentation and methodology [55]. The frits form the interface between the packed column and the open capillary on CEC columns and help to retain the stationary phase of the CEC column. However, frits are also responsible for variations in overall EOF and possibly even bubble formation. New approaches to fritless capillary columns have been introduced to improve the performance of CEC [75,76]. In the on-line temperature study, the temperature of water before, during, and after (Figure 17B) it passes through a frit is measured with a specially designed two-turn vertical solenoidal coil. This particular design represents one of the smallest solenoids used to date and was constructed to provide higher spatial accuracy, and it enables the measurement of the temperature of picoliter-to-nanoliter volumes. Using sleeve probe technology [50], the temperature-monitoring point can be moved as desired by sliding the column through the probe. The calculated temperature as a function of power dissipated throughout the capillary is shown in Figure 17C. The linear regression analysis of these plots indicates that the temperature is highest within the frit (24.6°C, 23.8°C), lowest before the frit (22.6°C), and intermediate after the frit (22.8°C). The particles in the frit obstruct the electrolyte and hence increase the resistance. Therefore, more power is dissipated in this region, reflected as increase in temperature. The water cools down as it exits but is still warmer than before the frit for several hundred microns.

In principle, a resonance frequency or difference in frequencies can be employed to perform NMR thermometry. The accuracy of temperature measurement depends on the accuracy of the thermocouple (0.2°C) in the NMR spectrometer. Generally, a precision of better than 0.2°C has been reported with these NMR thermometry measurements. Individual calibration plots may not be required when using aqueous buffers. Proper calibration procedures should be adopted with nonaqueous buffers and in the presence of organic modifiers.

4. FUTURE DIRECTIONS

Both CE-NMR and CEC-NMR have been established as powerful hyphenated separation techniques. The NMR sensitivity should be improved further in order for CE-NMR and CEC-NMR to become more widely applied techniques. One possible solution that can provide such an increase is the use of multiple RF coils on a single column, so that NMR spectra of a single migration band are recorded at multiple points. Fluidic systems containing multiple coils that allow individual bands to be parked and probed with NMR while continuing the separation are also expected to become available, greatly increasing the overall flexibility of the hyphenated system. Of course, higher-field-strength magnets will continue to become available, increasing the performance of NMR. Applications of CE-NMR and CEC-NMR to analyze peptides, proteins, and complex mixtures have not been reported extensively. For example, the separation and identification of molecular classes using pattern recognition rather than individual molecule identification can be performed. Such approaches can minimize long analytical procedures involved in metabolomics and proteomics. We expect the appearance of applications using capillary-scale separations hyphenated to NMR to increase greatly over the coming decade.

REFERENCES

1. Pullen FS, Swanson AG, Newman, MJ, Richards, DS. Rapid Commun Mass Spectrom 1995; 9:1003–1006.
2. Sandvoss M, Weltring A, Preiss A, Levsen K, Wuensch G. J Chromatogr 2001; 917:75–86.
3. Shockcor JP, Unger SE, Savina P, Nicholson JK, Lindon JC. J Chromatogr B 2000; 748:269–279.
4. Hjerten S. Chromatogr Rev 1967; 9:122–219.
5. Righetti PG, Gelfi C. Forensic Sci Int 1998; 92:239–250.
6. Liu SR, Shi YN, Ja WW, Mathies RA. Anal Chem 1999; 71:566–573.
7. Thormann W, Caslavska J. Electrophoresis 1998; 19:2691–2694.
8. Issaq HJ, Chan KC, Muschik GM, Janini GM. J Liq Chromatogr 1995; 18:1273–1288.
9. Miyawa JH, Lloyd DK, Alasandro MS. HRC—J Chromatogr B, 1998; 21:161–168.
10. Wang J, Schaufelberger DE, Guzman NA. J Chromatogr Sci. 1998; 36:155–160.
11. Behnke B, Metzger JW. Electrophoresis 1999; 20:80–83.
12. Olson DL, Peck TL, Webb AG, Magin RL, Sweedler JV. Science 1995; 270:1967–1970.
13. Wu NA, Peck TL, Webb AG, Magin RL, Sweedler JV. Anal Chem 1994; 66: 3849–3857.
14. Lacey ME, Subramanian R, Olson DL, Webb AG, Sweedler JV. Chem Rev 1999; 99:3133–3152.
15. Wada H. R & D of 1 GHz Class NMR Magnets, Experimental NMR Conference; Asilomar, CA, 2002.
16. Long HW, Gaede HC, Shore J, Reven L, Bowers CR, Kritzenberger J, Pietrass T, Pines A, et al. J Am Chem Soc 1993; 115:8491–8492.
17. Gitti R, Wild C, Tsiao C, Zimmer K, Glass TE, Dorn HC. J Am Chem Soc 1988; 110:2294–2296.

18. Stevenson S, Dorn HC. Anal Chem 1994;66:2993–2999.
19. Black RD, Early TA, Roemer PB, Mueller OM, Mogrocampero A, Turner LG, Johnson GA. Science 1993; 259:793–795.
20. Watanabe N, Niki E. Proc Jpn Acad Ser B 1978; 54:194–199.
21. Bayer E, Albert K, Nieder M, Grom E, Keller TJ. J Chromatogr 1979; 186:497–507.
22. Chin J, Fell JB, Jarosinski M, Shapiro MJ, Wareing JR. J Org Chem 1998; 63:386–390.
23. Lindon JC, Farrant RD, Sanderson PN, Doyle PM, Gough SL, Spraul M, Hofmann M, Nicholson JK. Magn Reson Chem 1995; 33:857–863.
24. Lenz EM, Greatbanks D, Wilson ID, Spraul M, Hofmann M, Troke J, Lindon JC, Nicholson JK. Anal Chem 1996; 68:2832–2837.
25. Lacey ME, Tan ZJ, Webb AG, Sweedler JV. J Chromatogr A 2001; 922:139–149.
26. Godejohann M, Preiss A, Mugge C, Wunsch G. Anal Chem 1997; 69:3832–3837.
27. Johnson CS, He Q. In: Advances in Magnetic Resonance. Vol. 13. New York: Academic Press, 1989:131–159.
28. Wu N, Peck TL, Webb AG, Magin RL, Sweedler JV. J Am Chem Soc 1994;116: 7929–7930.
29. Pusecker K, Schewitz J, Gfrorer P, Tseng LH, Albert K, Bayer E. Anal Chem 1998; 70:3280–3285.
30. Gfrorer P, Tseng LH, Rapp E, Albert K, Bayer E. Anal Chem 2001; 73:3234–3239.
31. Webb AG. Prog NMR Spectrosc 1997; 31:1–42.
32. Hoult DI, Richards RE. J Magn Reson 1976; 24:71–85.
33. Crouch RC, Martin GE. Magn Reson Chem 1992; 30:S66–S70.
34. Peck TL, Magin RL, Lauterbur PC. J Magn Reson Ser B 1995; 108:114–124.
35. Rogers JA, Jackman RJ, Whitesides GM, Olson DL, Sweedler JV. Appl Phys Lett 1997; 70:2464–2466.
36. Peck TL, Magin RL, Kruse J, Feng M. IEEE Trans Biomed Eng 1994; 41:706–709.
37. Li Y, Wolters AM, Malawey PV, Sweedler JV, Webb AG. Anal Chem 1999; 71: 4815–4820.
38. Webb AG, Sweedler JV, Raftery D. Parallel NMR detection. In: Albert K, ed. On-line LC-NMR and Related Techniques. Chichester, UK: Wiley, 2002:259–277.
39. Macnaughtan MA, Hou T, MacNamara E, Santini RE, Raftery D. J Magn Reson 2002; 156:97–103.
40. Hou T, Smith J, MacNamara E, Macnaughtan M, Raftery D. Anal Chem 2001; 73: 2541–2546.
41. Wolters AM, Jayawickrama DA, Larive CK, Sweedler JV. Anal Chem 2002; 74: 2306–2313.
42. Hou T, Smith J, MacNamara E, Macnaughtan M, Raftery D. Anal Chem 2001; 73: 2541–2546.
43. Wolters AM, Jayawickrama DA, Webb AG, Sweedler JV. Anal Chem 2002; 74: 5550–5555.
44. Olson DL, Lacey ME, Webb AG, Sweedler JV. Anal Chem 1999; 71:3070–3076.
45. Albert K, Bayer E. High-performance liquid chromatography proton nuclear magnetic resonance on-line coupling. In: Patonay G, ed. HPLC Detection, Newer Methods. New York: VCH, 1992:197–227.

46. Olson DL, Lacey ME, Sweedler JV. Anal Chem 1998; 70:645–650.
47. Schefer AB, Albert K. Capillary Separation Techniques. In: Albert K, ed. On-line LC-NMR and Related Techniques. New York: Wiley, 2002:237–246.
48. Schlotterbeck G, Ross A, Hochstrasser R, Senn H, Kuhn T, Marek D, Schett O. Anal Chem 2002; 74:4464–4471.
49. Gfrorer P, Schewitz J, Pusecker K, Bayer E. Anal Chem 1999; 71:315A–321A.
50. Kautz RA, Lacey ME, Wolters AM, Foret F, Webb AG, Karger BL, Sweedler JV. J Am Chem Soc 2001; 123:3159–3160.
51. Pusecker K, Schewitz J, Gfrorer P, Tseng LH, Albert K, Bayer E, Wilson ID, Bailey NJ, et al. Anal Commun 1998; 35:213–215.
52. Schewitz J, Gfrorer P, Pusecker K, Tseng LH, Albert K, Bayer E, Wilson ID, Bailey NJ, et al. Analyst 1998; 123:2835–2837.
53. Lacey ME, Webb AG, Sweedler JV. Anal Chem 2000; 72:4991–4998.
54. Wolters AW, Jayawickrama DA, Larive CK, Sweedler JV. Anal Chem 2002; 74: 4191–4197.
55. Lacey ME, Webb AG, Sweedler JV. Anal Chem 2002; 74:4583–4587.
56. Trumbull JD, Glasgow IK, Beebe DJ, Magin RL. IEEE Trans Biomed Eng 2000; 47:3–7.
57. Stocker JE, Peck TL, Webb AG, Feng M, Magin RL. IEEE Trans Biomed Eng 1997; 44:1122–1127.
58. Gfrorer P, Schewitz J, Pusecker K, Tseng LH, Albert K, Bayer E. Electrophoresis 1999; 20:3–8.
59. Lacey ME, Tan ZJ, Webb AG, Sweedler JV. J Chromatogr A 2001; 922:139–149.
60. Tallarek U, Vergeldt FJ, Van As H. J Phys Chem B 1999; 103:7654–7664.
61. Tallarek U, Van Dusschoten D, Van As H, Guiochon G, Bayer E. Angew Chem Int Ed 1998; 37:1882–1885.
62. Tallarek U, Baumeister E, Albert K, Bayer E, Guiochon G. J Chromatogr A 1995; 696:1–18.
63. Tallarek U, Bayer E, Van Dusschoten D, Scheenen T, Van As H, Guiochon G, Neue UD. AIChE J 1998; 44:1962–1975.
64. Vestermark A. Sci Tools 1970; 17:24–25.
65. Vestermark A. Ann NY Acad Sci 1973; 209:470–474.
66. Everaerts FM, Beckers JL, Verheggen T. P. E. M. Isotachophoresis. Theory, Instrumentation and Applications. Amsterdam, The Netherlands: Elsevier, 1976.
67. Veraart JR, Gooijer C, Lingeman H. Chromatographia 1997; 44:129–134.
68. Nishikawa T, Kambara H. Electrophoresis 1996; 17:1115–1120.
69. Terabe S, Otsuka K, Ando T. Anal Chem 1985; 57:834–841.
70. Burgi DS, Salomon K, Chien RL. J Liq Chromatogr 1991; 14:847–867.
71. Watzig H. Chromatographia 1992; 33:445–448.
72. Davis KL, Liu KLK, Lanan M, Morris MD. Anal Chem 1993; 65:293–298.
73. Kuroda A, Abe K, Tsutsumi S, Ishihara Y, Suzuki Y, Satoh K. Biomed Thermol 1993; 13:43–62.
74. Hindman JC. J Chem Phys 1966; 44:4582–4592.
75. Baltussen E, van Dedem GWK. Electrophoresis 2002; 23:1224–1229.
76. Ceriotti L, de Rooij NF, Verpoorte E. Anal Chem 2002; 74:639–647.

14

Applications of Capillary Electrochromatography

V. T. Remcho, Stacey L. Clark, and Angela Doneau
Oregon State University, Corvallis, Oregon, U.S.A.

Gabriela S. Chirica
Sandia National Laboratories, Livermore, California, U.S.A.

1. INTRODUCTION

Capillary electrochromatography (CEC) is an emerging technique that employs electroosmotic flow to transport analytes through the interstitial and intraparticular space of a selective chromatographic medium. An orthogonal technique that combines capillary electrophoresis (CE) and micro high-performance liquid chromatography (μHPLC), CEC engenders not only the advantages and limitations of its parent techniques, but also reveals new features that exceed the mere sum of inherited characteristics.

The phenomenon of electroosmotic flow (EOF), which is generated when an electrical field is applied across media having ionizable groups, has been extensively exploited in the past severals years in the capillary format in CE, due to more efficient heat transfer. Its intrinsically flat flow profile generates high efficiencies, to date comparable only to those of capillary gas chromatography. EOF controls the migration velocities of neutral analytes; in the case of charged compounds the total velocity is an additive function of electroosmotic and electrophoretic mobility. Simultaneously, chromatographic retention specific to HPLC is superimposed on the electro-driven transport typical of CE. Just as in μHPLC, a wide selectivity range is available and can be finely tuned by using mixtures of organic solvents and aqueous buffers. Mobile-phase additives, such as ion-pairing reagents, surfactants, and crown ethers, can further enhance selectivity opening the door for more applications (Table 1).

Table 1 Applications of CEC Classified by Type of Compounds Separated

Compound	Stationary phase	References
Acids		
p-Hydroxy benzoic acid, bumetanide, flurbiprofen	3-µm CEC Hypersil C18	[40]
4(4-Chloro-2-methylphenoxy) butyric acid	3-µm CEC Hypersil C18	[4]
Benzoic acids including positional isomers	5-µm Spherisorb SAX	[41]
Benzoic acids	5-µm Spherisorb SCX and 5-µm Spherisorb C18	[169]
Benzoic acid, salicylic acid, phenol	3-µm C6/SAX	[76]
Inorganic ions		
Iodide, iodate, perrhenate	5-µm Nucleosil SB or 5-µm TSK	[73]
Sulfite, sulfate, thiosulfate	IC-Anion-SW	[74]
Nitrate, iodide in the presence of acids & bases	3-µm C6/SAX	[76]
Bromate, bromide, chromate, iodate, iodide, nitrate, nitrite, thiocyanate	3-µm SAX or OT-Dionex AS5A particles electrostatically bound to the wall	[170]
Bromate, bromide, chromate, iodate, iodide, nitrate, nitrite, thiocyanate	9-µm Dionex IonPac AS9-HC	[75]
Fluoride, chloride, bromide, iodate, bromate, chlorate, sulfate, sulfite	Dionex IonPac AS5A particles electrostatically bound to the wall	[78]
Arsenic, chromium, selenium	Polyamine bonded phase	[79]
Pollutants		
PAHs	3-µm C18 and 1-µm silica	[80]
PAHs	1.5-µm nonporous C18	[31]
PAHs	Methacrylate monolith	[26,81]
PAHs	Entrapped 5-µm C18	[20–22]
PAHs	7-µm 4000 Å Nucleosil C18	[33]
Triazine herbicides	3-µm Hypersil C18	[82]
	3-µm Hypersil C8	
	3-µm Spherisorb C6/SCX	

Cinosulfuron and byproducts	3-µm C18	[171]
Carbonyl 2,4-dinitrophenylhydrazones	3-µm Spherisorb ODS 1	[172]
Insecticidal pyrethrin esthers	3-µm Hypersil C 18	[83]
Pyrethroid insecticides	5-µm Nucleosil C18	[84]
	5-µm Zorbax C8	
	5-µm J.T. Baker C18	
Carbamate and pyrethroid insecticides	5-µm Nucleosil C18	[87]
	5-µm Zorbax C8	
	5-µm J.T. Baker C18	
Pirimicarb and azoxystrobin pesticides	3-µm Hypersil C18	[86]
Phenols	3-µm C18	[173]
Mono and dichlorophenols	Silica with methylated β-CDs	[174]
Pentachlorophenol	3-µm Spherisorb ODS 1	[88]
Phenols in tobacco smoke	3-µm Hypersil C 18	[56]
Nitroaromatic and nitramine explosives	1.5-µm nonporous C18	[89]
Nitroaromatic and nitramine explosives	3-µm Hypersil C 18	[175]
	1.5-µm nonporous C18	
Amino acids, peptides, proteins		
PTH amino acids	3.5-µm Zorbax C18	[91]
PTH amino acids	6-µm Zorbax C18 sintered	[72]
Dansylated amino acids	10-µm ECTFE	[92]
Tryptophan, tyrosine	OT CEC DNA aptamers	[176]
NDA aminoacids	Methacrylate monolith	[28]
PTH aminoacids		
Peptides (2–7 amino acids)		
Trp-Arg, Arg-Trp	OT CEC DNA aptamers	[177]
Dipeptides	Methacylate monolith	[178]
Tripeptides	OT CEC porphyrin	[179]
Peptides (2–3 amino acids)	Methacrylate monolith	[180]
Peptides (2–5 amino acids)	Methacrylate monolith	[181,182]

TABLE 1 (*Continued*)

Compound	Stationary phase	References
Peptides (5 amino acids)	3-μm Hypersil C18	[95,107]
Peptides (4–5 amino acids)	8-μm gigaporous polymeric SCX	[183]
Peptides (4 amino acids)	Polyacrylamide poly(ethyleneglycol) monolith	[184]
Synthetic peptides	3-μm Hypersil C18	[94]
Synthetic peptides	5-μm Spherisorb SCX	[185]
Dipeptides		
Basic peptides	5-μm Spherisorb Si	[186]
Angiotensines	Acrylate monolith	[97]
Angiotensines	PS/DVB monolith	[29]
Lysozymes, angiotensins	OT CEC C18	[187]
Cytochrome c digest	3-μm Vydac C18	[188]
Chicken albumin digest, angiotensines		
β-Lactoglobulin	3-μm Vydac C18	[189]
Tryptic digest of transferrin	OT CEC Si	[99]
Tryptic digest of ovalbumin	Monolith	[190]
Map of cytochrome c	1.5-μm Gromsil ODS2	[93]
Tryptic digest of cytochrome c	3-μm Spherisorb SAX/C6	[94]
Carbonic anhydrase, α-lactalbumin, trypsin inhibitor, ovalbumn, conalbumin, hemoglobin variants	5-μm Spherisorb S5-W SAX	[191]
Ribonuclease, insulin, α-lactalbumin	Acrylate monolith	[97]
Lysozyme, α-chymotrypsinogen, ribonuclease A, cytochrome c	C18 monolith	[192]
Cytochrome c, lysozyme, myoglobin, ribonuclease A	OT CEC C8 diol	[193]
Cytochrome c mixture (horse, tuna, chicken, and bovine)	OT CEC cholesteryl	[101]
Cytochrome c mixture (horse, tuna, chicken, and bovine)	OT CEC C18	[194]
Trypsinogen, α-chymotrypsinogen, ribonuclease A, cytochrome c	OT CEC polyaspartic acid	[103]

Lysozyme, α-chymotrypsinogen, ribonuclease A, cytochrome c	OT CEC DVB	[195]
Carbohydrates		
Sucrose, saccharin	3-μm C18	[113]
Sucralose and related carbohydrates	3-μm C18	[196]
Glucose-maltohexose	Polyacrylamide poly(ethyleneglycol) monolith	[112]
p-Nitrophenyl-labeled glucopyranosides, maltooligosaccharides α- and β-anomers of glucopyranoside	5-μm Zorbax C18	[109]
Phenyl-methyl-pyrazolone-labeled monosaccharides	5-μm Hypersil ODS I	[110]
Phenyl-methyl-pyrazolone-labeled mono and di-saccharides	5-μm Hypersil ODS I	[108]
Phenyl-methyl-pyrazolone-labeled aldopentose and monosaccharides	Aminopropylated Si Octadecylammonium Si	[111]
Nucleotides		
AMP, ADP, and ATP	5-μm Nucleosil C18	[114]
Adenosine, cytidine, uridine, guanosine, thymidine	3-μm Hypersil phenyl	[116]
Adenosine, cytidine, inosine, uridine, guanosine, thymidine	3-μm CEC Hypersil C18	[115]
Thymine, cytosine, adenine, guanine, adenosine	OT-cholesteryl-undecanoate and cyanopentoxybiphenyl derivatized	[197]
Nucleic acids, mono-, di-, and tri-phosphonucleotides	10-μm ODSS	[119]
Dinucleotides	5-μm ODSS	
t RNAs	2-μm nonporous ODSS	
Purine and pyrimidine bases and their nucleotides	10-μm ODSS	[117,118]
Primicarb and related pyrimidines	3-μm Hypersil C18	[198]
Synthetic nucleoside	3-μm Spherisorb ODS1	[40]
PAH-DNA adduct products of in-vitro reactions	3-μm ODS	[120,121]
Miscellaneous		
Flavonoids (hespederin, hesperetin)	3-μm Hypersil C8	[160]
Flavonoids (hop acids)	3-μm Hypersil C18	[199]

TABLE 1 *(Continued)*

Compound	Stationary phase	References
Flavones in citrus essential oils	3-µm Spherisorb ODS 1	[162]
Antraquionones in rhubarb	3-µm Hypersil C18	[161]
Antraquionones in rhubarb	5-µm C18	[277]
Triglycerides of vegetable and fish oil	3-µm CEC Hypersil C18	[54,55,278]
Fatty acids of vegetable and fish oil	3-µm CEC Hypersil C18	[55]
Unsaturated fatty acids methyl esters	3-µm GROM-SIL ODS	[217]
Glycosphingolipids	5-µm Porous ODSS	[279]
Cannabinnoids	3-µm Hypersil C18	[66]
	3-µm Hypersil C8	
Retinyl esters	5-µm Nucleosil C18	[280]
Retinyl esters	7-µm Nucleosil C18	[163]
Retinyl esters	5-µm C30	[281]
Carotenoid isomers	3-µm C30	[164]
N-nitrosodiethanolamine in cosmetics	OT CEC C18	[282]
Aloins, and related constituents of aloe	3-µm C18	[283]
Food colorants and aromatic glucoronides	5-µm Nucleosil C18	[128]
Azo and antraquinone textile dyes	3-µm Hypersil C18	[129]
Alkaloids	7-µm Nucleosil C18	[215]
Fullurenes C_{60} and C_{70}	3-µm Vydac C18	[284]
Polystyrene standards	Methacrylate monolith	[285]
Celluloses	5-µm Nucleosil silica	[168]
Pharmaceuticals		
Steroids		
Fluticasone propionate and synthesis impurities	3-µm Hypersil C18	[122]
	3-µm Spherisorb ODS-I	[47]

Triamcinolone, hydrocortisone, prednidsolone, cortisone, methylprednisolone, betamethasone, dexamethasone, adrenosterone, fluocortolone, triamcinolone acetonide	3-µm Hypersil C18	[39,56]
Tipredane and related substances	3-µm Spherisorb ODS-I	[40,200]
Corticosteroids	3-µm Hypersil C18	[201]
Corticosterone, testosterone, androsten-3, 17-dione, androstan-3,17-dione, pregnan-3, 20-dione	6-µm Zorbax ODS	[36]
Aldosterone, hydrocortisone, testosterone	3-µm Hypersil C18	[202]
Digoxigenin, gitoxigenin, cinobufatalin, digitoxigenin, cinobufagin, bufalin	3-µm Spherisorb ODS-I	[203]
Hydrocortisone, testosterone, 17-α-methyltestosterone, progesterone	1.5-µm Nonporous Chromspher-ODS	[204]
Estriol, hydrocortisone, estradiol, estrone, testosterone, 17-α-methyltestosterone, 4-pregnen-20α-ol-3-one, progesterone	1.8-µm Zorbax ODS	[205]
Deesterified steroid, budesonide, steroid A	3-µm Spherisorb ODS-I	[206]
Hydrocortisone, prednisolone, hydrocortisone 21-acetate, testosterone	Poly (AMPS-co-IPPAm) hydrogel	[207]
Hydrocortisone, prednisolone, betamethasone, betamethasone dipropionate, clobetasol butyrate, fluticasone propionate, clobetase butyrate, betamethasone-17-valerate	3-µm Hypersil C18	[208]
Aldosterone, dexamethasone, β-estradiol, testosterone	3-µm Hypersil C18 1.5-µm NPS ODS 2	[209]
Desogestrel and analogs, tibolon and analogs	3-µm Hypersil C18	[210]
Dexamethasone, betamethasone valerate, fluticasone propionate	3-µm Hypersil C18/SCX	[211]
Three ingredients of hydrocortisone creme	3-µm Spherisorb ODS-I, sol-gel bonded	[212]
Neutral and conjugated steroids	Macroporous monolith, C12	[213]

TABLE 1 (*Continued*)

Compound	Stationary phase	References
Corticosteroids and esters (hydrocortisone, hydrocortisone 17-butyrate, hydrocortisone 21-acetate, hydrocortisone 17-valerate, hydrocortisone 21-caprylate, hydrocortisone 21-cypionate, hydrocortisone 21-hemisuccinate)	3-μm Spherisorb ODS-I	[214]
Cholesterol and ester derivatives	3-μm Hypersil C18	[123]
Corticosteroids (ouabain, strophantidin, 4-pregnene-6b,11b,21-triol-3,20-dione)	3 mm Spherisorb small pore ODS/SCX, sol-gel bonded	[215]
Bile acids and conjugates	Macroporous monolith, amino (normal phase)	[216]
	Macroporous monolith, C12 (reverse phase)	
Estrogens (estriol, estradiol, equiline, estrone)	3-μm GROM-SIL ODS-0 AB	[217]
Benzodiazepines		
Nitrazepam and diazepam	3-μm Hypersil C18	[39]
Cloxazolam, nitrazepam, clotiazepam, diazepam	Etched cholesterol-modified open-tubular column	[218]
Oxazepam, lorzepam, temazepam, diazepam, tofizepam	3-μm Phenyl	[219]
Nitrazepam, nimetazepam, estazolam, brotizolam, clonazepam, axazolam, haloxazolam, cloxazolam, medazepam	6.5-μm Cholesteryl	[124]
	Bonded-silica open-tubular, cholesteryl-modified capillary wall	
Temazepam, oxazepam, clonazepam, diazepam, nitrazepam	Open-tubular fused silica etched with NH_4HF_2, chemically bonded cholesteryl-10-undecanoate	[102]
Flunitrazepam, temazepam, diazepam, oxazepam, lorazepam, clonazepam, nitrazepam	PEM-coated capillary	[125]
Nonsteroidal anti-inflammatory drugs		
Ketoprofen, naproxen, fluribiprofen, indomethacin, ibuprofen	3-μm CEC Hypersil C18/SCX	[220]

Indoprofen, suprofen, tiaprofen, ketoprofen, naproxen, fenoprofen, carprofen, flurbiprofen, cicloprofen, ibuprofen	5-µm LiChrospher 100 RP-18	[221]
Etodolac and five metabolites	5-µm LiChrospher 100 RP-18	[126]
Acetaminophen, caffeine	5-mm Nucleosil C18, sol-gel bonded	[22]
Ibuprofen, indoprofen, fenoprofen, ketoprofen, suprofen, diclofenac, metenamic acid	Acrylamide-based monoliths	[222]
Ibuprofen, naproxen, ketoprofen, suprofen	Methacrylate-based macroporous SAX monolith	[127]
Tricyclic antidepressants		
Bendroflumethiazide, nortriptyline, chlomipramine, methdilazine, imipramine, desipramine	3-µm Spherisorb ODS-I	[47]
	3-µm Spherisorb SCX	
Nortriptyline, N-methyl-amitriptyline, amitriptyline, imipramine, clomipramine, N,N-dipropyl-protriptyline	3-µm Spherisorb ODS-I	[223]
Nortriptyline, amitriptyline, *N*-methyl-amitriptyline, desipramine, imipramine, chlomipramine, *N,N*-dimethyl-protriptyline, *N,N*-dipropyl-protriptyline	5-µm Nucleosil SCX	[128]
	5-µm Zorbax SCX	
Nortriptyline, imipramine, amitriptyline, clomipramine	3-µm Spherisorb SCX	[129]
Nortriptyline, N-methyl-amitriptyline, amitriptyline	Continuous bed, polyacrylamide with various contents of isopropyl and sulfonate ligands	[130]
Nortriptyline, doxepin, imipramine, amitriptyline, trimipramine, clomipramine	Nortriptyline MIP	[131]
Various pharmaceuticals		
Prostaglandins and relates impurities	3-µm Spherisorb ODS-I	[122]
	1.8-µm Zorbax SBC8	
Neutral related *S*-oxidation compounds	3-µm Hypersil C18	[224]
2-Phenylmethyl-1-naphthol	3-µm Hypersil C18	[225]
p-Hydroxybenzoic acid, bumetanide, flurbiprofen	3-µm Hypersil C18	[40]
	3-µm Hypersil C18/SCX	[220]
Thomapyrin, containing acetaminophen, caffeine, and acetylsalicylic acid	3-µm GROM-SIL 100 ODS-0 AB	[132]
Antiviral drug suramin	5-µm Nucleosil 100 C18	[114]

Table 1 (*Continued*)

Compound	Stationary phase	References
Amino group-containing drugs: codeine phosphate, ephedrine hydrochroride, thebaine, berberine, hydrochloride, jatrorrizine hydrochloride, cocaine hydrochloride	3-μm Micra bare silica	[226]
Isradepin and by-products	3-μm Hypersil C18	[70]
Morphine alkaloids	5-μm Nucleosil 100 C18	[114]
Antiepileptic drugs: ethosuccinimide, primidon, CBZ-10,11 diol, CBZ-10,11-epoxid, phenytoin, carbamazepine (CBZ)	3-μm Spherisorb ODS-I	[160]
Macrocyclic lactone, S541 factor B from *Streptamyces* S541	3-μm CEC Hypersil	[227]
Sulfanilamide, sulfaflurazol, sulfadicramide	5-μm Nucleosil 100–5C8	[228]
Cardiac glycosides: digoxigenin, digoxin, digitoxigenin	1.5-μm NPS ODS II	[209]
Tetracyclines	Etched and modified C18 open-tubular column	[229,230]
2-Phenylethylamine derivatives: epinephrine, DOPA, 2-amino-3-hydroxy-3-phenyl-propanol, ephedrine	5-μm Nucleosil 5C8	[160]
Vitamin D2 and D3	3-μm GROM-SIL 100 ODS-0 AB	[217]
Doxorubicin	3-μm Luna C18	[231]
Multicomponent aminoglycoside antibiotic (teicoplanin, six components)	3-μm Hypersil C18	[232]
Thalidomide and hydroxylated metabolites	5-μm Aminopropyl coated with amylose and cellulose derivatives	[233]
Thalidomide and metabolites	5-μm LiChrospher 100 RP-18	[234]
Methylamphetamine and impurities	1.5-μm NPS ODS II	[235]
Related opiate compounds (morphine, hydromorphone, nalorphine, codeine, oxycodone, diacetylmorphine)	1.5-μm NPS ODS II	[236]
Theophylline, caffeine, sulfanilamide	3-μm C18	[113]

Theophylline, caffeine, aminophylline, theobromine, β-hydrocyethyltheophylline, phenylbutazone, hydro-chloro-thiazide, acetominophen	3-μm silica	[237]
Fluvoxamine and possible isomers	3-μm Spherisorb ODS-I	[238]
Hydroquinone and ethers	5-μm LiChrospher 100 RP-18	[239]
Serotonin and metabolites	Open-tubular, fused silica etched with NH_4HF_2, chemically bonded 4-cyano-4′-pentoxybiphenyl	[102]
Neurotransmittors (5-hydroxytryptamine, dopamine, nor-epinephrine, epinephrine, DOPA)	Open-tubular, chemically modified wall with macro-cyclic dioxopolyamine (dioxo[13]aneN_4)	[240]
Cardiac glycosides (digitoxigenine, digitoxigenine-bis-digitoxoside, digitoxigenine-mono-digitoxine)	1-μm C8	[241]
Clenbuterol, salbutamol, methadone	3-μm Hypersil MOS	[242]
Neostigmine, salbutamol, fenoterol	5-μm Nucleosil C18	[243]
Salbutamol, salmeterol	3-μm Spherisorb C18 3-μm Spherisorb C6/SCX 3-μm Spherisorb SCX	[129]
Carbovir, ranitidine, ondansetron, imipramine, amitripty-line, clomipramine	3.5-μm Symmetry Shield RP-8	[244]
Benzylamine, nortriptyline, diphenhydramine, terbutalin, procainamide	3-μm Hypersil C18	[245]
Benzylamine, nortriptyline, diphenhydramine, pro-cainamide	3-μm Hypersil silica 3-μm Hypersil BDS silica 3-μm Hypersil silica	[45]
Basic and acidic drugs (metoclopramide, timolol, pro-cain, ambroxol, antipyrine, naproxen)	3-μm Spherisorb ODS-I, Hypersil C18, Hypersil C8	[246]
Amphetamine, methamphetamine, procaine, cocaine, quinine, heroin, noscapine, phenobarbital, methaqua-lone, diazepam, testosterone, testosterone propio-nate, cannabinol, delta-9-THC, delta-9-THC acid-A	3-μm Hypersil C8	[66]

Synergic aspects of CEC include the independence of electroosmotic flow on channel size (within practical limits) and lack of backpressure. This allows for longer columns and/or sorbents with smaller pores/channels to favor a rapid mass transfer process, as in the case of smaller-diameter packing and cast-in polymeric rods. Higher efficiencies, improved resolution, and larger peak capacity have been reported. For example, Figure 1 (from Ref. 1) illustrates the superior performance of CEC versus CE and μHPLC for the analysis of a drug mixture [1]. Commonly, reduced plate heights as low as 1–1.5 have been observed. In addition, EOF's independence of particle diameter offers efficiencies that are essentially constant over a wide range of flow velocities. The result is that fast separations are possible (EOF velocities of 1 mm/s) without compromising the resolution of analytes. Finally, as a miniaturized technique, analysis of low sample volumes is possible, as well as increased mass sensitivity and less packing and solvent consumption. It is therefore evident that CEC would be particularly suited to a large number of practical applications for liquid-phase

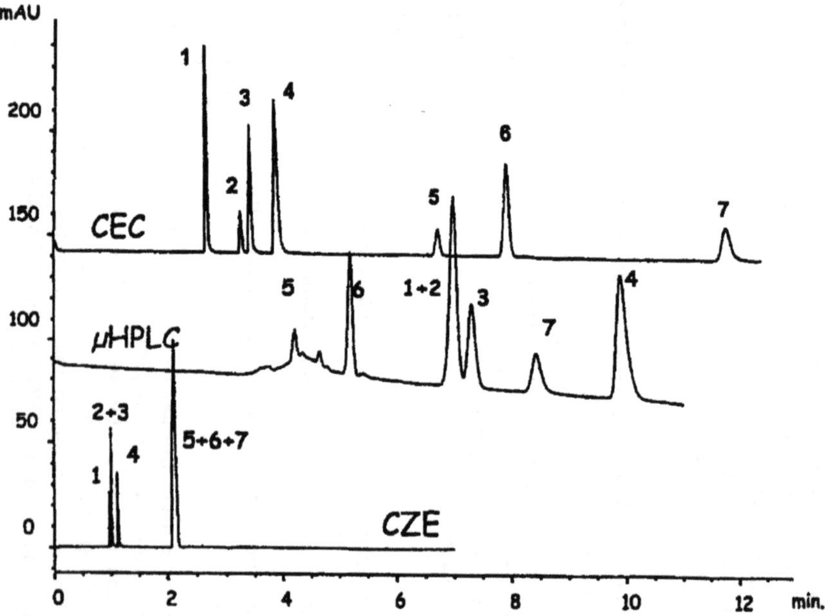

FIGURE 1 Comparison of separations of a drug mixture by CEC, HPLC, and CZE. Column, Spherisorb ODS-1, 3 μm, 250(335) × 0.1 mm. Voltage, 25 kV, pressure (HPLC), 200 bar. CZE, uncoated fused silica capillary 250(35) × 0.075 mm. (Reprinted from Ref. 1, with permission.)

separations. A number of excellent review articles detail assiduous efforts to improve the technique and find an ever-larger number of applications that take advantage of its attractive features [2–9].

Additional features that make CEC very attractive for a large number of applications include the ability to couple CEC to mass spectrometry (MS) or nuclear magnetic resonance (NMR), the possibility to preconcentrate samples at the head of the column, and the advantage of being able to stop the flow with minimal effect on the peak broadening. For a large variety of compounds that lack chromophores or for which derivatization is not desired, MS or NMR offers not only a detection option but also unequivocal characterization or structural elucidation. Due to compatibility of CEC flow rates with those required by a miniaturized electrospray ionization (ESI) source, there have been numerous reports of successful coupling of ESI-MS with CEC [7]. Developments of a few groups in the field of NMR miniaturization aim to use NMR as a detection method for CE and CEC. Their results indicate that in the near future, chemical, dynamic, and spatial structural determination of properties of organic compounds separated by CEC might be available [10]. Sample loadability onto a CEC column can be significantly increased by employing on-column focusing. As analytes dissolved in a mobile phase of lower elution strength (low percentage of organic buffer) are introduced into the column, they preconcentrate at the head of the column and elute only when the mobile phase contains a higher percentage of organic buffer [11]. Finally, due to the fact that electroosmotic-driven flow has no pressure drop associated with it, turning the voltage off can stop the migration of analytes and minimal effect on peak broadening and resolution is observed [12]. This so-called stop-flow or peak-parking feature is especially of use with detection systems that provide slower, although more complex, data acquisition, as is the case with an ion-trap mass spectrometer (Figure 2, from Ref. 11).

In practice, however, due mainly to a number of technical difficulties, most of which are attributable to lack of column ruggedness, it has yet to be shown that the theoretically predicted potential of CEC will fulfill its promise and result in a widespread routine technique. In the past 5–8 years, there has been a sustained effort in the separations community to understand theoretical aspects, based on which new developments arose in the field of column technology [9,13,14], application [4,5], and detection [7]. A number of these observations are summarized here, with an emphasis on their bearing on extending the application range of CEC.

Originally, CEC employed the same type of columns as μHPLC, namely, conventional HPLC silica-based particles packed into fused silica capillaries. Porous plugs attached to the fused silica walls—frits—hold the sorbent material within the confines of the capillary tubing. Just as in μHPLC, frit fragility significantly decreases the lifetime of the column. Moreover, nonspecific interactions

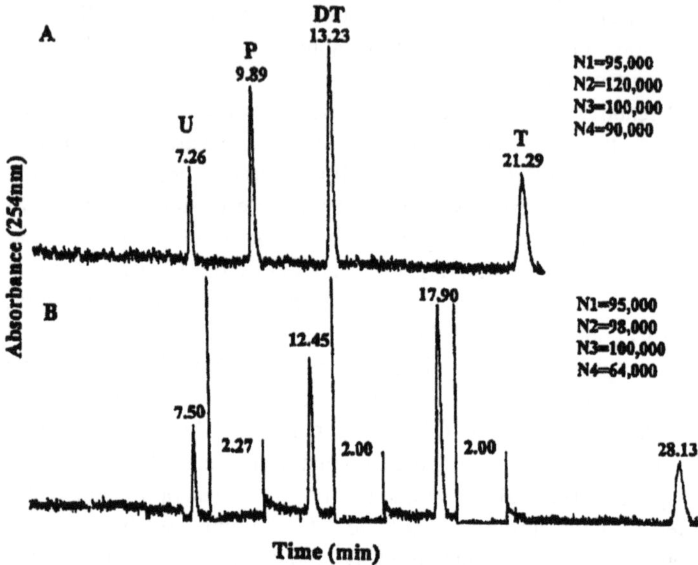

FIGURE 2 Comparison of retention times and separation efficiencies when the CEC system is operated in two different modes: (A) continuous mode; (B) stop-flow mode. (Reprinted from Ref. 11, with permission.)

occurring at the surface of the frits, differential flow at the interface between frits and the chromatographic bed, and, most important, the lack of reproducible frit fabrication procedures have had a deleterious effect on column-to-column reproducibility.

These circumstances have prompted the exploit of open-tubular (OT) type columns [15]. In such columns the capillary wall is activated and subsequently treated to generate a polymeric coating, molecular monolayer, or porous layer of stationary phase. Tan and Remcho developed the first OT CEC polymer coatings in 25-μm-i.d. capillaries. The polymethacrylate coatings provided a hydrophobic surface which promoted the separation of benzoates with efficiencies greater than 250,000 plates/m [16]. Later, the same authors extended the technology to the synthesis of a porous layer of molecularly imprint polymer for the separation of D and L isomers of dansyl phenylalanine [17] (Figure 3, from Ref. 17). Matyska and Pesek demonstrated that significant increase in the surface area could be achieved by chemically etching the surface of the capillary prior to attachment of the porous polymeric layer [18]. To date, several types of organic modifiers, including octadecyl, diol, cholestryl, and chiral selectors, have been attached to etched capillary surfaces and tested in the OT CEC format

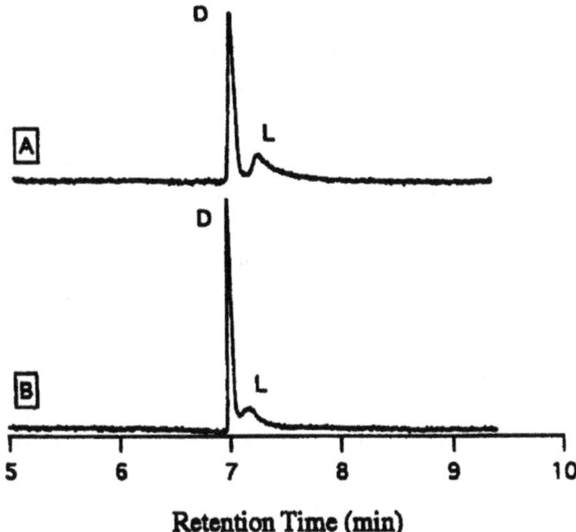

FIGURE 3 Effect of mobile-phase composition on the OT-LC separation of a mixture of D- and L-dansylphenylalanines. Capillary: L-dansylphenylalanine MIP, 25 mm × 100 cm (total length), 85 cm (effective length). Injection: 80 mbar, 3 s. Separation: pressure-driven flow at 350 mbar. Detection: on-column at 280 nm. Mobile phase: (A) 100% acetonitrile; (B) 99.5% acetonitrile/0.5% acetic acid. (Reprinted from Ref. 17, with permission.)

[19]. The main advantage of OT CEC is that separation efficiency can be doubled using this type of column. The trade-off is that the OT columns can easily be overloaded and therefore require a sensitive detection system. The small diameter of these columns precludes the use of UV detection, and fluorometric detection or mass spectrometry (MS) needs to be used. The use of fused silica capillaries with a "bubble cell" at the detection window has been reported as an alternative to employ UV detection. This features limit, to a certain extent, the range of practical applications of OT CEC.

Meanwhile, a further technological advance has been achieved with the design of new types of fritless columns. Specifically, continuous beds in the form of immobilized packed beds and monolithic porous polymer columns have been developed [13]. One of the immobilization designs proposed by Remcho et al. involved the entrapment of a conventional packed bed in a highly cross-linked porous silica or organic-based matrix that holds the sorbent in place, rendering the frits superfluous [20–22]. The latter approach involves in-situ polymerization of a monomeric mixture that generates a solid porous polymer in which individual components and/or subsequent derivatization dictate the retention mechanism. Both silica-based [23] and organic-based [24,25] monolithic columns have been prepared. The continuous-bed designs allow for column-length customization and seem more suited to analyzing real samples, as sections of the capillary contaminated or clogged by "dirty samples" could be cut off. Applications of monolithic columns range from polycyclic aromatic hydrocarbons (PAHs) [20–22,26,27] to benzene derivatives [24,25], to nonsteroidal antiinflamatory drugs [22] (Figure 4, from Ref. 22), to peptides [28,29]. The real potential of these columns has yet to be demonstrated.

The high retention, high sample capacity, and, in particular, the wide choice of selectivity given by conventional silica-based HPLC sorbents has made packed-column CEC the separation format of choice for most CEC applications. One critical element of sorbents employed in all types of CEC columns is the requirement for ionizable groups that are able to generate EOF as high voltages are applied at both ends of the columns. Silanol groups are an intrinsic part of all silica-based materials; a sufficient number of residual silanol groups are also present in surface-derivatized silica sorbents. At pHs higher than 2, residual free

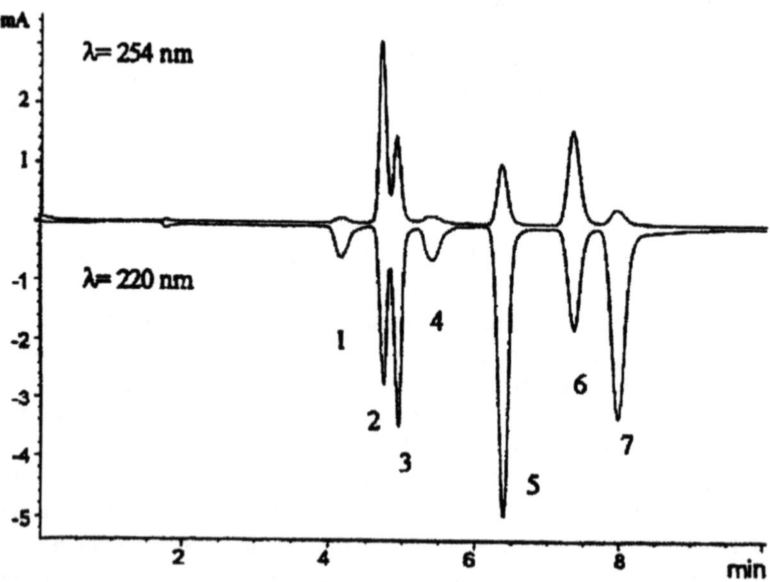

FIGURE 4 Nonsteroidal anti-inflammatory drugs, acetaminophen, and caffeine separated by CEC on a 75-mm-i.d. column packed with 5-mm ODS Nucleosil particles immobilized within a polymer matrix. Mobile phase: 70% acetonitrile/30% acetate, 10 mM (pH 3.0). UV detection at 254 and 220 nm; 20 kV applied, effective length 17 cm, total length 26 cm. Analytes: (1) unknown impurity; (2) acetaminophen; (3) caffeine; (4) aspirin; (5) naproxen; (6) flurbiprofen; (7) ibuprofen. (Reprinted from Ref. 22, with permission.)

silanol groups are ionized and therefore able to support EOF. Velocities of 1 mm/s can be achieved with silicate-based materials as mobile phases of pH 8 and 30 kV are employed. There is, however, a practical upper pH limit for silicate materials which is set by stability of the silica material itself. For applications which require operation at pH below 2 or greater than 8, a different chemistry is needed.

As CEC emerged, neutral molecules were extensively analyzed to illustrate its high efficiency and high speed advantage, or to provide a reference for column evaluation, especially as new column designs or sorbents were proposed. In the process, extensive theoretical treatment and experimental work enabled scientists to better understand and exploit CEC. The main goals, staggering efficiencies and short separation times, have been achieved. For example, Dittmann et al. [30] and Dadoo et al. [31] reported reduced plate heights as low as 1.2, for separation on PAHs using columns packed with 2.5-μm CEC Hypersil C18 and 1.5-μm Micra Scientific C18 nonporous silica, respectively. Later, Dadoo et al. exemplified the high speed advantage of CEC by separating five PAHs in less than 5 s (linear velocity of 20 mm/s) as 28 kV was applied across a 10-cm-long column packed with 1.5-μm Micra Scientific nonporous C18 silica beads. High efficiencies similar to those obtained in packed beds have also been reported for monolithic columns. Peters et al. prepared polymers that delivered flow velocities of 2 mm/s and efficiencies of 110,000 plates/m for the unretained compound [24]. Reduced plate heights for retained compounds varied from 1.2 to 1.9 in a column packed with 5-μm Nucleosil C18 particles subsequently immobilized by a silicate matrix [20].

Macroporous packing materials, particles of 5–10 μm with pore diameters ranging from 500 to 4000 Å, offer an alternative pathway to improving efficiency in CEC. The analytes can sample two distinct regions of pore space: the interparticle region (between adjacent particles of the packing material) and the intraparticle region (inside the particles). Remcho and co-workers showed, both in theory and in practice [32–34], that by adjusting the ionic strength of the buffer the "through-pore" or perfusive regime can be achieved in CEC and increased separation efficiencies can be obtained. Reduced plate height of 0.35–0.4 were demonstrated [34,35].

Simultaneously, instrumental development was addressed and stepwise or continuous mobile-phase gradient systems emerged; voltage and temperature programming was investigated, as more means to adjust selectivity, improve peak capacity, and further reduce analysis time [8]. Among several designs proposed for mobile-phase gradient generation, the one most widely used was that proposed Huber et al. [36] and Lister et al. [37]. Their method employs two μHPLC gradient pumps that deliver the mobile phase through a flow injection interface. They demonstrated increased resolution and significant reduction of analysis times for phenylthiohydantoin (PTH) amino acids and PAHs, respectively. Ericson et al. utilized the flow injection interface in conventional gradient elution (analytes migrate from the inlet of the column toward the outlet) and in counterflow mode (analytes injected at the outlet; their electrophoretic mobility is larger than the electroosmotic flow) to separate charged proteins [38]. Taylor et al. resolved a mixture of 9 corticosteroids and 5 thiazide diuretics using UV and ESI-MS, respectively [39] (Figure 5, from Ref. 39).

It was soon realized that in order for CEC to be recognized as more than just another iteration of HPLC and CE, additional benefits needed to be discovered. Today it is generally accepted that the future of CEC lies in its potential for tackling unique applications, such as complex mixtures of neutral, ionic, and charged molecules in real-life samples. This becomes possible as the advantage of large selectivity and high peak capacity is fully exploited. However, the analysis of charged compounds is a difficult challenge, as nonspecific interactions of basic compounds with the silica-based stationary phase or retarded migration of acidic molecules has been observed. This challenge turned into an excellent opportunity, as CEC became a prolific field for exploration and development of novel types of sorbents.

When negatively charged, the velocity of acidic compounds is smaller than that of EOF, since their electrophoretic mobility opposes the electroosmotic mobility. In such cases, electrokinetic injection must be avoided since very little or no sample is introduced into the column. Moreover, in instances when the magnitude of electrophoretic mobility is larger than that of the electroosmotic mobility, the analyte is not transported along the column. The obvious solution is to operate in ion-suppressed mode, i.e., at low pH, where acidic compounds are not ionized. The drawback is extensive analysis time, since the majority of silanol groups are also not ionized and EOF is significantly smaller.

In accordance with this prediction, Euerby et al. observed a slow EOF at pH 2.5 while separating a mixture of bumetanide, flurbiprophen, and *p*-hydroxybenzoic acid using a 3-μm CEC Hypersil C18-packed capillary [40]. The running time was 13 min and efficiencies varied from 97,000 to 174,000.

Ye et al. compared the performance of C18 and strong anion-exchange (SAX) silica-based packed columns for the separation of substituted benzoic acids [41]. The positively charged surface of the SAX packing produces a reverse EOF (direction from cathode to anode), and electrokinetic injection could now be employed. Long analysis times (~14 min) were noted at pH 2.2 with a C18-packed column. By contrast, at the same operating pH, a column packed with strong anion-exchange material resolved the analytes in 3 min with excellent efficiencies (220,000–310,000 plates/m). As expected, chromatographic retention on the two phases differed: in the case of C18 medium it was a function of hydrophobicity, while for the SAX medium the analytes elute in order of increasing pK_a. As expected, capacity factors were strongly dependent on the ionic strength of the buffer. Complete resolution of positional isomers of bromobenzoic acid was also achieved (Figure 6).

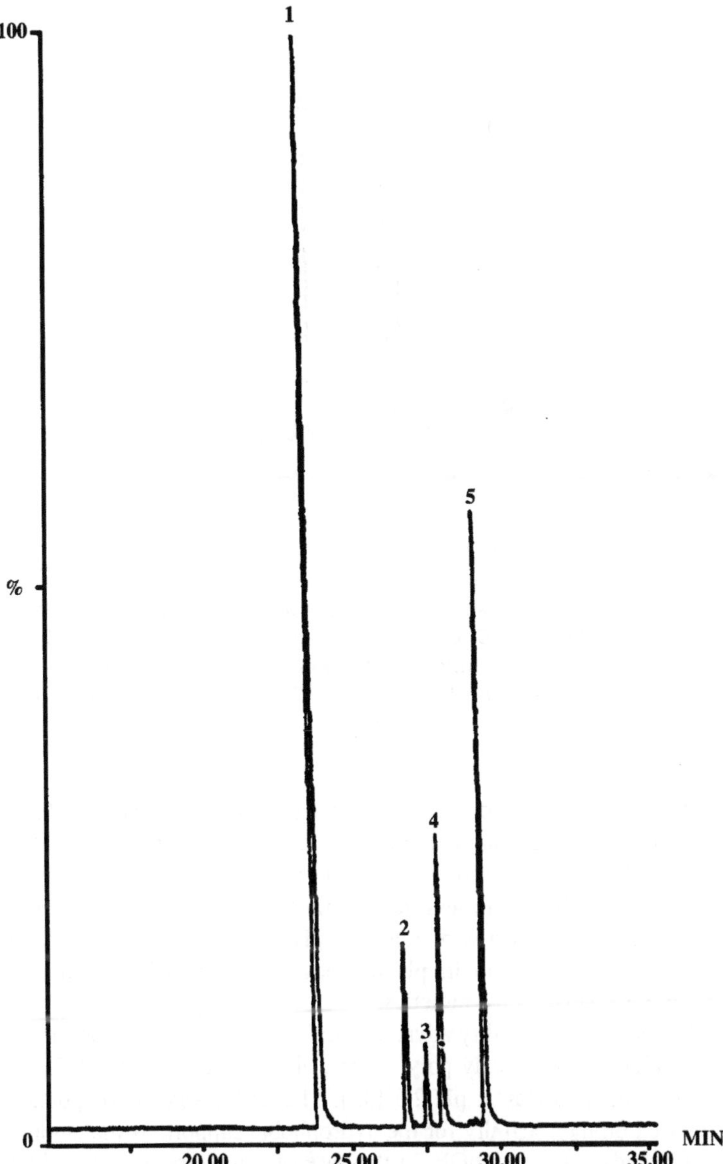

FIGURE 5 CEC-ESI-MS chromatogram recorded in selected ion monitoring mode (TIC) of a mixture of thiazide diuretics (100 μg/mL) using a step gradient. Voltage: 30 kV, HPLC injection volume: 5 μL, flow rate: 10 μL/min for 3 min, then 100 μL/min. Gradient: initial, ammonium acetate 5 mM in acetonitrile water (1/1) held for 3 min, then ramped to 80% acetonitrile in 0.1 min, maintained for 35 min. Column: Hypersil ODS, 3 μm, 46 cm fully packed. (1) hydroflumethiazide, (2) methylclothiazide, (3) metolazone, (4) epitiside, (5) bendrofluazide. (Reprinted from Ref. 39, with permission.)

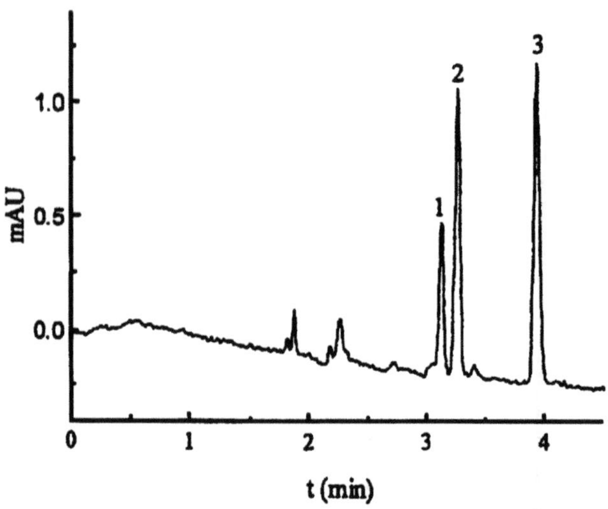

FIGURE 6 Chromatogram for separation of positional isomers of bromobenzoic acid in strong anion-exchange CEC. Experimental conditions: mobile phase, 50% acetonitrile in 10 mM phosphate buffer pH 2.2. Peaks: (1) *p*-bromobenzoic acid (pK_a 4.00); (2) *m*-bromobenzoic acid (pK_a 3.81); (3) *o*-bromobenzoic acid (pK_a 2.85). (Reprinted from Ref. 41, with permission.)

A new type of mixed-mode stationary phase which incorporates both strong anion-exchange (triehtylammonium methylstyrene) moieties and C18 (octadecylacrylate) ligands was used by Scherer et al. to separate acetylsalicylic acid and three of its metabolites in less than 2 min [42]. A rapid reversed EOF (1.8 mm/s) was generated over a wide range of pH, from 3 to 9. It was found that EOF velocity shows variation with pH, the magnitude of which is a function of the content of SAX versus C18 moieties.

Liu et al. used the pH stability advantage offered by the organic polymer-based stationary phases, specifically poly(styrene-divinylbenzene) (PSDVB), to separate weakly acidic analytes at pH 2.5 [43]. The undissociated compounds are separated based on their affinity for the hydrophobic support, as the anionic (sulfonic acid) ligands generated EOF. In the case of a mixture of linear alkyl chain carboxylic acids, the SCX column delivered significantly better separation efficiency, while severe tailing and longer retention was obtained with the non-derivatized PSDVB sorbent.

The progress of CEC has been closely observed in the pharmaceutical community. Since most drugs are basic compounds, a significant effort has been invested in understanding the behavior of basic analytes in CEC. It is believed that the difficulties associated with irreproducibility and severe peak tailing stem from mixed-mode interactions between the analytes and alkyl ligands (hydrophobic interaction) and underivatized silanol groups (ion exchange and hydrogen bonding). End-capped sorbents are not a solution, due to their inability to generate sufficient EOF. Consequently, the separation of basic molecules in the ion-suppressed mode requires pH higher than 8 which is detrimental to silica particles.

Wei et al. studied the separation of amines (aniline, ephedrine, codeine, cocaine, thebaine) and quaternary ammonium compounds (berberine, jatrorrhizine) on a bare silica stationary phase [44]. A thorough study of the separation mechanism revealed a complex multifunctional mechanism. Contributions from differential electrophoretic migration were superimposed on hydrophobic, cation-exchange, and normal-phase interactions. Retention was highly dependent on the pH (optimal pH 8.3), ionic strength, and the amount of organic modifier. As the content of acetonitrile exceeded 80%, retention was consistent with a normal-phase mechanism.

The chromatographic performance and separation mechanism seems to be sensitive not only to the type of stationary phase but also to the purity of the packing material, in particular to the metal content. At low pH, McKeown et al. observed differences in ion-exchange activity and EOF which affected the resolution and elution order between the columns [45]. The study concluded that better peak shapes could be achieved by employing silica packings with lowest metal content. However, the efficiency was significantly lower than in the case of the less pure phases.

Another way to improve peak tailing in normal or reversed-phase CEC is to add a silanol-masking agent, an amine, such as triethylamine, triethanolamine, or hexylamine, which scavenges the free silanol group. Lurie et al. used a pH 2.5 eluent, 3-μm C 8 packed column, hexylamine, and a step gradient in an attempt to reduce tailing and improve resolution of 17 drugs with polarities ranging from strongly acidic to weakly acidic, neutral, weakly basic, and strongly basic [46] (Figure 7, from Ref. 46). High-organic-content mobile phases were used to counteract the smaller EOF characteristic of lower pH. The importance of using the second-generation silica-based stationary phases, the new type of silica packing with low impurity content, was also stated by Dittman et al. [1]. They have shown a significant improvement in peak shape and efficiency, which led to complete separation of the seven basic solutes using a hexylamine-containing eluent. The significant differences observed in the elution order of the CEC, CE, and μHPLC separations speak once more for the orthogonality of the techniques.

An interesting (if not fully understood) phenomenon has been observed by Smith et al. as strong cation-exchange (SCX) phases were used to separate tricyclic antidepressants [47]. A certain focusing effect, which is not reproducible, produced peaks of staggering efficiencies, millions of plates per meter, the

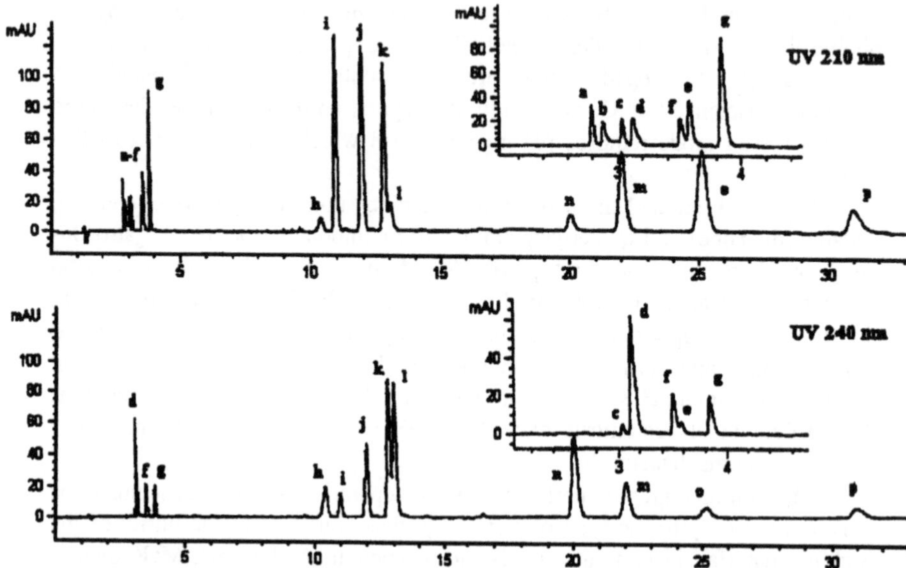

FIGURE 7 CEC step gradient of basic, neutral, and acidic compounds. Initial conditions (for first minute): acetonitrile/25 mM phosphate buffer pH 2.5 (60/40) with 2-mL/min hexylamine. Final conditions: acetonitrile/25 mM phosphate buffer pH 2.5 (75/25) with 2-mL/min hexylamine. A voltage of 25 kV and a temperature of 20°C was used. Analytes: (i) phenobarbital, (j) diazepam, (l) testosterone, (m) cannabinol, (n) testosterone propionate, (o) Δ9-tetrahydrocannabinol, and (p) Δ9-tetrahydrocannabinolic acid. (Reprinted from Ref. 46, with permission.)

largest efficiency numbers reported to date in CEC. Euerby et al. [40] and, more recently, Moffatt et al. [48] also observed such sharp peaks that are reminiscent of capillary GC.

Svec et al. also made use of the stability of methacrylate matrices to a very wide pH range and synthesized strong cation-exchange monoliths with sulfonic groups [49]. The columns were operated at pH 2.4 and 12, as acidic and, respectively, basic eluents were separated. Figure 8 (from Ref. 49) demonstrates the versatility of one column that, given different conditions, can perform fast and efficient separations on a wide pH range.

The ultimate challenge for any chromatographic technique is to separate neutral, ionic, acidic, and basic compounds found in real samples in the same run. Toward this goal, a series of papers published simultaneous separations of acidic, neutral, and basic compounds. Klamfl et al. used a mixed-mode (C6/SAX) silica-based packing to combine reverse EOF, hydrophobic interactions, as well as electrophoretic migration for the separation of a lab-made sample

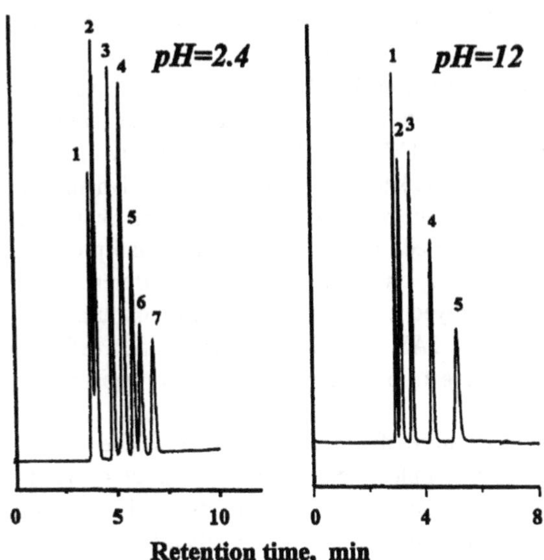

FIGURE 8 Electrochromatographic separation of aromatic acids (a) and anilines (b) on monolithic capillary columns [58]. Conditions: butyl methacrylateethylene dimethacrylate stationary phase with 0.3 wt% 2-acrylamido-2-methyl-1-propane-sulfonic acid; pore size, 750 nm; UV detection at 215 nm; voltage, 25 kV; pressure in vials, 0.2 MPa; injection, 5 kV for 3 s. (a) Capillary column, 30 cm (25 cm active length) × 100 μm i.d.; mobile phase, acetonitrile-5 mmol/L phosphate buffer, pH 2.4 (60/40, v/v). Peaks: 3,5-dihydroxybenzoic acid (1), 4-hydroxybenzoic acid (2), benzoic acid (3), 2-toluic acid (4), 4-chlorobenzoic acid (5), 4 bromobenzoic acid (6), 4-iodobenzoic acid (7). (b) Capillary column, 28 cm (25 cm active length) × 100 μm i.d.; mobile phase, acetonitrile/10-mmol/L NaOH, pH 12 (80/20, v/v). Peaks: 2-aminopyridine (1), 1,3,5-collidine (2), aniline (3), *N*-ethylaniline (4), *N*-butylaniline (5). (Reprinted from Ref. 49, with permission.)

containing nitrate and iodide ions, benzoic acid, salicylic acid, pyridine, aniline, 4 methoxy phenol, phenol, 4 chlorophenol, 4 nitrotoluene, and toluene [50]. Separation efficiencies varied from 75,000 plates/m for salicylic acid to 197,000 plates/m for phenol. The propensity of ion exchange versus hydrophobic interactions was tuned by varying the concentration of the competing ion versus the amount of organic modifier. Complex pH effects have been observed which result from pH influence on the dissociation rates of analytes reflected in both the variation of their effective mobilities (observed mobilities minus EOF) and the strength of ionic interactions with the stationary phase.

A simple approach to preparation of mixed-mode chromatographic beds was adopted by Zhang et al. [51]. They packed a C18/SCX column by mechanically mixing C18 particles with SCX particles. Separation of acidic, neutral, and basic compounds was achieved. It was also pointed out the advantage of facile adjustment of C18 to SCX content in the packed bed, which allows manipulation of hydrophobic and ion-exchange interactions as dictated by the sample.

Ye et al. found a parallel between CE and CEC, in dynamically modified capillary separations [52]. A column packed with bare silica was dynamically modified with cetyltrimethylaammoinium bromide (CTAB), a long-chain quaternary ammonium salt that was adsorbed onto the silica surface and produced a hydrophobic layer. The same additive, but with a completely different effect, was used Wu et al. [53]. They employed a methacrylate-based monolithic column and added in the mobile phase CTAB or sodium dodecyl sulfate (SDS) to generated EOF. Ten analytes, ranging from acidic to basic, were separated in both cases.

Various research groups have successfully employed CEC in the analysis of compounds in real-life biological samples. For example, the group of Sandra separated triglycerides, free fatty acids, and fatty acid esters [54,55] in vegetable oils and margarine (Figure 9). They employed 3-μm C18 silica-based packing and a nonaqueous eluent obtained by mixing acetonitrile, isopropanol, and hexane to which was added ammonium acetate. The free fatty acids, as well as methyl and phenacyl ester derivatives, were separated by CEC. It has to be pointed out that the more efficient CEC procedure afforded baseline separations of some components that co-elute in HPLC. This was the first report on analysis of biosamples in which CEC performed better than other separation techniques. Saeed et al. evaluated CEC for the separation of mono- and dihydroxy phenols in tobacco smoke [56]. They concluded that CEC is much faster that the GC or HPLC methods.

As early as 1997, Taylor et al. [57] demonstrated the gradient separation of corticosteroids in extracts of equine urine and plasma (Figure 10). The sample were purified using solid-phase extraction and automated dialysis, respectively. A reproducibility study revealed that peak broadening occurred only after the analysis of 200 urine extracts. Later, Stead et al. observed that on-line sample concentration could be easily achieved, as longer injection times had minimal influence on peak shape [58]. They demonstrated that the CEC separation of steroids in plasma was superior to HPLC. Several other groups also reported successful CEC separations of drugs and major metabolites in extracts of urine and plasma [59–62].

Due to the inherent complexity of biological samples and to the lack of chromophores in the case of various biomolecules, mass spectrometry would be the ultimate technique to complement CEC in analysis of real-life biological mixtures. In the past 10 years, this aspect has been recognized and numerous groups have worked on developing suitable interfaces for coupling CEC to MS. Mixtures of neutral isomeric compounds derived from the in-vitro reaction of

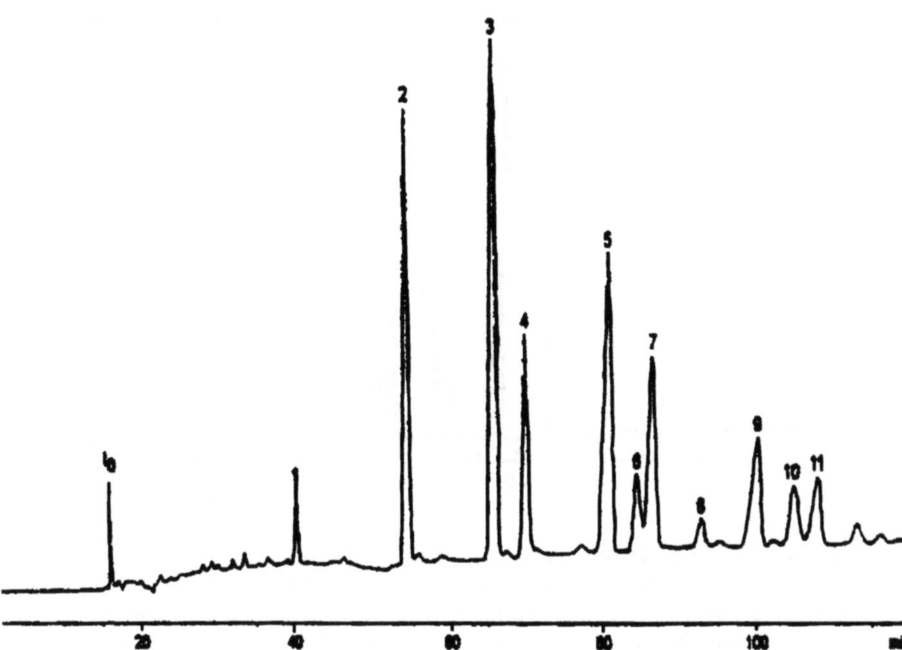

FIGURE 9 Separation of triglycerides of argan oil (*Argania spinosa*). Column: 40 cm $L \times 0.1$ mm i.d., FSOT, Hypersil C18 3 μm; mobile phase: 50 mM NH_4Ac in acetonitrile/isopropanol/*n*-hexane (57/38/5). Detection: UV at 200 nm. Temperature: 20°C. Voltage: 30 kV. Injection: 10 kV during 3 s. Peaks: 1 = LLnLn, 2 = LLL, 3 = OLL, 4 = PLL, 5 = OLO, 6 = SLL, 7 = PLO, 8 = PLP, 9 = OOO, 10 = SLO, and 11 = POO. (Reprinted from Ref. 55, with permission.)

two carcinogenic PAHs with calf thymus deoxyribonucleic acid (DNA) were analyzed on a capillary packed with 3-μm C18 particles [62]. The separation in CEC mode was faster and more efficient than in the HPLC mode. In a later study, a dilute DNA adduct mixture (10^{-6} M) were detected by employing on-column focusing and coupling the capillary to an ESI-MS [63,64] (Figure 11, from Ref. 63). Que et al. demonstrated the separation and identification of various bile acid mixtures (compounds that lack chromophores) using CEC coupled to electrospray ion-trap mass spectrometry [65]. High column efficiencies (610,000 plates/m) and very low detection limits (40 fmol) were achieved.

An excellent demonstration of the concept of "niche" application in CEC was offered by Lurie et al. in a study of CEC feasibility for analysis of seized drugs [66]. In the past, standard separation methods such as GC, GC/MS, HPLC, and CE have been employed in the analysis of cannabinoids, weakly acidic

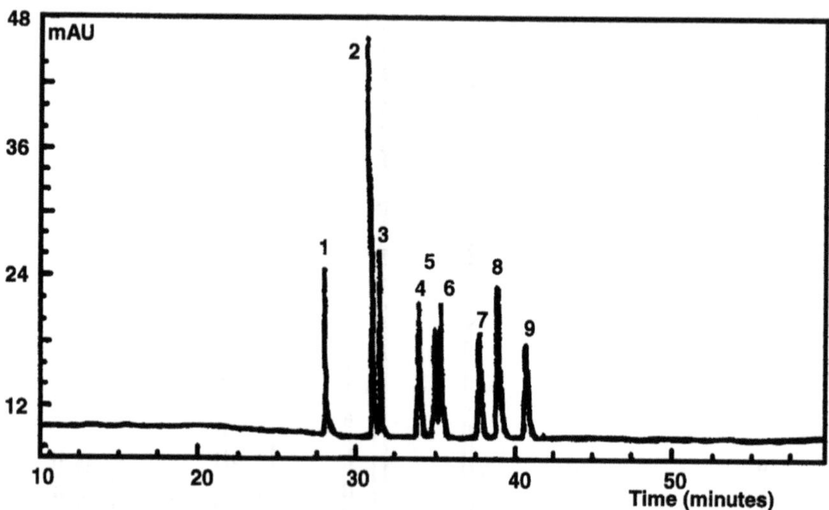

FIGURE 10 CEC-UV chromatogram (240 nm) of a mixture of 10 corticosteroids (100 μg/mL) using a linear gradient elution program. Voltage = 30 kV, HPLC injection volume = 10 μL, flow-rate = 10 μL/min for 3 min, then decreased to 100 μL/min. Gradient program = initial: ammonium acetate, 5 mM, in acetonitrile/water (17/83), held for 3 min, then ramped to 38% acetonitrile at 15 min and maintained to end of run. Column = Hypersil ODS, 3 μm, 42 cm total length, 30 cm packed length, 30.1 cm to window, 1 = triamcinolone, 2 = hydrocortisone and prednisolone co-eluting, 3 = cortisone, 4 = methylprednisolone, 5 = betamethasone, 6 = dexamethasone, 7 = adrenosterone, 8 = fluocortolone, 9 = triamcinolone acetonide. (Reprinted from Ref. 57, with permission.)

compounds that are major constituents of marijuana, hashish, and hash oil. None of these methods was able to meet the combined requirements for complete sample profiling (small amount of sample, satisfactory resolution, no thermal degradation, no sample derivatization). Using a 3-μm Hypersil C18 stationary phase at pH 2.57, complete separation of 7 standard cannabinoids was achieved in 26 min. Although this is a fairly long analysis time, due to the small dependence on efficiency of flow velocity, even the later-eluting compounds showed narrow peak shapes, and were baseline resolved. More important, as shown in Figure 12, the standard cannabinoids could be easily identified in the hashish and marijuana extracts.

In CEC, reproducibility has been the Achilles' heel. Whether it stems from irreproducible column manufacturing procedures, short column lifetime, instrument limitations, or sample injection and formation of air bubbles, this problem has prevented the rapid advance of CEC as a routine technique. The

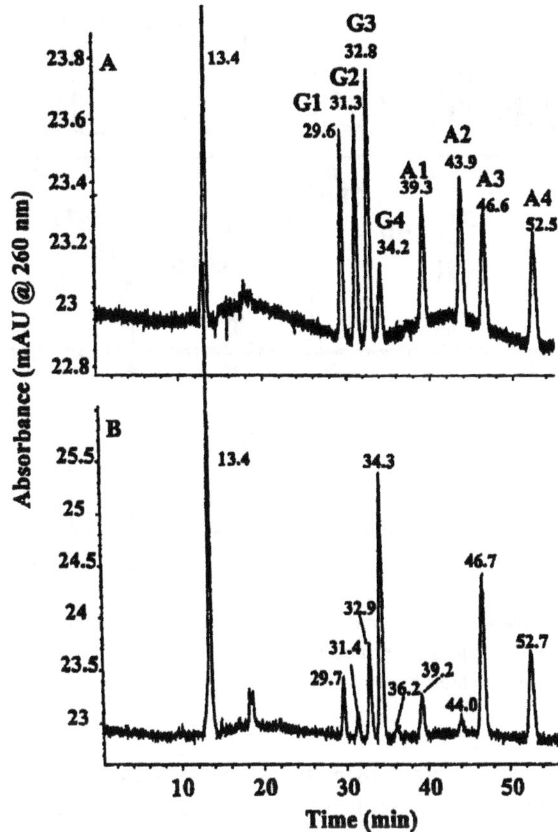

FIGURE 11 Separation of an adduct mixture formed from the in-vitro reactions of *anti*-5,6-dimethylchrysene 1,2-dihydrodiol 3,4-epoxide with calf thymus DNA with a ternary mobile phase consisting of 41% MeOH, 16% CAN, and 6 mM NH4OAc, (A) standards, (B) adducts from DNA reacted in vitro. (Reproduced from Am. Lab., 1999; 30: 15–29. Reprinted from Ref. 63, with permission.)

reproducibility issue is addressed in papers that explore theoretical and practical aspects of CEC in various depths. In a study published in 1996, Dulay et al. emphasized that column inhomogeneity and air bubble formation are major drawbacks in obtaining reproducible separations [67]. On a column packed with a mixture of C18 (major component) and bare silica particles (added to stabilize the bed and thereby reduced air bubble formation), they performed 150 consecutive runs over a few weeks. The reported RSDs for retention time and peak heights were 4% and 5%, respectively. Lelievre et al. used a home-made CEC

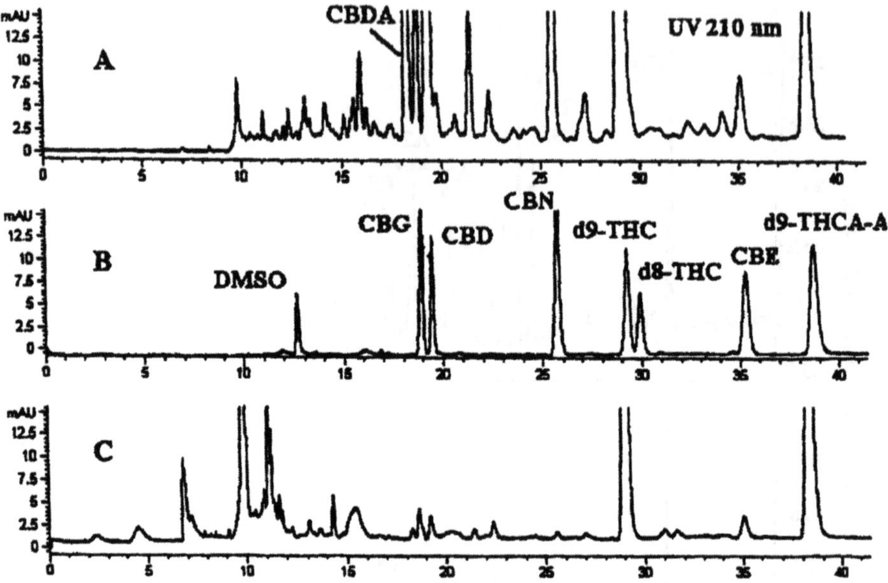

FIGURE 12 CEC of (A) concentrated hashish extract, (B) standard mixture of cannabinoids, and (C) concentrated marijuana extract. Conditions identical to Table 2. (Reprinted from Ref. 66, with permission.)

system and packed columns with 3-μm C18 particles and measured relative standard deviation (RSD) values of 3.5% for the EOF and 3.6% for the capacity factor of the most retained compound for run-to-run analysis [68]. For day-to-day analyses, four replicates, they recorded 4.1% RSD in t_0 and 6.3% RSD in the capacity factor of the most-retained compound. Robson et al. packed the capillary columns using supercritical carbon dioxide [69]. Retention times varied by less than 0.2% while peak area and peak height had 2.7% and 2.4% RSD, respectively. Over 200 runs were performed and the retention times varied by less than 2% RSD over 2 months.

While run-to-run reproducibility seems to approach values acceptable for routine analysis, batch-to-batch reproducibility has experienced large variation. Yamamoto et al. prepared three columns packed with 3-μm Hypersil C18 [70]. They studied the variation of retention factors for seven compounds and observed standard deviations of 9%. Walhagen et al. have published a more recent study dedicated entirely to the lifetime and reproducibility issues in CEC [71]. An HP^{3D}CE instrument produced by Agilent Technologies was used for these experiments. The capillary columns, supplied by the same company, were packed with 3-μm Hypersil C8 or C18. The columns were preconditioned according to the procedure suggested by the manufacturer. A sample containing uracil and three alkylbenzene was used as a test mixture for column evaluation. The RSDs for the retention factors of the retained analytes were 4% (four columns tested) for the C8 packing and 6% for the C18 packing. Efficiencies were better than 100,000 plates/m for the C8 columns and 125,000 plates/m for the C18 columns. The authors stated that, depending on the type of eluent and nature of analytes tested in these columns, 500 to 1000 repetitive runs could be obtained.

Indeed, the nature of the sample seems to be a critical parameter in the reproducibility equation. Short-term and long-term reproducibility of retention times and peak areas for CEC separation of cannabinoids have been measured by Lurie et al. [66]. The RSDs (seven replicates) for retention times are around 0.5% for short-term and vary from 1.49% to 2.03% for the long-term study, while peak areas showed up to 39% variation. Significantly improved reproducibility values were obtained when one compound, cannabinol, was used as a reference. The RSDs for relative peak areas varied from 2.3% to 3.95% for short-term and 1.65% to 8.1% for the long-term study.

In the case of continuous-bed or monolithic columns, due to the absence of frits there are virtually no bubble formation and bed rearrangement issues; thus it is expected that column-to-column reproducibility would be greatly improved. Asiae et al. showed that for sintered and subsequently reoctadecylated octadecylsilyl (ODS) packing, EOF varied by 2–3% within 30 days after 300 chromatographic runs [72]. The lifetime of the columns exceeded 300 runs. The packed beds immobilized within a methacrylate matrix showed less than 5% RSD (20 replicates) for capacity factors of retained compounds, for selected matrix compositions. Reproducibility issues in the case of porous polymer monoliths were investigated by Ngola et al. [26]. They noticed that their polymeric materials withstand exposure to various mobile phases and can be used for about 40 runs. Extended use is limited by the increased fragility of Teflon-coated capillaries rather than the stability of the polymer. Teflon coating allowed UV initiation of the polymerization reaction, which rendered a more uniform stationary phase. The authors reported run-to-run reproducibility of 0.6% RSD, and 10% RSD for batch-to-batch reproducibility.

2. INORGANIC ANIONS AND CATIONS

The first report on the use of ion-exchange chromatography in CEC mode was published by Li et al. [73]. They studied the feasibility of CEC for analysis of I^- and IO_3^-, as common constituents of nuclear wastes at Hanford, Washington, and other U.S. Department of Energy sites, and ReO_4^-, a surrogate for TcO_4^-. The CEC column was packed with silica-based strong anion exchanger (5-μm SB

Nucleosil). The ions were easily separated by CE and IE-CEC using pH 2.6 buffers. The separations were based on distinct mechanisms, a fact demonstrated by the differences in the elution order (Figure 13, from Ref. 73). The advantage of higher loading capacities was immediately apparent in the case of CEC. Unexpected, however, was the observation that CEC efficiencies could be higher than in the case of CE. A possible explanation would be the occurrence of a chromatofocusing effect in which analytes were retained in a fine band at the head of the column.

A comparison study of gradient liquid chromatography (LC) and pressurized IE CEC (EOF is amended by pressure-driven flow) was performed using organic polymer-based anion-exchange resins (5-μm TSK IC-Anion-SW). Kitagawa separated lanthanide cations by isocratic pressurized IE-CEC [74]. He obtained significant reduction in the analysis time of the light elements. Taking advantage of the electrophoretic mobility of cations, and combined electrophoretic mobility and ion-exchange interaction with the stationary phase in the case of anions, simultaneous separation of anions and cations was demonstrated.

The spectrum of CEC columns used to separate ionic compounds has been extended by the group of Haddad [75–77]. In a series of papers focused on the use of ion-exchange types of stationary phases to the packed particulate organic- or silica-based columns, they added wall-coated latex particles and continuous-rod organic-polymer capillaries. They demonstrated that the interplay of electrophoretic mobility, electroosmotic flow, and ion-exchange interactions afford adjustment of selectivity. The result is a separation mechanism appreciably different from CE or ion chromatography (IC), which provides superior performance in comparison to the parent techniques. For example, the optimal conditions for a Dionex AS9-HC-packed column lead to the separation of eight UV-absorbing anions in less than 3 min by pressurized and voltage-programmed CEC (Figure 14, from Ref. 75). Further changes in the mobile-phase composition by altering the nature [78] or the concentration [75] of the eluent ions were tools that afforded further manipulation of the ion-exchange contribution to the separation.

Recently, Breadmore et al. developed a theoretical model that describes analyte behavior in IE CEC [77]. The model includes the contribution of the CE component through the electrophoretic mobility of the analyte and the ionic strength of the buffer, as well as the ion-exchange component which acts through changes in the composition of background electrolyte. Excellent agreement ($r^2 > 0.98$) between experimental and theoretical data was reported for both packed and open-tubular CEC systems.

Detection is another aspect of IE CEC that has received attention, since only a few inorganic anions absorb UV radiation. Indirect UV detection was employed in some cases and it was found that it interfered with baseline stability

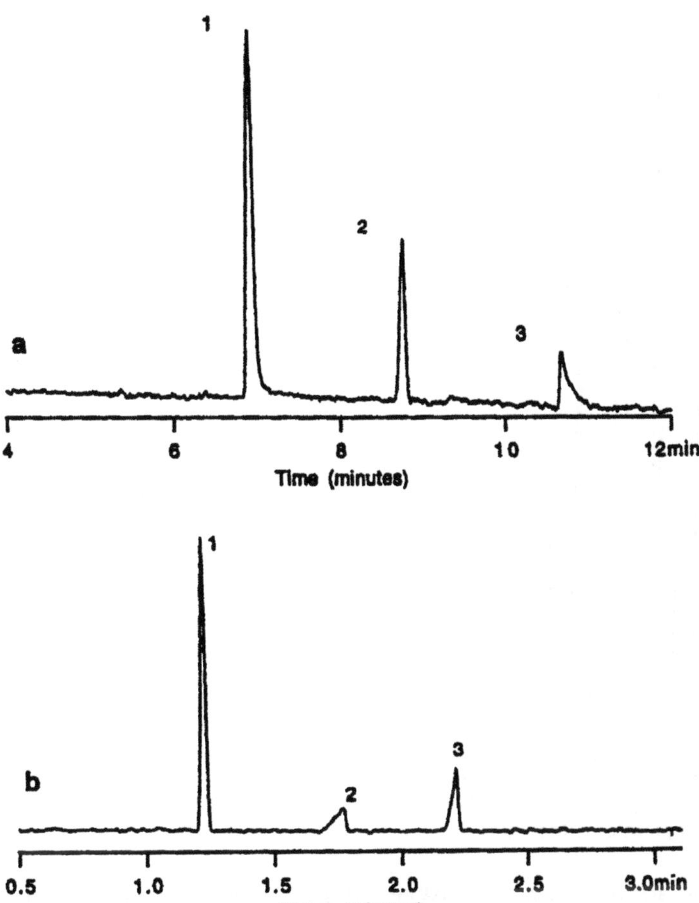

FIGURE 13 (a) Separation of mixture of iodide, iodate, and perrhenate on an IE-CEC column containing Nucleosil 5 μm SB. $L_{Tot} = 60$ cm, $L_{Bed} = 40$ cm, i.d. = 75 μm. Applied potential: –30 kV. UV absorption at 190 nm. Mobile phase: 5 mM phosphate buffer (pH 2.6). Solutes: (1) iodide; (2) iodate; (3) perrhenate. (b) Separation of mixture of iodide, iodate, and perrhenate on an open-tubular capillary. $L_{Tot} = 60$ cm, $L_{Det} = 40$ cm, i.d. = 75 μm. Applied potential: –30 kV. UV absorption at 190 nm. Mobile phase: 5 mM phosphate buffer (pH 2.6). Solutes: (1) iodide; (2) perrhenate; (3) iodate. (Reprinted from Ref. 73, with permission.)

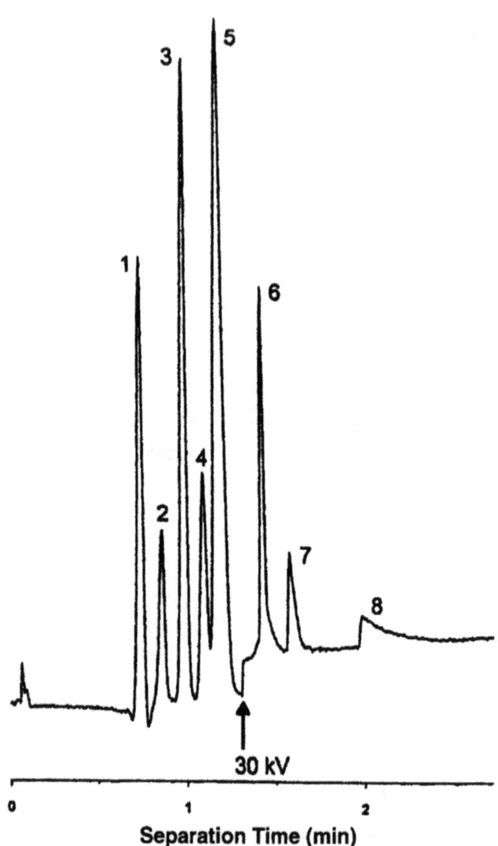

FIGURE 14 Separation of the test mixture using a step voltage gradient and a short packed column. Capillary: 75 mm i.d., 8 cm packed with Dionex AS9-HC (8.5 cm to detector, 34.5 cm total). Mobile phase: 2.5 mM hydrochloric acid (titrated to pH 8.05 with Tris). Flow is a combination of 10-bar pressure and EOF with −30 kV added at 1.3 min. All other conditions as given in Figure 1. (Reprinted from Ref. 75, with permission.)

[78]. Conductivity detection performed with a contactless conductivity detector through the packed bed was shown to be a more versatile and sensitive method than indirect UV detection [77]. Chen et al. employed inductively couple plasma mass spectrometry (ICP MS) as a detection tool for the simultaneous analysis of various ionic species of arsenic, chromium, and selenium [79]. The separation was achieved on an OT CEC column prepared by bonding a macrocyclic polyamine medium on the walls of the capillary.

3. POLLUTANTS

PAHs were the most common test mixture used for system testing in CEC. Facile UV detection combined with their nonpolar character provided, in most situations, a "clean" hydrophobic separation mechanism and conditions which clearly illustrate the advantages of CEC over capillary HPLC. For example, a sample of PAHs was used by Remcho et al. [21] to illustrate the effect that the entrapment matrix has on the performance of immobilized bed columns in CEC versus μHPLC mode (Figure 15, from Ref. 21). Dadoo et al. [31] also used

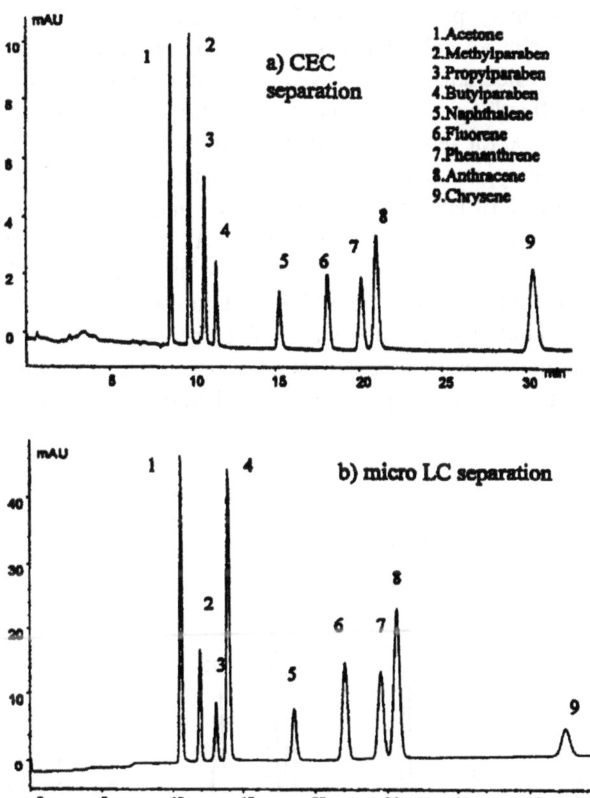

FIGURE 15 CEC and micro-LC separations of nine analytes achieved on the same column entrapped with composition O (see Table 1 for detailed description). (a) The CEC separation was achieved by applying 10 kV on 26-cm-long packed bed. (b) The micro-LC separation was performed on the same instrument (HP^{3D}CE by applying 9-bar gas pressure on the inlet vial (L_{eff} = 24 cm). Conditions: mobile phase, 80% acetonitrile/20% Tris buffer, 25 mM, pH 8.0. (Reprinted from Ref. 21, with permission.)

PAHs to demonstrate the excellent potential of CEC by resolving in under 10 min in isocratic mode 16 PAHs classified as priority pollutants by the U.S. Environmental Protection Agency. Yan et al. employed laser-induced fluorescence (LIF) for the detection of PAHs [80]. The limits of detection (LOD) for individual PAHs ranged between 1 nM and 10 pM, as the linear response spanned 4 orders of magnitude in concentration. A sample of 16 PAHs was also tested by Ngola et al. [26] on a new hydrophobic monolith (Figure 16, from Ref. 26). The synthetic procedure was readily transferable to the chip format and the first CEC separations on a chip were reported [26,81] (Figure 17).

A mixture of nine triazine herbicides was used by Dittmann et al. to compare different types of stationary phases: C18, C8, and C6/SAX [82]. Good separations were obtained with all columns as some selectivity changes were observed. Quality control considerations have driven the need for better analytical methods for fast, cost-effective determination of pesticides and insecticides. Henry et al. described a method for separation and quantitation of six insecticidally active phyrethrin esters in typical plant extracts and commercial formulation [83]. They employed a 3-μm C18 Hypersil-packed capillary and a ternary mobile phase, containing acetonitrile, aqueous buffer, and TFA. A novel concept referred to as surfactant-mediated (SM) CEC was introduced by the group of El Rassi [84]. It employs standard packed-column CEC and consists of adding a charged surfactant to the mobile phase. The hydrophobic analyte, in this case pyrethroid insecticides, partitions between the surfactant, which acts as a pseudo-stationary phase, and the fixed stationary phase of the packed chromatographic

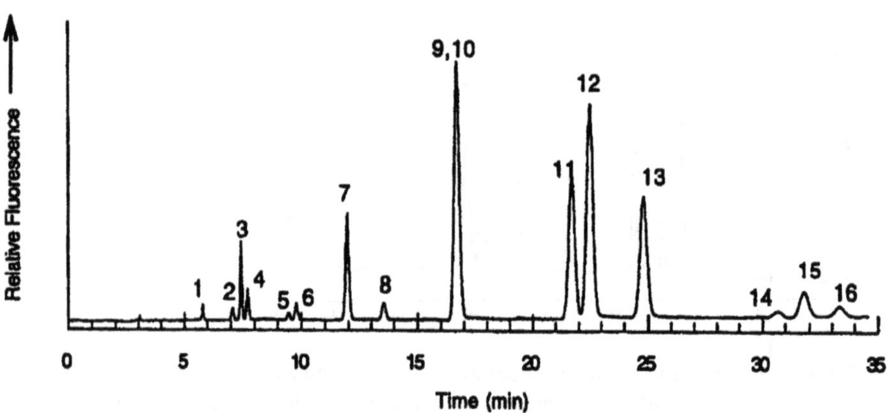

FIGURE 16 Electrochromatographic separation of 16 PAHs on the butyl stationary phase in 75/25 v/v acetonitrile/5 mM tris, pH 8, at field strength of 833 V/cm. (Reprinted from Ref. 26, with permission.)

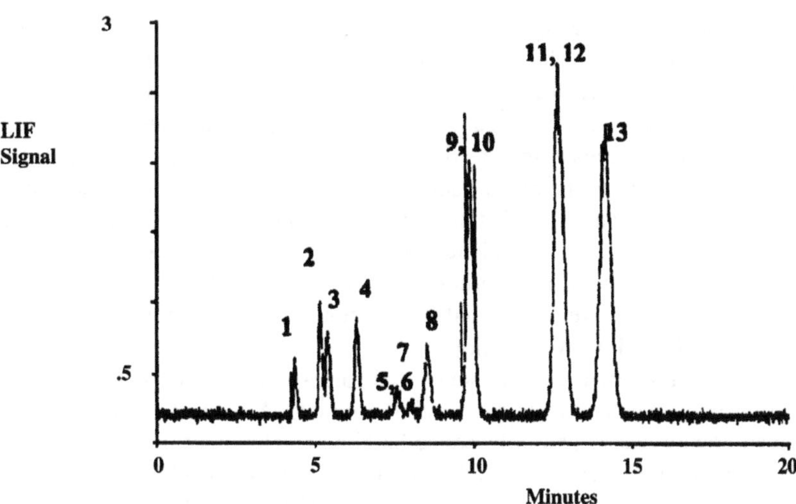

FIGURE 17 ChEC separation of 13 PAHs: (1) naphthalene, (2) acenaphthylene, (3) acenaphthene, (4) fluorene, (5) phenanthrene, (6) anthracene, (7) fluoranthene, (8) pyrene, (9) benz[*a*]anthracene, (10) chrysene, (11) benz[*b*]fluoranthene, (12) benzo[*k*]fluoranthene, (13) benzo[*a*]pyrene Length to detection 7 cm. Max voltage 2 kV. Running buffer was 80/20 (%, v/v) acetonitrileaqueous (20 mmol/L Tris pH 8.5) Injection: 2 kV, 5 min. (Reprinted from Ref. 81, with permission.)

bed. As a result, the geometric isomers and diastereoisomers of the pyrethroid insecticides are better resolved.

Environmental analytical chemists have found CEC interesting not only for its ability to analyze nonvolatile compounds but also for the possibility of performing on-line preconcentration of the sample. Trace enrichment is critical for environmental sample analysis since, in most instances, compounds are considered contaminants when present in very small concentrations. Furthermore, trace enrichment affords the use of UV detection, significantly decreasing the cost of sample analysis. A popular preconcentration approach is to generate analyte stacking at the head of the column by injecting solutes from an aqueous mixture. Yang et al. investigated on-line concentration using prolonged injection, which allows the detection of herbicides at concentrations as low as 10 μM [85]. It was discovered that introducing a plug of water at the head of the column, prior to analyte injection, increased the concentration ratio by 3 orders of magnitude.

Cooper et al. found that extended injection times doubled retention times and diminished EOF [86]. Another trace enrichment approach advocated by Tegeler et al. proposed sequential use of frontal and elution chromatography on a column prepared by coupling together two segments packed with different types of sorbents [87]. The sample was loaded on the first segment on the capillary, the "preconcentration segment," eluted on the second segment, the "separation column," and detected using UV absorbance. The method delivered 800- to 1000-fold increase in sensitivity for carbamate and pyrethroid insecticides. The feasibility of the method for "real sample" analysis was demonstrated with tap and lake water samples spiked with insecticides. During a study by Cassells et al., CEC was used to better understand the feasibility of direct toxicity tests performed with a bacteria-based soil toxicity kit [88]. Soil samples spiked with toxicants were extracted with various solvents and tested quantitatively using CEC with UV detection.

Another important environmental and forensic application is the identification of explosives and their degradation products. A method based on liquid chromatography was unable to produce baseline resolution of 14 nitroaromatic and nitramine explosive compounds. The problem was solved as Bailey et al. tested the separation of the same mixture in CEC mode (column packed with 1.5-μm C18 silica) [89]. A baseline resolution of all 14 compounds was realized in 7 min; in a second run, performed in less than 2 min, only 2 analytes of 14 co-eluted. To stabilize EOF and reduce peak tailing, SDS was added to the mobile phase. As in previous surfactant-mediated studies mentioned, changes in the elution order of the compounds were observed, suggesting a more complex separation mechanism (Figure 18, from Ref. 89).

Cooper et al. [86] analyzed environmental matrices derived from soil, plant, and animal extracts to study an insecticide, a fungicide, and their metabolites. The clean-up steps consisted simply of removal of the solid debris by centrifugation and filtration. The authors recognized that while retention time reproducibility was satisfactory for a study of pesticides metabolism, in order to achieve the sensitivity required for this type of analysis a sample preconcentration step was required before the CEC separation.

4. BIOMOLECULES

4.1. Amino Acids, Peptides, and Proteins

The wide spectrum of polarities encountered in amino acids and the structural complexity derived from the secondary and tertiary structures of proteins were features exploited extensively by CE and HPLC to achieve separation of complex mixtures. Since CEC combines the migrational differences of one with the selectivity range of the other, and has the additional capability of modulating the characteristics of the stationary phase, it is expected that the speed and resolving power of CEC would greatly benefit the field of proteomics.

As early as 1995, twelve dansyl-derivatized amino acids were resolved by

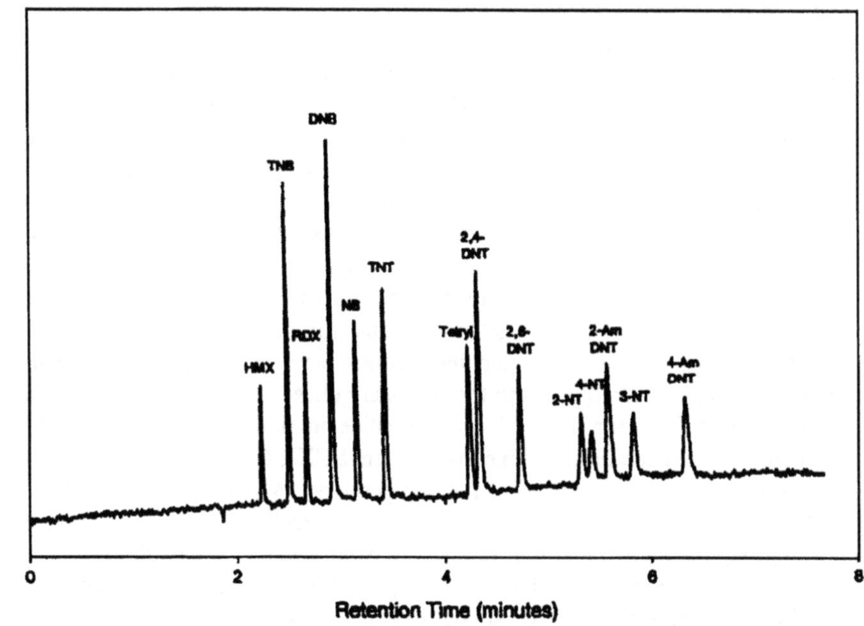

FIGURE 18 CEC separation of explosives with addition of SDS to mobile phase. Column: 34 cm × 75 mm i.d., 21 cm packed with 1.5-µm nonporous ODS II particles. Mobile phase: 20% methanol, 80% 10 mM MES, 5 mM SDS. Running potential: 12 kV (480 V/cm in packed portion). Injection: 2 s at 2 kV of 12.5 mg/L (each component) sample. (Reprinted from Ref. 89, with permission.)

CEC. The separation performed by Fujimoto et al. was achieved on a polyacrylamide gel column in 17 min [90]. Phenylthiohydantoin (PTH)-labeled amino acids were separated by Horvath's group by gradient CEC on a 3-µm Zorbax C18 column and, later, by employing an immobilized packed-bed capillary [91]. Alicea-Maldonado and Colon compared the separation of three dansylated amino acids by CE and CEC using a fluoropolymer as the stationary phase [92]. Elution of positively charged amino acids after the EOF marker suggested the interaction of solutes with the fluoropolymer.

N-alkyl silica-based stationary phases typically employed in HPLC separation of peptides and proteins have been investigated to a large extent in CEC mode. In most cases, gradient elution was required to optimize the resolution and analysis time. Tryptic digests maps of cytochrome c by pressurized CEC were better resolved than by µHPLC [93]. The experiments were conducted at low pH in the presence of trifluoroacetic acid (TFA) to prevent tailing of basic peptides, on a column packed with 1.5-µm C18 silica particles. It has been pointed out that the presence of TFA and the need to use high field strength to enhance the low EOF inherent to acidic conditions had the disadvantage of generating high currents, which in turn increased bubble generation.

A step forward from the use of reversed-phase material is the employment of mixed-mode packings. The field has been explored for both silica-based sorbents and organic-based materials; in the latter case a polymer completely void of hydrophobic character was not yet synthesized. Once again, the driving force behind the use of mixed-mode packing is the need to generate fast and stable EOF in both acidic and basic conditions. Moreover, the ionization sulfonic or amino groups, which are mostly responsible for EOF generation is less dependent on pH, as in the case of silanol groups. Ion-pairing agents are typically used to downplay the coulombic interactions between the oppositely charge moieties in peptides/proteins and the stationary phase. Adam et al. used pressurized gradient CEC for the analysis of peptide mixtures and tryptic digest of cytochrome c on capillaries packed with 3-μm SCX/C6 Waters Spherisorb [94]. Baseline resolution was achieved for two peptides, [Met^5] enkefalin and [Leu^5] enkefalin, which differ by only one amino acid residue. In a later study, the same group has shown that the retention of linear and cyclic peptides is dictated by the composition of the eluent, more specifically, the modifier content, pH, buffer concentration, and ionic strength [95]. At low organic content, retention originated from a chromatographic mechanism that involved predominantly hydrophobic interactions. As the concentration of organic modifier increased, the migration was governed mostly by the electrophoretic migration of individual peptides. Secondary interactions between the analytes and silanol groups and the strong cation-exchange groups were suppressed at low pH and increased ionic strength of the buffer. The addition of an ion-pairing reagent changed the elution order, indicating its strong effect on the surface charge density of the sorbent. Therefore, with commercially available particulate sorbents a thorough method development step is required to determined optimal separations conditions for a given mixtures of peptides. The researcher often resorts to incorporating mobile-phase additives, which are undesirable if subsequent coupling to MS is sought.

At another side of the method development spectrum are scientists who aimed to tailor stationary phases to specific applications. In this direction, Shediac et al. demonstrated the versatility of monolithic stationary phases through the one-step synthesis of polymer monoliths tunable for both charge and hydrophobicity [28]. A systematic study of the effect of buffer composition and degree of polymer hydrophobicity revealed optimal conditions for complete resolution of analytes of interest without the need for mobile-phase additives. Figure 19 (from Ref. 28) illustrates the CEC separation of 20 PTH-amino acids on a negatively charged lauryl stationary phase followed by UV detection. Likewise, another negatively charged lauryl polymer was employed in the separation of naphtha-

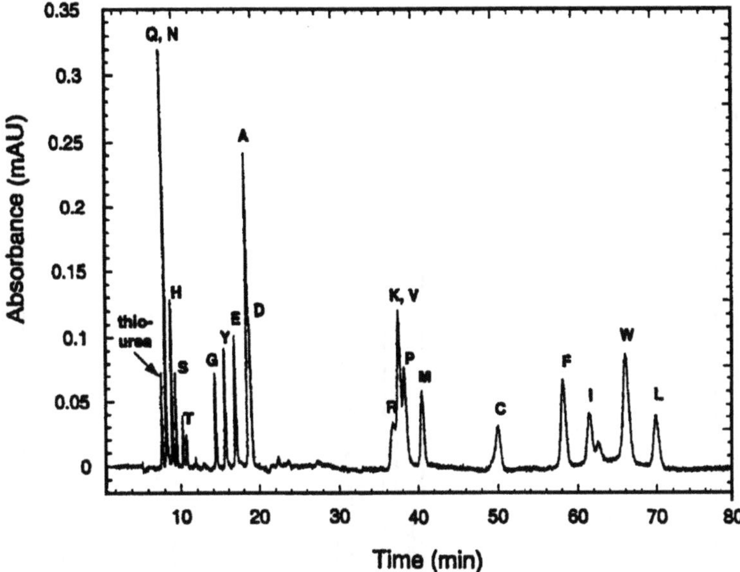

FIGURE 19 Electrochromatographic separation of 20 PTH-amino acids on negatively charged lauryl stationary phase in acetonitrile/25 mM phosphate pH 7.3 (40/60, v/v). UV detection at 214 nm. Field strength, 175 V/cm. Capillary: total length = 28.5 cm; length detector = 17.5 cm; i.d. = 100 μm. (Reprinted from Ref. 28, with permission.)

lene-2,3-dicarboxyaldehyde (NDA)-labeled amino acids followed by LIF detection, while basic bioactive peptides were resolved on positively charged butyl stationary phase and detected by measuring the UV absorbance. Analysis time was significantly reduced by increasing the surface charge of the monolith, to augment EOF, as well as by incorporating cellulose ester acrylate to improve resolution of individual peptides. A more recent study illustrated the power of tuning the selectivity of monoliths for the separation of tryptic digest fragments [96] (Figure 20, from Ref. 96).

Porous monolithic columns were prepared in a two-step process by in-situ copolymerization of glycidylmethacrylate, methyl methacrylate, and ethylene glycol methacrylate, followed by the reaction of epoxide groups with N-ethylbutylamine to generate a charged surface that supports EOF at low pH [97]. Mixtures of angiotensin-type peptides and small proteins were separated isocratically. The effect of organic modifier content, pH, and salt concentration on EOF and retention was investigated. The authors found that the elution order for proteins was consistent with that observed in reversed-phase chromatography, while an opposite trend was noted for peptide separation (Figure 21, from Ref.

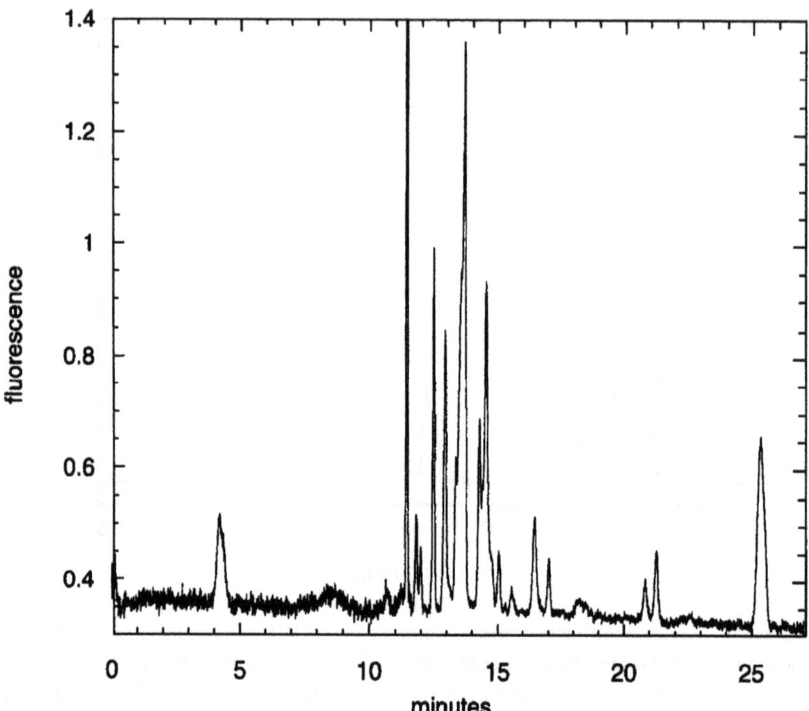

Figure 20 Tryptic digest of cytochrome c labeled with NDA 100-μm-i.d. negative lauryl column buffer–25/75, ACN/25 mM PB, pH 6.9, 5-s injection at 66 V/cm, run at 167 V/cm. (Reprinted from Ref. 96, with permission.)

97). The difference in behavior was ascribed to increased chromatographic retention and lower electrophoretic migration in the case of protein separations. Further evaluation of these phases lead to the observation that high resolution and efficiency separations, such as tryptic digest mapping, could be achieved at elevated temperatures (55°C) [98].

The group of Matyska and Pesek etched open-tubular columns with ammonium hydrogen difluoride and compared their performance to unetched fused silica for the separation of tryptic digests and protein mixtures [99]. The etched capillaries resolved 36 peaks in the analysis of a tryptic digest of transferrin, compared to 26 peaks separated by the unetched capillaries (Figure 22, from Ref. 99). An increase in the number of interacting sites due to the etching procedure is believed to be responsible for the improvement in resolution power. The group also reported on the utility of attaching C18, diol, cyanopentoxy, or cholesteryl moieties to the etched capillary wall [100–102]. Figure 23 (from

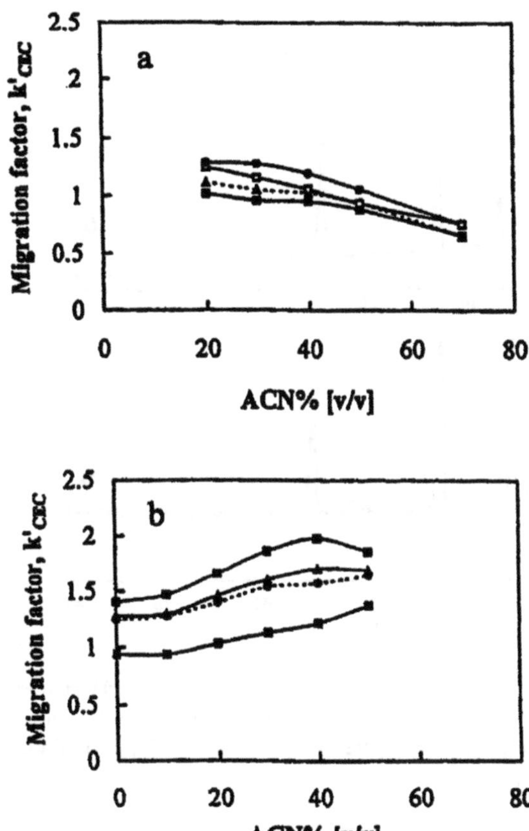

FIGURE 21 Plots of the migration factor, k'_{CEC}, of proteins and peptides against the acetonitrile concentration in the eluent. Column, 39 cm (effective length 29 cm) × 50 μm i.d., fused silica capillary with porous methacrylic monolith having tertiary amino functions; applied voltage, −25 kV; detection, 214 nm. (a) Proteins, conditions as in Figure 6: (■) ribonuclease A, (▲) insulin, (□) α-lactalbumin, (○) myoglobin. (b) Peptides, conditions as in Figure 9: (■) angiotensin II, (●) bradykinin, (▲) angiotensin I, (□) [Sar[1],Ala[8]]angiotensin II. (Reprinted from Ref. 97, with permission.)

Ref. 102) illustrates the difference in selectivity ascribed by three different ligands for the separation of four cytochrome C derivatives (chicken, tuna, horse, bovine). The improved chromatographic performance of etched and subsequently derivatized capillaries was ascribed to the larger number of bonding sites of the organic coating to the activated surface, which effectively block the access to the silanol groups.

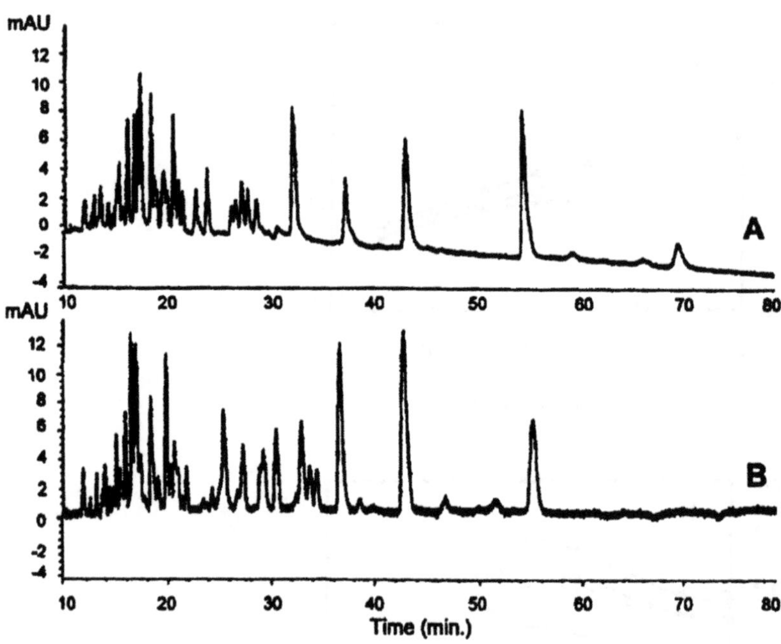

FIGURE 22 Electropherograms for the Lys-C digest of transferring on (A) etched and (B) unetched capillaries. Conditions: 50 mM phosphate buffer, pH 3.0; field 300 V/cm; temperature, 28°C; *l*, 56 cm. (Reprinted from Ref. 99, with permission.)

Xu and Regnier developed a stationary phase for cation-exchange chromatography in OT CEC by immobilizing poly(aspartic acid) [103]. The retention mechanism of proteins was similar to that encountered in HPLC and, more important, in the case of OT CEC gradient elution was not necessary. Chromatographic retention was a function of the pH and ionic strength of the mobile phase and the isoelectric point (pI) of the proteins. The authors advocated that the low phase ratio specific to OT CEC was responsible for the high efficiencies and isocratic elution mode.

Numerous researchers have dedicated their efforts to understanding the relationship between peptide and protein molecular properties (molecular mass and charge, hydrophobicity, surface charge anisotropy, surface area) and their retention in CEC as a function of field strength, gradient steepness, temperature, and variables related to surface characteristics of the stationary phase. So far, however, no reliable and comprehensive theory is available to model and predict peptide and protein retention, and the need to interface CEC with mass spec-

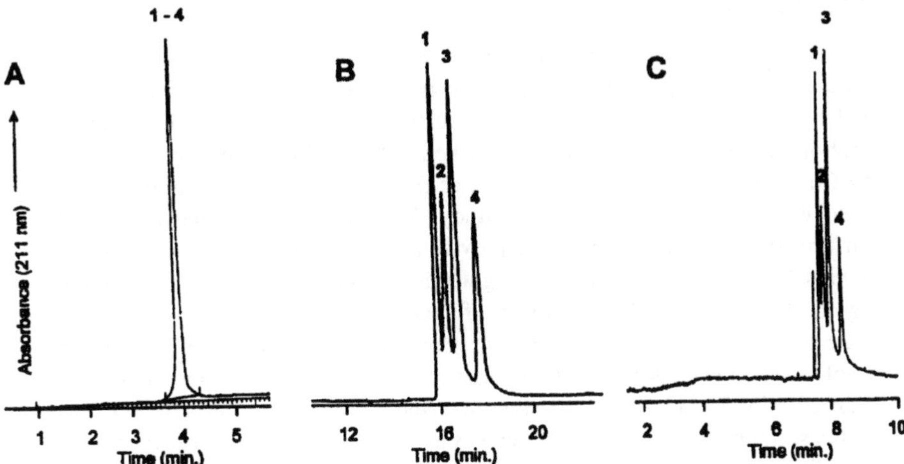

FIGURE 23 Separation of a cytochrome c mixture on 50-μm-i.d. capillaries: (A) bare capillary; (B) etched cholesteryl-modified capillary; (C) etched cyanopentoxy-modified capillary. Conditions: $V = 15$ kV; and mobile phase 100% aqueous pH 3.0 buffer (60 mM citric acid and 50 mM β-alanine). Dimensions: cholesteryl capillary, $L = 70$ cm and $l = 45$ cm. Bare and cyanopentoxy capillaries, $L = 50$ cm, and $l =$ 25 cm. Solutes: 1, chicken; 2, tuna; 3, horse; 4, bovine. (Reprinted from Ref. 102, with permission.)

trometry is even more necessary. Besides inherent difficulties associated with general design issues related to physical coupling of CEC to MS, the analysis of peptides and proteins by CEC MS include additional challenges such as incompatibilities between some mobile phases optimized for particular separations, specifically eluents that contain ion-pairing reagents, with the ionization process, and lower flow rates typically associated with operation at low pH.

CEC-MS has been successfully carried out by a number of groups for the analysis of numerous compounds including amino acids [104] and peptides [105,106]. More recently, the group of van der Greef has demonstrated remarkable progress in the coupling of CEC to an ion-trap MS via a nanoelectrospray interface [107]. Their new interface design obviates the need for any sheath flow to stabilize the spray and therefore affords analysis of small peptides in the attomole range. To date, this is the lowest amount analyzed using CEC-MS. The separation column was a 3-μm Hypersil C18-packed capillary. It was observed that the extent of nonspecific interactions between peptides and free silanols was considerably reduced, as smaller amounts of sample were injected.

4.2. Carbohydrates

Carbohydrates lack UV sensitivity and require derivatization with chromophore-containing reagents in order to be analyzed by UV detection. *p*-Nitro-phenol and 1-phenyl-3-methyl-5-pyrazolone (PMP) are commonly used for the derivatization of carbohydrates. Columns packed with ODS particles were employed by Gucek et al. in the CEC separation of five mono- and disaccharides derivatized with PMP [108]. In another study, Yang et al. recognized that a light surface coverage with ODS would provide enough selectivity for the separation of closely related *p*-nitrophenyl-α-D glucopiranosides and *p*-nitrophenyl-maltooligosaccharides, as the larger number of underivatized silanol groups would generate a faster EOF [109]. Figure 24 (from Ref. 109) illustrates the baseline resolution achieved with a mobile phase of low-content organic modifier in less than 20 min. The elution order was the same as in reversed-phase HPLC. The in-house-fabricated stationary phase was also able to distinguish between the α- and β-anomers of *p*-nitrophenyl-glucopyranoside. The difference in migration of the two anomers was due to the addition of borate in the mobile phase, which complexed both the α- and the β-anomers; as the borate β-anomer is a stronger complex, it eluted earlier (Figure 25, from Ref. 109).

The work of Suzuki et al. focused on investigating the separation mechanism on a similar reversed-phase stationary phase for different carbohydrates [110]. Using phosphate and borate buffers, common eluents for similar HPLC separations, the authors observed broad peaks and variable retention times. By replacing the eluent with HEPES buffer, the peaks were better resolved, but the separation was not reproducible. A lower-conductivity buffer containing 50 mM *N* (2 hydroxyethyl) piperazine-2′-(2ethanesulfonic acid) (pH 6) and acetonitrile produced reproducible separations of the PMP derivatives of monosaccharides found in glycoproteins (fucose, galactose, mannose, *N*-acetylgalactosamine and *N*-acetylglucosamine) and epimeric aldopentoses (arabinose, lyxose, ribose, and xylose). RSDs recorded for retention times and relative peak areas (internal standard 3-*O*-methyl-glucose) were 0.6% and 5%, respectively.

A later study by Suzuki et al. advocated the use of in-column derivatization with aminopropyl functionalities [111]. The group compared the performance of three CEC columns. The first column was a made of fused silica capillary packed with commercially available aminopropyl-derivatized particles (Develosil); the second column was obtained by packing Develosil in fused silica tubing onto which aminopropyl moieties had been attached, and, the third column was a bare silica-packed capillary which was treated after packing with octadecyltrimethoxysilylpropylammoniumchloride as a silylating reagent. The result was a gradual improvement in separation efficiency, peak shape, and elution time (Figure 26, from Ref. 111). The authors proposed that the opposition of wall-generated EOF with the EOF generated by the aminopropyl groups,

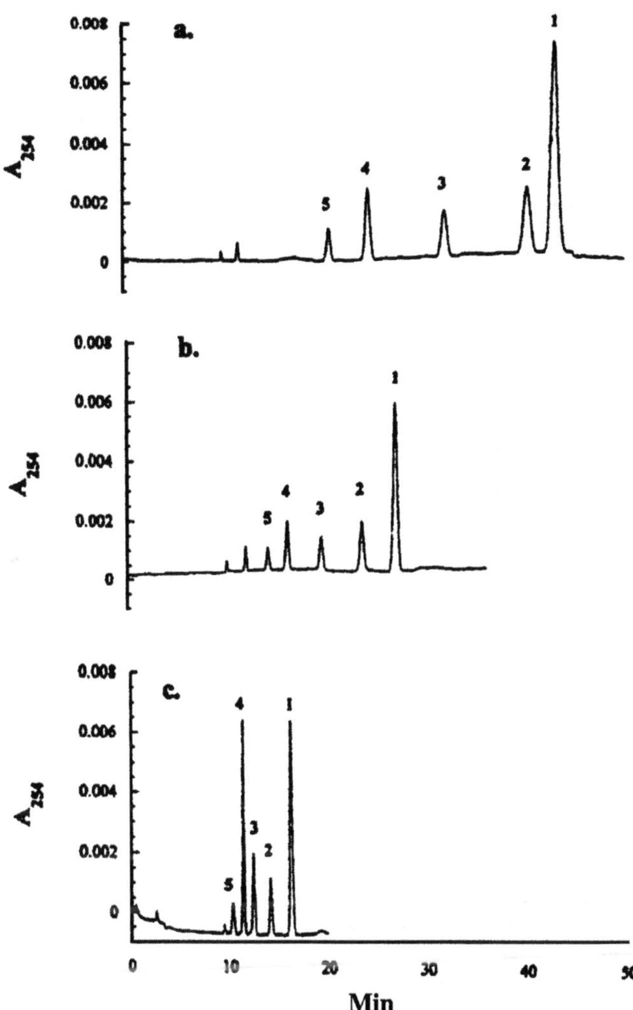

FIGURE 24 Electrochromatograms of *p*-nitrophenyl-α-D-glucopyranosides and maltooligosaccharides. Mobile phases: (c), 40% v/v of 5 mM NaH_2PO_4 (pH 6.0), 40% v/v H_2O, and 20% v/v acetonitrile; (b) 42.5% v/v of 5 mM NaH_2PO_4 (pH 6.0), 42.5% v/v H_2O, and 15% v/v acetonitrile; (a) 45% v/v of 5 mM NaH_2PO_4 (pH 6.0), 45% v/v H_2O, and 10% v/v acetonitrile; voltage, 10 kV; other conditions as in Figure 2. Solutes: 1, *p*-nitrophenyl-α-D-glucopyranoside; 2, *p*-nitrophenyl-α-D-maltoside; 3, *p*-nitrophenyl-α-D-maltotrioside; 4, *p*-nitrophenyl-α-D-maltotetraoside; 5, *p*-nitrophenyl-α-D-maltopentaoside. (Reprinted from Ref. 109, with permission.)

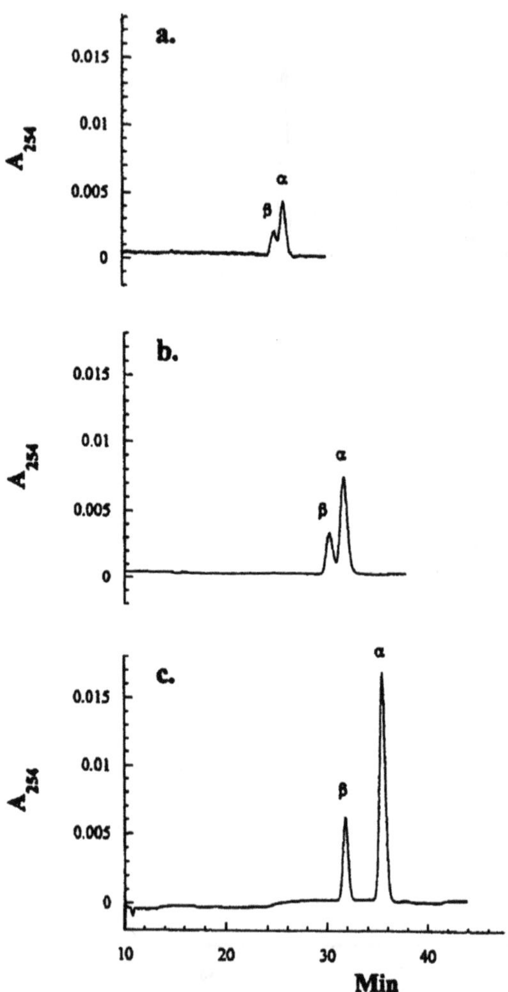

FIGURE 25 Electrochromatograms of β- and α-anomers of *p*-nitrophenylglucopyranoside. Mobile phase: (a) 42.5% v/v of 30 mM H_3BO_3 (pH 7.0), 42.5% v/v H_2O, and 15% v/v acetonitrile; (b) 42.5% v/v 50 mM H_3BO_3 (pH 7.5), 42.5% v/v H_2O, and 15% v/v acetonitrile; (c) 45% v/v of 50 mM H_3BO_3 (pH 7.5), 45% v/v H_2O, and 10% v/v acetonitrile; detection wavelength, 280 nm; other conditions as in Figure 24. Elution order: β-anomer first and α-anomer second. (Reprinted from Ref. 109, with permission.)

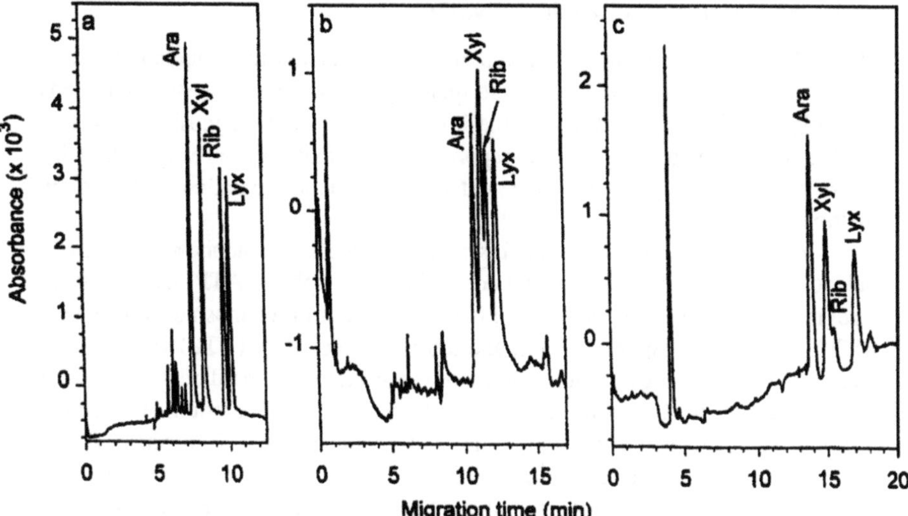

FIGURE 26 Comparison of the separate of PMP derivatives of aldopentose isomers on various columns: (a) a column prepared by in-column 3-aminopropylation of Nucleosil silica gel; (b) a column prepared by packing a commercial sample of amino silica, Develosil NH_2, in an APTMS-treated capillary; (c) a column prepared by packing Develosil-NH_2 in an uncoated capillary. Eluent, (25 mM HEPES–NaOH, pH 6.0)–acetonitrile (2/1, v/v); sample concentration, 50 nmol in 100 μL of eluent; injection, −2 kV for 3 s (from the cathodic end); applied voltage, −20 kV; detection, UV absorption at 245 nm. Peaks: Ara = D-arabinose, Xyl = D-xylose, Rib = D-ribose, Lyx = D-lyxose, all as PMP derivatives. (Reprinted from Ref. 111, with permission.)

and the nonspecific interaction between analytes and free silanol groups, were responsible for the poor performance of the first two columns.

The versatility of the monolithic columns was demonstrated by Palm and Novotny for the separation of mono- and oligosaccharides [112]. The compounds were derivatized with 2-aminobenzamide followed by reductive amination for LIF detection. The separation of the derivatized maltooligosaccharides was performed at pH 4. The derivatization reaction mixture was injected without any sample preparation into a macroporous polyacrylaminde/poly(ethylene glycol) column derivatized with butyl chains and vinylsulfonic acid.

Another detection scheme for carbohydrate analysis was used by Guo et al. [113]. The coupling of condensation nucleation light-scattering detection with pressurized CEC was investigated. The method afforded the separation of sucrose and saccharin with high sensitivity and no derivatization was required.

4.3. Nucleotides and Their Derivatives

There has been much interest in the analysis of nucleosides, nucleotides, and modified nucleosides. Qualitative and quantitative analysis of these compounds in biological fluids provides valuable information for the diagnosis of diseases and metabolic disorders. Several reports indicate that CEC is very suited for these types of analyses, as faster and more efficient separation methods are demonstrated.

As early as 1991, Verheij et al. separated adenosine monophosphate (AMP), adenosine diphosphate (ADP), and adenosine triphosphate (ATP) using voltage programming in pressurized CEC [114]. The separation of inosine triphosphate (ITP), uridine diphosphate (UDP), and uridine triphosphate (UTP) was detected using MS in the negative-ion continuous fast-atom bombardment (CF-FAB) mode. Helboe et al. optimized the condition for CEC separation of cytidine, uridine, inosine, guanosine, thymidine, and adenosine on a C18 phase [115]. They achieved baseline separation in 10 min under isocratic conditions with no mobile-phase additive. Calhours et al. studied the influence of pH, ionic strength of the buffer, and organic modifier content on the separation of nucleosides on a phenyl-bonded silica-packed column [116].

Numerous CEC studies pointed out the fact that stationary phases designed for HPLC might not be useful for CEC separations. As a consequence, novel stationary phases that allow better control of EOF and provide selectivity tailored for specific CEC applications have been developed. To the list of such phases, one should add the octadecylsulfonated silica (ODSS) phase developed by the group of El Rassi [117]. This phase consists of a charged hydrophilic sublayer bearing strong sulfonic acid groups, and a covalently attached nonpolar hydrophobic top layer of octadecyl moieties. The resulting EOF was higher than with ODS-silica phases and was relatively constant over a wide range of pH, a fact which allowed, to a certain extend, independent control of EOF and electrophoretic migration. A mixture of nucleosides and bases was separated on both silica-based C18 and ODSS phases [118]. The elution order differed in the two cases, indicating different retention mechanisms. The separation was further improved upon addition of tetrabutylammonium bromide to the mobile phase. Figure 27 (from Ref. 119) illustrates the excellent separation performance of the novel stationary phase as 12 mono-, di-, and triphosphate nucleotides are baseline resolved in less than 15 min [119]. It was also shown that nonporous ODSS-packed columns separated large nucleic acids (t-RNAs) that could not be resolved by capillary zone electrophoresis (Figure 28, from Ref. 119).

The remarkable progress achieved by Vouros et al. [120,121] in the CEC analysis of isomeric polyaromatic hydrocarbon deoxyribonucleic acid (DNA) adducts derived from in-vitro reactions of carcinogenic hydrocarbons with calf thymus DNA has been illustrated in a previous section. This work represents an

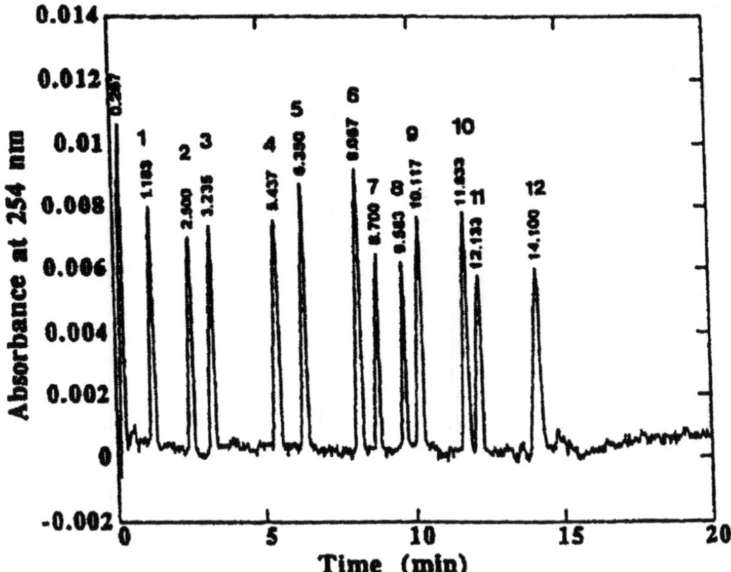

FIGURE 27 Typical electrochromatogram of mono-, di-, and triphosphate nucleotides. Capillary column, packed with 10-μm ODSS stationary phase, 20.5/27 cm × 100 μm i.d.; mobile phase, hydroorganic eluent containing 9.75 mM phosphate, 3.25 mM tetrabutyl-ammonium bromide and composed of 35% (v/v) acetonitrile and 65% (v/v) aqueous sodium phosphate, pH 6.50; running voltate, 20 kV, electrokinetic injection, 1.0 kV for 2 s. Solutes: 1, CMP; 2, UMP; 3, AMP; 4, GMP; 5, CDP; 6, UDP; 7, ADP; 8, CTP; 9, GMP; 10, UTP; 11, ATP; 12, GTP. (Reprinted from Ref. 119, with permission.)

excellent example of combining most of the attractive features of CEC, namely, speed, selectivity, gradient capability, on-line preconcentration capability, and relatively facile coupling to MS detection, to analyze complex biological samples.

5. PHARMACEUTICALS

Capillary electrochromatography has been applied to the analysis of a wide range of structurally diverse pharmaceutical compounds (Table 1).

Steroids of endogenous and synthetic origin have been often used to evaluate CEC columns and systems performance. For example, in one of the first papers on CEC and pharmaceutical compounds, published in 1994, Smith and Evans [122] presented an efficient separation of the corticosteroid fluticasone propionate from related impurities. On a 3-μm Spherisorb ODS-1-packed capil-

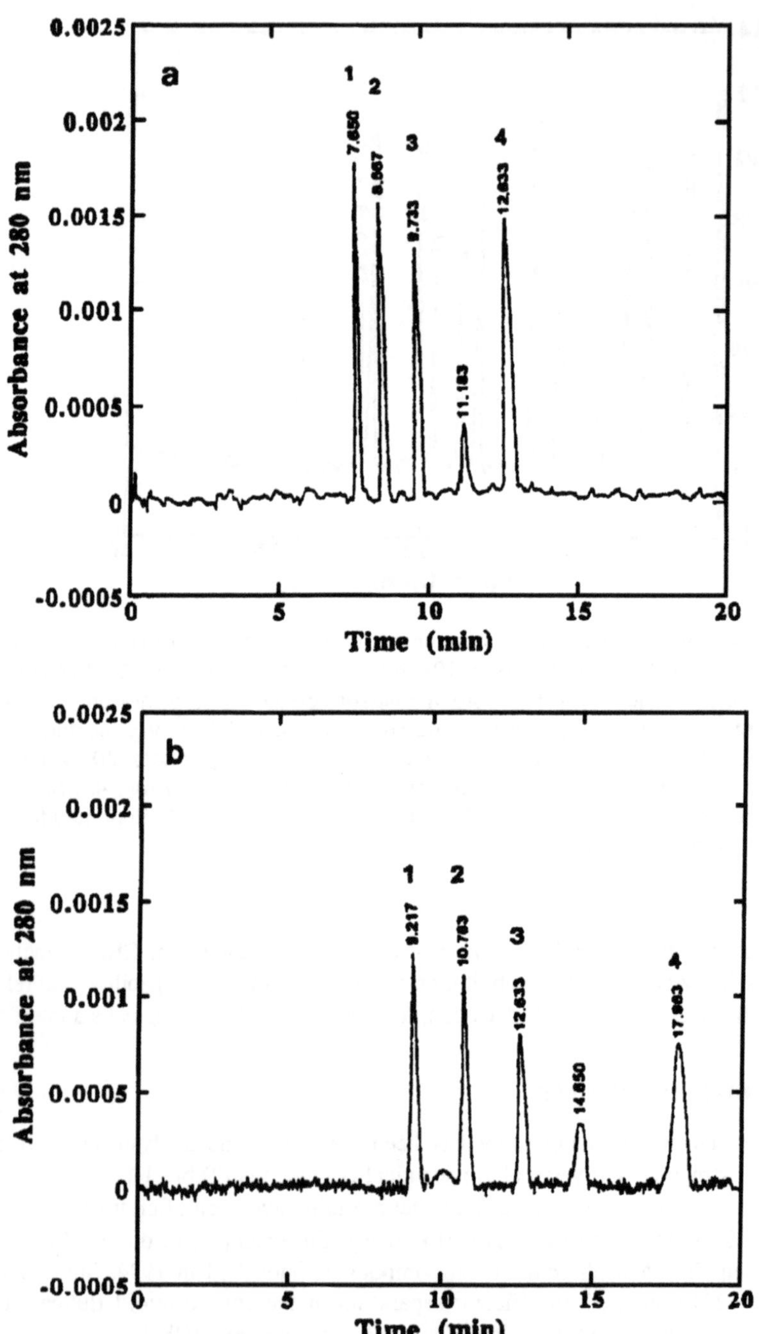
a
0.0025
0.002
0.0015
0.001
0.0005
0
-0.0005
Absorbance at 280 nm
1
2
3
4
7.650
8.567
9.733
11.183
12.633
0
5
10
15
20
Time (min)
b
0.0025
0.002
0.0015
0.001
0.0005
0
-0.0005
Absorbance at 280 nm
1
2
3
4
9.217
10.783
12.633
14.650
17.983
0
5
10
15
20
Time (min)

lary, the separation was achieved in 120 min. One year later, the same group reported that by increasing both the pH of the mobile phase and the organic content, it was possible to reduce the run time to 10 min [47].

Taylor et al. [39,56] analyzed a mixture of corticosteroids in extracts of equine urine and plasma. Gradient elution was used to facilitate trace enrichment at the head of the column. Huber et al. [36] presented another illustration for the use of capillary electrochromatography with gradient elution. Five steroid hormones were separated by using a capillary column packed with 6-μm Zorbax ODS stationary phase.

Cholesterol is one of the main components of cell membranes and has several functions in the body, including the synthesis of certain hormones such as vitamin D and bile acid. Over the years, gas chromatography has been used to characterize cholesterol and its derivatives. Thiam et al. [123] developed an isocratic CEC method that allows baseline separation of a complex mixture of cholesterol and 12 ester derivatives in less than 40 min. The use of a polymeric surfactant, poly(sodium N-undecanoyl-L-glycinate), in the CEC buffer reduced migration time and improved resolution of the analytes.

Another group of pharmaceutical compounds investigated using CEC is benzodiazepines. Benzodiazepines (pK_a values around 6–7) comprise a group of substances with anxiolytic and antihypnotic effects. Taylor and Teale [39] analyzed a mixture of nitrazepam and diazepam by CEC ES MS using a gradient system. Both linear and step gradients on C18 columns provided peak width reduction, but the analyses run times were lengthened compared to the corresponding isocratic system.

In a study presented by Jinno et al. [124], packed column capillary electrochromatography, open-tubular CEC, and microcolumn liquid chromatography using a cholesteryl silica bonded phase have been studied to compare the retention behavior for benzodiazepines. The results indicated that CEC was a promising method, as it yielded better resolution and faster analysis than microcolumn LC for benzodiazepines. Similar selectivity to HPLC was noted, except for a few solutes that were charged under the separation conditions. Columns packed with the ODS and cholesteryl phases were compared and showed totally different migration orders of the analytes. The retention on the cholesteryl silica sta-

---

FACING PAGE

FIGURE 28 Capillary electrochromatography of four t-RNAs. Capillary column, packed with 2-μm nonporous ODSS stationary phase, 20.5/27 cm × 100 μm i.d.; running voltage, 20 kV, electrokinetic injection, 1 kV for 2 s; mobile phase in (a), hydroorganic eluent containing 15 mM phosphate and composed of 40% (v/v) methanol. Solutes: 1, t-RNA^{Glu-}; 2 t-RNA^{Val-}; 3, t-RNA^{Lys-}; 4, t-RNA^{Phe}. (Reprinted from Ref. 119, with permission.)

tionary phase is based on the molecular structure of the analytes rather than the hydrophobicity, which governs the retention on the C18 phase.

Recently, Kapnissi et al. [125] reported the separation of seven benzodiazepines using open-tubular capillary electrochromatography (OT CEC). In their approach, fused silica capillaries were coated with thin films of physically adsorbed charged polymers by use of a polyelectrolyte multilayer (PEM) coating procedure. The PEM coating is constructed in situ by alternating rinses with positively and negatively charged polymers, where the negatively charged polymer is a molecular micelle. The coating was found to be remarkably stable, with excellent performance for more than 200 runs.

The separation of nonsteroidal anti-inflammatory drugs (NSAIDs) has recently attracted considerable interest. Nonsteroidal anti-inflammatory drugs are agents that, in addition to having anti-inflammatory action, also have analgesic, antipyretic, and platelet-inhibitory properties. They are used primarily in the treatment of chronic arthritic conditions and certain soft tissue disorders associated with pain and inflammation.

Strickmann et al. [126] used on-line coupling of CEC with electrospray ionization (ESI) mass spectrometry (MS) for the qualitative investigation of the biotransformation of the NSAID etodolac. The drug and its metabolites were analyzed in urine extracts. This hyphenated technique proved to be a useful tool for bioanalytical investigations.

Lammerhofer et al. [127] demonstrated the use of a strong anion-exchange stationary phase, prepared in a monolithic format, for the separation of a mixture of four NSAIDs—ibuprofen, naproxen, ketoprofen, and suprofen. The separation, presented in Figure 29, was achieved in 13 min with high column efficiencies of up to 231,000 plates/m.

CEC analysis of tricyclic antidepressants (TCAs) has seen much attention. TCAs are notorious for giving severe peak tailing when analyzed using standard reverse-phase stationary phases. Smith and Evans [47] reported an elegant solution to this problem by using a strong cationic exchanger. Astonishing peak efficiencies up to 8×10^6 plates/m were obtained. Unfortunately, as mentioned earlier, the focusing mechanism that occurs on this type of stationary phase is not reproducible and has not yet been fully explained.

The separation of TCAs and related quaternary ammonium compounds on different strong cation exchangers was studied by Enlund et al. [128]. Four cation-exchange materials, possessing propanesulfonic acid ligands, were prepared from different 5-μm bare-silica particles ranging from 80 to 800 Å in pore size. The best separation was produced on the small-pore materials, but the efficiency and symmetry were similar on all stationary phases compared.

In the SCX systems no amine additives are necessary to improve peak shape, so this raises the possibility of MS detection. A mixture of TCAs was analyzed by Spikmans et al. [129] using an automated capillary electrochroma-

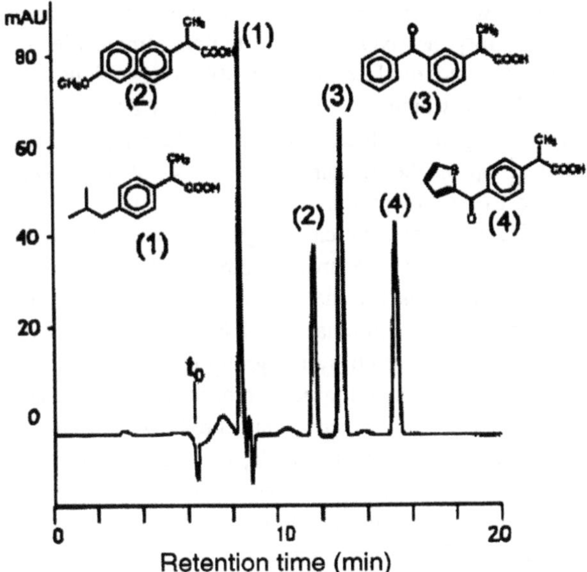

FIGURE 29 Separation of the nonsteroidal anti-inflammatory drugs ibuprofen (peak 1), naproxen (2), ketoprofen (3), and suprofen (4) in anion-exchange CEC mode using a strong anion-exchange monolithic column. Conditions: on-column alkylated monolith prepared from mixtures consisting of 8% 2-dimethylaminoethyl methacrylate, 24% 2-hydroxyethyl methacrylate, 8% ethylene dimethacrylate, 20% cyclohexanol, 40% 1-dodecanol; UV-initiated polymerization at room temperature for 16 h; $d_{p,mode} = 1423$ nm. Column dimensions: inner diameter 0.1 mm, total length 335 mm, effective length 250 mm. Mobile phase: 0.4 mol/L acetic acid and 4 mmol/L triethylamine in acetonitrile/methanol (60/40), voltage −25 kV, injection −5 kV for 5 s, temperature 50°C, UV detection at 250 nm. (Reprinted from Ref. 127, with permission.)

tography tandem mass spectrometry system. The separation was performed on an SCX and a mixed-mode (C6/SCX) column. Detection limits for all compounds analyzed were in the low nanomolar range.

Enlund et al. [130] have designed and evaluated polyacrylamide-based monolithic columns. On the best stationary phase, which has a molar ratio of 1:80 between the two functional ligands, vinylsulfonic acid and isopropylacrylamide, efficiencies of up to 200,000 plates/m were obtained for the separation of three tricyclic antidepressants.

A group of structurally similar tricyclic antidepressants was analyzed by Vallano and Remcho [131] using a molecular imprint polymer (MIP)-based chromatographic sorbent. MIPs were prepared via bulk polymerization using one of the antidepressants, nortriptyline (NOR), as the template molecule. As illustrated in Figure 30, with applied field strength of 900 V/cm, NOR was separated from the other five TCAs, which co-eluted, in less than 2.5 min. It is interesting to note that even amitripyline, which differs from the template only by the presence of a methyl group on the pendant amine, was not recognized by the MIP, and was separated from nortriptyline.

Finally, the utility of CEC in pharmaceutical analysis has been expanded by its combination with nuclear magnetic resonance spectroscopy. NMR is one of the most powerful analytical methods for the identification and structural elucidation of organic compounds. The combination of NMR detection with CEC for pharmaceutical analysis was reported by the group of Bayer [132]. A solution of the analgesic Thomapyrin, containing acetaminophen, caffeine, and acetylsalicylic acid, was investigated using isocratic and gradient continuous-flow CEC NMR. When operating in the isocratic mode, the components were poorly separated (Figure 31A). This was attributed to the relatively high concentration of the sample in methanol-d_4 and the large volume injected, necessary for NMR detection. The use of gradient elution led to more intense NMR signals, as illustrated in Figure 31B. The appropriate individual rows extracted from the gradient CEC NMR contour plot are shown in Figure 32.

6. CHIRAL COMPOUNDS

Enantioseparations are of importance in chemistry, the pharmaceutical industry, and the biological sciences. The first chromatographic separations of enantiomers were accomplished using gas chromatography (GC) [133]. Since that time, enantioseparations have been accomplished using high-performance liquid chromatography, supercritical fluid chromatography, capillary electrophoresis, and capillary electrokinetic chromatorgaphy.

The first electrically driven enantioseparations involved the addition of a chiral selector to the mobile phase in CE. This selector is usually a complexing agent and acts as a pseudo-stationary phase. The separation is accomplished by the difference in the distibution equilibria between the pseudo-stationary phase and the enantiomers [134]. The most common additives incorporated into these CE experiments were cyclodextrins and cyclodextrin derivatives [135–138]. However, these experiments required the replacement of the chiral selector after each electrophoretic run.

With the advent of CEC, the need to replace the additive was avoided. Enantiomoseparations are usually achieved in CEC experiments by immobilizing the chiral selector directly onto the inner capillary wall or onto particles which are then packed into the capillary, thus resulting in a chiral stationary

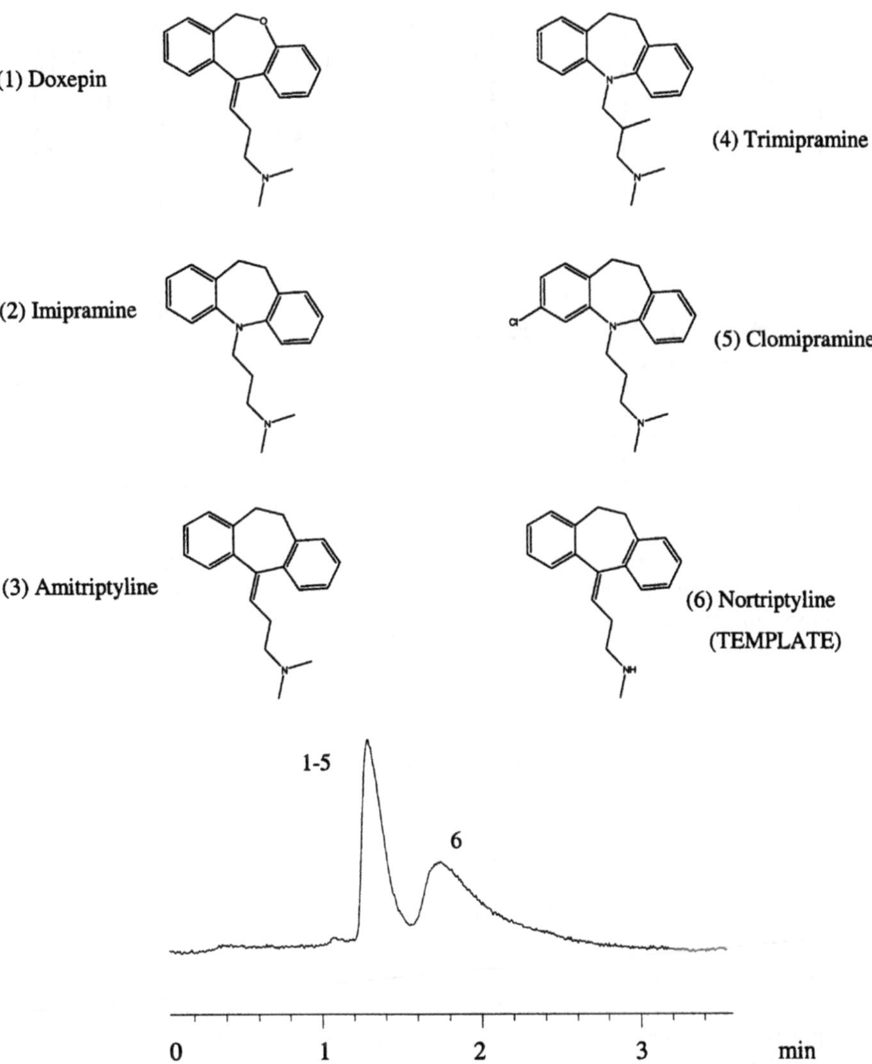

FIGURE 30 MIP-CEC separation of a simulated combinatorial library consisting of several tricyclic antidepressants. Conditions: capillary i.d. 100 mm; L_{tot} 33 cm; L_{bed} 22.5 cm; eluent: acetonitrile/10 mM Na acetate pH 3.0 (98/2) with 0.02% trifluoracetic acid and 0.015% triethylamine (v/v); voltage + 30 kV constant; injection: + 2 kV, 2 s; column temperature: 50°C. (Reprinted from Ref. 131, with permission.)

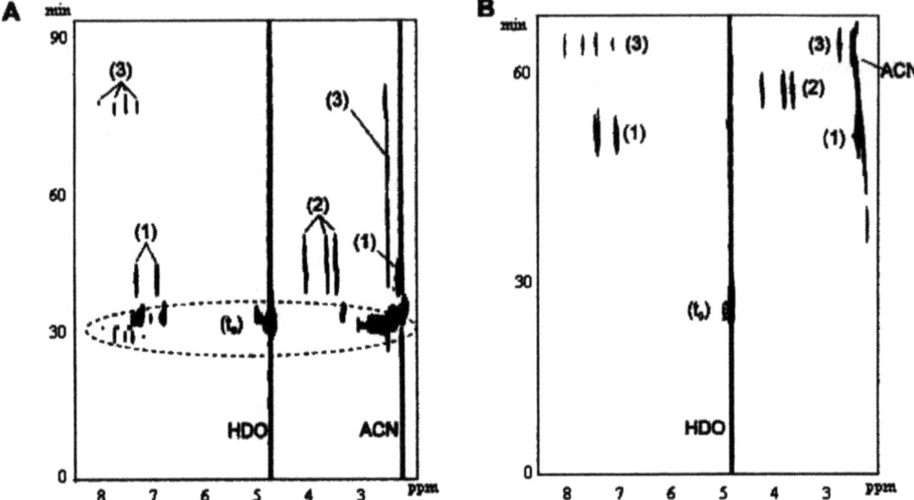

FIGURE 31 (A) Isocratic CEC-NMR chromatogram of the separation of Thomapyrin. (B) Gradient CEC-NMR chromatogram of the separation of the analgesic. 1, acetaminophen; 2, caffeine; 3, acetylsalicilic acid. (Reprinted from Ref. 132, with permission.)

phase (CSP). Table 2 shows different types of CSPs which have been utilized in CEC experiments.

6.1. Cyclodextrins and Cyclodextrin Derivatives

Cyclodextrins (CDs) are chiral compounds which interact with enantiomers via diastereomeric interactions. The separation is achieved because of the difference in stabilities of the resulting diastereomeric complexes formed between each enantiomer and the CD. In the first CEC experiments incorporating CDs, dimethylpolysiloxane containing chemically bonded permethylated β- or γ-CD (Chirasil-DEX) was chemically bonded to the inner walls of fused silica capillaries [139,140]. Electoosmotic flow is generated in these capillaries in the same manner as in fused silica capillaries. The Chirasil-DEX does not mask all the silanol groups, so while EOF is decreased, it is not entirely diminished by the coating. Since that time, CDs or CD derivatives have been bonded to silica particles which were then packed into capillaries, and the CD has been incorporated into continuous polymer beds known as monoliths. Table 3 shows some different CSPs, enantiomers separated, resolution, and the number of theoretical plates per meter.

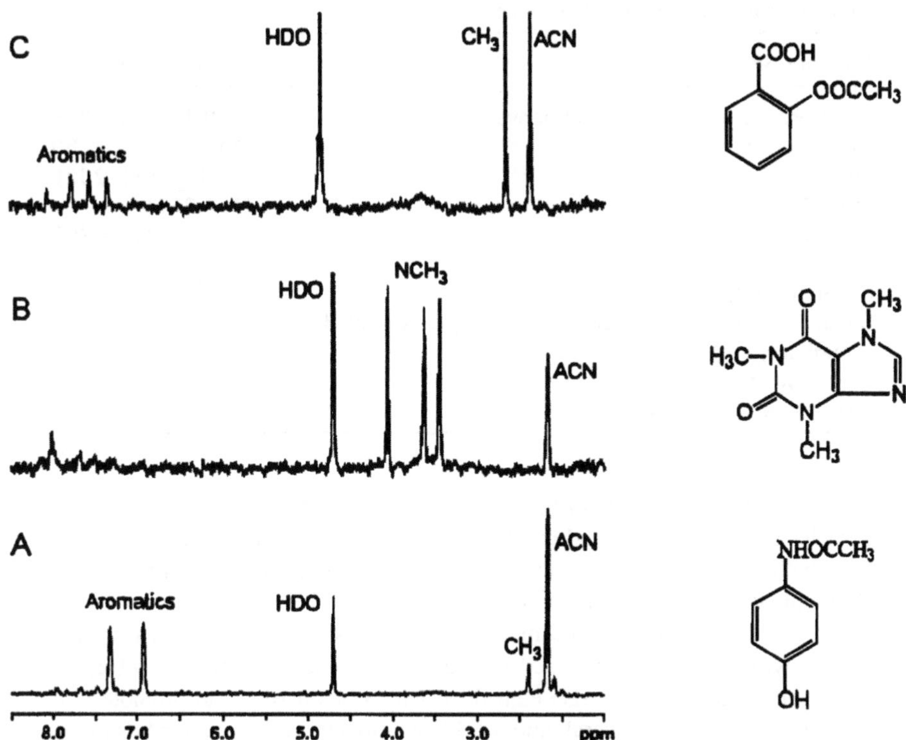

FIGURE 32 Single rows extracted from the continuous-flow CEC separation shown in Figure 32B: (A) acetaminophen (at 48–52 min); (B) caffeine (at 54–58 min); (C) acetylsalicylic acid (at 62–65 min). (Reprinted from Ref. 132, with permission.)

6.2. Biomolecules

Biomolecules are usually incorporated as stationary phases in the form of proteins. Proteins did not enjoy popularity as buffer additives in CE experiments due to high background absorbance caused by the protein. However, due to the versatility of protein-based stationary phases, the use of these molecules as CSPs has been investigated.

Li and Lloyd were among the first to report the use of immobilized proteins as a CSP in CEC [141]. They packed silica particles derivatized with α_1-acid glycoprotein (AGP) into CE capillaries. Chiral sites are present in the peptide sequence of AGP, as well as in the carbohydrate units present. The presence of the chiral centers sets up the same type of diastereomeric interactions as with CDs, as well as a host of other interactions such as hydrogen bonding. At the

TABLE 2 Chiral Stationary Phases used in CEC

Type of CSP	References
Cyclodextrins and cyclodextrin derivatives	[139,140,247–255]
Biomolecules	[141,142,256–259]
MIPs	[20,143,260–262]
"brush" type	[144,263]
Ion exchangers	[145–148,264,265]
Antibiotics	[154,266–270]
Chiral polyacrylamide	[271]
Chiral polymethacrylate[poly(dephenyl-2-pyridylmethyl-methacrylate]	[157,158]
Chiral ligand exchanger	[159]
Polysaccharide derivatives	[156,272–276]

normal operating pH range of these capillaries (3.0–7.5), both the surface of the particles and the immobilized protein carry a negative charge. Thus, EOF arises from both surfaces. The applicability of these capillaries was investigated using 19 chiral compounds, 10 of which were resolved (Table 3).

Kato et al. [142] created monolithic columns of sol-gel which were embedded with particles derivatized with an amino acid derivative to act as the CSP. The capillaries were also embedded with silica particles to promote EOF through the capillary. Several amino acid enantiomers were separated using the resulting capillaries. Detection was achieved by derivatizing the analytes with a fluorophore and measuring the fluorescence intensity. Figure 33 shows the resulting electrochromatogram of two pairs of enantiomers. While this method requires more sample preparation in order to derivatize the analytes, fluorescence detection increases the detection limit of the technique. The derivatization of the analytes did not seem to alter the enantiospecificity of the CSP, and they found the resolution and plate height of the embedded CEC capillaries superior to HPLC columns packed with particles derivitized with the CSP.

6.3. Molecularly Imprinted Polymers

Molecularly imprinted polymers (MIPs) allow for predetermined selectivity of enantiomers. MIPs are prepared by polymerizing a mixture of functional monomer(s) and cross-linking monomer in the presence of a template molecule. The template molecule remains in a "pocket" by its interaction with a functional monomer through hydrogen bonding. This allows the MIP to be found at the surface of the polymer. When polymerization is complete and the template molecule is removed, the polymer "remembers" the template molecule.

TABLE 3 Some Examples of Chiral Stationary Phases Used in CEC Applications, Resulting Resolution, and Efficiency

Chiral selector	Molecule used	CEC mode	Enantiomers separated	R_s or (α)	N_1/N_2	Refs.
Cyclodextrins and cyclodextrin derivitives	Chirasil-DEX	Open tubular	Phenylethanol	(1.05)	300,000/—	[139]
	Chirasil-DEX	Open tubular	Ibuprofen	—	—	[140]
			Etodolac	—	—	
			Phenylethanol	(1.17)	30,500/—	
			Flurbiprofen	(1.04)	31,162/—	
			Cicloprofen	(1.06)	19,828/—	
	Permethyl-β-cyclodextrin	Packed bed	Mephobarbital	(1.31)	17,565/17,695	[251]
			Hexobarbital	(1.25)	20,068/21,890	
			Pentobarbital	(1.06)	25,215/27,849	
			1-Methyl-5-(2-propyl)-5-(n-propyl)-barbituric acid	(1.07)	21,013/15,894	
			5-Ethyl-1-methyl-5-*n*-propyl)barbituric acid	(1.24)	18,734/19,714	
			Benzoin	(1.11)	18,739/16,156	
			α-Methyl-α-phenylsuccinimide	(1.09)	35,075/28,354	
			Glutethimide	(1.11)	17,425/15,543	
			MTH-proline	(1.17)	23,341/21,351	
			Methyl mandelate	(1.09)	43,509/40,653	
	β-Cyclodextrin	Continuous bed	Aspartic acid	5.08	78,000/70,000	[253]
			Glutamic acid	4.38	83,000/78,000	

TABLE 3 (*Continued*)

Chiral selector	Molecule used	CEC mode	Enantiomers separated	R_s or (α)	N_1/N_2	Refs.
			Serine	4.30	151,000/130,000	
			Valine	6.03	103,000/93,000	
			Norvaline	3.55	89,000/75,000	
			Leuine	4.47	65,000/50,000	
			Norleucine	2.94	95,000/95,000	
			Threonine	7.10	144,000/111,000	
			Methionine	2.70	95,000/93,000	
			Tryptophan	1.36	80,500/60,000	
			α-Amino-N-butyric acid	2.60	63,000/60,000	
			Phenylalanine	1.32	41,000/21,000	
			Phenylmercapturic acid	1.64	48,500/41,000	
			Warfarin	1.20	45,000/30,000	
			2-Phenoxypropionic acid	1.07	37,000/24,000	
			N-FMOC-DL-valine	1.54	46,000/38,000	
			Benzoin	1.78	26,000/23,000	
			1-(1-Naphthalene)ethanol	0.40	58,000/23,000	
	Sulfated β-cyclodextrin (SCD)	Packed bed	Tryptophan	2.6)	133,000/128,000	[254]
			Praziquantel	1.79	45,000/36,000	
			Atropine	11.23	365,000/288,000	
			Metoprolol	0.82	266,000/97,000	
			Verapamil	2.45	331,000/324,000	

Biomolecules	α-Acid glycoprotein	Packed bed	Disopyramide	—	—	[141]
			Pentobarbital			
			Hexobarbital			
			Cyclophosphamide			
			Benzoin			
	Amino acid derivative	Continuous bed	Alanine	2.00	—	[142]
			Glutamine	1.48	—	
			Glutamic acid	1.18	—	
			Isoleucine	2.60	—	
			Methionine	2.52	—	
			Phenylalanine	2.88	—	
			Proline	1.20	—	
			Serine	1.85	—	
			Threonine	1.74	—	
			Valine	4.45	—	
			2,3-Diaminopropionic acid	2.33	—	
			2-Aminobutyric acid	2.77	—	
	DNA Aptamers	Open tubular	Tryptophan	—	—	[258]
MIPs	L-Dansylphenylalanine	Continuous bed	Dansylphenylalanine	—	—	[20]
"Brush" type	(3R,4S)-Whelk-O,(S)-Naproxen-derived	Packed bed	10 chirally active compounds	2.63 to 30.95	121,000 to 20,0000/ 24,000 to 189,000	[144,263]
Ion exchangers	*tert*-Butylcarbamoyl quinine (anion exchanger)	Packed bed	N-derivized amino acids	0.7 to 12.5	2,724 to 29,851/ 4,484 to 22,766	[145,146]

Table 3 (*Continued*)

Chiral selector	Molecule used	CEC mode	Enantiomers separated	R_s or (α)	N_1/N_2	Refs.
	O-[2-(Methacryloyoxy)ethylcarbonyl]-10,11-dihydroquinidine (anion exchanger)	Continuous bed	N-derivitized amino acids	0.98 to 15.11	54,313 to 242,400/ 45,093 to 194,647	[147,148]
			2-(4-Chloro-2-methylphenoxy)-propionic acid	1.87	84,060/73,000	
			2-(2,4,5-Trichlorophenoxy)propionic acid	4.03	153,320/123,067	
	N-(4-Allyloxy-3,5-diclorobenzoyl)-1-amino-3-methylbutane	Packed bed	Talinolol	1.68	80,800/82,000	[265]
			Bupranolol	1.21	90,200/117,500	
			Bunitrolol	1.11	58,000/91,400	
			Celiprolol	1.20	97,000/99,500	
			Penbutolol	1.46	94,000/98,000	
			tert-Butyl-atenolol	1.75	90,200/85,400	
			tert-Butyl-metoprolol	2.02	119,300/126,000	
			tert-Butyl-propranolol	1.95	110,100/105,100	
			Atenolol	1.33	94,400/95,300	
			Metoprolol	1.39	81,500/100,400	
			Propranolol	0.67	58,100/67,600	
			Acebutolol	0.70	64,200/63,200	
			Practolol	1.30	91,400/93,200	
			Alprenolol	0.63	78,000/90,600	
			Pindolol	0.95	88,300/101,400	
			Mepindolol	0.68	82,200/73,000	
			Metipranolol	0.59	53,400/69,200	

Normetoprolol	0.61	65,200/29,600
Ethoxymethyl-metoprolol	1.40	105,200/116,300
Ethoxyethyl-metoprolol	1.25	84,000/96,800
4-Methoxy-propranolol	0.62	108,600/76,100
4-Hydroxy-propranolol	0.42	—
O-Allyl-propranolol	0.96	84,000/115,400
Oxprenolol	0.79	52,200/106,200
Carazolol	0.59	126,400/73,400
Sotalol	1.49	100,500/108,200
Nifenalol	1.11	33,400/53,100
Propafenone	0.64	55,600/62,400
Norpropafenone	0.73	82,400/99,800
Bamethan	1.05	51,000/37,500
Ephedrine	0.41	—
Mefloquine	3.16	61,600/49,000
Quinidine/quinine	4.53	62,800/138,400
Oxyphencyclimine	2.15	134,700/138,400
Etidocaine	1.51	254,000/360,200
Bupivacaine	0.90	192,000/187,800
Troeger's base	0.77	150,000/100,000
Promethazine	1.46	361,000/375,800
Dixyrazine	0.38	—
Pheniramine	0.58	103,800/50,800
Doxylamine	0.79	3,900/14,200
Benzetimide	0.52	192,600/65,000

TABLE 3 (*Continued*)

Chiral selector	Molecule used	CEC mode	Enantiomers separated	R_s or (α)	N_1/N_2	Refs.
			Camylofine	0.48	40,000/14,000	
			Lorazepam	0.70	68,600/50,600	
			O-(*tert*-Butyl-carbamoyl)mefloquine	15.28	85,200/53,400	
Antibiotics	Vancomycin	Packed bed	Pindolol	2.74	138,700/123,724	[154]
			Alprenolol	2.21	73,660/68,336	
			Atenolol	2.11	52,380/53,152	
			Fenoterol	1.11	27,364/22,056	
			Metoprolol	1.22	79,364/26,256	
			Sotalol	1.66	36,572/31,692	
			Propranolol	1.63	31,872/25,704	
			Bupivacaine	0.92	44,516/26,308	
			Labetalol	2.2	64,252/16,428	
			Verapamil	1.90	106,428/81,264	
			Terbutaline	2.42	121,496/111,456	
			Thalidomide	13.8	189,804/113,776	
			Ketamine	0.5	113,492/39,616	
	Vancomycin	Continuous bed	Thalidomide	2.5	120,000/80,000	[270]
			Warfarin	0.6	30,000/20,000	
			Coumachlor	1.0	9,000/9,000	
			Felodipine	0.6	8,000/5,000	
Chiral polymethacrylate [poly(diphenyl-2-pyridylmethylmethacrylate)]		Packed bed	Benzoin acetate	—	—	[157]
			Methylbenzoin	—	5,410/4,280	
			Troger's base	—	—	
			Trans-stillbene oxide	—	—	

		Packed bed	Benzoin	1.93	11,200/10,800	[158]
			Methylbenzoin	1.57	12,100/10,125	
			Isopropylbenzoin	3.22	8,775/8,600	
			Benzoinacetate	2.95	12,725/12,350	
			Benzoin oxime	0.35	3,540/3,820	
			trans-stilbene oxide	4.57	22,975/9,700	
			1,1′-Binaphthyl-2,2-diol	2.20	8,475/8,800	
Chiral ligand exchanger	N-(2-hydroxy-3-allyloxypropyl)-L-4-hydroxyproline	Continuous bed	Phenylalanine	2.11	—	[159]
Polysaccharide derivatives	Cellusose tris(3,5-dimethylphenyl-carbamate)	Packed bed	2-(Benzylsulfinyl)-benzamide	3.57	122,967/121,265	[156,273]
			2-(Benzylsulfinyl)-benzoic acid benzyl ester	8.01	114,038/69,157	
			Etozolin	5.85	192,217/159,756	
			Piprozolin	3.57	122,967/121,265	
Cation exchangers	N-(4-Allyloxy-3,5-diclorobenzoyl)-1-amino-3-methylbutane	Packed bed	Talinolol	1.68	80,800/82,000	[260]
			Bupranolol	1.21	90,200/117,500	
			Bunitrolol	1.11	58,000/91,400	
			Celiprolol	1.20	97,000/99,500	
			Penbutolol	1.46	94,000/98,000	
			tert-Butyl-atenolol	1.75	90,200/85,400	
			tert-Butyl-metoprolol	2.02	119,300/126,000	
			tert-Butyl-propranolol	1.95	110,100/105,100	
			Atenolol	1.33	94,400/95,300	
			Metoprolol	1.39	81,500/100,400	
			Propranolol	0.67	58,100/67,600	

Table 3 (*Continued*)

Chiral selector	Molecule used	CEC mode	Enantiomers separated	R_s or (α)	N_1/N_2	Refs.
			Acebutolol	0.70	58,100/100,400	
			Practolol	1.30	91,400/93,200	
			Alprenolol	0.63	78,000/90,600	
			Pindolol	0.95	88,300/101,400	
			Mepindolol	0.68	82,200/73,000	
			Metipranolol	0.59	53,400/69,200	
			Normetoprolol	0.61	65,200/29,600	
			Ethoxymethyl-metoprolol	1.40	105,200/116,300	
			Ethoxyethyl-metoprolol	1.25	84,000/96,800	
			4-Methoxy-propranolol	0.62	108,600/76,100	
			4-Hydroxy-propranolol	0.42	—	
			O-Allyl-propranolol	0.96	84,000/115,400	
			Oxprenolol	0.79	52,200/106,200	
			Carazolol	0.59	126,400/73,400	

Sotalol	1.49	100,500/108,200
Nifenalol	1.11	33,400/53,100
Propafenone	0.64	55,600/62,400
Norpropafenone	0.73	82,400/99,800
Bamethan	1.05	51,000/37,500
Ephedrine	0.41	—
Mefloquine	3.16	61,600/49,000
Quinidine/quinine	4.53	62,800/138,400
Oxyphencyclimine	2.15	134,700/138,400
Etidocaine	1.51	254,000/360,200
Bupivacaine	0.90	192,000/187,800
Troeger's base	0.77	150,000/100,000
Promethazine	1.46	361,000/375,800
Dixyrazine	0.38	—
Pheniramine	0.58	103,800/50,800
Doxylamine	0.79	3,900/14,200
Benzetimide	0.52	192,600/65,000
Camylofine	0.48	40,000/14,000
Lorazepam	0.70	68,600/50,600
O-(*tert*-Butyl-carbamoyl)mefloquine	15.28	85,200/53,400

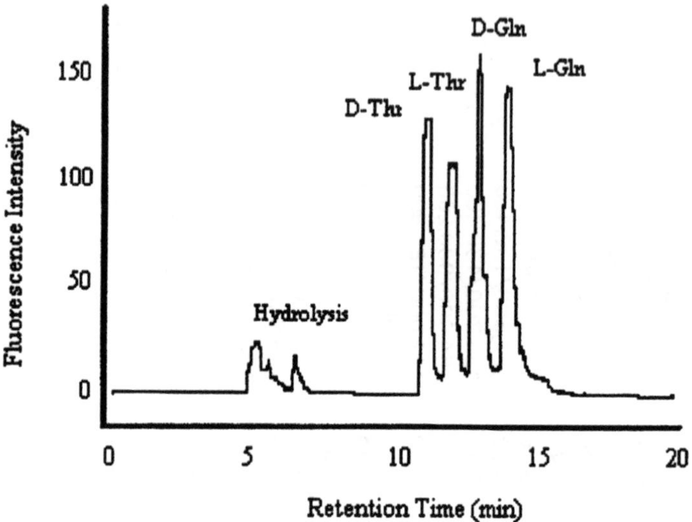

FIGURE 33 Electropherogram of threonine and glutamine enantiomers using an amino acid derivative CSP immobilized onto silica particles. Conditions: fused silica capillary, 30 cm × 75 mm i.d., packed segment is 15 cm in length, mobile phase is 30/70 5 mM phosphate (pH 2.5)/acetonitrile, field strength is 0.83 kV/cm. For resulting resolution, see Table 2. (Reprinted from Ref. 142, with permission.)

Peters et al. [143] used a valine-based chiral selector as the template molecule to prepare monolithic capillaries. These capillaries were used to successfully separate enantiomers of N-(3,5-dinitrobenzoyl)leucine. However, they found that the hydrophobicity of the monomers had a direct effect on the resolution and efficiency of the capillaries. The peaks tailed drastically due to reverse-phase interactions between the enantiomers and the monolith. They found that increasing the hydrophilicity of the monolith by the hydrolysis of the epoxide functionalities of the glycidyl methacrylate moieties resulted in a much more efficient separation.

Chirica and Remcho [20] created silicate-entrapped capillaries with predetermined selectivity by entrapping molecularly imprinted polymeric packing. The packing contained the chiral selector, and the silicate generated EOF. The resulting capillary was used to separate dansylphenylalanine enantiomers. Figure 34 shows a CEC separation of the enantiomers on the silicate-entrapped capillaries and an HPLC separation using a column packed with the MIP packing.

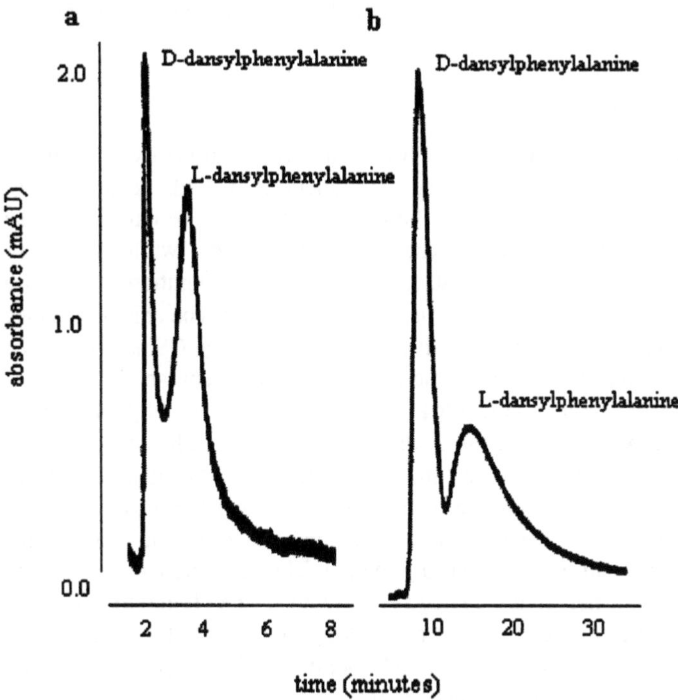

FIGURE 34 Separation of dansylphenylalanine on MIP CSP. (a) Silicate-entrapped CEC capillary. Conditions: i.d., 75 μm; L_{tot} 25 cm; L_{eff} 17 cm; applied potential 30 kV; electrokinetic injection, 10 kV for 10 s; UV detection at 280 nm, mobile phase is 80/20 acetonitrile/100 mM acetate (pH 3.0). (b) HPLC separation. Conditions: column, 15 cm × 0.46 cm; 1.0-mL/min flow rate, 2% acetic acid in acetonitrile, 30-μL injection, UV detection at 280 nm. (Reprinted from Ref. 20, with permission.)

6.4. "Brush"-Type Phases

"Brush" or Pirkle-type CSPs show promise in the enantioseparation of many classes of compounds, and also have shown to exhibit the high efficiencies expected in CEC separations. The high efficiency of enantioseparations of these CSPs is thought to be due to more favorable mass transfer kinetics between the analytes and these CSPs. Cavender et al. [144] used these CSPs bonded to silica particles to separate 10 chirally active compounds. They used (S)-Naproxen-derived CSP as well as the more widely used (3R, 4S)-Whelk-O. They found that while the (S)-Naproxen-derived CSP gave rise to more reproducible EOF, (3R,4S)-Whelk-O gave more efficient separations. The reproducibility of the

EOF with the former CSP was due to its "double-tethered" state when immobilized onto silica gel.

6.5. Ion Exchangers

An anion-exchange CSP usually involves a quinine functionality and one or more chiral centers. These CSPs carry two functions in CEC separations of enantiomers: chiral selectivity and the generation of EOF. The chirality of the CSP provides for enantioselective complexation interaction, while the anion-exchange ability provides an enantioselective molecular recognition driven by a coulombic interaction resulting in an ion-exchange process. The anion-exchange site is also responsible for generating EOF. Cation-exchange CSPs work in the same way, except that the exchange site carries a negative charge.

The group of Lindner [145,146] used *tert*-butylcarbamoylquinine as an anion-exchange-type CSP. This molecule was coated onto silica particles and packed into capillaries. The resulting stationary phase has a "zwitterionic" characteristic due to the positive charge of the CSP itself, and the negative charge of residual silanols. This characteristic allows control over the direction of the EOF (anodic versus cathodic) by varying the pH. The resulting capillaries were used for separation of negatively charged enantiomers. They found that the slow ion-exchange kinetics of the CSP gave rise to long runtimes; but they also found this to be avoidable by nonaqueous CEC conditions. This type of stationary phase has also been incorporated into an organic monolith and was successful in both the generation of EOF and enantioseparations [147,148].

6.6. Antibiotics

The antibiotics employed as CSPs generally have macrocyclic structures. These molecules are relatively large and contain several chiral centers and cavities that exhibit enantioselectivity via inclusion-complexation interactions with analytes. Antibiotics have been known to display a broad degree of enantioselectivity [149–153]. Vancomycin and teicoplanin are the most widely employed antibiotics for this use. Karlsson et al. [154] used a vancomycin CSP in both polar organic mode and reverse-phase CEC to separate a variety of enantiomers (Table 3).

6.7. Polysaccharide Derivatives

Polysaccharide derivatives used as CSPs interact with chiral analytes in much the same manner as cyclodextrins. These molecules have been shown to exhibit high chiral recognition ability in HPLC [155]. The main advantage of CEC over HPLC is the enhanced efficiency. In chiral separations, slow mass transfer kinetics between the CSP and chiral analytes have somewhat diminished the efficiency advantage of the technique. The goal of using polysaccharide derivatives as CSPs was to use a smaller amount of CSP, thus enhancing the mass transfer kinetics.

The group of Blaschke [156] investigated this idea using cellulose tris(3,5-dichlorophenylcarbamate) coated onto silica gel. Figure 35 shows the result of the study when differing amounts of the CSP were coated onto the particles. The CSP was studied using nonaqueous CEC, which also enhances the chiral recognition ability of the polysaccharide derivative. Their results agreed with theory: less CSP results in more efficient separations. The high degree of chiral recognition of the CSP is also exhibited in their results.

6.8. Other CSPs

Many other molecules have been incorporated into CEC for use as CSPs. Table 2 lists some the more common ones that have been used. The use of the positively charged polymer poly(diphenyl-2-pyridylmethylmethacrylate) as a CSP has been thoroughly investigated by Blaschke et al. [157,158]. This CSP carries a negative charge. The polymer has been coated onto silica particles which are underivatized and those derivitized with an aminopropyl group to investigate the generation of EOF. With the use of these CSPs it is possible to generate anodic or cathodic EOF, and they have been investigated concerning the promise of improving the efficiency of enantioseparations in CEC.

There have also been examples of ligand-exchange CSPs. Schmid et al. [159] used a ligand-exchange monomer as a chiral selector. The chiral selector, monomer, cross-linker, and charged monomer were polymerized to produce monolithic capillaries capable of chiral recognition and generation of EOF. The separation is achieved due to the differences in the stability between the ternary mixed copper complexes formed by the enantiomers and the CSP.

7. MISCELLANEOUS

A large number of analytes separated by CEC are not among the main categories mentioned earlier in this text. Some of these applications are listed in Table 1, and supplemental information can be found by the interested reader in excellent review articles published by Sandra et al. [4,5].

Additional noteworthy applications of CEC include natural products such as the plant flavonoids hesperetin and hesperidin [160], anthraquinones extracted from rhubarb and from Chinese medicine [161], and heterocyclic compounds present in oils of bergamot, mandarin, and sweet orange [162]. The CEC analysis of retinyl esters has been investigated by Roed et al. in nonaqueous mode for the separation of liver extracts of arctic seal [163]. Carotenoid isomers were also separated on C30 stationary phases by nonaqueous CEC [164]. It was found that CEC offered increased resolution compared to HPLC, and in CEC

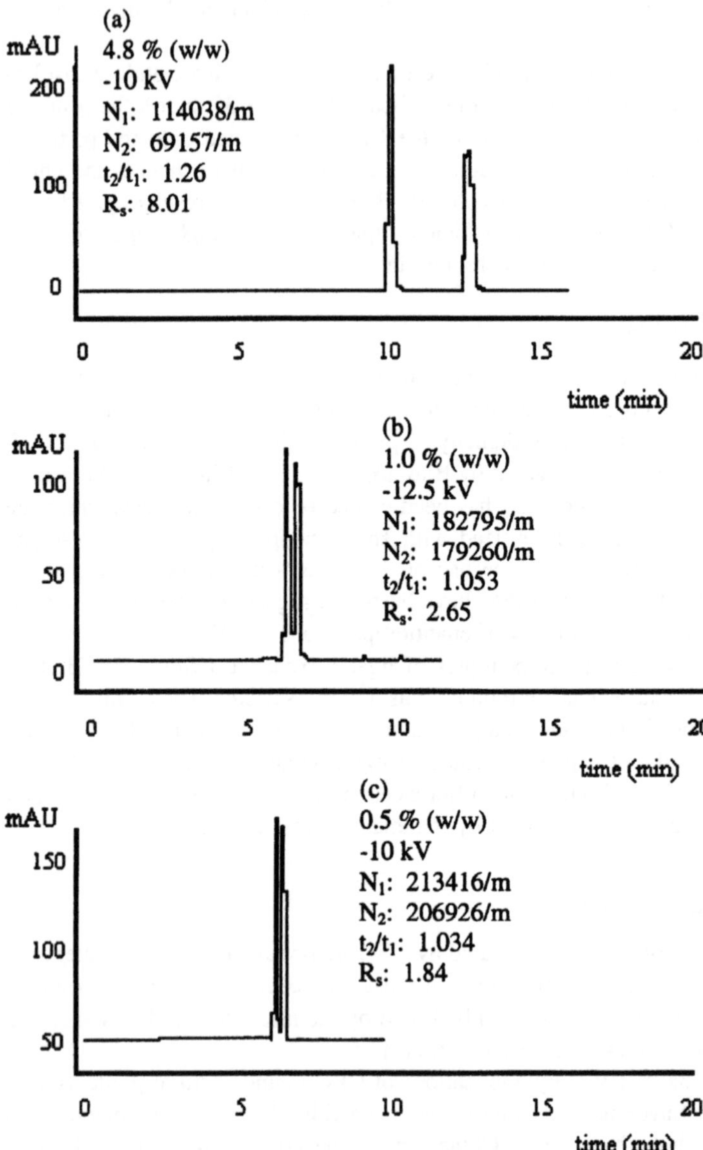

FIGURE 35 CEC enantioseparations of 2-(benzylsulfinyl)benzamide in capillaries packed with derivitized with differing amounts of the polysaccharide derivative cellulose tris(3,5-dichlorophenylcarbamate). The stationary phase contained (a) 4.8%, (b) 1.0%, and (c) 0.5% (w/w) of the chiral selector. (Reprinted from Ref. 156, with permission.)

format the C30 phases delivered efficiencies similar to those encountered with standard C18 phases. The separation of 17 carotenoid standards afforded identification of the peaks obtained in a gradient-elution CEC separation of a composite food extract [164]. Kvasnickova et al. developed a polymer-based stationary phase for the analysis of lignans from seeds of *Schisandra chinensis* [165]. The results of quantitative CEC analyses were in good agreement with those determined by similar HPLC analyses.

The feasibility of CEC for the analysis of dyes was also investigated in conjuncture with MS detection. Hugener et al. separated five water-soluble dyes by pressurized CEC and identified the peaks using ESI MS [166]. Lord et al. looked at the separations of nonionic textile dyes, a topic of great interest for the textile industry as well as archeology and forensic science [167]. The authors demonstrated the advantages of CEC, rapid and efficient separations, as well as the capability of using MS for analyte detection.

Another excellent demonstration of the wide application range of CEC has been the microanalytical determination of the molecular mass distribution of celluloses in the investigation of cellulose-based objects of cultural and historical value [168]. Stol et al. employed a column packed with bare silica particles of 5 μm diameter with a nominal pore size of 300 Å to achieve the separation of 2-kDa to 500-kDa celluloses by size exclusion (SE) CEC. The higher speed and smaller sample size advantages of SE CEC compared to SE HPLC distinguished CEC as the method of choice for the study of cellulose-based objects of archeological importance.

REFERENCES

1. Dittman MM, Masuch C, Rozing G. J Chromatogr A 2000; 887:209–221.
2. Altria KD. J Chromatogr A 1999; 856:443–463.
3. Colon LA, Burgos G, Maloney TD, Contron JM, Rodriquez RL. Electrophoresis 2000; 21:3965–3993.
4. Dermaux A, Sandra P. Electrophoresis 1999; 20:3027–3065.
5. Vanhoenacker G, Van der Bosch T, Rozing G, Sandra P. Electrophoresis 2001; 22:4064–4103.
6. Unger KK, Huber M, Walhagen K, Hennessy TP, Hearn MTW. Anal Chem 2002; 74:200A–207A.
7. Choudhary G, Apffel A, Yin H, Hancock W. J Chromatogr A 2000; 887:85–101.
8. Rimmer CA, Piraino SM, Dorsey JG. J Chromatogr A 2000; 887:115–124.
9. Zou H, Huang X, Ye M, Luo Q. J Chromatogr A 2002; 954:5–32.
10. Gfrorer P, Schewitz J, Puseker K, Bayer E. Anal Chem 1999; 71:315A–321A.
11. Ding J, Vourous P. Anal Chem 1997; 69:379–384.
12. Ding J, Vourous P. J Chromatogr A 2000; 887:103–113.
13. Tang Q, Lee ML. Trends Anal Chem 2000; 19:648–698.
14. Svec F, Peters EC, Sykora D, Frechet JMJ. J Chromatogr A 2000 887:3–29.

15. Jinno K, Sawada H. Trends Anal Chem 2000; 19:664–675.
16. Tan JZ, Remcho VT. Anal Chem 1997; 69:581–586.
17. Tan JZ, Remcho VT. Electrophoresis 1998; 19:2055–2060.
18. Pesek JJ, Matyska MT. J Chromatogr A 1996; 736:255–264.
19. Pesek JJ, Matyska MT. J Chromatogr A 2000; 887:31–41.
20. Chirica GS, Remcho VT. Electrophoresis 1999; 20:50–56.
21. Chirica GS, Remcho VT. Electrophoresis 2000; 21:3093–3101.
22. Chirica GS, Remcho VT. Anal Chem 2000; 72(15):3605–3610.
23. Minakuchi H, Nakanishi K, Soga N, Ishizuka N, Tanaka N. J Chromatogr A 1997; 762:135.
24. Peters EC, Petro M, Svec F, Frechet JMJ. Anal Chem 1998; 70:2288–2295.
25. Peters EC, Petro M, Svec F, Frechet JMJ. Anal Chem 1998; 70:2296–2302.
26. Ngola SM, Fintschenko Y, Choi WY, Shepodd TJ. Anal Chem 2001; 73:849–856.
27. Dulay MT, Quirino JP, Bennett BD, Kato M, Zare RN. Anal Chem 2001; 73: 3921–3926.
28. Shediac R, Ngola SM, Throckmorton DJ, Anex DS, Shepodd TJ, Singh AK. J Chromatogr A 2001; 925:251–263.
29. Gusev I, Huang X, Horvath C. J Chromatogr A 1999; 855:273–290.
30. Dittmann MM, Rozing G. J Chromatogr A 1996; 744:63–74.
31. Dadoo R, Zare RN, Yan C, Anex DS. Anal Chem 1998; 70:4787–4792.
32. Li DM, Remcho VT. J Microcol Sep 1997; 9:389–397.
33. Vallano PT, Remcho VT. Anal Chem 2000; 72:4255–4265.
34. Vallano PT, Remcho VT. J Phys Chem B 2001; 105:3223–3228.
35. Stol R, Kok WT, Poppe H. J Chromatogr A 1999; 853:45–54.
36. Huber C, Choudhary G, Horvath C. Anal Chem 1997; 69:4429–4436.
37. Lister AS, Rimmer CA, Dorsey JG. J Chromatogr A 1998; 828:105–112.
38. Ericson C, Hjerten S. Anal Chem 1999; 71:1621.
39. Taylor MR, Teale P. J Chromatogr 1997; 768:89–95.
40. Euerby MR, Gilligan D, Johnson CM, Roulin SCP, Myers P, Bartle KD. J Microcol Sep 1999; 11:305–311.
41. Ye M, Zou H, Liu Z, Ni J. J Chromatogr A 2000; 887:223–231.
42. Scherer B, Steiner F. J Chromatogr A 2001; 924:197–209.
43. Liu Y, Pietrzyk J. Anal Chem 2000; 72:5930–5938.
44. Wei W, Luo GA, Hua GY, Yan C. J Chromatogr A 1998; 817:65–74.
45. McKeown AP, Euerby MR, Johnson JM, Koeberlee M, Lomax H, Ritchie H, Ross P. Chromatographia 2000; 52:777–786.
46. Lurie IS, Conver TS, Ford VL. Anal Chem 1998; 70:4563–4569.
47. Smith NW, Evans MB. Chromatographia 1995; 41:197–203.
48. Moffatt F, Cooper PA, Jessop KM. Anal Chem 1999; 71:1119–1124.
49. Svec F, Peters EC, Sykora D, Frechet JMJ. J Chromatogr A 2000; 887:3–29.
50. Klampfl C, Hilder EF, Haddad PR. J Chromatogr 2000; 888:267–274.
51. Zhang LH, Zhang YK, Shi W, Zou HF. J High Resolution Chromatogr 1999; 22: 666–670.
52. Ye M, Zou H, Liu Z, Ni J, Zhang Y. J Chromatogr 1999; 855:137–145.
53. Wu R, Zou H, Ye M, Lei Z, Ni J. Electrophoresis 2001; 22:544–551.

54. Sandra P, Dermaux A, Ferraz V, Dittman MM, Rozing G. J Microcol Sep 1997; 9:409–419.
55. Dermaux A, Sandra P, Ksir M, Zarrouck FF. J High Resolution Chromatogr 1998; 21:545–548.
56. Saeed M, Depala M, Craston DH, Anderson IGM. Chromatographia 1999; 49: 391–398.
57. Taylor MR, Teale P, Westwood SA, Perrett D. Anal Chem 1997; 69:2554–2558.
58. Stead DA, Reid RG, Taylor RB. J Chromatogr 1998; 798:259–267.
59. Paterson C, Boughtflower R, Higton D, Palmer E. Chromatographia 1997; 46: 599–604.
60. Roed L, Lundanes E, Greibrokk T. J Microcol Sep 1999; 11:421–430.
61. Sirimanne SR, Barr JR, Patterson DG Jr. J Microcol Sep 1999; 11:109–166.
62. Ding JM, Szeliga J, Dipple A, Vouros P. J Chromatogr 1997; 781:327–334.
63. Ding J, Vouros P. J Chromatogr A 2000; 887:103–113.
64. Ding J, Vouros P. Anal Chem 1997; 69:379–384.
65. Que AH, Konse T, Baker AG, Novotny MV. Anal Chem 2000; 72:2703–2710.
66. Lurie IS, Meyers RP, Conver T. Anal Chem 1998; 70:3255–3260.
67. Dulay MT, Chan Y, Rakewstraw DJ, Zare RN. J Chromatogr A 1996; 725:361–366.
68. Lelievre F, Yan C, Zare RN, Gareil P. J Chromatogr A 1996; 723:145–156.
69. Robson MM, Roulin S, Shariff SM, Raynor MW, Bartle KD, Clifford AA, Myers P, Euerby MR, Johnson CM. Chromatographia 1996; 43:313–321.
70. Yamamoto H, Baumann J, Erni F. J Chromatogr A 1992; 593:13–319.
71. Walhagen K, Unger KK, Hearn MTW. J Chromatogr A 2000; 894:35–43.
72. Asiaie R, Huang X, Farnan D, Horvath C. J Chromatogr A 1998; 806:251–263.
73. Li D, Knobel HH, Remcho VT. J Chromatogr B 1997; 695:169–174.
74. Kitagawa S, Tsuji A, Watanabe H, Nakashima M, Tsuda T. J Microcol Sep 1997; 9:347–356.
75. Hilder E, Klampfl CW, Haddad PR. J Chromatogr A 2000; 890:337–345
76. Klampfl CW, Hilder E, Haddad PR. J Chromatogr A 2000; 888:267–274.
77. Breadmore MC, Hilder E, Macka M, Haddad PR. Trends Anal Chem 2001; 20: 355–364.
78. Boyce MC, Breadmore M, Macka M, Doble P, Haddad PR. Electrophoresis 2000; 21:3073–3080.
79. Chen WH, Lin SY, Liu CY. Anal Chim Acta 2000; 410:25–35.
80. Yan C, Dadoo R, Zhao H, Zare RN, Rakestraw DJ. Anal Chem 1995; 67:2026–2029.
81. Fintschenko Y, Choi WY, Ngola SM, Shepodd TT. Fresenius J Anal Chem 2001; 371:174–181.
82. Dittman MM, Rozing R. J Microcol Sep 1997; 9:399–408.
83. Henry CW, McCarroll ME, Warner IM. J Chromatogr A 2001; 905:319–327.
84. Tegeler T, El Rassi Z. Electrophoresis 2002; 23:1217–1223.
85. Yang CM, El Rassi Z. Electrophoresis 1999; 20:2337–2342.
86. Cooper PA, Jessop KM, Moffatt F. Electrophoresis 2000; 21:1574–1579.
87. Tegeler T, El Rassi Z. J Chromatogr A 2002; 945:267–279.

88. Cassells NP, Lane CS, Depala M, Saeed M, Craston DH. Chemosphere 2000; 40: 609–618.
89. Bailey CG, Yan C. Anal Chem 1998; 70:3275–3279.
90. Fujimoto C, Kino J, Savada H. J Chromatogr A 1995; 716:107–113.
91. Huber C, Choundhary G, Horvath C. Anal Chem 1997; 69:4429–4436.
92. Alicea-Maldonado R, Colon L. Electrophoresis 1999; 20:37–42.
93. Behnke B, Metzger JW. Electrophoresis 1999; 20:80–83.
94. Adam T, Unger KK. J Chromatogr A 2000; 894:241–251.
95. Walhagen K, Unger KK, Hearn MTW. Anal Chem 2001; 73:4924–4936.
96. Throckmorton D, Shediac R, Shepodd T, Singh A. J Chromatogr 2001; 1:251–265.
97. Zhang S, Huang X, Zhang J, Horvath C. J Chromatogr A 2000; 887:465–477.
98. Zhang S, Zhang J, Horvath C. J Chromatogr A 2001; 914:189–200.
99. Pesek J, Matyska M, Swedberg S, Udivar S. Electrophoresis 1999; 20:2343–2348.
100. Matyska MT, Pesek JJ, Boysen RI, Hearn MTW. Anal Chem 2001; 73:5116–5125.
101. Matyska MT, Pesek JJ, Boysen RI, Hearn MTW. J Chromatogr A 2001; 924: 211–221.
102. Matyska MT, Pesek JJ. Anal Chem 1999; 71:5508–5514.
103. Xu W, Regnier F. J Chromatogr A 1999; 853:243–256.
104. Choudhary G, Horvath C, Banks JF. J Chromatogr A 1998; 828:469.
105. Apffel A, Yin H, Hancock WS, McManigill D, Frenz J, Wu SL. J Chromatogr A 1999; 832:149.
106. Huang P, Xiaoying J, Chen Y, Srivasan JR, Lubman DM. Anal Chem 1999; 71: 1786–1791.
107. Gucek M, Gaspari M, Walhagen K, Vreeken RJ, Verheij ER, van der Greef J. Rapid Commun Mass Spectrom 2000; 14:1448–1454.
108. Gucek M, Pilar B. Chromatographia 2000; 51:S139–S142.
109. Yang C, El Rassi Z. Electrophoresis 1998; 19:2061–2067.
110. Suzuki S, Yamamoto M, Kuwahara Y, Makiura K, Honda S. Electrophoresis 1998; 19:2682–2688.
111. Suzuki S, Kuwahara Y, Makiura K, Honda S. J Chromatogr A 2000; 873:247–256.
112. Palm A, Novotny M. Anal Chem 1997; 69:4499–4507.
113. Guo W, Koropchak JA, Yan C. J Chromatogr A 1999; 849:587–597.
114. Verheij ER, Tjaden UR, Niessen WMA, van der Greej J. J Chromatogr A 1991; 554:339–349.
115. Helboe T, Hansen SH. J Chromatogr A 1999; 836:315–324.
116. Calhours X, Morin P, Agrofoglio L, Dreux M. J High Resolution Chromatogr 2000; 23:138–142.
117. Zhang M, El Rassi Z. Electrophoresis 1998; 19:2068–2072.
118. Zhang M, El Rassi Z. Electrophoresis 1999; 20:31–36.
119. Zhang M, Yang C, El Rassi Z. Anal Chem 1999; 71:3277–3282.
120. Ding J, Vorous P. J Chromatogr A 2000; 887:103–113.
121. Ding J, Vourous P. Anal Chem 1997; 69:379–384.
122. Smith NW, Evans MB. Chromatographia 1994; 38:649–657.

123. Thiam S, Shamsi SA, Henry CW III, Robinson JW, Warner IM. Anal Chem 2000; 72:2541–2546.
124. Jinno K, Sawada H, Catabay AP, Watanabe H, Sabli NBH, Pesek JJ, Matyska MT. J Chromatogr A 2000; 887:479–487.
125. Kapnissi C, Akbay C, Schlenoff JB, Warner IM. Anal Chem 2002; 74:2328–2335.
126. Strickmann DB, Blaschke G. J Chromatogr B 2000; 748:213–219.
127. Lammerhofer M, Svec F, Frechet JMJ, Lindner W. J Chromatogr A 2001; 925: 265–277.
128. Enlund AM, Isaksson R, Westerlund D. J Chromatogr A 2001; 918:211–220.
129. Spikmans V, Lane SJ, Tjaden UR, van der Greef J. Rapid Commun Mass Spectrom 1999; 13:141–149.
130. Enlund AM, Ericson C, Hjerten S, Westerlund D. Electrophoresis 2001; 22:511–517.
131. Vallano PT, Remcho VT. J Chromatogr A 2000; 887:125–135.
132. Gfrorer P, Schewitz J, Pusecker K, Tseng L-H, Albert K, Bayer E. Electrophoresis 1999; 20:3–8.
133. Gil-Av E, Feibush B, Charles-Sieger R. Tetrahedron Lett 1966; 10:1009–1015.
134. Snopek J, Jelinek I, Smolkova-Keulemansova E. J Chromatogr 1988; 452:571.
135. Terabe S, Ozake H, Otsuka K, Ando T. J Chromatogr 1985; 332:211.
136. Nishi H, Fukuyama T, Terabe S. J Chromatogr 1991; 553:503.
137. Knopek J, Smolkova-Keulemansova E. J Chromatogr 1988; 438:211.
138. Fanali S. J Chromatogr 1991; 545:437.
139. Mayer S, Schurig V. J High Resolution Chromatogr 1992; 15:129–131.
140. Mayer S, Schurig V. J Liq Chromatogr 1993; 16:915–931.
141. Li S, Lloyd DK. Anal Chem 1993; 65:3684–3690.
142. Kato M, Dulay MT, Bennett B, Chen J-R, Zare RN. Electrophoresis 2000; 21: 3145–3151.
143. Peters EC, Lewandowski K, Petro M, Svec F, Fréchet MJ. Anal Comm 1998; 35: 83–86.
144. Cavender DM, Wolf C, Spence PL, Pirkle WH, Derrico EM, Rozing GP. J Chromatogr A 1997; 782:175–179.
145. Tobler E, Lämmerhofer M, Lindner W. J Chromatogr A 2000; 875:341–352.
146. Lämmerhofer M, Tobler E, Lindner W. J Chromatogr A 2000; 887:421–437.
147. Lämmerhofer M, Peters EC, Yu C, Svec F, Fréchet JMJ, Lindner W. Anal Chem 2000; 72:4614–4622.
148. Lämmerhofer M, Svec F, Fréchet JMJ. Anal Chem 2000; 72:4623–4628.
149. Armstrong DW, Tang YB, Chen SS, Zhou YW, Bagwill C, Chen JR. Anal Chem 1994; 66:1473–1484.
150. Armstrong DW, Gasper MP, Rundlett KL. J Chromatogr A 1995; 689:285–304.
151. Desiderio C, Fanali S. J Chromatogr A 1998; 818:281–282.
152. Medvedovici A, Sandra P, Toribio L, David F. J Chromatogr A 1997; 785:159–171.
153. Dönnecke J, Svensson LA, Gyllenhaal O, Karlsson KE, Karlsson A, Vessman J. J Microcol Sep 1999; 11:521–533.
154. Karlsson C, Karlsson L, Armstrong DW, Owens PK. Anal Chem 2000; 72:4394–4401.

155. Okamoto Y, Kawashima M, Hatada K. J Chromatogr A 1986; 363:173.
156. Girod M, Chankvetadze B, Okamoto Y, Blaschke G. J Sep Sci 2001; 24:27–34.
157. Krause K, Chankvetadze B, Okamoto Y, Blaschke G. Electrophoresis 1999; 20: 2772–2778.
158. Krause C, Chankvetadze B, Okamoto Y, Blaschke G. J Microcol Sep 2000; 12: 398–406.
159. Schmid MG, Grobuischek N, Tuscher C, Gübitz G, Végvári Á, Egidijus M, Maruška A, Hjertén S. Electrophoresis 2000; 21:3141–3144.
160. Eimer T, Unger KK, van der Greef J. Trends Anal Chem 1996; 15:463–468.
161. Li Y, Liu H, Ji X, Li J. Electrophoresis 2000; 21:3109–3115.
162. Cavazza A, Bartle KD, Dugo P, Mondello L. Chromatographia 2001; 53:57–62.
163. Roed L, Lundanes E, Greibrokk T. J Chromatogr A 2000; 890:347–353.
164. Sander LC, Pursch M, Maerker B, Wise SA. Anal Chem 1999; 71:3477–3483.
165. Kvasnickova L, Glatz Z, Sterbova H, Kahle V, Slanina J, Musil P. J Chromatogr A 2001; 916:265–271.
166. Hugener M, Tinke AP, Tjaden UR, Niessen WMA, van der Greef J. J Chromatogr A 1993; 647:375–385.
167. Lord GA, Gordon DB, Tetler LW, Carr CM. J Chromatogr A 1995; 700:27–33.
168. Stol R, Pedersoli JL, Poppe H, Kok WT. Anal Chem 2002; 74:2314–2320.
169. Zhang LH, Zhang YK, Shi W, Lei ZD, Zou HF. J High Resolution Chromatogr 1999; 22:666–670.
170. Breadmore MC, Hilder EF, Macka M, Avdalovic N, Haddad PR. Electrophoresis 2001; 22:503–510.
171. Mayer M, Rapp E, Marck C, Bruin GJM. Electrophoresis 1999; 20:43–49.
172. Engelhardt H, Hafner FT. Chromatographia 2000; 52;769–776.
173. Fung YS, Long YH. J Chromatogr A 2001; 907:301–311.
174. Araki T, Chiba M, Tsunoi S, Tanaka M. Anal Science 2000; 16:412–424.
175. Hilmi A, Luong JHT. Electrophoresis 2000; 21:1395–1404.
176. Kotia RB, Li L, McGown LB. Anal Chem 2000; 72:827–831.
177. Charles JAM, McGown LB. Electrophoresis 2002; 23:1599–1604.
178. Wu R, Zou H, Fu H, Jin W, Ye M. Electrophoresis 2002; 23:1239–1245.
179. Charvatova J, Kral V, Deyl Z. J Chromatogr B 2002; 770:155–163.
180. Wu R, Zou H, Ye M, Lei Z, Ni J. Anal Chem 2001; 73:4918–4923.
181. Yu C, Svec F, Frechet JMJ. Electrophoresis 2000; 21:120–127.
182. Svec F, Peters E, Sykora D, Frechet JMJ. J High Resolution Chromatogr 2000; 23:3–18.
183. Ludtke S, Adam T, Unger KK. J Chromatogr A 1997; 786:229–235.
184. Palm A, Novotny MV. Anal Chem 1997; 69:4499–4507.
185. Ye ML, Zou HF, Liu Z. J Chromatogr A 2000; 869:385–394.
186. Ye ML, Zou HF, Liu Z, Ni JY, Zhang YK. J Chromatogr A 1999; 855:137–145.
187. Pesek JJ, Matyska M, Williamsen EJ, Evanchic M, Hazari V, Konjuh K, Takhar S, Tranchina R. J Chromatogr A 1997; 786:219–228.
188. Wu JT, Huang P, Li MX, Lubman DM. Anal Chem 1997; 69:2908–2913.
189. Huang P, Wu JT, Lubman DM. Anal Chem 1999;70:3003–3008.
190. He B, Ji J, Regnier F. J Chromatogr A 1999; 853:257–262.
191. Zhang J, Huang X, Zhang SH, Horvath C. Anal Chem 2000; 72:3022–3029.

192. Ericson C, Hjerten S. Anal Chem 1999; 71:1621–1627.
193. Pesek JJ, Matyska M, Sandoval JE, Williamsen EJ. J Liq Chromatogr Relat Technol 1996; 19:2843–2865.
194. Pesek. JJ, Matyska MT, Cho J. J Chromatogr A 1999; 845:237–246.
195. Huang X, Zhang J, Horvath C. J Chromatogr A 1999; 858:91–101.
196. Zhao RR, Johnson BP. J Liq Chromatogr Relat Technol 2000; 23:1851–1857.
197. Matyska MT, Pesek JJ, Katrekar A. Anal Chem 1999; 71:5508–5514.
198. Moffatt F, Cooper PA, Jessop KM. J Chromatogr A 1999; 855:215–226.
199. Vanhoenacker G, Dermaux A, De Keukeleire D, Sandra P. J Sep Sci 2001; 24: 55–58.
200. Euerby MR, Johnson CM, Bartle KD, Myers P, Roulin SCP. Anal Commun 1996; 33:403–405.
201. Carlsson A, Petersson P, Walhagen A. In: Sandra P, Ed. CD-ROM Proceedings of the 20th International Symmposium on Capillary Chromatography, Riva del Garda, Italy, IOPMS, Kortrijk, Belgium 1998.
202. Gordon DB, Lord GA, Jones DS. Rapid Commun Mass Spectrom 1994; 8:544–548.
203. Lord GA, Gordon DB, Myers P, King BW. J Chromatogr A 1997; 768:9–16.
204. Seilar RM, Kok WT, Kraak JC, Poppe H. Chromatographia 1997; 46:131–136.
205. Seilar RM, Kok WT, Kraak JC, Poppe H. J. Chromatogr A 1998; 808:71–77.
206. Euerby MR, Johnson CM, Cikalo M, Bartle KD. Chromatographia 1998; 47:135–140.
207. Fujimoto C, Fujise Y, Matsuzawa E. Anal Chem 1996; 68:2753–2757.
208. Frame LA, Robinson ML, Lough WJ. J Chromatogr A 1998; 798:243–249.
209. Mayer M, Rapp E, Marck C, Bruin GJM. Electrophoresis 1999; 20:43–49.
210. Hilhorst MJ, Somsen GW, de Jong GH. Chromatographia 2001; 53:190–196.
211. Spikmans V, Lane SJ, Smith NW. Chromatographia 2001; 51:18–24.
212. Adam T, Unger KK. GIT Labor Fachz 1999; 43:1056–1061.
213. Que AH, Palm A, Baker AG, Novotny MV. J Chromatogr A 2000; 887:379–391.
214. Djordjevic NM, Fitzpatrick F, Houdiere F, Lerch G, Rozing GP. J Chromatogr A 2000; 887:245–252.
215. Tang QL, Lee ML. J High Resolution Chromatogr 2000; 23:73–80.
216. Que AH, Konse T, Baker AG, Novotny MV. Anal Chem 2000; 72:2703–2710.
217. Rentel C, Gfroerer P, Bayer E. Electrophoresis 1999; 20:2329–2336.
218. Catabay AP, Sawada H, Jinno K, Pesek JJ, Matyska MT. J Capillary Electrophoresis 1988; 5:89–95.
219. Cahours X, Morin P, Dreux M. J Chromatogr A 1999; 845:203–216.
220. Euerby MR, Johnson CM, Bartle KD. LC-GC Int. 1998; 11:39–44.
221. Desiderio C, Fanali S. J Chromatogr A 2000; 895:123–132.
222. Hoegger D, Freitag R. J Chromatogr A 2001; 914:211–220.
223. Endund AM, Westerlund D. J Chromatogr A 2000; 895:17–25.
224. Miyawa JH, Alasandro MS, Riley CM. J Chromatogr A 1997; 769:145–153.
225. Miyawa JH, Lloyd KD, Alasandro MS. J High Resolution Chromatogr 1998; 21: 161–168.
226. Wei W, Luo GA, Hua GY, Yan C. J Chromatogr A 1998; 817:65–74.

227. Lane SJ, Boughtflower R, Paterson C, Underwood T. Rapid Commun Mass Spectrom 1996; 10:733–736.
228. Dekkers SEG, Tjaden UR, van der Greef J. J Chromatogr A 1995; 712:201–209.
229. Pesek JJ, Matyska MT. J Chromatogr A 1996; 736:313–320.
230. Pesek JJ, Matyska MT. J Capillary Electrophor 1997; 5:213–217.
231. Nagaraj S, Karnes HT. Biomed Chromatogr 2000; 14:234–242.
232. Taylor RB, Vorarat S, Reid RG, Boyle SP, Moody RR. J Capillary Electrophor Microchip Technol 1999; 6:131–136.
233. Meyring M, Chankvetadze B, Blaschke G. J Chromatogr A 2000; 876:157–167.
234. Meyring M, Strickmann D, Chankvetadze B, Blaschke G, Desiderio C, Fanali S. J Chromatogr B 1999; 723:255–264.
235. Lurie IS, Bailey CG, Anex DS, Bethea MJ, McKibben TD, Casale JF. J Chromatogr A 2000; 870:53–68.
236. Lim JT, Zare RN, Bailey CG, Rakestraw DJ, Yan C. Electrophoresis 2000; 21: 737–742.
237. Lai EPC, Dabek-Zlotorzynska E. Electrophoresis 1999; 20:2366–2372.
238. Hilhorst MJ, Somsen GW, de Jong GH. J Chromatogr A 2000; 872:315–321.
239. Desiderio C, Ossicini L, Fanali S. J Chromatogr A 2000; 887:489–496.
240. Guan N, Zeng ZR, Wang YC, Fu EQ, Cheng JK. Anal Chim Acta 2000; 418: 145–151.
241. Luedtke S, Unger KK. Chimia 1999; 53:498–500.
242. Stol R, Mazereeuw M, Tjaden UR, van der Greef J. J Chromatogr A 2000; 873: 293–298.
243. Mazereeuw M, Spikmans V, Tjaden UR, van der Greef J. J Chromatogr A 2000; 872:315–321.
244. Smith NW. J Chromatogr A 2000; 887:233–243.
245. Gillot NC, Euerby MR, Johnson CM, Barrett DA, Shaw PN. Chromatographia 2000; 51:167–174.
246. Dittmann MM, Masuch K, Rozing GP. J Chromatogr A 2000; 887:209–221.
247. Armstrong DW, Tang Y, Ward T, Nichols M. Anal Chem 1993; 65:1114–1117.
248. Szeman J, Ganzler K. J Chromatogr A 1994; 668:509–517.
249. Szeman J, Ganzler K. J Chromatogr A 1994; 668:509–517.
250. Jakubetz H, Czesla H, Schurig V. J Micro Sep 1997; 9:421–431.
251. Wistuba D, Czesla H, Roeder M, Schurig V. J Chromatogr A 1998; 815:183–188.
252. Wei W, Luo G, Xiang R, Yan C. J Microcol Sep 1999; 11:263–269.
253. Koide T, Ueno K. J High Resolution Chromatogr 2000; 23:59–66.
254. Wistuba D, Schurig V. Electrophoresis 2000; 21:3152–3159.
255. Ye M, Zou H, Lei Z, Wu R, Liu Z, Ni J. Electrophoresis 2001; 22:518–525.
256. Yang J, Hage DS. Anal Chem 1994; 66:2719–2725.
257. Lloyd DK, Li S, Ryan P. J Chromatogr A 1995; 694:285–296.
258. Kotia RB, Li L, McGown LB. Anal Chem 2000; 72:827–831.
260. Liu Z, Otsuka K, Terabe S. J Sep Sci 2001; 24:17–26.
260. Bruggemann O, Freitag R, Whitcombe MJ, Vulfson EN. J Chromatogr A 1997; 781:43–53.
261. Schweitz L, Andersson LI, Nilsson S. Anal Chem 1997; 69:1179–1183.
262. Schweitz L, Spegel P, Nilsson S. Analyst 2000; 125:1899–1901.

263. Wolf C, Spence PL, Pirkle WH, Cavender DM, Derrico EM. Electrophoresis 2000; 21:917–924.
264. Lammerhofer M, Lindner W. J Chromatogr A 1998; 829:115–125.
265. Tobler E, Lammerhofer M, Wuggenig F, Hammerschmidt F, Lindner W. Electrophoresis 2002; 23:462–476.
266. Dermaux A, Lynen F, Sandra P. J High Resolution Chromatogr 1998; 21:575–576.
267. Carter Finch AS, Smith NW. J Chromatogr A 1999; 848:375–385.
268. Fanali S, Rudaz S, Veuthey J-L, Desiderio C. J Chromatogr A 2001; 919:195–203.
269. Desiderio C, Aturki Z, Fanali S. Electrophoresis 2001; 22:535–543.
270. Kornysova O, Owens PK, Maruska A. Electrophoresis 2001; 22:3335–3338.
271. Krause K, Girod M, Chankvetadze B, Blaschke G. J Chromatogr A 1999; 837: 51–63.
272. Meyring M, Chankvetadze B, Blaschke G. J Chromatogr A 2000; 876:157–167.
273. Chankvetadze B, Kartozia I, Breitkreutz J, Girod M, Knobloch M, Okamoto Y, Blaschke G. J Sep Sci 2001; 24:251–257.
274. Girod M, Chankvetadze B, Blaschke G. Electrophoresis 2001; 22:1282–1291.
275. Chankvetadze B, Kartozia I, Breitkreutz J, Okamoto Y, Blaschke G. Electrophoresis 2001; 22:3327–3334.
276. Chankvetadze L, Kartozia I, Yamamoto C, Chankvetadze B, Blaschke G, Okamoto Y. Electrophoresis 2002; 23:486–493.
277. Ding J, Ning B, Fu G, Lu Y, Dong S. Chromatographia 2000; 52:285–288.
278. Dermaux A, Medvedovichi A, Ksir MH, Talbi M, Sandra P. J Microcol Sep 1999: 11:451–459.
279. Zhang MQ, Ostrander GK, El Rassi Z. J Chromatogr A 2000; 887:287–297.
280. Roed L, Lundanes E, Greigrokk T. J Microcol Sep 2000; 12:561–567.
281. Roed L, Lundanes E, Greigrokk T. Electrophoresis 1999; 20:2373–2378.
282. Matyska M, Pesek J, Yang L. J Chromatogr A 2000; 887:497–503.
283. Girelli AM, Messina A, Ferrantelli P, Sinibaldi M, Tarola AM. Chromatographia, 2001; 53:S284–S289.
284. Whitaker KW, Sepaniak MJ. Electrophoresis 1994; 15:1341–1345.
285. Peters EC, Petro M, Svec F, Frechet JMJ. Anal Chem 1998; 70:2296–2302.

15

Clinical Applications of Microfluidic Devices

Joan M. Bienvenue, James Karlinsey,
James P. Landers, and Jerome P. Ferrance
University of Virginia, Charlottesville, Virginia, U.S.A.

1. INTRODUCTION

Much of the early work on the application of microfluidic devices to clinical analyses focused on electrophoretic microchips, particularly for DNA separations. Through this work, it was shown that the diagnostic ability of the microchip separations was equivalent to, but much faster than, current slab-gel methods for a number of clinical assays. A number of other microchip applications, including preconcentration, filtration, enzymatic reactions, simple flow analyses, or coupling with detection methods, are also now being applied to clinical applications with equivalent diagnostic ability. These methods are not yet being applied in the clinical laboratory, because many are not yet robust and rugged enough to move beyond the research stage. Instrumentation and microdevices are only beginning to be designed for commercial production, where testing and validation will have to take place before these methods are ready for routine clinical use.

The one exception to this is microchip electrophoretic separations, for which instrumentation and microdevices are now commercially available. Though not designed specifically for clinical applications, clinical analyses have been carried out on these instruments and have been shown to perform at the necessary sensitivity and selectivity. However, the major advantages of the microchip separations, fast separation times and reduced volumes, are not of themselves enough to move the clinical community to this new method. Although they are fast, microchip separations still only incrementally reduce the total analysis time for many analyses. The sample processing steps which precede separation remain the largest contributor to total analysis time and generate much larger volumes than are actually necessary for the microchip analysis. Some of these processing steps are the new applications being developed on microchips, with the intended goal of reducing the time, cost, and volumes of these processes to better match those of microchip separations.

With these processes also performed on microchips, the true advantage of microdevices begins to be evident. Traditional analyses require multiple steps in multiple instruments/devices, with sample handling between each step creating opportunities for sample loss or contamination. Microchips provide the ability to incorporate all of the processes onto a single device in an integrated fashion, such that one process leads directly into the next with little or no user intervention. This leads to the type of "sample in–answer out" processing that has been promoted for years as "lab-on-a-chip." The methods and techniques discussed in this chapter can be combined in various ways to form microdevices that perform a diversity of clinical analyses. These integrated devices can be designed to perform rapid, single sample analyses directly at the point of care, or parallel-processed to provide low-cost, high-throughput screening of patient samples in a centralized laboratory.

While the fully integrated device described is still a few years from being realized, this chapter reports on the processes already being developed for microfluidic devices having applications for clinical analyses. One particular omission from the research reported here is microarray-type microchip devices, which are currently available for both DNA and protein genotyping. Because most designs for these devices do not employ microfluidics, but act simply as microdetection devices, they have not been included in the scope of this chapter. What has been included is the wide variety of clinical analytes currently being investigated using microchips, along with the detection and separation technologies used to analyze them. We focus on three types of clinically relevant analytes: small molecules (including drugs, ions, neurotransmitters, carbohydrates, and amino acids), proteins and peptides, and DNA. The major focus is on DNA because of the wealth of separations and sample processing steps already reported on microdevices for DNA. In addition, we explore some of the challenges of integrating the individual processes into a single device. The examples and literature cited here are by no mean the complete sum of the current microchip work applicable to clinical analyses, as advances are continuing at an astonishing pace, but give some idea of the breadth and cutting-edge nature of the research.

2. SMALL MOLECULES

2.1. Drugs

Application of microdevices to the analysis of drugs can provide important benefits for both therapeutic agents and drugs of abuse. Rapid identification of a drug or its metabolites in blood or urine can accurately target treatments in drug overdoses or allergic reactions. Simple monitoring of drug concentrations is often important to maintain a specific therapeutic level in the blood, or to ensure that the drug is being properly metabolized. Microdevices for drug analysis will also find applications in pharmaceutical rather than clinical applications, with the requirements for high throughput in drug development being a major driving force. Pharmaceutical applications are more likely to benefit from the initial work, as mass spectrometry (MS) is used as the detection method for many of these devices. While mass spectrometry is widely used in the pharmaceutical industry, it is usually not applied in clinical laboratories.

Pharmaceutical application was first demonstrated by rapid MS determination of sildenafil from commercial tablets with no sample pretreatment using an octagonal microchip designed for high-throughput applications [1]. Eight microchannels, fabricated in a poly(methyl methacrylate) (PMMA) chip, ended in sharp tips designed to create an electrospray source. By rotating the position of the chip, sequential electrospray analysis was performed by applying a voltage in the appropriate sample well. In this instance, the low-cost, disposable microdevice served only as a multiplex delivery device, with no processing performed on the chip.

Plastic microdevices for high-throughput screening with MS detection were also prepared for detection of aflatoxins and barbiturates. These devices incorporated concentration techniques interfaced with electrospray ionization MS (ESI-MS) through capillaries [2]. The microfluidic device for aflatoxin detection employed an affinity dialysis technique, in which a poly (vinylidene fluoride) (PVDF) membrane was incorporated in the microchip between two channels. Small molecules were dialyzed from the aflatoxin/antibody complexes, which were then analyzed by MS. A similar device was used for concentrating barbiturate/antibody complexes using an affinity ultrafiltration technique. A barbiturate solution was mixed with antibodies and then flowed into the device, where uncomplexed barbiturates were removed by filtration. The antibody complex was then dissociated and electrokinetically mobilized for MS analysis. In each case, the affinity preconcentration improved the sensitivity by at least one to two orders of magnitude over previously reported detection limits.

Mass spectrometric detection has also been directly interfaced with microchip separations for drug detection. These studies, detecting imipramine and desipramine in fortified human plasma, show analysis of spiked analytes in clinical sample matrices for drug detection [3]. These widely used tricyclic antidepressants inhibit the reuptake of the neurotransmitters serotonin and norepinephrine in the central nervous system. Unfortunately, the 5-mg/mL detection limit found for these antidepressants with this method is not low enough to detect typical clinical levels of the drugs. Combinatorial library characterization and preclinical drug delivery studies should benefit, however, since the concentrations utilized are considerably higher than those encountered in pharmacokinetic studies.

More traditional fluorescence detection on microchips has been applied to drugs that are naturally fluorescent. The chiral forms of the antibiotic gemifloxacin were separated in a poly(dimethylsiloxane) (PDMS) microchip using a chiral crown ether as a selector [4]. Analysis of the drug in urine samples exhibited problems due to the high concentration of sodium ions, which interfere with binding to the crown ether. This was negated by addition of EDTA as a chelating agent to reduce binding inhibition effects. The limit of detection was only 3 μM due to the high background fluorescence contributed by the PDMS when excited with 325-nm light. However, this limit of detection was still found to be acceptable when compared to the 4 μM limit found using the traditional UV detection method.

Drugs that are not naturally fluorescent can also be detected by laser-induced fluorescence (LIF) if they are tagged with a fluorescent compound. Amphetamine and its analogs, used as central nervous system stimulants and drugs of abuse, were derivatized with fluorescein isothiocyanate (FITC) and analyzed in urine samples [5]. Urinary extracts, prepared by solid-phase extraction, and fortified urine were both subjected to the labeling procedure. The extracts exhibited a sensitivity of about 200 ng/mL, relevant for toxicological drug screening, but the limit for identification of drugs in fortified urine was only 10 μg/mL, too high for practical purposes. Higher efficiencies were found with microchip separations, compared to capillary separations, with no loss of accuracy or precision. Compounds from this same class were also labeled with 4-fluoro-7-nitrobenzofurazane (NBD-F) and separated using a micellar electrokinetic chromatography (MEKC) method in a channel microfabricated in glass (Figure 1) [6]. In addition, chiral separation of these compounds was achieved using γ-cyclodextrin (γ-CD) as the chiral selector and low concentrations of sodium dodecyl sulfate (SDS). Though labeled off-line in the study, the authors propose NBD-F as a possible on-chip pre- or postcolumn labeling reagent because of the rapid kinetics of the reaction.

Thermal lens microscopy (TLM) is another detection technique which has been applied for drug detection on microchips [7]. This detection method relies on the transfer of vibrational energy from absorbed laser light to the solution as heat. Heating of the solution changes the refractive index of the solution, causing a thermal lens effect that displaces the focal point of a probe laser. A schematic is given in Figure 2. This method was employed for sensitive detection of L-ascorbic acid and dehydroascorbic acid (DHAA) levels in urine, which can be used in the clinical diagnosis of diseases such as scurvy. An integrated flow injection analysis (FIA) system was constructed on a microdevice for determination of ascorbic acid by a redox-complexation reaction. The reaction was carried out directly on the quartz microchip by flowing the reactants into the channel using microsyringe pumps, allowing them to mix by molecular diffusion. The

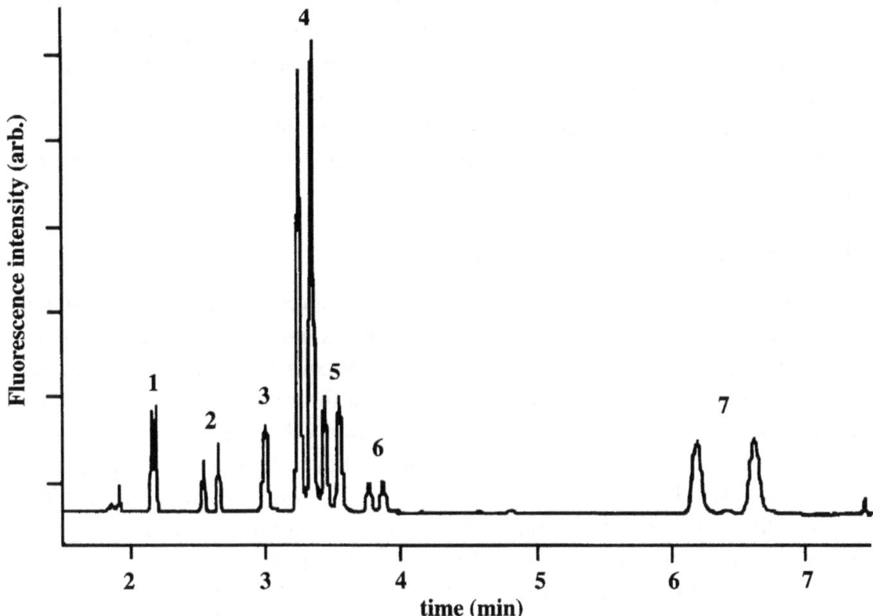

FIGURE 1 Separation of 1,(+/−)-norephedrine; 2, (+/−)-ephedrine; 3, (+/−)-cathinone; 4, (+/−)-amphetamine; 5, (+/−)-pseudoephedrine; 6, (+/−)-methcathinone, and 7, (+/−)-methamphetamine. Separation conditions: 160-mm separation length; 50 mM phosphate buffer, pH 7.35; 10 mM HS-γ-CD, 1.5 mM SDS; 8 kV. (Reprinted with permission from Ref. 6.)

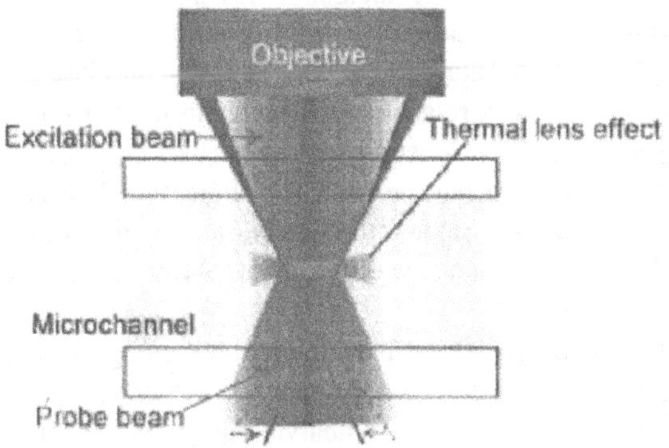

FIGURE 2 Schematic of thermal lens microscopy (TLM). (Reprinted with permission from Ref. 8.)

reaction product, ferroin, absorbs laser light and was detected by TLM in both pharmaceutical preparations and urine, with no interference from colored or other active analytes in the urine. The method showed a limit of detection of 0.1 μM ascorbic acid, with a linear range up to 40 μM, sufficient for clinical analysis. DHAA concentrations were determined by reacting the urine sample with a reducing agent, then measuring the ascorbic acid concentration again, and calculating the change.

2.2. Endogenous Small Molecules and Ions

TLM has also been demonstrated for the determination of catecholamines (CA) in pharmaceutical preparations [8]. Catecholamines, including epinephrine, norepinephrine, dopamine, and L-DOPA, are endogenous neurotransmitters useful for diagnosing diseases, but are also synthesized for treatment of asthma, myocardial infarction, and neural disorders. These four CA were independently detected in the microchip FIA system described above. An oxidation reaction in the microchannel produced aminochromes by reaction of each catecholamine with sodium metaperiodate; concentrations were determined in a reaction time of only 15 s. Linear correlations between the TLM signal and the concentrations of each CA were determined over the range of 5–20 μg/mL, values suitable for pharmaceutical analysis. This method is being integrated with a microchip separation domain since all four catecholamines are normally present in biological fluids. For clinical applications, however, this method would have to become much more sensitive, as plasma levels of epinephrine are in the low-pg/mL range and norepinephrine levels in plasma range from 100 to 350 pg/mL in normal individuals.

Epinephrine was also detected in a microchip FIA system which employed chemiluminescence (CL) as the detection method [9]. Epinephrine-containing samples were injected into a stream containing lucigenin flowing into the FIA chip. The epinephrine/lucigenin solution underwent a light-producing oxidation/reduction reaction when mixed with an alkaline solution on a silicon/glass hybrid microchip. The intensity was linearly related to the epinephrine concentration when the channel was designed to allow sufficient mixing of the reaction components. Again, this device could not yet be used for clinical analyses, as even in microchips with different mixing modes, the detection limit was found to be no better than 800 nM, while plasma levels of epinephrine are in the 0.1–0.3 nM range.

Neurotransmitters have also been detected on microchips using electrochemical detection, eliminating the need for on-chip reactions or derivitization. Amperometric detection, a current change when an analyte passes the detection electrodes, was demonstrated on a microchip for the determination of dopamine concentrations in standard solutions [10]. The microdevice developed in this study was a wide channel into which sample introduction through a capillary allowed spatial control of sample within the channel. The device was used for flow injection analysis, to detect dopamine concentration changes, and for separation of dopamine from catechol, a standard EC detection substrate. Spatially resolved detection in the microdevice was accomplished using parallel measurements at a microarray of 100 platinum detection electrodes. This allowed spatially resolved continuous electrophoretic separations to be performed on the device, but no detection limits were reported for this method to show that it was sensitive enough for clinical applications. Amperometric detection after separation of dopamine and catechol was also reported with thick-film carbon electrodes [11]. These electrodes were printed on ceramic strips which were placed at the exit from the separation channel. Because the detector was not part of the separation microdevice, the detection electrode was easily replaced. This system reported detection limits as low as 380 nM dopamine, due to the lower noise found with the thick electrodes.

Direct integration of the electrodes into the separation microdevice for electrochemical detection was also reported by two other groups. Photolithographic placement of a platinum electrode at the exit of an electrophoresis channel in glass microdevices allowed EC detection following separation of dopamine, epinephrine, and catechol with limits of detection ($S/N = 2$) of 3.7, 6.5, and 12 μM, respectively [12]. An alternative method, suggested as more flexible, formed electrodes by sputtering gold directly onto the channel outlet. This also provided electrochemical analysis of mixtures via the amperometric detection of L-DOPA, dopamine, and isoproterenol separated in the electrophoretic channel. A linear response was found for dopamine from 20 to 200 μM, with a 1.0 μM limit of detection [13]. While these methods do not yet provide the sensitivity necessary for clinical analysis of neurotransmitters, the electrochemical detection limits are sensitive enough for analysis of other small molecules.

Electrochemical detection was easily employed for the determination of uric acid in urine, abnormal concentrations of which have been linked to several disease states [14]. A glass/PDMS hybrid device with an off-chip platinum electrode was used to evaluate standard samples for both dopamine and uric acid. The linear responses for dopamine and uric acid were 1–165 and 15–110 μM, respectively, with a 1 μM limit of detection for both. Normal concentrations of uric acid in urine are 800–8000 μM, thus a 50 to 75-fold dilution was used with the urine samples analyzed to place them within the linear range of the detection method. Uric acid concentrations in these urine samples were confirmed using the clinically accepted method. This new method should allow clinical detection of both abnormally high and abnormally low uric acid concentrations in urine samples on a microdevice.

Conductivity detectors represent another method for EC detection that do not require species that are electroactive. With this method, changes in the conductivity of the migrating analyte zones from the background electrolyte are detected rather than the species themselves. This method was applied for determination of oxalate at clinically relevant concentrations in urine on a PMMA electrophoretic microchip [15]. Chloride, with an absolute ionic mobility close to the electrophoretic mobility of oxalate, posed a problem with the conductivity detector, as the chloride concentration in urine (50–150 mM) is approximately three orders of magnitude higher than the oxalate concentration. At a low pH (3.2–4.5), however, adequate resolution was attained to determine oxalate concentrations down to 80 nM, in the presence of more than a 10^4 excess of chloride (Figure 3).

Endogenous molecules and ions in urine were also studied using a number of other detection methods on microchips. Carnitine and selected acylcarnitines in human urine were purified using solid-phase extraction (SPE) before separation and detection on a microchip electrophoresis/mass spectroscopy device [16]. Quadrupole selective ion monitoring (SIM) provided both sufficient sensitivity and scan rate to detect the targeted carnitines, at a concentration range of 35–124 μM in the extracted urine. Laser-induced fluorescence detection has been employed for determination of carbohydrate levels in urine on microdevices, measurements that can be used to detect metabolic diseases [17]. Mono- and disaccharide standards were first labeled with 9-aminopyrene-1,4,6-trisulfonate (APTS) and then separated on an electrophoretic microchip to establish calibration curves. Urine samples from both normal and galactosemia patients were deproteinated and then evaluated using this technique to show the clinical utility of this analysis method. Absorbance detection, using a fiber-optic system, has also been employed for clinical evaluation of urine samples on microdevices. In this method, calcium ions in urine were determined by complexation with arsenazo III immobilized on polymeric beads [18]. Beads were placed in a reservoir fabricated into a channel at the detection point, and samples were electrokinetically mobilized through the channel. Calcium ion concentrations in urine samples were easily determined using calibration curves established for the device. These calcium measurements, requiring no sample pretreatment, were found to correlate well with clinical determinations.

2.3. Amino Acids

Amino acids (AA) have also been studied on microdevices for development of microclinical analysis devices. In urine, normal ranges for standard amino acids and their metabolites range from 0 to 24 mM, with abnormal concentrations indicative of a number of disease states. Plasma concentrations of certain amino acids can also be used for disease detection. Elevated homocysteine levels in plasma is an independent risk factor for cardiovascular disease. Microdevices employing end-column amperometric detection were used for the determination

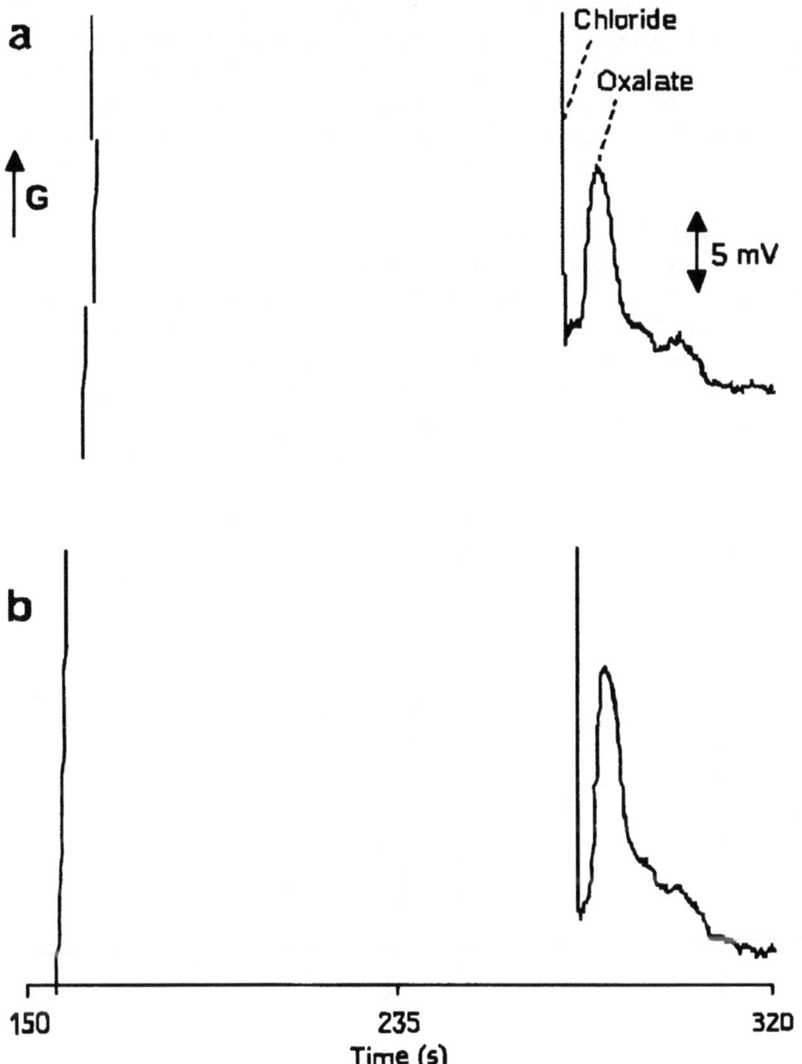

FIGURE 3 Repeatable separation of oxalate and chloride in a 100-fold diluted urine on the chip: (a) first and (b) tenth run from a series of 10 consecutive runs with one sample. The driving current was stabilized at 12 μA. G, increasing conductance. (Reprinted with permission from Ref. 15.)

of both total homocysteine (tHcy) and protein-bound homocysteine (pbHcy) in plasma [19]. Capillary separations, using off-column EC detection, were first optimized to provide a clinically appropriate linear range, then transferred to the microchip platform. Homocysteine and reduced glutathione, 200 μM each, were separated and detected on the microchip, but additional evaluation of the microchip system to determine the limits of detection for these compounds was not performed.

The electropherogram in Figure 4 shows the free-solution electrophoretic separation of four amino acids on a PMMA microchip employing a conductivity detector [20]. Baseline resolution was achieved for alanine, valine, glutamine, and tryptophan. A calibration plot was determined for alanine spanning a concentration range of 15–80 nM, with a limit of detection of 8.0 nM. Separation and detection of peptides, proteins, and oligonucleotides were also performed on this microdevice, using techniques that included free-solution electrophoresis, MEKC, and capillary electrochromatography, but no clinical samples were evaluated.

Harrison and co-workers were the first to show quantitative analysis of six fluorescein isothiocyanate (FITC)-labeled amino acids on a microdevice, with separations possible in less than 5 s [21]. The concentrations they examined

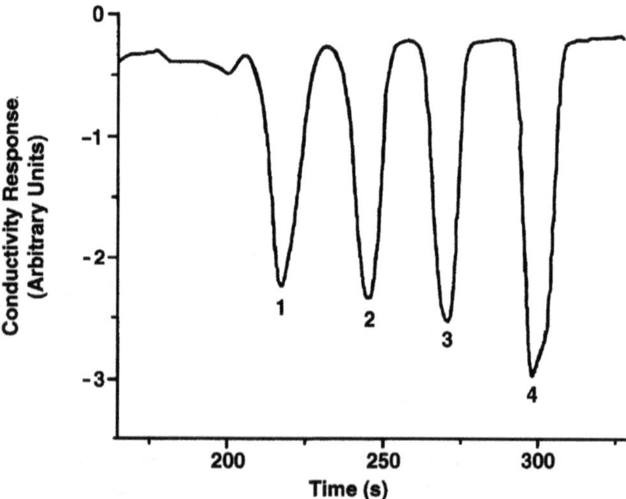

FIGURE 4 Separation of 100 μM amino acid mixture consisting of (1) alanine, (2) valine, (3) glutamine, and (4) tryptophan in an unmodified PMMA microchip using indirect, contact conductivity detection. Electrophoretic conditions: 3-s electrokinetic injection time; E= 150 V/cm for the electrophoresis. (Reprinted with permission from Ref. 20.)

were in the clinical range, but no clinical applications were investigated. The FITC labeling method has also been combined with MEKC on an electrophoretic microdevice. In this study both standards and urine samples were evaluated, with detection limits as low as 3.3 nM observed [22]. Micellar electrochromatography on microchips was also performed on amino acids labeled with tetramethylrhodamine isothiocyanate [23]. This study showed separation of 19 of the 20 naturally occurring amino acids in 165 s. Concentrations of each amino acid were 10 μM, well within the clinical range, but no clinical samples were evaluated.

The ability to detect amino acids with on-chip sample labeling was reported in a number of studies. Two studies looked at pre- and postcolumn labeling of amino acids directly on the microdevice using the reaction of *o*-phthaldialdehyde with amino acids to form fluorescent compounds [24,25]. These studies separated only three amino acids, however. For separation and detection of 19 of the standard amino acids, Munro and co-workers used indirect fluorescence detection to overcome these problems [26]. Employing a buffer containing a background fluorescent molecule, analyte zones were detected as a decrease in signal as they passed the detector. Using this method, they were able to detect elevated amino acid levels in urine resulting from metabolic disorders or kidney dysfunction.

The FITC labeling method was also applied to chiral separations of amino acids on a microchip to determine the enantiomeric ratios of amino acids found on a meteorite [27]. Since biotic amino acids are normally single enantiomers, chiral separations of amino acids are not truly clinical in nature, but illustrate the potential for chiral separations of small molecules of clinical interest. Mathies and co-workers used this technique to search for evidence of life in extraterrestrial environments. Enantiomeric forms of Val, Ala, Glu, and Asp could be discriminated by addition of α-, β-, or γ-cyclodextrin (CD) to the run buffer. Improved resolution with faster separations was found with respect to conventional CE. This method has been modified, by addition of SDS to the buffer, to perform cyclodextrin-modified micellar electrokinetic chromatography (CD-MEKC) [28]. Increasing the SDS concentration decreased the magnitude of electroosmotic flow (EOF), increasing the effective migration distance, and therefore the resolution on the microchips.

3. PROTEINS AND PEPTIDES

A significant amount of work on microdevices has recently focused on proteins and peptides. Some of this was initiated as part of the larger proteomics effort, but currently the work is centered largely on the transfer of current analytical methods to microchips. This includes enzymatic assays and immunoassays, both of which are routinely utilized in clinical laboratories. We have included these methods under separate subsections, but will first detail work in other areas of protein research on microchips.

3.1. Mass Spectroscopic Analyses

Mass spectroscopic analysis of proteins and peptide fragments has become an often-utilized technique over the past few years, and a number of microdevices have been reported for use with this analysis method. Though not yet used routinely in clinical analyses, detecting expression patterns and specific mutations in proteins may well be the best way to detect diseases and propose treatments specific for a given mutation. Transfer of this type of analysis to the clinical laboratory may await the development of microdevices for the sample preparation steps, which can interface directly with MS detectors.

Trypsin digestion of proteins has been carried out on a microdevice for producing peptide fragments that can be analyzed by MS [29]. This device provided the ability to cut a protein band directly from a polyacrylamide gel, then elute and digest the protein in a single step. While this device did not couple directly with the MS detector, other microdevices, similar to those reported for preparing and delivering small molecules for MS analysis, have been designed for proteins. Sample cleanup was addressed using a microdialysis microdevice that was integrated with an electrospray tip for analyzing sample by ESI-MS [30]. This greatly enhanced the signal-to-noise ratio while reducing sample waste and improving sensitivity. A more complex device provided dual microdialysis for removing both high- and low-molecular-weight species before the ESI-MS interface [31]. This technique allowed the molecular-weight range for analysis to be selected. Another method used for sample concentration was a microdevice designed to create gradients by controlling the electrokinetic mobilization of two solvents from different reservoirs [32]. The researchers constructed a solid-phase bed in their electrospray interface capillary, onto which they loaded a trypsin-digested sample. The microchip-generated gradient, with nanoliter-per-minute flow rates, then allowed sequential mobilization of peptides into the MS, separating and concentrating the analytes.

A number of reports have detailed the coupling of microchip CE directly with ESI-MS for proteins, giving sensitive detection of both proteins and peptides in nanomolar concentrations [33–35]. Proteins and peptides investigated included myoglobin, recombinant human growth hormone, ubiquitin, endorphin, figrinopeptide a, and trypsin-digested lectin. Samples deposited in different reservoirs on one of these microchips could be sequentially mobilized without cross-contamination. Membrane proteins were separated on a microdevice and analyzed at trace levels using both a triple quadrupole and a quadrupole time-of-flight mass spectrometer (QqTOF-MS). A concentration limit of 3.2–43.5 nM was found for different peptides achieved by selective ion monitoring [36].

While all of these devices used normal zone electrophoresis separation techniques, isoelectric focusing (IEF) methods on microchips have also been interfaced with ESI-MS for high-resolution separations of proteins [37]. Figure 5 shows the electropherogram and corresponding mass spectra generated by this device. For on-chip sample preconcentration before separation, a polarity-switching technique was employed to achieve subnanomolar detection limits for many peptide standards [38].

3.2. Size-Based Separations

Many popular protein sizing techniques used in clinical analyses have been adapted as capillary electrophoresis methods, and are now being translated to microdevices. SDS capillary gel electrophoresis has been performed in microfabricated channels, achieving similar separation efficiencies to conventional capillary-based SDS CE while accelerating protein separations by a factor of 20 [39]. Six fluorescently tagged proteins, with masses from 9 to 116 kDa, were separated in less than 35 s. The chip surface plays an important role in the efficiency of protein separations, with proteins being known to bind to both hydrophilic and hydrophobic surfaces. To prevent protein adsorption to the capillary walls, PDMS coatings were examined with different surfactants for detection of a number of proteins [40]. This work was transferred to a microdevice, showing detection of a fluorescently tagged immunoglobulin.

Labeling of proteins before separation causes a number of problems. For example, labeling can change the order of separation, and multiple labeling can cause additional peaks in the separation. As a result, Colyer and co-workers incorporated postseparation labeling of the proteins with 2-toluidinonapthalene-6-sulfonate (TNS) on microdevices, separating individual serum proteins (IgG, transferrin, α_1-antitrypsin, and albumin) to mimic the γ, β, α_1, and albumin zones of human serum [41]. While separation of serum proteins can provide information regarding various disease states, sensitivity limitations and differential complexation with the TNS prevented separation of actual clinical samples. Additionally, TNS/protein complexes require excitation at UV wavelengths for fluorescent emission. NanoOrange is another dye that forms highly fluorescent complexes when bound to hydrophobic protein regions, except its maximal excitation wavelength is 488 nm. This dye was used on a microdevice, for postcolumn labeling of the model serum proteins: α-lactalbumin, β-lactalbumin A, and β-lactalbumin B [42]. These proteins were eluted in order of decreasing pI, with the separation completed in 36 s.

Postcolumn labeling requires a more complex microdevice than the simple crossed-tee design. Direct incorporation of the fluorescent label into the run buffer would eliminate this problem, but would have to not interfere with the separations. This was shown by Jin et al., who carried out protein/SDS size-

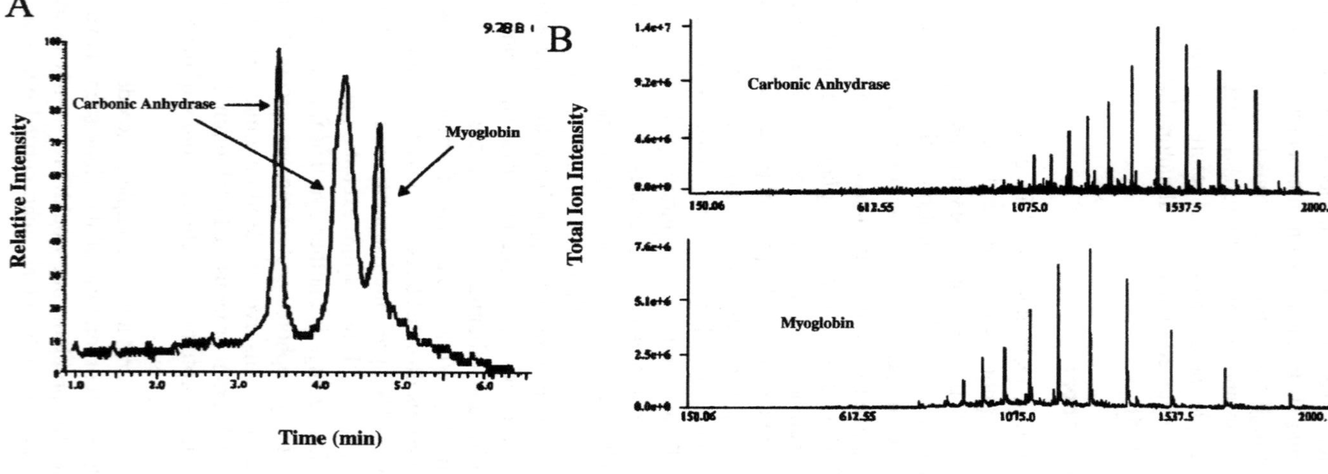

Figure 5 (A) IEF Electopherogram of carbonic anhydrase (pI 5.9 and 6.8) and myoglobin (pI 7.2) mixture containing 0.05 mg/mL for each protein, obtained at 0.05 µL/min with 0.5-psi nitrogen and 0.2 µL/min sheath liquid assistance. (B) Positive ESI mass spectra of carbonic anhydrase and myoglobin from peaks 1 and 3. (Reprinted with permission from Ref. 37.)

based separations on microdevices, with LIF detection possible simply by including NanoOrange into the run buffer [43]. A microchip separation of a size standard is shown in Figure 6. This work has been further extended to increase the sensitivity, lowering the limit of detection to less than 2 ng/mL [44] and applying it to the microchip separation of serum proteins under nondenaturing conditions [45]. Incorporating the dye directly into the run buffer is also employed in a commercial microchip instrument for protein separations [46]. This method requires a destaining step, however, to dilute the SDS concentration in the buffer and lower the background fluorescence. The microchip is designed to sequentially analyze 11 different samples with enough sensitivity to detect 30 nM carbonic anhydrase.

3.3. Enzymatic Assays

Enzymatic assays are often used clinically to detect the presence or concentration of specific molecules in biofluids. These types of assays can be performed

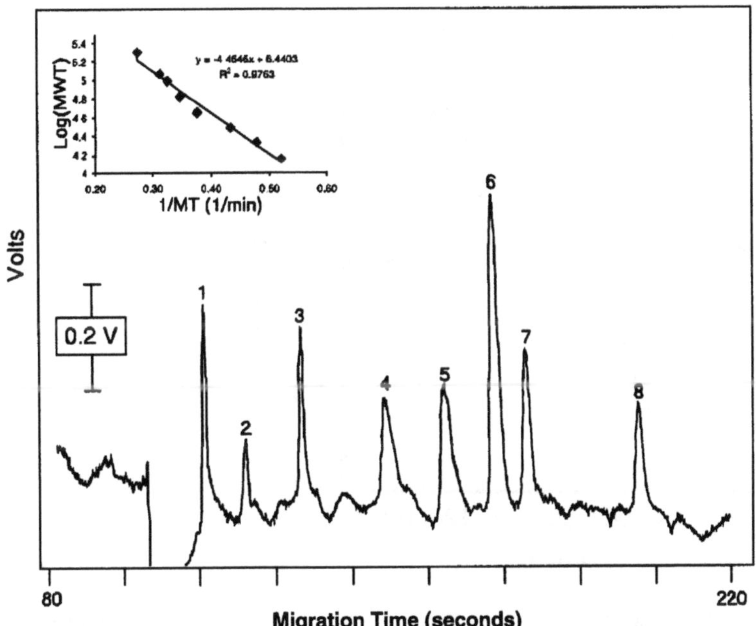

Figure 6 Microchip electrophoresis of protein-SDS size standard with laser-induced fluorescence detection (488/590 nm). NanoOrange stock solution (500×) was added into CE-SDS run buffer at 1× concentrations. Inset shows linear relationship between log (MWT) and 1/MT. (Reprinted with permission from Ref. 43.)

on microdevices using both immobilized and free enzymes, with a number of different detection methods. Reaction kinetics for the enzyme β-galactosidase (β-Gal) were assayed on a microchip with a substrate that hydrolyzed to form the fluorescent product resorufin [47]. Using electrokinetic flow, precise concentrations of substrate, enzyme, and inhibitor were mixed in nanoliter volumes with the cross intersections and mixing tees used for valving. The two fluid streams were mixed proportionately by varying the field strengths in the mixing channels providing the ability to change reagent concentrations dynamically in a single experiment, as demonstrated in Figure 7. Inhibition of β-Gal was also evaluated, by varying concentrations of different inhibitors. The microchip assays required only 20 min, and the amount of reagent consumed was reduced by four orders of magnitude over a conventional assay. A fluorogenic enzymatic assay was also demonstrated with human T-cell phosphatase (TCPTP) on a glass microchip [48]. The assay showed the ability to accurately control flow from a number of reservoirs with solutions of varying viscosities by application of external pressure sources.

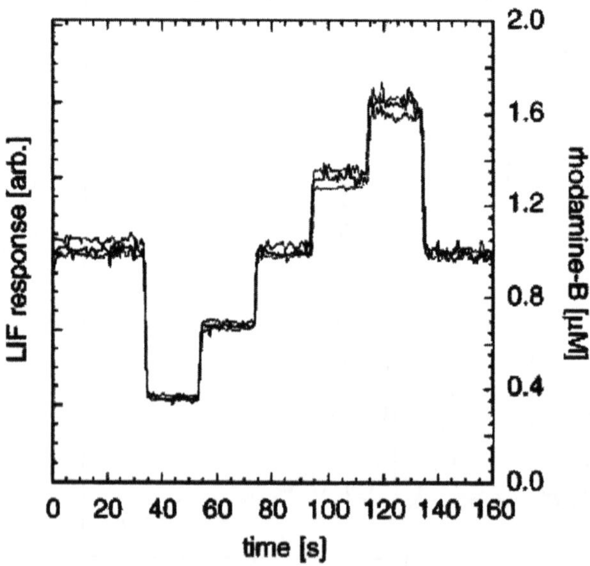

FIGURE 7 Fluorescence intensity versus time for three runs of a step-gradient increase from 0.32 to 1.68 μM rhodamine B. The left axis fluorescence signal was collected 20 mm downstream from the mixing cross, and the right axis shows the concentration of rhodamine B corresponding to each signal plateau. (Reprinted with permission from Ref. 47.)

A more complicated microchip was developed to perform a shift assay for detecting protein kinase A activity; this kinase transfers a phosphate group from an ATP to a serine residue [49]. In eukaryotic cells, kinases play a role in signal transduction and are believed to be involved in a number of diseases, including cancer. While kinase activity is traditionally measured using radiolabeled reagents, the microchip method relied on phosphorylation of a fluorogenic substance attached to a short peptide chain containing a serine residue. In the microdevice, the enzyme, substrate, and varying amounts of the inhibitor H-89 were mixed in a reaction chamber. After an incubation period, an aliquot of the mixture was injected into a separation channel, where the phosphorylated product could be separated from the unreacted substrate. This type of microchip shift assay greatly reduces the reagent consumption and provides the potential for high-throughput screening or kinetic measurements.

Silicon wafers, machined to form deep parallel channels, provided a lamellae structure with considerably increased reactor surface area for enzyme immobilization [50]. Because the enzyme is exposed in a thin layer, no substrate diffusion into the enzyme layers occurred, and the reactor displayed flow-independent enzyme activity using immobilized peroxidase and glucose oxidase. The peroxidase activity was detected at different flow rates by simply monitoring the absorbance of Trinder reagent, which produces a red color as the H_2O_2 is reacted. In the glucose oxidase microdevice, fluid from a microdialysis fiber passes over the immobilized enzyme, and an oxygen electrode, positioned at the microdevice outlet, measures the drop in the oxygen concentration. The dialysis probe was sequentially immersed into various glucose solutions, ranging from 0 to 100 mM, suggesting that this type of sampling system could be operated in an in-vivo environment. To increase the catalytic activity, porous silicon dice were incorporated into the reactor to further increase the surface area for immobilization [51]. The apparent enzyme activity increased, but a flow-rate dependency was introduced because of the need to saturate the entire porous region to obtain true enzyme measurements.

3.4. Immunoassays

Electrophoresis-based immunoassays have been successfully applied on the microchip format using a number of different assay methods. Relying only on the electrophoretic separation capabilities of microchips, competitive assays have been performed by a number of researchers. Cortisol in serum was analyzed by mixing serum with anti-cortisol antibody and fluorescently labeled cortisol [52]. The mixture was then separated, giving a sharp free labeled antigen peak and a slowly migrating diffuse antibody–antigen band in less than 30 s. Serum cortisol standards were used to create a calibration curve with a working range of 1–60 μg/dL, covering the range of clinical interest. A competitive immunoassay was also performed on a microchip for the determination of serum thyroxine (T4), used to assess thyroid function [53]. The antigen was completely separated from the antibody/antigen complex in less than 1 min, with linear T4 quantitation over the range of normal adult serum levels.

Chiem and Harrison used a simple crossed-tee chip design to carry out direct and competitive immunoassays [54]. Direct assays were performed by mixing the labeled theophylline or labeled bovine serum albumin with varying amounts of antibody. The mixtures were then separated on microchips to show the presence of the complex (Figure 8). Competitive assays for theophylline concentrations required mixing the serum or standard samples with the antibody and labeled antigen and allowing the mixture to incubate before the microchip separation. An integrated immunoreactor was then developed for the serum theophylline competitive immunoassay [55]. This microdevice included separate reservoirs for the sample, the labeled antigen, and the antibody, which were mixed together sequentially by application of the appropriate voltages to the reservoirs. The mixture incubated as it passed through a mixing coil, and was subsequently injected into the separation channel. Results were equivalent to

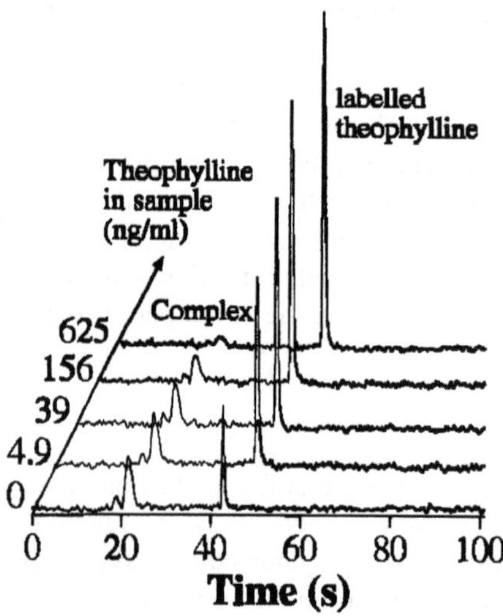

FIGURE 8 Series of on-chip electropherograms for a competitive Th assay using 10 μL of labeled theophylline, 1.5 μL of stock antitheophylline, and increasing amounts of unlabeled Th. A pH 8.0, 50 mM tricine/40 mM NaCl/0/0.1% Tween-20 buffer was used, with a 6-kV separation potential. (Reprinted with permission from Ref. 54.)

mixtures incubated off-chip, with a detection limit of 0.26 mg/L, and a linear range up to 40 mg/L.

A different kind of microfluidic device, called a T-sensor, was used to perform flow immunoassays [56]. This technique relied on low-Reynolds-number flow of two adjacent streams, one containing antibody and another fluorescent-labeled antigen. While no mixing of streams occurs, antigen diffuses into the antibody stream rather rapidly. As it complexes with an antibody, however, the diffusion slows and fluorescence accumulates at the edge of the antibody stream. This is detected at a set distance along the flow path to allow for sufficient diffusion of the antigen. Competitive assays can be performed by adding a specific concentration of labeled antigen directly to the sample and measuring a difference in fluorescence in the antibody stream. Many therapeutic drugs, drugs of abuse, and pesticides are within the sensitivity range of this technique, and there is no need to pretreat blood samples before measurement.

Kitamori's research group has successfully integrated immunosorbent assays into microdevices, using polystyrene beads as the immunosorbent. Beads are loaded into the chip through a channel terminating in a dam as shown in Figure 9. After analysis, the beads can be removed by reversing the flow. The initial work utilized adsorption of human secretory immunoglobulin A (s-IgA) onto the bead surface, which was then reacted with colloidal gold-conjugated

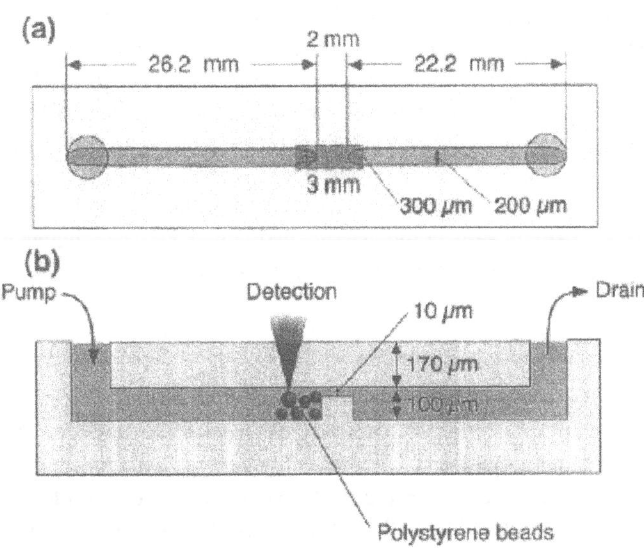

FIGURE 9 Layout of the glass microchip for immunosorbent assay: (a) overview; (b) cross section. (Reprinted with permission from Ref. 57.)

anti-s-IgA antibody and detected using TLM [57]. A conventional immunoassay could not detect 1-μg/mL s-IgA in a sample, which was well within the detection limit of this microchip bead immunoassay. Typically, concentrations in human saliva are approximately 200 μg/mL; thus, this system is suitable for practical clinical measurements. A commonly used marker of colon cancer, carcinoembryonic antigen, was also detected in this system using a three-antibody sandwich immunoassay [58]. Again, limits were much lower than with conventional procedures, so practical applications are possible. A similar device, prepared for detection of interferon-γ using the sandwich immunosorbent assay, was designed with the capability for simultaneous interrogation of four samples [59]. The device required a single pump with a common inlet for filling the channels with beads and addition of the antibody solutions. The detection limits in each channel were in good agreement, and, while the total assay time increased, the time for an individual sample was more comparable with traditional assays.

Another type of immobilized immunoassay, with the antibody directly immobilized on the microchip substrate, has been demonstrated on microdevices. These devices were used for heterogeneous competitive immunoassays for human immunoglobulin G (IgG), elevated levels of which may be associated with chronic infection or cancer [60]. A three-layer system was used to immobilize the antibodies in a specific region of the PDMS/glass device. An initial biotin-conjugated protein was adsorbed to the microchip surface which was then treated with neutravidin. Biotin-conjugated antibodies were then flowed into the microchip in the detection region, with biotin-conjugated dextran used to block the remaining sites. Because these microchips were designed to be disposable, anti-mouse IgG was co-immobilized with anti-human IgG to allow Cy3-labeled mouse IgG to be used as an internal standard. Fluorescence from the Cy3 label does not overlap with the fluorescence from the Cy5-labeled human IgG used for the competitive assay, allowing both immunoglobulins to be detected simultaneously. Analysis of patient sera could easily distinguish between patients with normal IgG serum levels (8–16 mg/mL) and those with elevated levels (>16 mg/mL).

4. DNA

One of the most important applications of capillary electrophoresis has been the separation of DNA for clinical purposes. Thus, it is logical that a major application of microchip technology is DNA separations. One of the first examples of a DNA separation on a microdevice with a clinical application was accomplished by Effenhauser et al. [61]. This research demonstrated separation of fluorescently labeled oligonucleotides, ranging in size from 10 to 25 base pairs, in only 45 s in a 3.8-cm channel length. Because the DNA fragments were so short, they were able to effect the separation without a sieving matrix. That same year, Woolley et al. [62] demonstrated gel electrophoresis in microchips for the separation of longer DNA fragments. Fragments (70 to 1000 bp) from a *Hae*III digest of ϕX174 DNA were separated using hydroxypropyl cellulose (HPC) as a replaceable sieving matrix in 120 s, as illustrated in Figure 10. The figure also illustrates the dependence of the separation on the choice of interca-

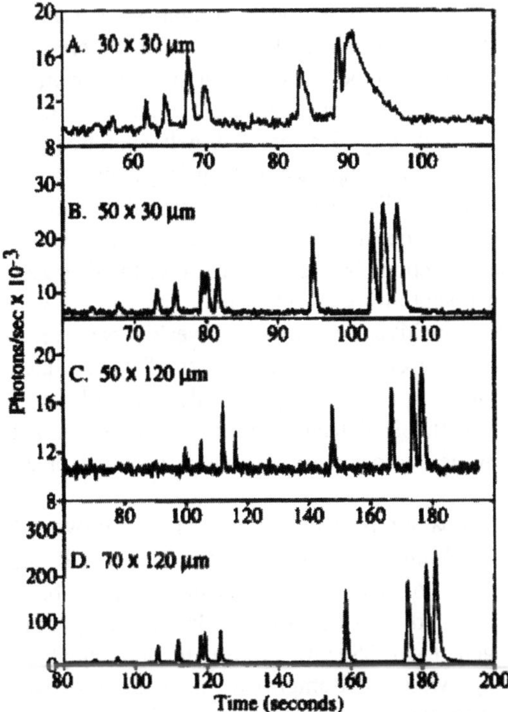

FIGURE 10 Electropherograms obtained with various channel sizes and matrix conditions for *Hae*III digests of ϕX174 phage DNA. (A) Separation channel, 30 μm; cross channel, 30 μm; sample (plug injected for 5 s), 100 ng/μL; separation in the presence of 1 μM TO. (B) Separation channel, 50 μm; cross channel, 30 μm; sample (plug injected for 1 s), 10 ng/μL; separation in the presence of 1 μM TO. (C) Separation channel, 50 μm; cross channel, 120 μm; sample (stack injected for 1 s), 10 ng/μL; separation in the presence of 0.1 μM TO6. (D) Separation channel, 70 μm; cross channel, 120 μm; sample (plug injected for 1 s), 10 ng/μL; separation in the presence of 0.1 μM TO6. Sensitivities of DNA detection with TO and TO6 are comparable at the concentrations used. (Reprinted with permission from Ref. 62.)

lating dye, the channel dimensions, and the separation conditions. The authors also showed a separation of polymerase chain reaction (PCR)-amplified HLA-DQ alpha alleles, further demonstrating the utility of microdevices for clinically relevant DNA separations.

4.1. DNA Separations of PCR Products

A number of groups have since demonstrated the diagnostic capabilities of a numerous types of microchip DNA separations. One common way microchip CE is used to provide clinically relevant information is through separations of PCR-amplified products. Often, this is accomplished by detecting the presence of a specific fragment size in the amplified product, as shown in work performed to detect herpes simplex virus (HSV) in serum [63]. HSV, a causative agent of encephalitis, normally requires PCR amplification, slab-gel separation, and Southern blot analysis for detection, but an amplified 111-bp fragment from the virus, indicative of viral infection, was easily separated and detected on a microchip using HEC as a sieving matrix. The increased speed of detection (100 s versus 1–2 days for conventional detection) is important because the disease is curable when detected early and potentially fatal when not. The authors were able to discern samples that were positive, negative, and weakly positive, as shown in Figure 11, demonstrating the diagnostic capabilities of these separations. This size separation technique was also employed for detection of hepatitis C virus on a plastic microchip with LIF detection in a few hundred seconds [64]. Again, the diagnostic utility of microchip separations is apparent in this work.

Another area in which size separations of PCR products have proven effective has been cancer diagnostics. Munro et al. [65] demonstrated that B- and T-cell lymphoproliferative disorders could be identified using an HEC-based microchip separation with no loss of diagnostic capacity. In this study, fragments from the T-cell receptor γ gene (150–250 bp) and from the immunoglobulin heavy-chain gene (80–140 bp) were PCR-amplified. Separations of each from normal controls show broad smears of DNA due to the polyclonal nature of these genes, which allows for an immune response to invading antigens. Lymphoproliferative disorders, however, result in an overexpression of a single clone, detected as an intense band in gels or a sharp peak in microchip separations, allowing for clear determination between the "normal" and diseased states.

While these examples illustrate diagnosis by identification of a single PCR-amplification product, microdevices have also been used for separation of clinically relevant multiplex amplifications. Duchenne muscular dystrophy (DMD) is caused by a number of mutations within the very large dystrophyn gene. These are identified by PCR amplification of several exon regions of the gene in two multiplex PCR reactions. Separation of these products on microchips has

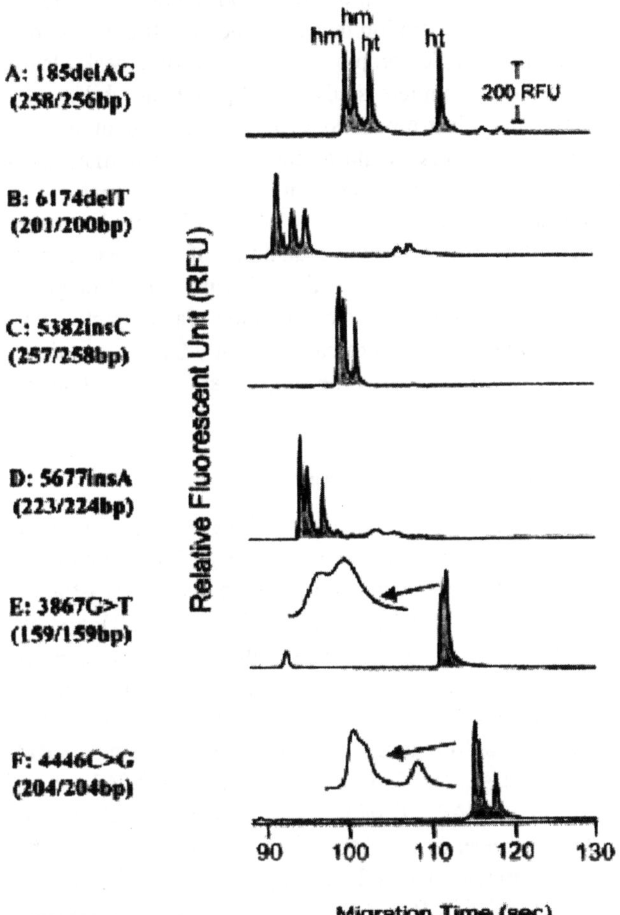

FIGURE 11 Fast mutation detection via heteroduplex analysis on a microfabricated electrophoretic chip. The heterozygous mutants were specified in the figure. The separation buffer was 2.5% HEC containing 10% glycerol in 1× TBE (pH = 8.6) for (a–d) and 4.5% HEC containing 10% glycerol and 15% urea in 1× TBE (pH = 8.6) for (e–f). The microchannel on the chip was coated with PVP (Mr 1,000,000) and detection mediated by laser-induced fluorescence (em/ex 520 nm/488 nm). The PCR products were injected into the channel for 100 s at 333 V/cm (effective microchannel length of 5.5 cm), and the separation voltage was 573 V/cm. (Reprinted with permission from Ref. 69.)

been explored by two different groups. The initial work [66] identified the mutation in a patient sample in a single separation that combined the fragments from the two multiplex reactions in a single sample. The absence of a peak clearly identified the mutation. A more complete study investigated the ability of a commercial microchip separation instrument to quantify the amount of each fragment present [67]. Female carriers of the disease contain a normal and a mutated copy of the gene, thus the relative amounts of each fragment are required to accurately diagnose the specific mutation. Using 15 control samples, normal ranges were established for the relative quantity of each exon in an amplified sample. Patient samples were then evaluated, with correct diagnosis of DMD in 35 patient samples. While the exact mutation was correctly diagnosed relative to standard Southern blot identification in only 28 of the patients, the mismatches were found to be related to the amplification rather than the microchip separations.

4.2. Mutation Genotyping Using Microchips

Mutations in DMD are normally deletions or duplications of large regions of the gene. More subtle mutations, which can lead to a variety of common disease states, can be detected on microdevices using techniques such as single-stranded conformational polymorphism (SSCP), allele-specific PCR (AS-PCR), and heteroduplex analysis (HDA). One of the first reports using these techniques describes a method for SSCP analysis of mutations in the tumor susceptibility genes BRCA1 and BRCA2 [68]. SSCP exploits electrophoretic mobility differences due to mutation-induced conformational changes in single-stranded DNA. This work focused on the 185delAG and 5382insC mutations in BRCA1, as well as the 6147delT mutation in BRCA2, all of which have been correlated with increased incidence of breast cancer in the Ashkenazi Jewish population. DNA fragments encompassing each mutation were PCR-amplified using labeled primers, then denatured and separated. SSCP profiles, clearly identifying each mutation, could be generated in less than 120 s. This research presents the possibility of genetically screening select populations using high-throughput microchip systems for mutation detection.

This same group also utilized heteroduplex analysis and allele-specific PCR to investigate mutation detection in the BRCA1 and BRCA2 tumor susceptibility genes [69,70]. In HDA, duplex DNA is formed by denaturing the double-stranded PCR product and allowing the strands to reanneal. When mutations are present in one of the genes, homoduplexes are formed between like strands, but heteroduplex DNA is formed between wild-type and mutant PCR fragment strands. These heteroduplexes can be easily identified by their change in electrophoretic mobility, resulting in multiple peaks in the subsequent microchip separation. This technique proved highly effective for detecting mutations in the microchip format, being sensitive enough to identify mutations present in concentrations as low as 1–10% of the total DNA concentration in only 130 s. In subsequent research, HDA was used in concert with AS-PCR, a technique using three primers in which one primer is constructed to amplify a particular fragment only if the mutation is present. The resulting amplified products were then separated using the previously described microchip HDA procedure, and detection of three heterozygous mutations (an insertion, deletion, and substitution mutation) was accomplished in 180 s or less, as seen in Figure 11.

Another important method for detecting single base changes at specific locations in DNA is by single nucleotide polymorphism (SNP) analysis. A large number of SNPs have been identified and can be used for genotyping. Associations between specific disease susceptibility and specific SNP mutations are already being investigated. An early paper detailing a microdevice SNP separation detected SNP sites in the p53 tumor suppressor gene in 100 s, results which were confirmed by sequencing with commercial instrumentation [71]. Mathies and co-workers expanded upon this work, using multiplex microchip separations to study three common variants of the HFE locus associated with the genetic disease hereditary hemochromatosis [72,73]. The assay used AS-PCR with fluorescently labeled energy-transfer allele-specific primers for the three variants, the amplification products of which were separated on a 96-channel radial-array microdevice in less than 10 min. These results demonstrated that high-throughput screening of SNP mutations for clinical samples is possible.

SNP analysis on microdevices has also been achieved using pyrosequencing. In this method, a series of enzymes is used to generate light following addition of a base to the extending DNA chain. This technique was performed on a microdevice by immobilizing the enzymes required for the pyrosequencing process on polystyrene beads [74]. These beads were flowed into a filter-type holder fabricated in a silicon/glass device and then loaded with the sample DNA. By flowing nucleotides sequentially over the beads, the SNP sequence could be determined. This work also explored variations in the p53 tumor suppressor gene, again illustrating the application of microdevice SNP detection for clinical targets.

4.3. DNA Sequencing

The rapid sequencing of the human genome was most notably accomplished through a conversion from traditional slab-gel analysis to capillary electrophoretic analysis. Microchips have the capability to further improve analysis in this area, and a variety of groups are working toward optimization of throughput, efficiency, resolution, and performance of microdevices for DNA sequencing. In an early work detailing sequencing on the microchip platform, Wooley et al. [75] described separation of sequencing extension fragments using a polyacrylamide sieving matrix with both one- and four-color detection. Another study, by Ehrlich and co-workers [76], quantitatively assessed linear polyacrylamide (LPA) as a sieving matrix for sequencing in microdevices, detailing how a variety of conditions (injection size, device length, channel folding, etc.) affect resolution of the DNA fragments. Separations of 350 bases in only 7 min were reported in 10-cm-long channels. The group further assessed LPA as a sieving matrix in microdevices by optimizing the buffer composition, LPA composition, temperature, and electric field strength, achieving 580-bp read lengths on an 11.5-cm single-channel device [77]. Longer read lengths could be achieved by reducing the voltage; however, this nearly doubled the separation time, from 18 to 30 min. This same group further increased the sequence read length to 800 bp by increasing the channel length while at the same time multiplexing the separations using a 48-channel array [78]. However, the 40-cm channel lengths created a device much larger than is normally acceptable for microscale analysis.

Research has also focused on improving high throughput by employing shorter microchannels for decreased analysis times [79,80]. One report presents sequencing up to 500 bases in a 10-cm separation channel using an LPA sieving matrix and single-color detection. Further work describes a 16-channel microchip fabricated on a 10-cm-diameter wafer with an effective channel length of 7.6 cm. Four-color confocal fluorescent detection allowed assignments up to 450 bases in all 16 channels in 15 min with this device. Using an automated base-calling program, up to 543 bases have been called at an accuracy of greater than 99%, in less than 18 min. The authors describe the effects of temperature, template concentration, chip design, and injection time (Figure 12) on their separations.

All of these groups are continuing to refine their current microdevice designs to further improve the detection, read lengths, speed, and throughput of their systems. In terms of the actual separation, one group has shown that further increases in read lengths could be accomplished by increasing the effective channel length, and compared this to the results possible in standard capillary separations [81]. This study looked at 48- and 50-cm-long microchannels fabricated in standard borofloat glass and sequencing results from capillaries with the same separation conditions. Single-base resolution of over 600 bases with an accuracy of 98% was achieved in the microchips, identical to that of the capillary. Read lengths could indeed be increased by increasing the channel length; however, the length of time needed for separation was on the order of hours. While the above research describes a variety of work using linear polyacrylamides for high-resolution separations, a number of other novel sieving matrices for high-resolution separations of DNA are also being investigated. A variety of detection systems are also being developed for this work.

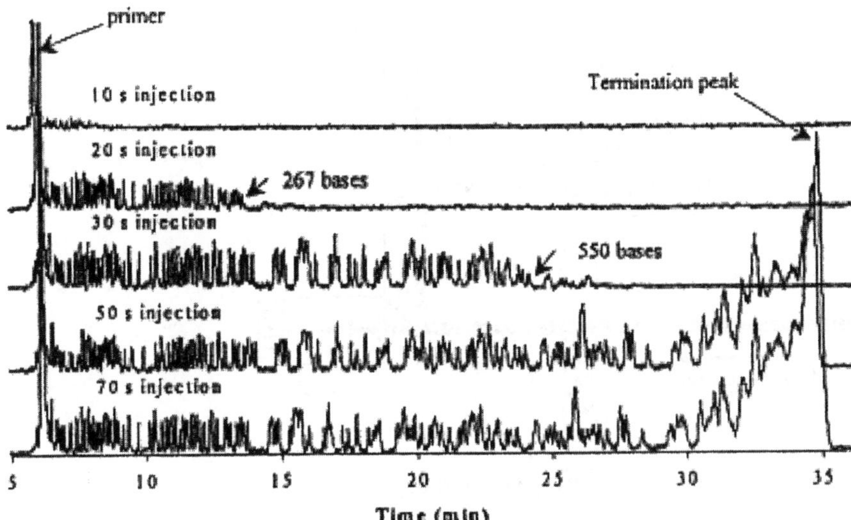

FIGURE 12 Effect of injection times on steady injection states of DNA fragments of different sizes. The arrows point to the last peaks that reached steady injection states. The separations were performed at ambient temperature by using 4% LPA and an electric field strength of ~0.180 V/cm for injection and separation. Four-color sequencing samples were used, but only the data for T-terminated traces are presented. The data shown were obtained from a channel with an effective separation distance of 7.46 cm. (Reprinted with permission from Ref. 80.)

For applications such as four-color DNA sequencing, multiple fluorophores must be detected independently at different wavelengths. Techniques reported for multicolor detection include diffraction gratings, prisms, and filter wheels. Mathies and co-workers employ a series of dichroic beam splitters and filters to select out the desired wavelengths (Figure 13) [82,83]. However, the optics must be changed to use different fluorophores, splitting of the fluorescent signal decreases the sensitivity, detection is limited to the designed number of components, and a separate detector is required for each of the four detection wavelengths. Acoustooptic tunable filter (AOTF) technology is being used for multicolor detection by Landers and co-workers. An AOTF is an electronically tunable spectral bandpass filter that allows passage of particular wavelengths through a crystal based on the acoustic wave propagating through the crystal. This technology has a narrow bandwidth, allows rapid changes between filtered wavelengths, has no moving parts, and employs only a single detector for nearly simultaneous detection of multiple wavelengths. Another group, developing a

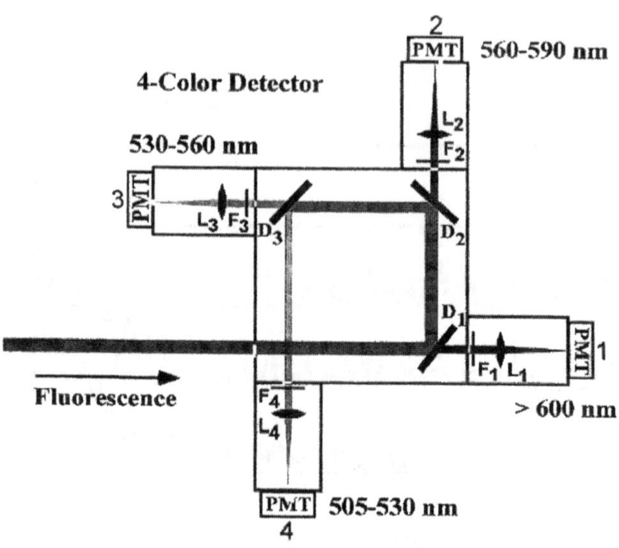

FIGURE 13 Four-color detector. Pinholes and apertures are indicated by blank white spaces between compartments. D, dichroic beamsplitter; F, filter; L, lens; PMT, photomultiplier tube. Fluorescence is directed sequentially starting at D_1. The fluorescent light is then split, filtered, and focused onto each PMT. The filter bandwidths for each channel, as defined by the dichroics and filters, correspond to the four common DNA sequencing dye emission windows. (Reprinted from permission from Ref. 83.)

48-channel device for sequencing, employed a highly sensitive charge-coupled device (CCD) detector allowing for simultaneous detection of many fluorescent dyes [84]. Separations of single-stranded DNA fragments up to 500 bases in length were demonstrated in their system, including accurate sizing of GeneCalling fragments.

The CCD device, besides allowing multiple-wavelength detection, also provides a means to easily image the entire array of channels. Any multiplex channel array requires the ability to detect very rapidly from all of the separation channels. With the CCD camera, the excitation laser is spread over the entire array to excite all of the channels simultaneously. In contrast, Mathies has applied both translation of the microchip and scanning of the laser across the channels to excite each channel individually. Rapid scanning is required to collect enough data points for each peak in every channel, but cross-talk is observed as the laser sweeps from one channel to the next. For the circular chip design with 96 channels, they have developed a spinning objective-type system to move the entire detection system from one channel to the next. Another approach, one that uses no moving parts and allows direct addressing of each channel, relies on acoustooptic technology similar to the AOTF [85,86]. This system, called an acoustooptic deflector, bends laser light passing through it to different extents based on the frequency of the acoustic wave propagating in the crystal. By changing the acoustic wave, the laser can be directed to the exact center of each channel, without having to scan between channels as with the translational approaches.

5. DNA PROCESSING ON MICRODEVICES

Clinical analysis of DNA typically requires purification of DNA from a cell or biofluid and amplification of DNA for high-sensitivity or sequencing applications. As mentioned in the introduction, faster separations must be accompanied by faster sample preparation steps if this technology is going to be accepted and implemented. Thus, application of microdevices to both of these processes is already being explored by a number of researchers, with the goal of reducing the processing time and volumes for complete total clinical analysis to a minimal amount of time. In addition, work has also been carried out on integration of these processing steps into a single device that could carry out the entire analysis. We report here the progress achieved on development of microdevices for each step and also advances made toward developing integrated devices.

5.1. DNA Extraction and Purification

DNA extraction and purification were traditionally accomplished using organic extraction and ultracentrifugation-based procedures, which are both time-consuming and not easily transferable to the microscale. Newer methods employ solid-phase extraction (SPE) on silica surfaces, glass fibers, modified magnetic beads, and ion-exchange resins—techniques that save time and are also more amenable to chip applications.

The SPE process for purifying DNA on silica surfaces was first miniaturized by packing a small bed of silica beads in a capillary connector [87,88]. DNA adsorbs to the silica surface in the presence of a chaotropic salt such as guanidine, and contaminants such as protein are removed by washing with an isopropanol solution. Purified DNA is then eluted in low-salt buffer, ready for PCR amplification or other analysis steps. Using this process, DNA was extracted from whole blood in about 10 min with a 70% efficiency. Lysing the cells in the guanidine load solution was the only off-line preparation required. Attempts to transfer this method to a microdevice showed some promise, but efficient extraction of DNA in a silica bead-packed channel was not reproducible due to difficulties in packing the beds. High back pressures also resulted during repeated use of the devices, due to compression of the bed. An alternative process is to use the silica surface of the microdevice itself for the DNA adsorption; however, the surface area is not large enough to provide sufficient capacity. To overcome this restriction, silicon microchips were reactive ion-etched to produce pillars throughout the channel to increase the surface available for DNA adsorption [89]. These devices, though suitable for DNA extraction, are expensive to produce, and therefore are not practical for clinical use.

Silica bead-packed microchips were further investigated, this time using a sol-gel matrix to hold the beads in place. This eliminated the problems due to bed compression, and produced reproducible extractions with >60% efficiency on single microchips and between microchips [90]. The adsorbed DNA was eluted in a very small volume of buffer (~10 μL), making this an ideal method for integration into a lab-on-a-chip or micro total analysis system (μ-TAS) device. In subsequent efforts, flow rates of the load, wash, and elution solutions through these devices were increased to allow faster purifications. Using the increased flow rates, DNA was purified from cultures of both *Salmonella typhimurium* and *Bacillus anthracis*, then PCR-amplified to demonstrate the ability to detect these bacteria [91]. This work was further modified to eliminate the need for the tedious bead-packing step, using the sol-gel itself as the silica matrix, since as a precursor liquid the matrix was easily formed in place. This work demonstrated reproducible extraction efficiencies of >90% on the sol-gel-filled microdevices, and the technique has been applied to whole blood for purification of genomic DNA and to cerebral spinal fluid for extraction of viral DNA [92].

5.2. PCR Amplification

Since its introduction into molecular biology by Mullis et al. [93] in 1986, the polymerase chain reaction has revolutionized clinical diagnostics. Amplification of DNA or RNA now permits genetic analysis of DNA present in very low copy numbers, previously too minimal for detection. Conventional PCR thermocyclers use a heating block to temperature cycle solutions in small polypropylene tubes. The first microfabricated device for PCR, made by Wilding and co-workers [94,95], applied this same temperature-cycling principle, using a silicon-glass device with a Peltier heater and cooler to perform amplification of λ phage and *Campylobacter jejuni* bacterial DNA in a 10-μL reaction volume. The same group also integrated a cell-sorting step into the microdevice, isolating white blood cells from whole blood using reactive ion-etched silicon micropost filters to selectively capture the white cells [96]. The PCR reaction mixture was added directly to the cells in a filter chamber and thermocycled, with amplification times comparable to those of commercial, bench-top thermocyclers. At the same time, Wooley et al. [97] reported on a PCR microdevice consisting of a polysilicon heater with a polypropylene insert, epoxied to a microchip. This device demonstrated more rapid heating and cooling of the PCR reaction mixture than other microdevices, with heating and cooling rates of 10 and 2.5°C/s, respectively. The β-globin gene from *Salmonella* DNA was amplified using 30 cycles in only 15 min on this device.

A unique approach to PCR amplification on microdevices was reported by Kopp et al. [98], who used continuous flow through a serpentine channel fabricated in a microchip to effect the temperature changes necessary for thermocycling. Three heating devices incorporated into the microchip provided distinct temperature zones, as seen in Figure 14. The PCR reaction mix was flowed through the channel by hydrostatic pumping, with the reaction times at each temperature set by the path length and the flow rate of the solution. The amount of product obtained was limited by the fixed number of temperature cycles, set by the microchip design, thus restricting the flexibility of the chip.

As with traditional thermocyclers, the microchip PCR methods discussed to this point utilize contact methods, i.e., the heating/cooling device is in direct contact with the microchip or microchamber. These methods limit the efficiency and effectiveness of thermocycling, due to the thermal load of the device itself. A more efficient way to handle thermocycling of the PCR solution is heating only the solution itself, not the microchip. Such a "noncontact" method has been described by two groups [99,100], who used IR light excitation of a vibrational band of water to heat the PCR reaction mixture. This allowed for faster temperature cycling and reduced volumes of PCR mixture, making the amplification process faster and more efficient. In a capillary design, nanoliter volumes of PCR mixture could be thermocycled with cycle times of the order of 3 s using this heating method [101]. In the microchip format, this IR-mediated thermocycling has been applied to glass, glass/PDMS hybrid, and polyimide microdevices [102]. The increased efficiency and reduced amplification times with microchip PCR were illustrated by a 15-cycle amplification of a 500-bp fragment of λ-phage DNA in only 240 s in a polyimide device.

6. INTEGRATED MICRODEVICES

With the goal of a micro total analysis system on a single microdevice (lab-on-a-chip) in mind, several research groups have begun the quest to integrate processing steps and separations into a single device. Mathies and co-workers were the first to achieve integration of PCR and separation in a single device [97]. As described above, this design used a silicon heater-enclosed polypropylene PCR chamber, epoxied to a microchip as a reservoir for the separation channel etched into the microchip. After thermocycling in the PCR chamber, the amplified β-globin gene product from *Salmonella* DNA was separated in ~100 s. This demonstrated the feasibility of PCR/CE coupled analysis in microdevices. Ramsey and co-workers [103], made the first attempt at PCR/CE on a

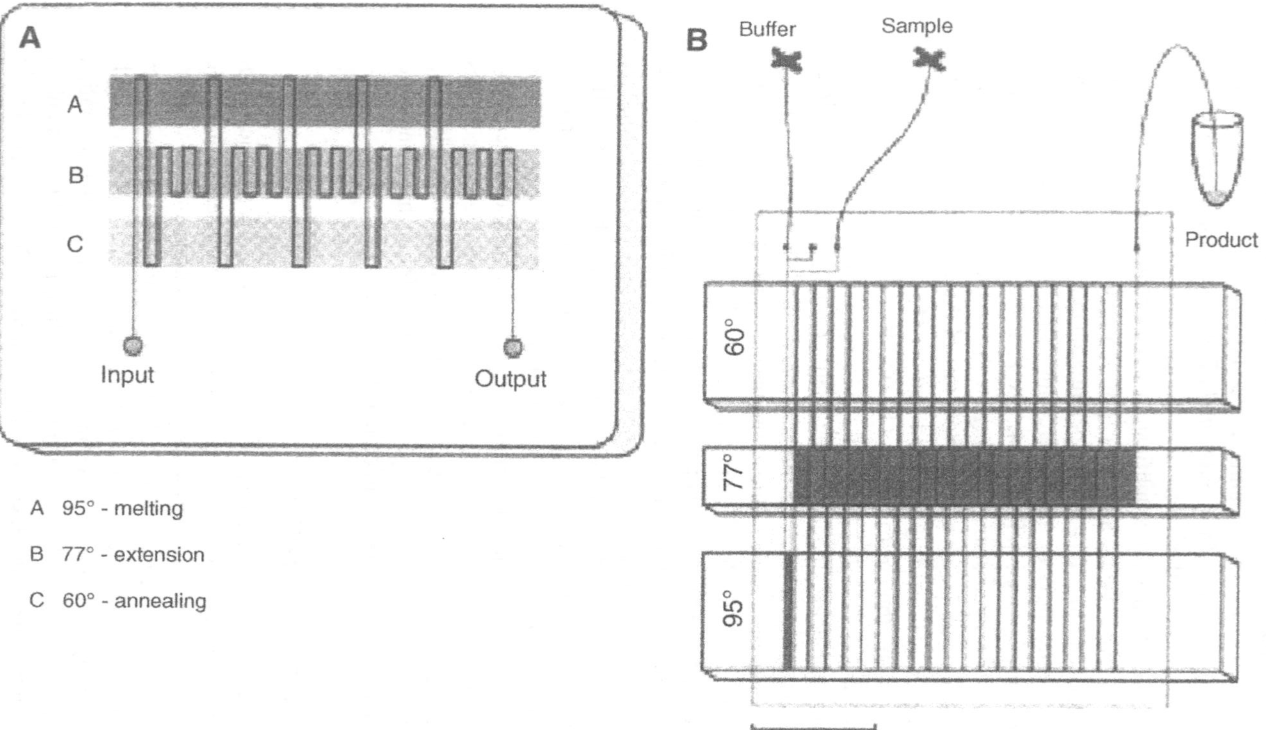

Figure 14 (A) Schematic of a chip for flow-through PCR. Three well-defined zones are kept at 95, 77, and 60°C by means of thermostated copper blocks. The sample is hydrostatically pumped through a single channel etched into the glass chip. The channel passing through the three temperature zones defines the thermal cycling process. (B) Layout of the device used in the study. The device has three inlets on one side of the device and an outlet on the opposite side. Only two inlets were used, one carrying the sample, the other bringing a constant buffer flow. The chip incorporates 20 identical cycles, except for the first one, which includes a threefold increase in DNA melting time. (Reprinted with permission from Ref. 98.)

single device—an improvement on the stand-alone PCR chamber epoxied to the microchip. They constructed three PCR reservoirs on a microdevice, all of which were connected to a single separation channel. Thermocycling was carried out by placing the entire device in a conventional thermocycler, with amplification of separate DNA targets in each of the three reservoirs. The product from each reservoir was then separated concurrently in the same separation channel by simultaneously applying the same voltage to all three PCR reservoirs. This same group also reported the integration of cell lysis, PCR, and CE on a multi-functional microchip [104]. *Escherichia coli* cells loaded into a reservoir were lysed by thermocycling, then thermocycled again to perform the PCR. As before, the thermocycling, which took 3.75 h, was carried out in a commercial thermocycler and then the products were separated by electrophoresis. This integrated chip represented a step forward, but long cycle times due to thermocycling of the entire microchip represent a serious disadvantage of this method.

This same group later detailed a more elegant method for DNA amplification and subsequent electrophoretic separation. This system employs Peltier thermoelectric elements above and below the microchip to heat the PCR chamber using radiative heat, as shown in Figure 15 [105]. Rapid cooling is accomplished by cooling fans and integral heat sinks, resulting in cycle times of approximately 1.25 min—a vast improvement on prior work. After amplification, the DNA was concentrated in the injection region using a porous, semiperme-

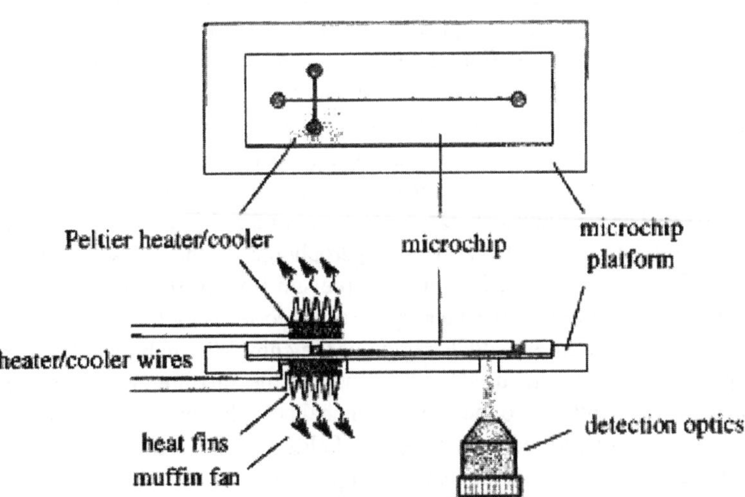

FIGURE 15 Schematic of the dual Peltier assembly for rapid thermal cycling followed by electrophoretic analysis on-chip. (Reprinted with permission from Ref. 105.)

able polysilicate membrane to separate a side channel from the separation channel. This membrane allowed conduction of ionic current during injection from the PCR chamber to the side channel, but the pores were too small to allow passage of the DNA itself, thereby concentrating the DNA in the injection region. Once concentrated, the DNA was injected into the separation channel and electrophoretically separated using a PDMA sieving matrix.

Mathies and co-workers developed a more complicated microfluidic device capable of integrated PCR/CE, with on-chip PCR in a 280-nL chamber etched in the glass substrate [106]. Advancements included microfluidic valves and hydrophobic vents connected to the PCR chamber for sample introduction and to prevent sample evaporation during thermal cycling, as demonstrated in previous work [107]. Here, the microfluidic valve and hydrophobic vent technology was developed with 50–100-nL dead volumes to enable the precise loading and containment of submicroliter samples in the PCR chamber. A thin-film resistive heater and a miniature thermocouple below the chamber are used to cycle the temperature, with cycle times on the order of 30 s. Following amplification, the valve manifold is removed from the chip to provide access for platinum electrodes and for filling the electrophoresis reservoirs. PCR amplification and separation of a 136-bp amplification product derived from the M13/pUC19 cloning vector and a 231-bp product amplified from a human genomic DNA control sample was achieved. Though complex, this device represents one of the most functional and integrated CE/PCR microdevices reported to date.

Burns and co-workers have developed a device for PCR amplification and electrophoretic separation that employs microfabricated fluidic channels, heaters, temperature sensors, and fluorescence detectors to analyze nanoliter volumes of DNA samples, operating as a single, closed system [108]. As shown in Figure 16, the microdevice includes an injector, a sample mixing and positioning system, a temperature-controlled reaction chamber, an electrophoretic separation channel, and an integrated fluorescence detector, allowing for rapid, elegant analysis of DNA samples. A DNA solution is placed in one reservoir and the PCR mixture in a second reservoir. Hydrophobic patches, positioned just beyond vent lines in each injection channel, prevent flow of solutions by capillary action. Pressure through the vent splits off precise nanoliter drops, which mix together and move into the PCR region of the chip. Microfabricated heaters and temperature sensors are used to thermocycle the mixture, which is then hydrodynamically injected into the separation channel. The fluorescently labeled DNA is then electrophoresed under applied fields of <10 V/cm and detected using an integrated photodiode beneath the separation channel. This device was used to successfully amplify a 106-bp fragment of DNA and represents the first rapid and fully coupled nanoliter-volume DNA amplification and separation system.

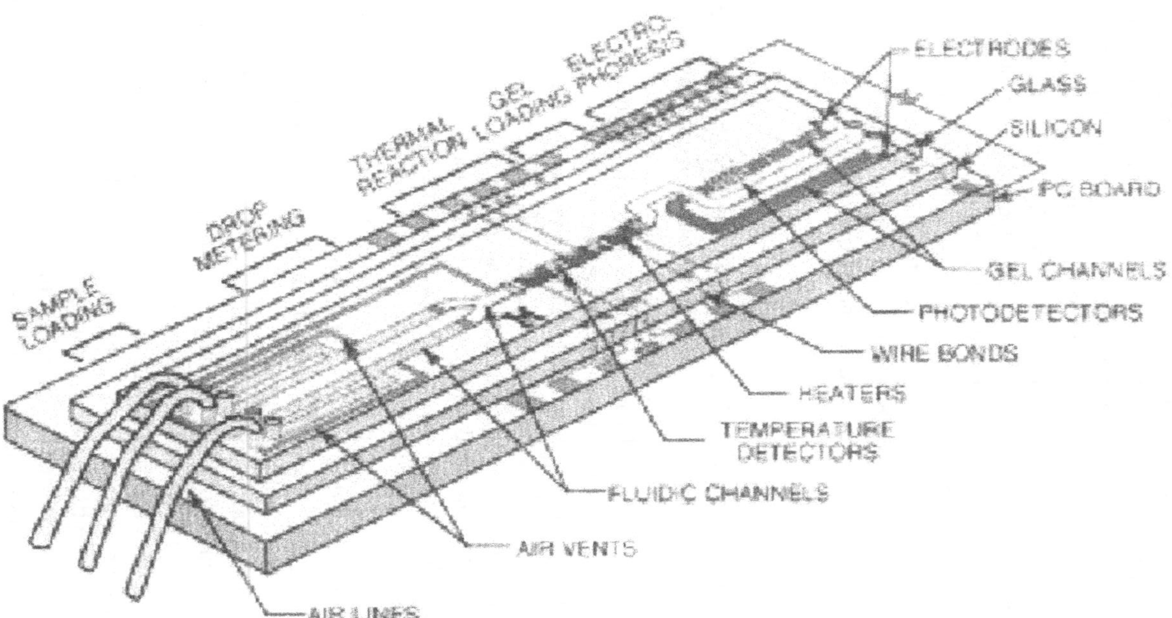

Figure 16 Schematic of integrated device with two liquid samples and electrophoresis gel present. The only electronic component not fabricated on the silicon substrate, excluding those for control and data processing, is an excitation source placed above the electrophoresis channel. (Reprinted with permission from Ref. 108.)

While these devices have been looking to integrate PCR and DNA separation in a single microdevice, a polycarbonate microfluidic device for automated multistep genetic analysis has also been reported [109]. This device can accomplish extraction and concentration of nucleic acids, chemical amplification, serial enzymatic reactions, metering, mixing, and DNA hybridization, all in one instrument smaller than a credit card. The authors reported detection of mutations in a 1.6-kb region of the HIV genome with samples containing low copy numbers. The elements in this device are amenable to parallel and integrated analysis and would be useful for low-cost genetic analysis for point-of-care clinical diagnostics. Unfortunately, this device does not incorporate a separation step, the main analytical method for clinical laboratories.

7. CONCLUSIONS

A large number of issues remain to be addressed in the development of microdevices for clinical applications. As stated in the introduction, two embodiments of the clinical microdevice can be foreseen: one, a single analysis chip designed for rapid analysis at the point of care; and a second, high-throughput device designed for use in a centralized laboratory. These devices could be disposable, possibly preloaded with the necessary reagents for a given analysis, but reusable devices provide advantages in terms of validating results and economics, particularly with complex devices requiring numerous microfabrication steps. A second issue with respect to the final microdevice design is modular versus fully integrated devices. Modular devices provide the opportunity to select the best substrate for each processing step, the ability to store samples for possible reanalysis at a later point, and the possibility of using a single device to extract DNA which can then be used in a number of amplification microdevices to test for various mutations or SNPs. In the end, all of these devices may well find their own niche in the clinical laboratory.

At the moment, a large number of researchers are moving the various processing steps required for clinical analyses to microchip formats in both stand-alone and integrated devices. Microchip separations are furthest along and have been shown to have equivalent diagnostic ability with conventional methods, but in much shorter times, for a large number of clinical assays. While most microchip PCR amplifications have not yet been directed toward clinically important species, this will begin to change as the clinical community realizes the power that microdevices provide and becomes more involved in the application of these devices to real samples. Overall, progress continues to be made in all microchip arenas, as evidenced by the ever-increasing number of microchip publications. With this much effort directed toward a common goal, the clinical lab-on-a-chip may come to fruition in the not-so-distant future.

REFERENCES

1. Yuan CH, Shiea J. Sequential electrospray analysis using sharp-tip channels fabricated on a plastic chip. Anal Chem 2001; 73:1080–1083.
2. Jiang Y, Wang PC, Locascio LE, Lee CS. Integrated plastic microfluidic devices with ESI-MS for drug screening and residue analysis. Anal Chem 2001; 73:2048–2053
3. Deng Y, Zhang H, Henion J. Chip-based quantitative capillary electrophoresis/mass spectrometry determination of drugs in human plasma. Anal Chem 2001; 73:1432–1439.
4. Cho SI, Lee KN, Kim YK, Jang J, Chung DS. Chiral separation of gemifloxacin in sodium-containing media using chiral crown ether as a chiral selector by capillary and microchip electrophoresis. Electrophoresis 2002; 23:972–977.
5. Ramseier A, von Heeren F, Thormann W. Analysis of fluorescein isothiocyanate derivatized amphetamine and analogs in human urine by capillary electrophoresis in chip-based and fused-silica capillary instrumentation. Electrophoresis 1998; 19: 2967–2975.
6. Wallenborg SR, Lurie IS, Arnold DW, Bailey CG. On-chip chiral and achiral separation of amphetamine and related compounds labeled with 4-fluoro-7-nitrobenzofurazane. Electrophoresis 2000; 21:3257–3263.
7. Sorouraddin HM, Hibara A, Proskurnin MA, Kitamori T. Integrated FIA for the determination of ascorbic acid and dehydroascorbic acid in a microfabricated glass-channel by thermal-lens microscopy. Anal Sci 2000; 16:1033–1037.
8. Sorouraddin HM, Hibara A, Kitamori T. Use of a thermal lens microscope in integrated catecholamine determination on a microchip. Fresenius J Chem 2001; 371:91–96.
9. Kamidate T, Kaide T, Tani H, Makino E, Shibata T. Effect of mixing modes on chemiluminescent detection of epinephrine with lucigenin by an FIA system fabricated on a microchip. Anal Sci 2001; 17:951–955.
10. Gavin PF, Ewing AG. Characterization of electrochemical array detection for continuous channel electrophoretic separations in micrometer and submicrometer channels. Anal Chem 1997; 69:3838–3845.
11. Wang J, Tian BM, Sahlin E. Micromachined electrophoresis chips with thick-film electrochemical detectors. Anal Chem 1999; 71:5436–5440.
12. Woolley AT, Lao K, Glazer AN, Mathies RA. Capillary electrophoresis chips with integrated electrochemical detection. Anal Chem 1998; 70:684–688.
13. Wang J, Tian BM, Sahlin E. Integrated electrophoresis chips/amperometric detection with sputtered gold working electrodes. Anal Chem 1999; 71:3901–3904.
14. Fanguy JC, Henry CS. The analysis of uric acid in urine using microchip capillary electrophoresis with electrochemical detection. Electrophoresis 2002; 23:767–773.
15. Zuborova M, Masar M, Kaniansky D, Johnck M, Stanislawski B. Determination of oxalate in urine by zone electrophoresis on a chip with conductivity detection. Electrophoresis 2002; 23:774–781.
16. Deng Y, Henion J, Li J, Thibault P, Wang C, Harrison DJ. Chip-based capillary electrophoresis/mass spectrometry determination of carnitines in human urine. Anal Chem 2001; 73:639–646.

17. Easley C, Jin LJ, Landers JP, Ferrance JP. Carbohydrate detection on microchips for clinical detection of galactosemia. J Chrom A 2003; in press.
18. Malcik N, Ferrance JP, Caglar P, Landers JP. Development of a microchip sensor for calcium detection in biological fluids. In preparation.
19. Pasas SA, Lacher NA, Davies MI, Lunte SM. Detection of homocysteine by conventional and microchip capillary electrophoresis/electrochemistry. Electrophoresis 2002; 23:759–766.
20. Galloway M, Stryjewski W, Henry A, Ford SM, Llopis S, McCarley RL, et al. Contact conductivity detection in poly(methyl methacylate)-based microfluidic devices for analysis of mono- and polyanionic molecules. Anal Chem 2002; 74: 2407–2415.
21. Harrison DJ, Fluri K, Seiler K, Fan ZH, Effenhauser CS, Manz A. Micromachining a miniaturized capillary electrophoresis-based chemical-analysis system on a chip. Science 1993; 261:895–897.
22. von Heeren F, Verpoorte E, Manz A, Thormann W. Micellar electrokinetic chromatography separations and analyses of biological samples on a cyclic planar microstructure. Anal Chem 1996; 68:2044–2053.
23. Culbertson CT, Jacobson SC, Ramsey JM. Microchip devices for high-efficiency separations. Anal Chem 2000; 72:5814–5819.
24. Jacobson SC, Koutny LB, Hergenröder R, Moore AW, Ramsey JM. Microchip capillary electrophoresis with an integrated postcolumn reactor. Anal Chem 1994; 66:3472–3476.
25. Jacobson SC, Hergenröder R, Moore AW, Ramsey JM. Precolumn reactions with electrophoretic analysis integrated on a microchip. Anal Chem 1994; 66:4127–4132.
26. Munro NJ, Huang Z, Finegold DN, Landers JP. Indirect fluorescence detection of amino acids on electrophoretic microchips. Anal Chem 2000; 72:2765–2773.
27. Hutt LD, Glavin DP, Bada JL, Mathies RA. Microfabricated capillary electrophoresis amino acid chirality analyzer for extraterrestrial exploration. Anal Chem 1999; 71:4000–4006.
28. Rodriguez I, Jin LJ, Li SF. High-speed chiral separations on microchip electrophoresis devices. Electrophoresis 2000; 21:211–219.
29. Jin LJ, Ferrance JP, Sanders JC, Landers JP. A microchip-based proteolytic digestion system driven by electroosmotic pumping. Lab Chip 2003; 3:11–18.
30. Xu N, Lin Y, Hofstadler SA, Matson D, Call CJ, Smith RD. A microfabricated dialysis device for sample cleanup in electrospray ionization mass spectrometry. Anal Chem 1998; 70:3553–3556.
31. Xiang F, Lin Y, Wen J, Matson DW, Smith RD. An integrated microfabricated device for dual microdialysis and on-line ESI-ion trap mass spectrometry for analysis of complex biological samples. Anal Chem 1999; 71:1485–1490.
32. Figeys D, Aebersold R. Nanoflow solvent gradient delivery from a microfabricated device for protein identifications by electrospray ionization mass spectrometry. Anal Chem 1998; 70:3721–3727.
33. Xue Q, Foret F, Dunayevskiy YM, Zavracky PM, McGruer NE, Karger BL. Multichannel microchip electrospray mass spectrometry. Anal Chem 1997; 69: 426–430.

34. Figeys D, Ning Y, Aebersold R. A microfabricated device for rapid protein identification by microelectrospray ion trap mass spectrometry. Anal Chem 1997; 69: 3153–3160.
35. Li J, Thibault P, Bings NH, Skinner CD, Wang C, Colyer C, et al. Integration of microfabricated devices to capillary electrophoresis-electrospray mass spectrometry using a low dead volume connection: application to rapid analyses of proteolytic digests. Anal Chem 1999; 71:3036–3045.
36. Li J, Kelly JF, Chernushevich I, Harrison DJ, Thibault P. Separation and identification of peptides from gel-isolated membrane proteins using a microfabricated device for combined capillary electrophoresis/nanoelectrospray mass spectrometry. Anal Chem 2000; 72:599–609.
37. Wen J, Lin Y, Xiang F, Matson DW, Udseth HR, Smith RD. Microfabricated isoelectric focusing device for direct electrospray ionization-mass spectrometry. Electrophoresis 2000; 21:191–197.
38. Li J, Wang C, Kelly JF, Harrison DJ, Thibault P. Rapid and sensitive separation of trace level protein digests using microfabricated devices coupled to a quadrupole—time-of-flight mass spectrometer. Electrophoresis 2000; 21:198–210.
39. Yao S, Anex DS, Caldwell WB, Arnold DW, Smith KB, Schultz PG. SDS capillary gel electrophoresis of proteins in microfabricated channels. Proc Natl Acad Sci USA 1999; 96:5372–5377.
40. Badal MY, Wong M, Chiem N, Salimi-Moosavi H, Harrison DJ. Protein separation and surfactant control of electroosmotic flow in poly(dimethylsiloxane)-coated capillaries and microchips. J Chromatogr A 2002; 947:277–286.
41. Colyer CL, Mangru SD, Harrison DJ. Microchip-based capillary electrophoresis of human serum proteins. J Chromatogr A 1997; 781:271–276.
42. Liu Y, Foote RS, Jacobson SC, Ramsey RS, Ramsey JM. Electrophoretic separation of proteins on a microchip with noncovalent, postcolumn labeling. Anal Chem 2000; 72:4608–4613.
43. Jin LJ, Giordano BC, Landers JP. Dynamic labeling during capillary or microchip electrophoresis for laser-induced fluorescence detection of protein-SDS complexes without pre- or postcolumn labeling. Anal Chem 2001; 73:4994–4999.
44. Giordano BC, Jin LJ, Landers JP. Optimization of buffer conditions for dynamic labeling of proteins during capillary and microchip electrophoresis. 2003; In preparation.
45. Giordano BC. 2002. Personal communication.
46. Bousse L, Mouradian S, Minalla A, Yee H, Williams K, Dubrow R. Protein sizing on a microchip. Anal Chem 2001; 73:1207–1212.
47. Hadd AG, Raymond DE, Halliwell JW, Jacobson SC, Ramsey JM. Microchip device for performing enzyme assays. Anal Chem 1997; 69:3407–3412.
48. Kerby M, Chien RL. A fluorogenic assay using pressure-driven flow on a microchip. Electrophoresis 2001; 22:3916–3923.
49. Cohen CB, Chin-Dixon E, Jeong S, Nikiforov TT. A microchip-based enzyme assay for protein kinase A. Anal Biochem 1999; 273:89–97.
50. Laurell T, Rosengren L. A micromachined enzyme reactor in (110)-oriented silicon. Sens Actuators, B 1994; 19:614–617.

51. Laurell T, Drott J, Rosengren L, Lindstrom K. Enhanced enzyme activity in silicon integrated enzyme reactors utilizing porous silicon as the coupling matrix. Sensors Actuators, B 1996; 31:161–166.
52. Koutny LB, Schmalzing D, Taylor TA, Fuchs M. Microchip electrophoretic immunoassay for serum cortisol. Anal Chem 1996; 68:18–22.
53. Schmalzing D, Koutny LB, Taylor TA, Nashabeh W, Fuchs M. Immunoassay for thyroxine (T4) in serum using capillary electrophoresis and micromachined devices. J Chromatogr, B: Biomed Sci Appl 1997; 697:175–180.
54. Chiem N, Harrison DJ. Microchip-based capillary electrophoresis for immunoassays: analysis of monoclonal antibodies and theophylline. Anal Chem 1997; 69: 373–378.
55. Chiem NH, Harrison DJ. Microchip systems for immunoassay: an integrated immunoreactor with electrophoretic separation for serum theophylline determination. Clin Chem 1998; 44:591–598.
56. Hatch A, Kamholz AE, Hawkins KR, Munson MS, Schilling EA, Weigl BH, et al. A rapid diffusion immunoassay in a T-sensor. Nat Biotechnol 2001; 19: 461–465.
57. Sato K, Tokeshi M, Odake T, Kimura H, Ooi T, Nakao M, et al. Integration of an immunosorbent assay system: analysis of secretory human immunoglobulin A on polystyrene beads in a microchip. Anal Chem 2000; 72:1144–1147.
58. Sato K, Tokeshi M, Kimura H, Kitamori T. Determination of carcinoembryonic antigen in human sera by integrated bead-bed immunoassay in a microchip for cancer diagnosis. Anal Chem 2001; 73:1213–1218.
59. Sato K, Yamanaka M, Takahashi H, Tokeshi M, Kimura H, Kitamori T. Microchip-based immunoassay system with branching multichannels for simultaneous determination of interferon-gamma. Electrophoresis 2002; 23:734–739.
60. Linder V, Verpoorte E, de Rooij NF, Sigrist H, Thormann W. Application of surface biopassivated disposable poly(dimethylsiloxane)/glass chips to a heterogeneous competitive human serum immunoglobulin G immunoassay with incorporated internal standard. Electrophoresis 2002; 23:740–749.
61. Effenhauser CS, Paulus A, Manz A, Widmer HM. High-speed separation of antisense oligonucleotides on a micromachined capillary electrophoresis device. Anal Chem 1994; 66:2949–2953.
62. Woolley AT, Mathies RA. Ultra-high-speed DNA fragment separations using microfabricated capillary array electrophoresis chips. Proc Natl Acad Sci USA 1994; 91:11348–11352.
63. Hofgartner WT, Huhmer AF, Landers JP, Kant JA. Rapid diagnosis of herpes simplex encephalitis using microchip electrophoresis of PCR products. Clin Chem 1999; 45:2120–2128.
64. Chen YH, Wang WC, Young KC, Chang TT, Chen SH. Plastic microchip electrophoresis for analysis of PCR products of hepatitis C virus. Clin Chem 1999; 45: 1938–1943.
65. Munro NJ, Snow K, Kant JA, Landers JP. Molecular diagnostics on microfabricated electrophoretic devices: from slab gel—to capillary—to microchip-based assays for T- and B-cell lymphoproliferative disorders. Clin Chem 1999; 45:1906–1917.

66. Kricka LJ, Wilding P. Micromachining. A new direction for clinical analyzers. Pure Appl Chem 1996; 68:1831–1836.
67. Ferrance J, Snow K, Landers JP. Evaluation of microchip electrophoresis as a molecular diagnostic method for Duchenne muscular dystrophy. Clin Chem 2002; 48:380–383.
68. Tian H, Jaquins-Gerstl A, Munro N, Trucco M, Brody LC, Landers JP. Single-strand conformation polymorphism analysis by capillary and microchip electrophoresis: a fast, simple method for detection of common mutations in BRCA1 and BRCA2. Genomics 2000; 63:25–34.
69. Tian H, Brody LC, Landers JP. Rapid detection of deletion, insertion, and substitution mutations via heteroduplex analysis using capillary- and microchip-based electrophoresis. Genome Res 2000; 10:1403–1413.
70. Tian H, Brody LC, Fan S, Huang Z, Landers JP. Capillary and microchip electrophoresis for rapid detection of known mutations by combining allele-specific DNA amplification with heteroduplex analysis. Clin Chem 2001; 47:173–185.
71. Schmalzing D, Belenky A, Novotny MA, Koutny L, Salas-Solano 0, El-Difrawy S, et al. Microchip electrophoresis: a method for high-speed SNP detection. Nucleic Acids Res 2000; 28:E43.
72. Medintz I, Wong WW, Sensabaugh G, Mathies RA. High speed single nucleotide polymorphism typing of a hereditary haemochromatosis mutation with capillary array electrophoresis microplates. Electrophoresis 2000; 21:2352–2358.
73. Medintz I, Wong WW, Berti L, Shiow L, Tom J, Scherer J, et al. High-performance multiplex SNP analysis of three hemochromatosis-related mutations with capillary array electrophoresis microplates. Genome Res 2001; 11:413–421.
74. Andersson H, van der Wijngaart W, Stemme G. Micromachined filter-chamber array with passive valves for biochemical assays on beads. Electrophoresis 2001; 22:249–257.
75. Woolley AT, Mathies RA. Ultra-high-speed DNA sequencing using capillary electrophoresis chips. Anal Chem 1995; 67:3676–3680.
76. Schmalzing D, Adourian A, Koutny L, Ziaugra L, Matsudaira P, Ehrlich D. DNA sequencing on microfabricated electrophoretic devices. Anal Chem 1998; 70: 2303–2310.
77. Salas-Solano O, Schmalzing D, Koutny L, Buonocore S, Adourian A, Matsudaira P, et al. Optimization of high-performance DNA sequencing on short microfabricated electrophoretic devices. Anal Chem 2000; 72:3129–3137.
78. Koutny L, Schmalzing D, Salas-Solano O, El-Difrawy S, Adourian A, Buonocore S, et al. Eight hundred-base sequencing in a microfabricated electrophoretic device. Anal Chem 2000; 72:3388–3391.
79. Liu S, Shi Y, Ja WW, Mathies RA. Optimization of high-speed DNA sequencing on microfabricated capillary electrophoresis channels. Anal Chem 1999; 71:566–573.
80. Liu S, Ren H, Gao Q, Roach DJ, Loder RT Jr, Armstrong TM, et al. Automated parallel DNA sequencing on multiple channel microchips. Proc Natl Acad Sci USA 2000; 97:5369–5374.
81. Backhouse C, Caamano M, Oaks F, Nordman E, Carrillo A, Johnson B, et al.

DNA sequencing in a monolithic microchannel device. Electrophoresis 2000; 21: 150–156.

82. Shi Y, Simpson PC, Scherer JR, Wexler D, Skibola C, Smith MT, et al. Radial capillary array electrophoresis microplate and scanner for high-performance nucleic acid analysis. Anal Chem 1999; 71:5354–5361.
83. Medintz IL, Paegel BM, Blazej RG, Emrich CA, Berti L, Scherer, et al. High-performance genetic analysis using microfabricated capillary array electrophoresis microplates. Electrophoresis 2001; 22:3845–3856.
84. Simpson JW, Ruiz-Martinez MC, Mulhern GT, Berka J, Latimer DR, Ball JA, et al. A transmission imaging spectrograph and microfabricated channel system for DNA analysis. Electrophoresis 2000; 21:135–149.
85. Huang Z, Munro N, Huhmer AF, Landers JP. Acousto-optical deflection-based laser beam scanning for fluorescence detection on multichannel electrophoretic microchips. Anal Chem 1999; 71:5309–5314.
86. Huang Z, Jin L, Sanders JC, Zheng Y, Dunsmoor C, Tian H, et al. Laser-induced fluorescence detection on multichannel electrophoretic microchips using microporcessorembedded acousto-optic laser beam scanning. IEEE Trans Biomed Eng 2002; 49:859–866.
87. Tian H, Huhmer AF, Landers JP. Evaluation of silica resins for direct and efficient extraction of DNA from complex biological matrices in a miniaturized format. Anal Biochem 2000; 283:175–191.
88. Tian H, Brody LC, Mao D, Landers JP. Effective capillary electrophoresis-based heteroduplex analysis through optimization of surface coating and polymer networks. Anal Chem 2000; 72:5483–5492.
89. Christel LA, Petersen K, McWilliam W, Northrup MA. Rapid, automated nucleic acid probe assays using silicon microstructures for nucleic acid concentration. J Biomech Eng 1999; 121:272–279.
90. Wolfe KA, Breadmore MC, Ferrance JP, Power ME, Conroy JF, Norris PM, et al. Toward a microchip-based solid-phase extraction method for isolation of nucleic acids. Electrophoresis 2002; 23:727–733.
91. Breadmore MC, Wolfe KA, Arcibal IG, Leung WK, Dickson D, Giordano BC, et al. Microchip-based purification of DNA from biological samples. Anal Chem 2003; accepted.
92. Wu Q, Hassan B, Ferrance J Landers JP. Solid phase extraction of DNA from blood and cerebral spinal fluids on microdevices. 2003. In preparation.
93. Mullis K, Faloona F, Scharf S, Saiki R, Horn G, Erlich H. Specific enzymatic amplification of DNA in vitro: the polymerase chain reaction. Cold Spring Harbor Symp Quant Biol 1986; 51 Pt 1:263–273.
94. Shoffner MA, Cheng J, Hvichia GE, Kricka LJ, Wilding P. Chip PCR. I. Surface passivation of microfabricated silicon-glass chips for PCR. Nucleic Acids Res 1996; 24:375–379.
95. Cheng J, Shoffner MA, Hvichia GE, Kricka LJ, Wilding P. Chip PCR. II. Investigation of different PCR amplification systems in microbabricated silicon-glass chips. Nucleic Acids Res 1996; 24:380–385.
96. Wilding P, Kricka LJ, Cheng J, Hvichia G, Shoffner MA, Fortina P. Integrated

cell isolation and polymerase chain reaction analysis using silicon microfilter chambers. Anal Biochem 1998; 257:95–100.
97. Woolley AT, Hadley D, Landre P, deMello AJ, Mathies RA, Northrup MA. Functional integration of PCR amplification and capillary electrophoresis in a microfabricated DNA analysis device. Anal Chem 1996; 68:4081–4086.
98. Kopp MU, Mello AJ, Manz A. Chemical amplification: continuous-flow PCR on a chip. Science 1998; 280:1046–1048.
99. Northrup MA, Mariella RP, Carrano AV, Balch JW. U.S. Patent 5,589,136. 1996.
100. Oda RP, Strausbauch MA, Huhmer AF, Borson N, Jurrens SR, Craighead J, et al. Infrared-mediated thermocycling for ultrafast polymerase chain reaction amplification of DNA. Anal Chem 1998; 70:4361–4368.
101. Huhmer AF, Landers JP. Noncontact infrared-mediated thermocycling for effective polymerase chain reaction amplification of DNA in nanoliter volumes. Anal Chem 2000; 72:5507–5512.
102. Giordano BC, Ferrance J, Swedberg S, Huhmer AF, Landers JP. Polymerase chain reaction in polymeric microchips: DNA amplification in less than 240 seconds. Anal Biochem 2001; 291:124–132.
103. Waters LC, Jacobson SC, Kroutchinina N, Khandurina J, Foote RS, Ramsey JM. Multiple sample PCR amplification and electrophoretic analysis on a microchip. Anal Chem 1998; 70:5172–5176.
104. Waters LC, Jacobson SC, Kroutchinina N, Khandurina J, Foote RS, Ramsey JM. Microchip device for cell lysis, multiplex PCR amplification, and electrophoretic sizing. Anal Chem 1998; 70:158–162.
105. Khandurina J, McKnight TE, Jacobson SC, Waters LC, Foote RS, Ramsey JM. Integrated system for rapid PCR-based DNA analysis in microfluidic devices. Anal Chem 2000; 72:2995–3000.
106. Lagally ET, Medintz I, Mathies RA. Single-molecule DNA amplification and analysis in an integrated microfluidic device. Anal Chem 2001; 73:565–570.
107. Lagally ETS, P. C., Mathies, R. A. Monolithic integrated microfluidic DNA amplification n and capillary electrophoresis analysis system. Sens Actuators B 2000; B63:138–146.
108. Burns MA, Johnson BN, Brahmasandra SN, Handique K, Webster JR, Krishnan M, et al. An integrated nanoliter DNA analysis device. Science 1998; 282:484–487.
109. Anderson RC, Su X, Bogdan GJ, Fenton J. A miniature integrated device for automated multistep genetic assays. Nucleic Acids Res 2000; 28:E60.

Index